T0298361

Functional Analysis for the Applied Mathematician

Functional Analysis for the Applied Mathematician is a self-contained volume providing a rigorous introduction to functional analysis and its applications. Students from mathematics, science, engineering, and certain social science and interdisciplinary programs will benefit from the material. It is accessible to graduate and advanced undergraduate students with a solid background in undergraduate mathematics and an appreciation of mathematical rigor. Students are called upon to actively engage with the material, to the point of proving some of the basic results or their straightforward generalizations, both within the text and within the generous set of exercises.

Features:
- Replete with exercises and examples
- Suitable for graduate students and advanced undergraduates
- Develops the basics of functional analysis, exploring the interplay between algebraic linear space theory and topology
- Presents a variety of applications, often dealing with partial differential equations and their numerical approximation
- Doubles as a reference book with an extensive index listing the concepts and results

Textbooks in Mathematics
Series editors:
Al Boggess, Kenneth H. Rosen

Encounters with Chaos and Fractals, Third Edition
Denny Gulick and Jeff Ford

Differential Calculus in Several Variables
A Learning-by-Doing Approach
Marius Ghergu

Taking the "Oof!" Out of Proofs
Alexandr Draganov

Vector Calculus
Steven G. Krantz and Harold Parks

Intuitive Axiomatic Set Theory
José Luis García

Fundamentals of Abstract Algebra
Mark J. DeBonis

A Bridge to Higher Mathematics
James R. Kirkwood and Raina S. Robeva

Advanced Linear Algebra, Second Edition
Nicholas Loehr

Mathematical Biology: Discrete and Differential Equations
Christina Alvey and Daniel Alvey

Numerical Methods and Analysis with Mathematical Modelling
William P. Fox and Richard D. West

Business Process Analytics
Modeling, Simulation, and Design
Manuel Laguna and Johan Marklund

Quantitative Literacy Through Games and Gambling
Mark Hunacek

Measure Theory and Integral
Theory and Practice
John Srdjan Petrovic

Contemporary Abstract Algebra, Eleventh Edition
Joseph A. Gallian

Algebra, Second Edition
Groups, Rings, and Fields
Louis Halle Rowen and Uzi Vishne

Functional Analysis for the Applied Mathematician
Todd Arbogast and Jerry L. Bona

https://www.routledge.com/Textbooks-in-Mathematics/book-series/CANDHTEX-BOOMTH

Functional Analysis for the Applied Mathematician

Todd Arbogast
The University of Texas at Austin, USA

Jerry L. Bona
The University of Illinois at Chicago, USA

CRC Press
Taylor & Francis Group
Boca Raton London New York

CRC Press is an imprint of the
Taylor & Francis Group, an **informa** business
A CHAPMAN & HALL BOOK

Designed cover image: Elzbieta Krzysztof/Shutterstock

First edition published 2025
by CRC Press
2385 NW Executive Center Drive, Suite 320, Boca Raton FL 33431

and by CRC Press
4 Park Square, Milton Park, Abingdon, Oxon, OX14 4RN

CRC Press is an imprint of Taylor & Francis Group, LLC

ISBN: 978-1-032-79156-2 (hbk)
ISBN: 978-1-032-79460-0 (pbk)
ISBN: 978-1-003-49213-9 (ebk)

DOI: 10.1201/9781003492139

Typeset in Latin Modern font
by KnowledgeWorks Global Ltd.

Publisher's note: This book has been prepared from camera-ready copy provided by the authors.

Note on the cover image
"On the 100th anniversary of the most famous mathematical discussion at the Planty Gardens." Stefan Banach and Otton Nikodym had a habit of discussing mathematics on their evening walks in Planty Park, Kraków, Poland. One day in 1916, Hugo Steinhaus heard them discussing the Lebesgue integral, and joined the conversation. This sculpture by Stefan Dousa with Nikodym on the left and Banach on the right was unveiled in 2016. It commemorates the chance meeting, and, like Steinhaus, invites us all to join them in the advanced exploration of mathematics.

Contents

Preface   ix

Author Biographies   xiii

Synopsis   xv

CHAPTER 1 ▪ Preliminaries   1

1.1	ELEMENTARY OR POINT-SET TOPOLOGY	1
1.2	LEBESGUE MEASURE AND INTEGRATION	12
1.3	HOLOMORPHIC FUNCTIONS AND COMPLEX CONTOUR INTEGRATION	24
1.4	EXERCISES	32

CHAPTER 2 ▪ Normed Linear Spaces and Banach Spaces   35

2.1	BASIC CONCEPTS AND DEFINITIONS	35
2.2	LINEAR MAPS AND THE DUAL SPACE	42
2.3	SOME IMPORTANT EXAMPLES	47
	2.3.1 Finite-dimensional spaces	47
	2.3.2 The spaces ℓ^p	50
	2.3.3 The Lebesgue spaces $L^p(\Omega)$	53
2.4	HAHN-BANACH THEOREMS	59
2.5	APPLICATIONS OF THE HAHN-BANACH THEOREM	65
2.6	THE OPEN MAPPING THEOREM	71
2.7	THE UNIFORM BOUNDEDNESS PRINCIPLE	77
2.8	THE EMBEDDING OF X INTO ITS DOUBLE DUAL X^{**}	79
2.9	COMPACTNESS AND WEAK CONVERGENCE IN AN NLS	81
	2.9.1 The norm or strong topology	81
	2.9.2 The weak and weak-∗ topologies	82
2.10	THE DUAL OF AN OPERATOR	89
2.11	EXERCISES	93

CHAPTER 3 ■ Hilbert Spaces   103

3.1   BASIC PROPERTIES OF INNER-PRODUCTS   103
3.2   BEST APPROXIMATION AND ORTHOGONAL
      PROJECTION   107
3.3   DUALITY IN HILBERT SPACES   111
3.4   ORTHONORMAL SUBSETS AND BASES   113
3.5   WEAK CONVERGENCE IN A HILBERT SPACE   122
3.6   EXERCISES   124

CHAPTER 4 ■ Spectral Theory and Compact Operators   129

4.1   DEFINITIONS OF THE RESOLVENT AND SPECTRUM   130
4.2   BASIC SPECTRAL THEORY IN BANACH SPACES   132
4.3   COMPACT LINEAR OPERATORS ON A BANACH SPACE   134
4.4   BOUNDED SELF-ADJOINT LINEAR OPERATORS ON A
      HILBERT SPACE   143
4.5   COMPACT SELF-ADJOINT LINEAR OPERATORS ON A
      HILBERT SPACE   149
4.6   THE ASCOLI-ARZELÀ THEOREM   153
4.7   STURM-LIOUVILLE THEORY   155
      4.7.1   Sturm-Liouville problems and Green's functions   158
      4.7.2   Spectral properties of the solution operator   164
      4.7.3   Some applications   168
4.8   EXERCISES   171

CHAPTER 5 ■ Distributions   178

5.1   THE NOTION OF GENERALIZED FUNCTIONS   178
5.2   TEST FUNCTIONS   181
5.3   DISTRIBUTIONS   184
5.4   OPERATIONS WITH DISTRIBUTIONS   188
      5.4.1   Multiplication by a smooth function   188
      5.4.2   Differentiation   189
      5.4.3   Translations and dilations of $\mathbb{R}^d$ 192
      5.4.4   Convolutions   193
5.5   CONVERGENCE OF DISTRIBUTIONS AND
      APPROXIMATIONS TO THE IDENTITY   196

5.6 SOME APPLICATIONS TO DIFFERENTIAL EQUATIONS   199
    5.6.1   Ordinary differential equations   199
    5.6.2   Partial differential equations and fundamental
          solutions   203
5.7 LOCAL STRUCTURE OF $\mathcal{D}'$ 208
5.8 EXERCISES   209

CHAPTER 6 ■ The Fourier Transform   213

6.1 MOTIVATION FOR FOURIER ANALYSIS   213
6.2 THE $L^1(\mathbb{R}^d)$ THEORY   216
6.3 THE SCHWARTZ SPACE THEORY   221
6.4 THE $L^2(\mathbb{R}^d)$ THEORY   228
6.5 THE $\mathcal{S}'$ THEORY   231
6.6 SOME APPLICATIONS   238
    6.6.1   The heat equation   238
    6.6.2   The Schrödinger equation   239
    6.6.3   Signal processing and translation invariance   241
6.7 EXERCISES   246

CHAPTER 7 ■ Sobolev Spaces   252

7.1 DEFINITIONS AND BASIC PROPERTIES   253
7.2 EXTENSIONS FROM Ω TO $\mathbb{R}^d$ 259
7.3 THE SOBOLEV EMBEDDING THEOREM   265
7.4 COMPACTNESS   273
7.5 THE H^s SOBOLEV SPACES   275
7.6 TRACE THEOREMS   281
7.7 THE $W^{s,p}(\Omega)$ SOBOLEV SPACES   289
7.8 EXERCISES   290

CHAPTER 8 ■ Boundary Value Problems   295

8.1 SECOND ORDER LINEAR ELLIPTIC PDES   296
    8.1.1   Practical examples   297
    8.1.2   Boundary conditions (BCs)   299
8.2 VARIATIONAL PROBLEMS AND MINIMIZATION OF
    ENERGY   300
8.3 THE CLOSED RANGE THEOREM AND LINEAR
    OPERATORS BOUNDED BELOW   305

8.4	THE BABUŠKA-LAX-MILGRAM THEOREM	306
8.5	APPLICATION TO LINEAR ELLIPTIC PDES	311
8.5.1	The general Dirichlet problem	312
8.5.2	The Neumann problem with lowest order term	313
8.5.3	The Neumann problem with no zeroth order term	315
8.5.4	Elliptic regularity	317
8.6	GALERKIN METHODS	318
8.7	GREEN'S FUNCTIONS	321
8.8	EXERCISES	325

CHAPTER 9 ■ Differential Calculus in Banach Spaces — 334

9.1	DIFFERENTIATION	335
9.1.1	The chain rule	339
9.1.2	The Mean-Value Theorem	341
9.1.3	Partial differentiation	343
9.2	FIXED POINTS AND CONTRACTIVE MAPS	345
9.3	NONLINEAR EQUATIONS	348
9.3.1	Newton methods	349
9.3.2	The Inverse Function Theorem	351
9.3.3	The Implicit Function Theorem	355
9.4	HIGHER DERIVATIVES	358
9.5	EXTREMA	363
9.6	EXERCISES	367

CHAPTER 10 ■ The Calculus of Variations — 372

10.1	THE EULER-LAGRANGE EQUATIONS	373
10.2	CONSTRAINED EXTREMA AND LAGRANGE MULTIPLIERS	380
10.3	LOWER SEMICONTINUITY AND EXISTENCE OF MINIMA	384
10.4	EXERCISES	390

Bibliography	395
Index	397

Preface

In the Fall Semester of 1993, faculty from the Mathematics, Computer Science, Physics and several Engineering Departments at the University of Texas at Austin banded together in a new initiative, the Texas Institute for Computational and Applied Mathematics (TICAM). This was a degree-granting research program with its own curriculum in Computational and Applied Mathematics (CAM). The idea was to provide students with a broad interdisciplinary but high-level education in mathematics, computer science, and the engineering and scientific applications of these subjects. The first-year graduate mathematics offering in CAM, a year-long course taught from the Mathematics Department, was initially put on its feet by Bona and Professor Karen Uhlenbeck. Later, Arbogast joined in and for several years, Arbogast and Bona alternated teaching the two semesters of this course.

Over the years, TICAM broadened its scope to embrace all the sciences, engineering, finance, and medicine, and it was renamed the Institute for Computational Engineering and Sciences. To honor its founder, Professor J. Tinsley Oden, the name was modified in 2019 to the Oden Institute of Computational Engineering and Sciences. The graduate program also had a new moniker, becoming the Computational Science, Engineering, and Mathematics (CSEM) program in 2010.

The CSEM program attracts rather well-prepared students, so its first-year graduate courses in mathematics and computer science move along at a good clip. As the goal of the program is to produce students whose focus is on applications, but who can use expertly the tools of applied and computational mathematics, we put a lot of content into these courses. The material covered in the two-semester mathematics course did not settle down right away, but by the turn of the millennium, it was reaching a steady state. Because of the research interests among the institute's faculty, the mathematics course was designed to prepare students well for further study of functional analysis, partial differential equations, and modern numerical analysis. As we went along, we noticed that students were being asked to acquire at least three different books as references to the material developed in the course. This led the two of us to start producing course notes, and those became the basis for this book.

The text begins with a short chapter reminding the reader of some prerequisite material. It includes essentially all of the elements of point-set topology, Lebesgue integration theory, and complex analysis required by the subsequent development. Especially the topology will intrude often and ubiquitously into the rest of the text. However, it should be emphasized that a previous course in measure theory is not required. Aside from proving certain technical properties of the Lebesgue and Sobolev spaces, all that finds use in the body of the text are the three main limit theorems, and these arise mostly in examples. A bit of complex contour integration is also included for the student who may encounter in a later chapter the need to compute certain Fourier transforms. The well-prepared student will know most or all of the material in the first chapter. Others seem to assimilate this material as the course develops. We do not cover this chapter explicitly within the CSEM course.

Chapters 2–4 cover basic functional analysis of Banach and Hilbert spaces, exploring the interplay between algebraic linear space theory, topology, and duality, often in the context of applications to analysis and partial differential equations. These chapters begin from normed linear spaces and end with basic spectral theory for compact operators and its applications, including an extended discussion of Sturm-Liouville theory. This is followed by a chapter on the elements of generalized functions or distribution theory. The first semester CSEM course is intended to cover Chapters 2–5, at a fast pace. Over the years, various people have taught the course, and some instructors find that they can cover only Chapters 2–4.

The second half of the book explores applications of the functional analysis developed in the first half. Chapter 6 presents Fourier analysis in both L^2 and the space of tempered distributions, while Chapter 7 explores Sobolev spaces, including the embedding theorems, compactness results, and trace theorems. This is material that we feel is essential to have under one's command when pursuing modern applied analysis.

The development then turns in Chapter 8 to linear boundary-value problems for second-order elliptic equations where the use of some of the methods and concepts introduced earlier are on display. Closed range and Lax-Milgram-type results are developed, as well as Galerkin methods and Green's functions.

The book closes with two chapters on nonlinear functional analysis. Calculus in Banach spaces, fixed point theorems, inverse and implicit function theorems, Newton and Newton-Kantorovich results, and extrema are discussed in Chapter 9. It segues naturally into the final chapter on calculus of variations. (It is often the case that not enough time remains in the semester to fully cover this chapter.)

The book is relatively self-contained throughout its development, providing a rigorous introduction to functional analysis and its applications, except at a few points which require advanced results from analysis or logic, or look ahead to more general results only stated without proof. It is not the most general development possible. Rather it is an introduction to the subject with an emphasis on the basics useful in applications. Students are called upon to actively participate in the course's development in the generous set of exercises both within the text and at the end of each chapter. Some of these problems are strictly for practice with the material, but others call upon students to prove some of the basic results or their straightforward generalizations. The extensive index may prove useful to students in navigating the concepts and results, as well as to those using the text as a reference book.

While the material is tailored to the CAM/CSEM curriculum, our course has attracted students from the Mathematics Department as well as occasional students from engineering and the sciences, especially physics. Other instructors have used our course notes for a one-semester course in functional analysis and its applications. The chapters on calculus in Banach spaces together with the chapter on calculus of variations provided the starting point for an advanced course in nonlinear functional analysis taught by one of us.

Over the years, many of our mathematics colleagues have taken turns with the course and the set of notes we compiled. We are especially grateful to Professors Irene Gamba, Luis Caffarelli, Alexis Vasseur, Thomas Chen, and others for their diligent efforts in improving the quality of the notes. We are also grateful to the many former graduate students, unfortunately too many to name individually, who have endeavored to learn the material, and provided exceptional feedback. We hope that students and instructors find the book a useful asset in their journey into mathematics and its applications.

Todd Arbogast
Jerry L. Bona

July 1, 2024

Author Biographies

Todd Arbogast, Professor of Mathematics, was born on December 9, 1957. He earned his Ph.D. from the University of Chicago in 1987 under the direction of Professor Jim Douglas, Jr. He has held faculty positions at Purdue University, Rice University, and the University of Texas at Austin, where he is a core member of the Oden Institute for Computational Engineering and Sciences. He is a Fellow of both the American Mathematical Society and the Society for Industrial and Applied Mathematics. His research contributes to the development and analysis of numerical algorithms for the approximation of partial differential systems, homogenization and multiscale modeling, and scientific computation, as well as applications of the same to the modeling and simulation of multiphase flow and transport through geologic porous media.

Jerry L. Bona, Professor of Mathematics, Statistics & Computer Science, was born on February 5, 1945. He earned his Ph.D. in 1971 from Harvard University under the supervision of Professor Garrett Birkhoff. His early work in the Fluid Mechanics Research Institute at the University of Essex with Professors Brooke Benjamin and J. J. Mahony resulted in a model equation for long waves in non-linear dispersive systems, known as the Benjamin-Bona-Mahony equation. He has held faculty positions at the University of Chicago, the Pennsylvania State University, the University of Texas at Austin, and the University of Illinois at Chicago. He is a fellow of the American Mathematical Society, the Society for Industrial and Applied Mathematics, the Centre de Recherche Mathématiques of the Université de Montréal, and the American Association for the Advancement of Science. His research is in fluid mechanics, oceanography, coastal engineering, mathematical aspects of biology, mathematical economics, and the associated theory of partial differential equations, computational mathematics, and numerical analysis.

.

Synopsis

Chapter 1, Preliminaries, is designed to assist the reader with basic prerequisite material that will find use throughout. Three topics are discussed. Well prepared students may wish to skip one or more of these topics and simply refer back to parts of the material as needed. The first topic is elementary or point-set topology, which is concerned with open and closed sets and concepts of convergence of sequences, limit points, continuity of functions, and compactness. The subject is discussed in a general setting, although a particular emphasis is on metric topologies. The second topic discussed is Lebesgue measure and integration. Certain technical details need to be formalized regarding how one measures the size of a set and integrates a function. The three main convergence theorems (monotone, Fatou, and dominated) are also discussed. Topic three provides some elementary complex analysis, including the basics of holomorphic or analytic functions, the residue theorem, and complex contour integration. The latter is useful for computing Fourier transforms and will come up in Chapter 6.

Chapter 2, Normed Linear Spaces and Banach Spaces, explores the interplay between vector space theory, topology, and duality. One can begin the subject of Functional Analysis from many starting points. This book starts the subject with finite and infinite dimensional normed linear spaces over the ground field of either real or complex numbers. Banach spaces find application in many branches of pure and applied mathematics, and it is important to have a basic understanding of their properties. Both Banach and Hilbert spaces (Chapter 3) are special examples of normed linear spaces. The norm gives both a measure of the size of a vector and induces a metric topology. Once the interaction of topology and vector space structure is understood, the three fundamental theorems of the theory, the Hahn-Banach theorem, the open mapping theorem, and the uniform boundedness principle are discussed in detail and many of their important corollaries are derived. Interest in all these results turns around mappings between normed linear spaces that interact well with their structure. These are the linear and continuous mappings. The dual space is a special collection of such mappings, the set of continuous linear mappings from a given space into the ground field. Such objects map an abstract vector into the space of real or complex numbers. The set of all these numbers characterizes the vector. Important examples of Banach spaces, the Lebesgue spaces of real or complex-valued integrable functions, are explored in some detail. The chapter concludes with the notion of weak convergence and compactness which will play roles many times in what follows.

Chapter 3, Hilbert Spaces, discusses linear spaces with an inner-product. An inner-product induces a norm, and hence a topology, and also leads naturally to the concepts of orthogonality, best approximation, and basis. Given a point, there is a closest point (or best approximation) within any closed convex set that minimizes the norm of the difference of the two points. If the convex set is a linear subspace, then the difference of the point and its best approximation is orthogonal to the subspace. This leads to a discussion of orthogonal projection operators. Duality in Hilbert spaces is simpler than it is in general Banach spaces, since the Reisz representation theorem specifies that the dual space of a Hilbert space is equivalent to the space itself. Every Hilbert space has an orthonormal basis of vectors that are both mutually orthogonal and normalized to each have size one. As an important application, one sees that Fourier series forms a basis for the Lebesgue space of order 2 of periodic functions. The chapter ends with a discussion of weak convergence in a Hilbert space, which is much simpler than that in Banach spaces.

Chapter 4, Spectral Theory and Compact Operators, is concerned with questions of invertibility of an operator mapping a Banach space to itself. Invertibility of an operator in infinite dimensions can be rather subtle. For example, spectral values may not have associated eigenvalues with corresponding eigenvectors. The theory discussed can be applied to gain insight into the structure of a particular operator. The theory is further developed for special operators known as compact and/or self-adjoint operators. In particular, compact self-adjoint operators on a Hilbert space provide an orthonormal basis of eigenvectors for the space which behaves nicely under the action of the operator. As a consequence of the Ascoli-Arzelà theorem, important examples of compact operators include many integral operators. The chapter ends with an extended application of spectral theory to boundary-value problems in the form of the classic Sturm-Liouville theory. The notion of a Green's function is introduced, which is an integral operator and thus compact. Cosine and sine series are extracted, and the technique of separation of variables is justified.

Chapter 5, Distributions, also called generalized functions, are a generalization of the notion of a function mapping some domain to the real or complex field. Distributions are defined via duality. The basic domain is not Euclidean space, but the space of test functions consisting of infinitely differentiable functions with compact support. These form a vector space with a special notion of convergence, which leads to a topology on the space. This topology does not arise from a metric, so some care is required in this chapter. The dual of the space of test functions is the space of distributions. So, a distribution is a linear mapping of the test functions into the real or complex numbers. Most of the functions appearing in analysis can be viewed as distributions with their action being defined by integration of its product with a test function. A distribution that is not given by a function is evaluation of a test function at a point. This is well-defined, and leads to the so-called delta function distribution, but it is certainly not defined by integration against a function. Many operations related to functions extend to distributions, including multiplication by a smooth function, translations and dilations, convolutions, and, most importantly, differentiation. A topology is given to the space of distributions, allowing discussion of the convergence of sequences of distributions. The chapter concludes with applications to ordinary and partial differential equations and the notion of a fundamental solution.

Chapter 6, The Fourier Transform, concerns decomposing a function into simple harmonic functions. The Fourier transform is the amplitude of each simple harmonic function in the decomposition. One can recover the function by inverting the process, at least for some functions. It is easy to define the Fourier transform on the Lebesgue space of order 1. It maps into but not onto the Lebesgue space of order infinity, so the inverse transform is not so easily understood. The natural space of definition is the Lebesgue space of order 2; however, it is not a simple task to define the transform here. Instead, the Fourier transform is defined on a class of functions, the Schwartz space of infinitely differentiable functions that decrease rapidly at infinity. On this space, interesting properties of the Fourier transform can be developed, including that it is invertible, that convolution transforms to multiplication, and that differentiation transforms to multiplication by a polynomial. By continuous extension of the operator from the Schwartz class, one obtains the Fourier transform on the Lebesgue space of order 2 with the same properties. The dual of the Schwartz space is the space of tempered distributions, a subset of the distributions, on which the Fourier transform can be defined. Some important problems in applied mathematics related to partial differential equations and translation invariant operators arising in signal processing are solved using what was learned about the Fourier transform.

Chapter 7, Sobolev Spaces, explores families of Banach spaces of functions that have (distributional) derivatives up to some order that lie in some Lebesgue space. The Sobolev embedding theorem shows that some of these spaces are embedded in others. This is fundamentally a trade-off between the number of derivatives and the order of the Lebesgue norm. In some important cases, these spaces are contained within the set of continuous functions, and so can viewed as having point values. The notion of fractional derivatives, which appears in Chapter 8, is introduced by way of the Fourier transform. Useful concepts for a function belonging to a particular Sobolev space are considered, such as its extension beyond the boundary of its domain and its trace, i.e., its value on the boundary. The compactness results of the Rellich-Kondrachov theorem are also developed. In the right circumstances, this allows extracting a convergent subsequence from a bounded sequence. An understanding of Sobolev spaces is essential to have under one's command when pursuing modern applied analysis.

Chapter 8, Boundary Value Problems, turns to a development of the techniques needed to solve linear elliptic boundary-value problems. The basic theory is developed to solve abstract variational problems and operator equations. Many of the concepts and methods introduced earlier are on display in this chapter. Indeed, it was attempts to solve such problems that led to many of these ideas. New concepts are also introduced, including the Poincaré inequality and results that relate properties of an operator to its dual operator, such as the closed range theorem, inf-sup, and coercivity conditions leading to Lax-Milgram-type existence and uniqueness results for variational problems. The theory is then brought to bear upon elliptic boundary-value problems with various boundary conditions in Sobolev spaces. Elliptic regularity, Galerkin methods, and Green's functions also arise in the discussion.

Chapter 9, Differential Calculus in Banach Spaces, begins the study of nonlinear functional analysis. Although many of the results will appear at first glance to be familiar from advanced calculus, the focus here is on abstract and possibly infinite dimensional Banach spaces. Elementary calculus in Banach spaces is developed first, including notions of differentiation, the chain rule, and a mean-value theorem. Attention is then turned to solving nonlinear equations. Developed are the Banach contraction mapping theorem for finding fixed points, Newton-like methods to find zeroes, and inverse and implicit function theorems. The chapter closes with a discourse on higher-order derivatives and extrema. Along the way, the ideas are applied to showing that nonlinear ordinary and partial differential equations have solutions, and that the eigenvalues of a matrix depend continuously on its coefficients, at least until a bifurcation occurs.

Chapter 10, The Calculus of Variations, is a short introduction to this classic subject. One is given a real-valued functional and asked to find functions that minimize the functional. On its face, one merely needs to use the previous chapter to find the critical points, and determine which are minima. However, when the functional is an integral operator of a certain form often appearing in applications, the search for a critical point is simplified and leads to the Euler-Lagrange equations. This system of second-order, nonlinear ordinary differential equations speeds the analysis. An exploration of minimization under constraints is also pursued, resulting in the theory of Lagrange multipliers. A brief foray into the theory's modern counterpart concludes the chapter, resulting in the elucidation of the concept of lower semicontinuity. Several classic problems are solved, including the isoperimetric problem, and the existence of geodesics on manifolds is established.

CHAPTER 1

Preliminaries

Discussed in this chapter are some pertinent aspects of topology, measure theory, and complex integration that are needed in the rest of the book. This material is treated as background, and well-prepared students may wish to skip one or more of these topics.

In the précis of these subjects offered here, a few of the proofs are presented. However, many proofs are omitted or relegated to the problem set at the end of the chapter. If no indication of proof is provided, either it follows directly from the definitions or the reader should provide the relatively simple proof so as to better engage the ideas.

1.1 ELEMENTARY OR POINT-SET TOPOLOGY

In applied mathematics, we are often faced with analyzing mathematical structures as they might relate to real-world phenomena. In applying mathematics, real phenomena or objects are conceptualized as abstract mathematical objects. Collections of such objects are called *sets*. The objects in a set of interest may also be related to each other; that is, there is some *structure* on the set. We call such a structured set a *space*.

Examples.

(1) A vector space (algebraic structure).

(2) The set of integers $\mathbb{Z}$ (number theoretical structure or arithmetic structure).

(3) The set of real numbers $\mathbb{R}$ or the set of complex numbers $\mathbb{C}$ (algebraic and topological structure).

We start the discussion of spaces by putting forward sets of "points" on which we can talk about the notions of *convergence* or *limits* and associated *continuity* of functions. A simple example is a set X with a notion of distance between any two points of X. A sequence $\{x_n\}_{n=1}^{\infty} \subset X$ converges to $x \in X$

DOI: 10.1201/9781003492139-1

if the distance from x_n to x tends to 0 as n increases. This definition relies on the following formal concept.

Definition. A *metric* or *distance function* on a set X is a function $d : X \times X \to \mathbb{R}$ satisfying:

(a) (positivity) for any $x, y \in X$, $d(x, y) \geq 0$, and $d(x, y) = 0$ if and only if $x = y$;

(b) (symmetry) for any $x, y \in X$, $d(x, y) = d(y, x)$;

(c) (triangle inequality) for any $x, y, z \in X$, $d(x, y) \leq d(x, z) + d(z, y)$.

A *metric space* (X, d) is a set X together with an associated metric $d : X \times X \to \mathbb{R}$.

Example. $(\mathbb{C}^d, |\cdot|)$ is a metric space, where for $x, y \in \mathbb{C}^d$, the distance from x to y is

$$|x - y| = \left(\sum_{i=1}^{d} |x_i - y_i|^2 \right)^{1/2}.$$

Similarly, $(\mathbb{R}^d, |\cdot|)$ is a metric space with the same metric.

It turns out that the notion of distance or metric is sometimes stronger than what actually appears in practice. The more fundamental concept upon which much of the mathematics developed here rests is that of limits. That is, there are important spaces arising in applied mathematics that have well-defined notions of limits, but these limiting processes are *not* compatible with any metric. We shall see such examples later; let it suffice for now to motivate introducing a weaker definition of limits.

A sequence of points $\{x_n\}_{n=1}^{\infty}$ can be thought of as converging to x if every "neighborhood" of x contains all but finitely many of the x_n, where a neighborhood is a subset of points containing x that we think of as "close" to x. Such a structure is called a *topology*. One studies this structure in the subject called *point-set topology*. It is formalized as follows.

Definition. A *topological space* $(X, \mathcal{T})$ is a nonempty set X of points with a family $\mathcal{T}$ of subsets, called *open*, with the properties:

(a) $X \in \mathcal{T}$, $\emptyset \in \mathcal{T}$;

(b) If $\omega_1, \omega_2 \in \mathcal{T}$, then $\omega_1 \cap \omega_2 \in \mathcal{T}$;

(c) If $\omega_\alpha \in \mathcal{T}$ for all α in some index set $\mathcal{I}$, then $\bigcup_{\alpha \in \mathcal{I}} \omega_\alpha \in \mathcal{T}$.

The family $\mathcal{T}$ is called a *topology* for X. Given $A \subset X$, we say that A is *closed* if its complement $A^c = X \setminus A = \{x \in X : x \notin A\}$ is open.

For the time being, the discussion is focused on a fixed but arbitrary topological space $(X, \mathcal{T})$.

Example. If X is any nonempty set, we can always define the two topologies:

(1) $\mathcal{T}_1 = \{\emptyset, X\}$, called the *trivial* topology;

(2) $\mathcal{T}_2$ consisting of the collection of all subsets of X, called the *discrete* topology.

Proposition 1.1. *The sets $\emptyset$ and X are both open and closed. Any finite intersection of open sets is open. Any intersection of closed sets is closed. The union of any finite number of closed sets is closed.*

Proof. We need only show the last two statements, as the first two follow directly from the definitions. Let $A_\alpha \subset X$ be closed for $\alpha \in \mathcal{I}$. Then one of De Morgan's laws yields

$$\left(\bigcap_{\alpha \in \mathcal{I}} A_\alpha \right)^c = \bigcup_{\alpha \in \mathcal{I}} A_\alpha^c \quad \text{is open.}$$

Finally, if $\mathcal{J} \subset \mathcal{I}$ is finite, the other De Morgan law gives

$$\left(\bigcup_{\alpha \in \mathcal{J}} A_\alpha \right)^c = \bigcap_{\alpha \in \mathcal{J}} A_\alpha^c \quad \text{is open.} \qquad \square$$

It is often convenient to define a simpler collection of open sets that immediately generates a topology.

Definition. Given a topological space $(X, \mathcal{T})$ and an $x \in X$, a *base for the topology at x* is a collection $\mathcal{B}_x$ of open sets containing x such that for any open $E \ni x$, there is $B \in \mathcal{B}_x$ such that

$$x \in B \subset E.$$

A *base for topology* $\mathcal{B}$ is a collection of open sets that contains a base at x for all $x \in X$.

Proposition 1.2. *A collection $\mathcal{B}$ of subsets of X is a base for a topology $\mathcal{T}$ if and only if (1) each $x \in X$ is contained in some $B \in \mathcal{B}$ and (2) if $x \in B_1 \cap B_2$ for $B_1, B_2 \in \mathcal{B}$, then there is some $B_3 \in \mathcal{B}$ such that $x \in B_3 \subset B_1 \cap B_2$. If (1) and (2) are valid, then*

$$\mathcal{T} = \{E \subset X : E \text{ is a union of subsets in } \mathcal{B}\}.$$

Proof. ($\implies$) Since X and $B_1 \cap B_2$ are open, (1) and (2) follow from the definition of a base at x.

($\impliedby$) Let $\mathcal{T}$ be defined as above. Then $\emptyset \in \mathcal{T}$ (the vacuous union), $X \in \mathcal{T}$ by (1), and arbitrary unions of sets in $\mathcal{T}$ are again in $\mathcal{T}$. It remains to show the intersection property. Let $E_1, E_2 \in \mathcal{T}$, and $x \in E_1 \cap E_2$ (if $E_1 \cap E_2 = \emptyset$, there is nothing to prove). Then there are sets $B_1, B_2 \in \mathcal{B}$ such that

$$x \in B_1 \subset E_1, \quad x \in B_2 \subset E_2,$$

so

$$x \in B_1 \cap B_2 \subset E_1 \cap E_2.$$

But, property (2) then provides a $B_3 \in \mathcal{B}$ such that

$$x \in B_3 \subset E_1 \cap E_2.$$

Thus $E_1 \cap E_2$ is a union of elements in $\mathcal{B}$, and is thus in $\mathcal{T}$. □

We remark that instead of using open sets, one can equally well consider *neighborhoods* of points $x \in X$, which are sets $N \ni x$ such that there is an open set E satisfying $x \in E \subset N$.

Theorem 1.3. *If (X,d) is a metric space, then $(X,\mathcal{T})$ is a topological space, where a base for the topology is given by*

$$\mathcal{T}_B = \{B_r(x) : x \in X \text{ and } r > 0\},$$

where

$$B_r(x) = \{y \in X : d(x,y) < r\}$$

is the ball of radius r about x.

Proof. Point (1) of Proposition 1.2 is clear. For (2), suppose $x \in B_r(y) \cap B_s(z)$. Then $x \in B_\rho(x) \subset B_r(y) \cap B_s(z)$, where $\rho = \frac{1}{2}\min(r - d(x,y), s - d(x,z)) > 0$. □

Thus metric spaces have a natural topological structure. However, not all topological spaces are induced as above by a metric, so the class of topological spaces is genuinely richer.

Definition. Let $(X,\mathcal{T})$ be a topological space. The *closure* of $A \subset X$, denoted $\bar{A}$, is the intersection of all closed sets containing A:

$$\bar{A} = \bigcap_{\substack{F \text{ closed} \\ F \supset A}} F.$$

Proposition 1.4. *The set $\bar{A}$ is closed, and it is the smallest closed set containing A.*

Proof. This follows by Proposition 1.1 and the definition. □

Definition. The *interior* of $A \subset X$, denoted A°, is the union of all open sets contained in A, *viz.*

$$A^\circ = \bigcup_{\substack{E \text{ open} \\ E \subset A}} E.$$

Proposition 1.5. *The set A° is open and is the largest open set contained in A.*

Proof. This also follows from Proposition 1.1 and the definition. □

Proposition 1.6. *It is always the case that*

$$A \subset \bar{A}, \quad \bar{\bar{A}} = \bar{A}, \quad A \subset B \implies \bar{A} \subset \bar{B}, \quad \overline{A \cup B} = \bar{A} \cup \bar{B},$$

$$and \quad A \ closed \iff A = \bar{A}.$$

Moreover,

$$A \supset A^\circ, \quad A^{\circ\circ} = A^\circ, \quad A \subset B \implies A^\circ \subset B^\circ, \quad (A \cap B)^\circ = A^\circ \cap B^\circ,$$

$$and \quad A \ open \iff A = A^\circ.$$

Proposition 1.7. *It holds that* $(A^c)^\circ = (\bar{A})^c$ *and* $(A^\circ)^c = \overline{(A^c)}$.

Proof. The first result follows because

$$x \notin (\bar{A})^c \iff x \in \bar{A} \iff x \in \bigcap_{\substack{F \ closed \\ F \supset A}} F$$

$$\iff x \notin \left(\bigcap_{\substack{F \ closed \\ F \supset A}} F \right)^c \iff x \notin \bigcup_{\substack{F^c \ open \\ F^c \subset A^c}} F^c = (A^c)^\circ.$$

The second result is the complement of the first since $(A^c)^c = A$. □

Definition. A point $x \in X$ is an *accumulation point* of $A \subset X$ if every open set containing x intersects $A \setminus \{x\}$. Also, a point $x \in A$ is an *interior point* of A if there is some open set E such that

$$x \in E \subset A.$$

Finally, $x \in A$ is an *isolated point* if there is an open set $E \ni x$ such that $E \cap A = \{x\}$.

Proposition 1.8. *For* $A \subset X$, $\bar{A}$ *is the union of the set of accumulation points of* A *and* A *itself, and* A° *is the union of the interior points of* A.

Definition. A set $A \subset X$ is *dense* in X if $\bar{A} = X$.

Definition. The *boundary* of $A \subset X$, denoted ∂A, is

$$\partial A = \bar{A} \cap \overline{A^c}.$$

Proposition 1.9. *If* $A \subset X$, *then* ∂A *is closed and*

$$\bar{A} = A^\circ \cup \partial A, \quad while \quad A^\circ \cap \partial A = \emptyset.$$

Moreover,

$$\partial A = \partial A^c = \{x \in X : every \ open \ E \ni x \ intersects \ both \ A \ and \ A^c\}.$$

Note that if x is an isolated point of $A \subset \mathbb{R}$, then $x \notin A^\circ$, so $\partial A \neq \partial(A^\circ)$ in general.

Definition. A sequence $\{x_n\}_{n=1}^\infty \subset X$ *converges* to $x \in X$, or has *limit* x, if given any open $E \ni x$, there is $N > 0$ such that $x_n \in E$ for all $n \geq N$ (i.e., the entire tail of the sequence is contained in E).

Proposition 1.10. *If* $\lim_{n \to \infty} x_n = x$, *then* x *is an accumulation point of the set consisting of the elements of the sequence* $\{x_n\}_{n=1}^\infty$.

Notice that in a general topological space, a point x can be an accumulation point of $\{x_n\}_{n=1}^\infty$ despite the fact that no subsequence $\{x_{n_k}\}_{k=1}^\infty$ converges to x.

Example. Let X be the set of nonnegative integers and $\mathcal{T}_B = \{\{0, 1, \ldots, i\}$ for each $i \geq 1\}$ be a base for a topology on X. Then $\{x_n\}_{n=1}^\infty$ with $x_n = n$ has 0 as an accumulation point, but no subsequence converges to 0.

If $x_n \to x \in X$ and $x_n \to y \in X$, it is possible that $x \neq y$.

Example. Let $X = \{a, b\}$ and $\mathcal{T} = \{\emptyset, \{a\}, \{a, b\}\}$. Then the sequence $x_n = a$ for all n converges to both a and b.

Definition. A topological space $(X, \mathcal{T})$ is called *Hausdorff* if given distinct points $x, y \in X$, there are disjoint open sets E_1 and E_2 such that $x \in E_1$ and $y \in E_2$.

Proposition 1.11. *If* $(X, \mathcal{T})$ *is Hausdorff, then every set consisting of a single point is closed. Moreover, limits of sequences are unique.*

Definition. A point $x \in X$ is a *strict limit point* of $A \subset X$ if there is a sequence $\{x_n\}_{n=1}^\infty \subset A \setminus \{x\}$ such that $\lim_{n \to \infty} x_n = x$.

As will be seen now, metric spaces are less susceptible to the sort of pathology that can obtain in a general topological space.

Proposition 1.12. *If* (X, d) *is a metric space and* $A \subset X$, *then every* $x \in \partial A$ *is either an isolated point or a strict limit point of* A *and* A^c.

Proposition 1.13. *If* (X, d) *is a metric space and* $\{x_n\}_{n=1}^\infty$ *is a sequence in* X, *then* $x_n \to x$ *if and only if, given* $\epsilon > 0$, *there is* $N > 0$ *such that*

$$d(x, x_n) < \epsilon \quad \forall \, n \geq N.$$

That is, $x_n \in B_\epsilon(x)$ *for all* $n \geq N$.

Proof. If $x_n \to x$, then for any fixed open set $E \ni x$, a suitable tail of the sequence lies in E. In particular, this holds for the open sets $B_\epsilon(x)$. Conversely, if E is any open set containing x, then since the open balls centered around x form a base for the topology, there is an $\epsilon > 0$ such that $B_\epsilon(x) \subset E$. By assumption, there is an N such that for $n \geq N$, $x_n \in B_\epsilon$ and so the tail $\{x_n : n \geq N\}$ lies in E. $\qquad \square$

Proposition 1.14. *Every metric space is Hausdorff.*

Proposition 1.15. *If (X, d) is a metric space and $A \subset X$ has an accumulation point x, then there is some sequence $\{x_n\}_{n=1}^{\infty} \subset A$ such that $x_n \to x$.*

Proof. Given an integer $n \geq 1$, there is some point $y \in A$ such that $y \in B_{1/n}(x)$, since x is an accumulation point. For each n, choose such a point and call it x_n. Then, the sequence $\{x_n\}_{n=1}^{\infty}$ has the property that $x_n \to x$. $\square$

Proposition 1.16. *Suppose that (X, d) is a metric space and $A \subset X$ is nonempty. Then $x \in \bar{A}$ if and only if there is some sequence $\{x_n\}_{n=1}^{\infty} \subset A$ such that $x_n \to x$.*

Proof. If $x \in \bar{A}$, then either $x \in A$ and $x_n = x$ provides the sequence, or x is an accumulation point and Proposition 1.15 provides the sequence. Conversely, if $x_n \in A$ and $x_n \to x$, then every ball about x contains at least one of the x_n, and so either x is in A or x is an accumulation point of A. $\square$

A natural way to define continuity of a function f mapping a topological space X to a topological space Y would be to demand that whenever $x_n \to x$ in X, then $f(x_n) \to f(x)$ in Y. In general topological spaces, this notion, which is called *sequential continuity*, is subject to non-intuitive pathology. The problems arising from limit processes in general topological spaces are avoided by using the following definition of continuity.

Definition. A function f mapping a topological space $(X, \mathcal{T})$ into a topological space $(Y, \mathcal{S})$ is *continuous* if the inverse image of every open set in Y is open in X.

We say that f is continuous at a point $x \in X$ if given any open set $E \subset Y$ containing $f(x)$, then $f^{-1}(E) = \{x \in X : f(x) \in E\}$ contains an open set D containing x. That is,

$$x \in D \quad \text{and} \quad f(D) = \{f(x) \in Y : x \in D\} \subset E.$$

It transpires that a map $f : X \to Y$ is continuous if and only if it is continuous at each point of X.

Proposition 1.17. *If $f : X \to Y$ and $g : Y \to Z$ are continuous, then $g \circ f : X \to Z$ is continuous.*

Proposition 1.18. *If f is continuous and $x_n \to x$, then $f(x_n) \to f(x)$.*

The converse of Proposition 1.18 is false in general. As mentioned already, when the hypothesis $x_n \to x$ always implies $f(x_n) \to f(x)$, the function f is deemed to be *sequentially continuous*. Proposition 1.18 states that continuous functions are always sequentially continuous.

Proposition 1.19. *If $f : X \to Y$ is sequentially continuous and if X is a metric space, then f is continuous.*

Proof. Let $E \subset Y$ be open and $A = f^{-1}(E)$. We must show that A is open. Suppose not. Then there is some $x \in A$ such that $B_r(x) \not\subset A$ for any $r > 0$. Thus, for $r_n = 1/n$, $n \geq 1$ an integer, there is some $x_n \in B_{r_n}(x) \cap A^c$. Since f is sequentially continuous and $x_n \to x$, then $f(x_n) \to f(x) \in E$. Because E is open, this in turn implies that $f(x_n) \in E$ for n sufficiently large, a contradiction since $x_n \in A^c$ implies $f(x_n) \notin E$. $\qquad\square$

Let $(X, \mathcal{T})$ and $(Y, \mathcal{S})$ be topological spaces and suppose there is a map $f : X \to Y$ that is injective (one-to-one), surjective (onto), and such that both f and f^{-1} are continuous. Then f and f^{-1} map open sets to open sets. That is, $E \subset X$ is open if and only if $f(E) \subset Y$ is open. Therefore $f(\mathcal{T}) = \mathcal{S}$ and, from a topological point of view, X and Y are indistinguishable. Any topological property of X is shared by Y, and conversely. For example, if $x_n \to x$ in X, then $f(x_n) \to f(x)$ in Y and, conversely, $y_n \to y$ in Y implies $f^{-1}(y_n) \to f^{-1}(y)$ in X.

Definition. A *homeomorphism* between two topological spaces X and Y is a one-to-one continuous mapping f of X onto Y for which f^{-1} is also continuous. If there is a homeomorphism $f : X \to Y$, we say that X and Y are *homeomorphic*.

It is possible to define two or more nonhomeomorphic topologies on any set X of at least two points. If $(X, \mathcal{T})$ and $(X, \mathcal{S})$ are topological spaces, and $\mathcal{S} \supset \mathcal{T}$, then we say that $\mathcal{S}$ is *stronger* than $\mathcal{T}$ or that $\mathcal{T}$ is *weaker* than $\mathcal{S}$.

Example. The trivial topology $\mathcal{T} = \{\emptyset, X\}$ on a set X is weaker than any other topology on X. The discrete topology on X consisting of all subsets of X is stronger than any other topology on X.

Proposition 1.20. *The topology $\mathcal{S}$ is stronger than $\mathcal{T}$ if and only if the identity mapping $I : (X, \mathcal{S}) \to (X, \mathcal{T})$ is continuous.*

Proposition 1.21. *Given a collection $\mathcal{C}$ of subsets of X, there is a weakest topology $\mathcal{T}$ containing $\mathcal{C}$.*

Proof. Since the intersection of topologies is again a topology (prove this),

$$\mathcal{C} \subset \mathcal{T} = \bigcap_{\substack{\mathcal{S} \supset \mathcal{C} \\ \mathcal{S} \text{ a topology}}} \mathcal{S}$$

is the weakest such topology (which is nonempty since the discrete topology is a topology containing $\mathcal{C}$). $\qquad\square$

Given a topological space $(X, \mathcal{T})$ and $A \subset X$, a topology $\mathcal{S}$ on A may be defined by restriction. Specifically, the *inherited* or *induced* topology $\mathcal{S}$ on A is

$$\mathcal{S} = \mathcal{T} \cap A \equiv \{E \subset A : \text{ there is some } G \in \mathcal{T} \text{ for which } E = A \cap G\}.$$

That $\mathcal{S}$ is a topology on A is easily verified. When viewed with its induced topology, A is called a *subspace* of X.

Given two topological spaces $(X, \mathcal{T})$ and $(Y, \mathcal{S})$, the *product topology* $\mathcal{R}$ on the Cartesian product

$$X \times Y = \{(x, y) : x \in X,\ y \in Y\}$$

has as a base for open sets the collection

$$\mathcal{R}_B = \{E_1 \times E_2 : E_1 \in \mathcal{T},\ E_2 \in \mathcal{S}\}.$$

It is easily verified that this is indeed a base for a topology. Moreover, $\mathcal{T}$ and $\mathcal{S}$ could be replaced by associated bases and the same topology $\mathcal{R}$ would obtain.

Example. If (X, d_1) and (Y, d_2) are metric spaces, then a base for $X \times Y$ is

$$\{B_r(x) \times B_s(y) \subset X \times Y : x \in X,\ y \in Y \text{ and } r, s > 0\},$$

wherein the balls are defined with respect to the appropriate metric d_1 or d_2. Moreover, the nonnegative function $d : (X \times Y) \times (X \times Y) \to \mathbb{R}$ defined by

$$d((x_1, y_1), (x_2, y_2)) = d_1(x_1, x_2) + d_2(y_1, y_2)$$

is a metric that induces the product topology.

Example. The real plane $\mathbb{R}^2$ has two equivalent and natural bases for the usual Euclidean topology, namely the set of all (open) discs, and the set of all (open) rectangles.

The construction just introduced can be generalized to define a topology on an arbitrary product of topological spaces. Let $\{(X_\alpha, \mathcal{T}_\alpha)\}_{\alpha \in \mathcal{I}}$ be a collection of topological spaces. Then $X = \times_{\alpha \in \mathcal{I}} X_\alpha$, defined to be the collection of all points $\{x_\alpha\}_{\alpha \in \mathcal{I}}$ with the property that $x_\alpha \in X_\alpha$ for all $\alpha \in \mathcal{I}$, has a *product topology* with base

$$\mathcal{T}_B = \Big\{ \underset{\alpha \in \mathcal{I}}{\times} E_\alpha : E_\alpha \in \mathcal{T}_\alpha\ \forall\ \alpha \in \mathcal{I}$$
$$\text{and } E_\alpha = X_\alpha \text{ for all but finitely many } \alpha \in \mathcal{I} \Big\}.$$

The *projection map* $\pi_\alpha : X \to X_\alpha$ is defined for $x = \{x_\beta\}_{\beta \in \mathcal{I}}$ by $\pi_\alpha x = x_\alpha$. Thus π_α provides the αth coordinate of x.

Remark. The notation $\{x_\alpha\}_{\alpha \in \mathcal{I}}$ is properly understood as a map $g : \mathcal{I} \to \bigcup_{\alpha \in \mathcal{I}} X_\alpha$ such that $g(\alpha) = x_\alpha \in X_\alpha$ for all $\alpha \in \mathcal{I}$. Then $X = \times_{\alpha \in \mathcal{I}} X_\alpha$ is the collection of all such maps, and $\pi_\alpha(g) = g(\alpha)$ is evaluation at $\alpha \in \mathcal{I}$. However, we will continue to use the more informal conception of the product X as consisting of "points" $\{x_\alpha\}_{\alpha \in \mathcal{I}}$.

Proposition 1.22. *Each π_α is continuous. Furthermore, the product topology is the weakest topology on X that makes each π_α continuous.*

Proof. If $E_\alpha \subset X_\alpha$ is open, then

$$\pi_\alpha^{-1}(E_\alpha) = \underset{\beta \in \mathcal{I}}{\times} E_\beta,$$

where $E_\beta = X_\beta$ for $\beta \neq \alpha$, is a basic open set and so is open. Finite intersections of these sets must be open, and indeed these form our base. It is thus clear that the product topology as defined must form the weakest topology for which each π_α is continuous. □

Proposition 1.23. *If X and Y_α, $\alpha \in \mathcal{I}$, are topological spaces, then a function $f : X \to \times_{\alpha \in \mathcal{I}} Y_\alpha$ is continuous if and only if $\pi_\alpha \circ f : X \to Y_\alpha$ is continuous for each $\alpha \in \mathcal{I}$.*

Proposition 1.24. *If X is Hausdorff and $A \subset X$, then A is Hausdorff (in the inherited or induced topology). If $\{X_\alpha\}_{\alpha \in \mathcal{I}}$ are Hausdorff, then $\times_{\alpha \in \mathcal{I}} X_\alpha$ is Hausdorff (in the product topology).*

Most topologies of interest here have an infinite number of open sets. In general, large collections of open sets work against drawing certain types of important conclusions. However, there is an important class of topological spaces that possess a helpful finiteness property.

Definition. Let $(X, \mathcal{T})$ be a topological space and $A \subset X$. A collection $\{E_\alpha\}_{\alpha \in \mathcal{I}} \subset \mathcal{T}$ is called an *open cover* of A if each E_α is an open set and $A \subset \bigcup_{\alpha \in \mathcal{I}} E_\alpha$. If every open cover of A contains a finite subcover (i.e., any open cover $\{E_\alpha\}$ has the property that a finite number of its members already covers A), then A is called *compact*.

An interesting point immediately arises: does the compactness of A depend upon its external environment? Another way to ask this is, if $A \subsetneq X$ is compact, is A compact when it is viewed as a subset of itself? That is to say, $(A, \mathcal{T} \cap A)$ is a topological space, and $A \subset A$, so is A also compact in this context? What about the converse? If A is compact in itself, is A compact in X? It is easy to verify that both these questions are answered in the affirmative. Thus, compactness is a property of a set and its topology, independent of any larger space within which it may live.

The Heine-Borel theorem states that every closed and bounded subset of $\mathbb{R}^d$ is compact, and conversely. The proof is technical and can be found in most introductory books on real analysis (see, e.g., [18] or [19]).

Proposition 1.25. *A closed subset of a compact space is compact. A compact subset of a Hausdorff space is closed.*

Proof. Let X be compact and $F \subset X$ closed. If $\{E_\alpha\}_{\alpha \in \mathcal{I}}$ is an open cover of F, then $\{E_\alpha\}_{\alpha \in \mathcal{I}} \cup F^c$ is an open cover of X. By compactness, there is a finite subcover $\{E_\alpha\}_{\alpha \in \mathcal{J}} \cup F^c$. But then $\{E_\alpha\}_{\alpha \in \mathcal{J}}$ covers F, so F is compact.

Suppose X is Hausdorff and $K \subset X$ is compact. (We often write $K \subset\subset X$ in this case, and read it as "K compactly contained in X.") It suffices to show that K^c is open. Fix $y \in K^c$. Since X is Hausdorff, for each $x \in K$, there are open sets E_x and G_x such that $x \in E_x$, $y \in G_x$, and $E_x \cap G_x = \emptyset$. The sets $\{E_x\}_{x \in K}$ form an open cover of K, so a finite subcollection $\{E_x\}_{x \in A}$ still covers K. Thus

$$G = \bigcap_{x \in A} G_x$$

is open, contains y, and does not intersect K. Since y is arbitrary, K^c is open and therefore K is closed. □

Proposition 1.26. *The continuous image of a compact set is compact.*

An amazing fact about compact spaces is contained in the following theorem. Its proof can be found in most introductory texts in analysis or topology (see [11], [14], [18], [21]).

Theorem 1.27 (Tychonoff theorem). *Let $\{X_\alpha\}_{\alpha \in \mathcal{I}}$ be an indexed family of compact topological spaces. Then the product space $X = \times_{\alpha \in \mathcal{I}} X_\alpha$ is compact in the product topology.*

A common way to use compactness in metric spaces is delineated in the following result, which also characterizes compactness in that context.

Proposition 1.28. *Suppose (X, d) is a metric space. Then X is compact if and only if X is sequentially compact, which means that every sequence $\{x_n\}_{n=1}^{\infty} \subset X$ has a subsequence $\{x_{n_k}\}_{k=1}^{\infty}$ which converges to some point in X.*

Proof. Suppose that X is both compact and contains a sequence $\{x_n\}_{n=1}^{\infty}$ with no convergent subsequence. This sequence cannot contain a point that repeats an infinite number of times. Hence, it may be assumed without loss of generality that no point of the sequence is repeated, For each n, let

$$\delta_n = \inf_{m \neq n} d(x_n, x_m).$$

If $\delta_n = 0$ for some n, then there are x_{m_k}, $k = 1, 2, \ldots$, such that $m_k > m_{k-1}$ for $k > 1$ and

$$d(x_n, x_{m_k}) < \frac{1}{k},$$

which is to say, $x_{m_k} \to x_n$ as $k \to \infty$, a contradiction. So $\delta_n > 0$ for all n, and

$$\bigcup_{n=1}^{\infty} B_{\delta_n}(x_n) \cup \left(\bigcup_{n=1}^{\infty} \overline{B_{\delta_n/2}(x_n)} \right)^c$$

is an open cover of X with no finite subcover, contradicting the compactness of X and establishing the forward implication.

Suppose now that every sequence in X has a convergent subsequence. Let $\{U_\alpha\}_{\alpha \in \mathcal{I}}$ be a minimal open cover of X. By this we mean that no U_α may be removed from the collection if it is to remain a cover of X. Thus for each $\alpha \in \mathcal{I}$, there exists $x_\alpha \in X$ such that $x_\alpha \in U_\alpha$ but $x_\alpha \notin U_\beta$ for all $\beta \neq \alpha$. If $\mathcal{I}$ is infinite, we can choose $\alpha_n \in \mathcal{I}$ for $n = 1, 2, \ldots$ and a subsequence that converges, *viz.*

$$x_{\alpha_{n_k}} \to x \in X \quad \text{as} \quad k \to \infty.$$

Now $x \in U_\gamma$ for some $\gamma \in \mathcal{I}$. But then there is an $N > 0$ such that for all $k \geq N$, $x_{\alpha_{n_k}} \in U_\gamma$, a contradiction. Thus any minimal open cover is finite, and so X is compact. □

1.2 LEBESGUE MEASURE AND INTEGRATION

The Riemann integral is quite satisfactory for continuous functions, or functions with not too many discontinuities, defined on bounded subsets of $\mathbb{R}^d$. However, it is not so satisfactory for discontinuous functions, nor can it be easily generalized to functions defined on sets outside $\mathbb{R}^d$, such as probability spaces. Measure theory resolves these difficulties. It seeks to provide a way of specifying the 'size' of a large collection of subsets of a set X. From such a well-defined notion of size, the integral of a class of functions much larger than continuous functions can be defined. Additionally, this more general notion of integration behaves much better with regard to taking limits. Indeed, the lack of certain types of limit theorems in Riemann integration theory was an impetus for the development of Lebesgue integration. We summarize the basic theory here, but omit most of the proofs. They can be found in most standard texts in real analysis (see e.g., [18], [19], [20]).

It turns out that a consistent measure of subset size cannot be defined for all subsets of a set X. One must either modify the notion of size or restrict to only certain types of subsets. The latter course appears a good one since, as we will see, the subsets of $\mathbb{R}^d$ that can be measured include any set that can be approximated well by disjoint unions of rectangles.

Definition. A collection $\mathcal{A}$ of subsets of a set X is called a *σ-algebra* on X if

(a) $X \in \mathcal{A}$;

(b) whenever $A \in \mathcal{A}$, $A^c \in \mathcal{A}$;

(c) whenever $A_n \in \mathcal{A}$ for $n = 1, 2, 3, \ldots$ (countably many A_n), then also $\bigcup_{n=1}^{\infty} A_n \in \mathcal{A}$.

Proposition 1.29. *For a σ-algebra $\mathcal{A}$, the following properties hold:*

(a) $\emptyset \in \mathcal{A}$;

(b) *If $A_n \in \mathcal{A}$ for $n = 1, 2, \ldots$, then $\bigcap_{n=1}^{\infty} A_n \in \mathcal{A}$;*

(c) *If $A, B \in \mathcal{A}$, then $A \setminus B = A \cap B^c \in \mathcal{A}$.*

The proof follows directly from the definition.

Definition. By a *measure* on a σ-algebra $\mathcal{A}$, we mean a countably additive function $\mu : \mathcal{A} \to R$, where either $R = \mathbb{R}^+ = [0, +\infty]$, giving what is called a *positive measure* (as long as $\mu \not\equiv +\infty$), or $R = \mathbb{C}$, giving a *complex measure*. *Countably additive* means that if $A_n \in \mathcal{A}$ for $n = 1, 2, \ldots$, and $A_i \cap A_j = \emptyset$ for $i \neq j$, then

$$\mu\left(\bigcup_{n=1}^{\infty} A_n\right) = \sum_{n=1}^{\infty} \mu(A_n).$$

That is to say, the size or *measure* of a set is the sum of the measures of any countable collection of disjoint subsets of the set that cover it entirely.

Proposition 1.30. *Let $\mathcal{A}$ be a σ-algebra and μ a measure on $\mathcal{A}$.*

(a) $\mu(\emptyset) = 0$.

(b) *If $A_n \in \mathcal{A}$, $n = 1, 2, \ldots, N$ are pairwise disjoint, then*

$$\mu\left(\bigcup_{n=1}^{N} A_n\right) = \sum_{n=1}^{N} \mu(A_n).$$

(c) *If μ is a positive measure and $A, B \in \mathcal{A}$ with $A \subset B$, then*

$$\mu(A) \leq \mu(B).$$

(d) *If $A_n \in \mathcal{A}$, $n = 1, 2, \ldots$, and $A_n \subset A_{n+1}$ for all n, then*

$$\mu\left(\bigcup_{n=1}^{\infty} A_n\right) = \lim_{n\to\infty} \mu(A_n).$$

(e) *If $A_n \in \mathcal{A}$, $n = 1, 2, \ldots$, $\mu(A_1)$ is finite (i.e., μ is a complex measure or μ is a positive measure and $\mu(A_1) < \infty$), and $A_n \supset A_{n+1}$ for all n, then*

$$\mu\left(\bigcap_{n=1}^{\infty} A_n\right) = \lim_{n\to\infty} \mu(A_n).$$

Proof. (a) Since $\mu \not\equiv +\infty$, there is an $A \in \mathcal{A}$ such that $\mu(A)$ is finite. Now, write $A = A \cup \emptyset$; these two sets are pairwise disjoint, so $\mu(A) = \mu(A) + \mu(\emptyset)$. It follows that $\mu(\emptyset) = 0$.

(b) Let $A_n = \emptyset$ for $n > N$. Then

$$\mu\left(\bigcup_{n=1}^{N} A_n\right) = \mu\left(\bigcup_{n=1}^{\infty} A_n\right) = \sum_{n=1}^{\infty} \mu(A_n) = \sum_{n=1}^{N} \mu(A_n).$$

(c) Let $C = B \setminus A$. Then $C \cap A = \emptyset$, so

$$\mu(A) + \mu(C) = \mu(C \cup A) = \mu(B),$$

and the fact that $\mu(C) \geq 0$ gives the result.

(d) Let $B_1 = A_1$ and $B_n = A_n \setminus A_{n-1}$ for $n \geq 2$. Then the B_n are pairwise disjoint, and, for any $N \leq \infty$,

$$A_N = \bigcup_{n=1}^{N} A_n = \bigcup_{n=1}^{N} B_n,$$

so

$$\mu\left(\bigcup_{n=1}^{\infty} A_n\right) = \mu\left(\bigcup_{n=1}^{\infty} B_n\right) = \sum_{n=1}^{\infty} \mu(B_n)$$

$$= \lim_{N\to\infty} \sum_{n=1}^{N} \mu(B_n) = \lim_{N\to\infty} \mu\left(\bigcup_{n=1}^{N} B_n\right) = \lim_{N\to\infty} \mu(A_N).$$

(e) Let $B_n = A_n \setminus A_{n+1}$ and $B = \bigcap_{n=1}^{\infty} A_n$. Then the B_n and B are pairwise disjoint,

$$A_N = A_1 \setminus \bigcup_{n=1}^{N-1} B_n \quad \text{and} \quad A_1 = B \cup \bigcup_{n=1}^{\infty} B_n.$$

In consequence of the countable additivity,

$$\mu(A_1) = \mu(B) + \sum_{n=1}^{\infty} \mu(B_n),$$

which is finite, so

$$\mu(B) = \mu(A_1) - \sum_{n=1}^{N-1} \mu(B_n) - \sum_{n=N}^{\infty} \mu(B_n) = \mu(A_N) - \sum_{n=N}^{\infty} \mu(B_n).$$

Since the series $\sum_{n=1}^{\infty} \mu(B_n)$ converges, the limit as $N \to \infty$ of the second term on the right-hand side of the last equation is zero and the result follows. □

Definition. A triple $(X, \mathcal{A}, \mu)$ consisting of a set X, a σ-algebra $\mathcal{A}$ of subsets of X, and a measure μ defined on $\mathcal{A}$ is called a *measure space*.

An important σ-algebra is one generated by a topology, namely the family $\mathcal{B}$ of all Borel sets associated with a topological space. This is the smallest σ-algebra that contains all the open sets and is closed under countable unions and the taking of complements.

Definition. The collection of *Borel sets* $\mathcal{B}$ of a topological space $(X, \mathcal{O})$ is the smallest family of subsets of X with the properties:

(a) each open set is in $\mathcal{B}$, i.e., $\mathcal{O} \subset \mathcal{B}$;

(b) if $A \in \mathcal{B}$, then $A^c \in \mathcal{B}$;

(c) if $\{A_n\}_{n=1}^{\infty} \subset \mathcal{B}$, then $\bigcup_{n=1}^{\infty} A_n \in \mathcal{B}$.

That there is such a smallest family follows from the facts that the family of all subsets satisfies (a)–(c), and if $\{\mathcal{B}_\alpha\}_{\alpha \in \mathcal{I}}$ is any collection of families satisfying (a)–(c), then $\bigcap_{\alpha \in \mathcal{I}} \mathcal{B}_\alpha$ also satisfies (a)–(c).

Note that closed sets are in $\mathcal{B}$, as well as countable intersections by De Morgan's law. Obviously, $\mathcal{B}$ is a σ-algebra. In this text, the only Borel sets we will come across are the Borel sets associated to Euclidean space $\mathbb{R}^d$ with its standard metric topology.

Remark. The above definition makes sense relative to the open sets in any topological space.

Theorem 1.31. *There exists a unique positive measure μ, called* Lebesgue measure, *defined on the Borel sets $\mathcal{B}$ of $\mathbb{R}^d$, which is such that μ assigns the obvious measure to rectangles and μ is translation invariant. More precisely, if $A \subset \mathcal{B}$ is a rectangle, i.e., there are numbers a_i and b_i such that*

$$A = \{x \in \mathbb{R}^d : a_i < x_i \text{ or } a_i \leq x_i \text{ and } x_i < b_i \text{ or } x_i \leq b_i \ \forall \ i\},$$

then $\mu(A) = \prod_{i=1}^{d}(b_i - a_i)$. And, if $x \in \mathbb{R}^d$ and $A \in \mathcal{B}$, then

$$\mu(x + A) = \mu(A),$$

where $x + A = \{y \in \mathbb{R}^d : y = x + z \text{ for some } z \in A\} \in \mathcal{B}$.

The construction of Lebesgue measure is somewhat tedious, and can be found in many texts in real analysis (see again [18], [19], [20]). An interesting point arising in this theorem is to determine why $x + A \in \mathcal{B}$ if $A \in \mathcal{B}$. This follows since the mapping $f(y) = y + x$ is a homeomorphism of $\mathbb{R}^d$ onto $\mathbb{R}^d$, and hence preserves the open sets which generate the Borel sets.

Another interesting point presents itself as soon as Lebesgue measure is constructed on the Borel sets. If $A \in \mathcal{B}$ is such that $\mu(A) = 0$, we say A is a *set of measure zero*. As an example, a $(d-1)$-dimensional hyperplane has d-dimensional measure zero. If we intersect the hyperplane with a set $A \subset \mathbb{R}^d$, the measure should still be zero; however, such an intersection may not be a Borel set. It seems natural to require that if $\mu(A) = 0$ and $B \subset A$, then μ applies to B and $\mu(B) = 0$.

Let the collection of all sets of measure zero be denoted

$$\mathcal{Z} = \{A \subset \mathbb{R}^d : \ \exists \ B \in \mathcal{B} \text{ with } \mu(B) = 0 \text{ and } A \subset B\}$$

and define the *Lebesgue measurable sets* $\mathcal{M}$ to be

$$\mathcal{M} = \{A \subset \mathbb{R}^d : \ \exists \ B \in \mathcal{B}, \ Z_1, Z_2 \in \mathcal{Z} \text{ such that } A = (B \cup Z_1) \setminus Z_2\}.$$

We leave it to the reader to verify that $\mathcal{M}$ is a σ-algebra.

Next extend $\mu : \mathcal{M} \to [0, \infty]$ by

$$\mu(A) = \mu(B)$$

where $A = (B \cup Z_1) \smallsetminus Z_2$ for some $B \in \mathcal{B}$ and $Z_1, Z_2 \in \mathcal{Z}$. In mathematics, every definition must be *well-defined*, meaning that all implied assertions hold true and that it is unambiguously interpreted (i.e., it has a unique interpretation, especially with respect to any possible choices that one may make). In this case, the domain and range of μ are properly formulated, and the value assigned to A is independent of the choice of its decomposition, as is easily verified since $\mu|_{\mathcal{Z}} = 0$.

Theorem 1.32. *There exists a σ-algebra $\mathcal{M}$ of subsets of $\mathbb{R}^d$ and a positive measure $\mu : \mathcal{M} \to [0, \infty]$ satisfying the following.*

(a) *Every open set in $\mathbb{R}^d$ is in $\mathcal{M}$.*

(b) *If $A \subset B \in \mathcal{M}$ and $\mu(B) = 0$, then $A \in \mathcal{M}$ and $\mu(A) = 0$.*

(c) *If A is a rectangle with x_i bounded between a_i and $b_i \geq a_i$, then $\mu(A) = \prod_{i=1}^{d}(b_i - a_i)$.*

(d) *The measure μ is translation invariant, i.e., if $x \in \mathbb{R}^d$ and $A \in \mathcal{M}$, then $x + A \in \mathcal{M}$ and $\mu(A) = \mu(x + A)$.*

Sets outside $\mathcal{M}$ exist, and are called *unmeasurable* or *nonmeasurable* sets. We shall not meet any in this course. Moreover, for most of the results in the present text, μ restricted to the Borel sets suffices.

Consider now functions defined on measure spaces, taking values in the extended real number system $\overline{\mathbb{R}} \equiv \mathbb{R} \cup \{-\infty, +\infty\}$, or in $\mathbb{C}$.

Definition. Suppose $\Omega \subset \mathbb{R}^d$ is measurable. A function $f : \Omega \to \overline{\mathbb{R}}$ is *measurable* if the inverse image of every open set in $\mathbb{R}$ is measurable. A function $g : \Omega \to \mathbb{C}$ is measurable if its real and imaginary parts are measurable.

Note that measurability depends on $\mathcal{M}$, but not on μ! It is enough to verify that the sets

$$E_\alpha = \{x \in \Omega : f(x) > \alpha\}$$

are measurable for all $\alpha \in \mathbb{R}$ to conclude that f is measurable.

Theorem 1.33.

(a) *If f and g are measurable, so are $f + g$, $f - g$, and fg. If f and g are also real-valued, then $\max\{f, g\}$ and $\min\{f, g\}$ are also measurable.*

(b) *If f is measurable and $g : \mathbb{R} \to \mathbb{R}$, or in case f is complex valued, $g : \mathbb{C} \to \mathbb{C}$ is continuous, then $g \circ f$ is measurable.*

(c) *If f is defined on $\Omega \subset \mathbb{R}^d$, f continuous and Ω measurable, then f is measurable.*

(d) *If $\{f_n\}_{n=1}^{\infty}$ is a sequence of real-valued, measurable functions, then*

$$\inf_n f_n, \quad \sup_n f_n, \quad \liminf_{n\to\infty} f_n, \quad and \quad \limsup_{n\to\infty} f_n$$

are measurable functions.

The last statement in Theorem 1.33 uses some important terminology. Given a nonempty set $S \subset \mathbb{R}$ (such as $S = \{f_n(x)\}_{n=1}^{\infty}$ for $x \in \Omega$ fixed), the *infimum* of S, denoted $\inf S$, is the greatest number $\alpha \in [-\infty, +\infty)$ such that $s \geq \alpha$ for all $s \in S$. The *supremum* of S, $\sup S$, is the least number $\alpha \in (-\infty, +\infty]$ such that $s \leq \alpha$ for all $s \in S$. Given a sequence $\{y_n\}_{n=1}^{\infty}$ (such as $y_n = f_n(x)$ for $x \in \Omega$ fixed),

$$\liminf_{n\to\infty} y_n = \sup_{n\geq 1} \inf_{m\geq n} y_m = \lim_{n\to\infty} \left(\inf_{m\geq n} y_m \right)$$

and, similarly,

$$\limsup_{n\to\infty} y_n = \inf_{n\geq 1} \sup_{m\geq n} y_m = \lim_{n\to\infty} \left(\sup_{m\geq n} y_m \right).$$

Corollary 1.34. *If f is real-valued and measurable, then so are*

$$f^+ = \max\{f, 0\}, \quad f^- = -\min\{f, 0\}, \quad and \quad |f|.$$

Moreover, if $\{f_n\}_{n=1}^{\infty}$ is a sequence of measurable functions which converges pointwise, then the limit function is measurable.

Remark. With these definitions, $f = f^+ - f^-$ and $|f| = f^+ + f^-$.

Definition. If X is a set and $E \subset X$, then the function $\mathcal{X}_E : X \to \mathbb{R}$ given by

$$\mathcal{X}_E(x) = \begin{cases} 1 & \text{if } x \in E, \\ 0 & \text{if } x \notin E, \end{cases}$$

is called the *characteristic function* of E. If $s : X \to \mathbb{R}$ has finite range, then s is called a *simple function*.

If the range of s is $\{c_1, \ldots, c_n\}$ and

$$E_i = \{x \in X : s(x) = c_i\},$$

then

$$s(x) = \sum_{i=1}^{n} c_i \mathcal{X}_{E_i}(x),$$

and s is measurable if and only if each E_i is measurable.
Every function can be approximated by simple functions.

Theorem 1.35. *Given any function $f : \Omega \subset \mathbb{R}^d \to \overline{\mathbb{R}}$, there is a sequence $\{s_n\}_{n=1}^{\infty}$ of simple functions such that*

$$\lim_{n \to \infty} s_n(x) = f(x) \quad \text{for every } x \in \Omega$$

(i.e., s_n converges pointwise to f). If f is measurable, then each s_n can be chosen measurable. Moreover, if f is bounded, then $\{s_n\}_{n=1}^{\infty}$ can be chosen so that the convergence is uniform. If $f \geq 0$, then $\{s_n\}_{n=1}^{\infty}$ may be chosen to be monotonically increasing at each point.

Proof. If $f \geq 0$, define for $n = 1, 2, \ldots$ and $i = 1, 2, \ldots, n2^n$,

$$E_{n,i} = \left\{ x \in \Omega : \frac{i-1}{2^n} \leq f(x) < \frac{i}{2^n} \right\},$$

$$F_n = \{ x \in \Omega : f(x) \geq n \}.$$

Then

$$s_n(x) = \sum_{i=1}^{n2^n} \frac{i-1}{2^n} \mathcal{X}_{E_{n,i}}(x) + n\mathcal{X}_{F_n}$$

has the desired properties. In the general case, let $f = f^+ - f^-$ and approximate f^+ and f^- as above. □

With these prefatory notions in hand, it is now straightforward to define the Lebesgue integral. Let $\Omega \subset \mathbb{R}^d$ be measurable and suppose $s : \Omega \to \mathbb{R}$ is a measurable, simple function given as

$$s(x) = \sum_{i=1}^{n} c_i \mathcal{X}_{E_i}(x).$$

The Lebesgue integral of s over Ω is taken to be

$$\int_{\Omega} s(x)\, dx = \sum_{i=1}^{n} c_i \mu(E_i).$$

If $f : \Omega \to [0, \infty]$ is measurable, define

$$\int_{\Omega} f(x)\, dx = \sup_{s} \int_{\Omega} s(x)\, dx,$$

where the supremum is taken over all measurable simple functions satisfying $0 \leq s(x) \leq f(x)$ for $x \in \Omega$. Note that the integral of f may be $+\infty$.

If f is measurable and real-valued, then $f = f^+ - f^-$, where $f^+ \geq 0$ and $f^- \geq 0$. In this case, define

$$\int_{\Omega} f(x)\, dx = \int_{\Omega} f^+(x)\, dx - \int_{\Omega} f^-(x)\, dx,$$

provided at least one of the two integrals on the right is finite.

Finally, if f is complex-valued, apply the above construction to the real and imaginary parts of f, provided both the integrals of these are finite.

Definition. We say that a real-valued measurable function f is *integrable* if the integrals of f^+ and f^- are both finite. If only one is finite, then f is not integrable; however, in that case the value $+\infty$ or $-\infty$ is assigned to the integral, depending on which of f^+ and f^- has a finite integral.

Proposition 1.36. *The real-valued measurable function f is integrable over Ω if and only if*

$$\int_\Omega |f(x)|\, dx < \infty.$$

Definition. Let $\Omega \subset \mathbb{R}^d$ be measurable. The class of all integrable functions on Ω is denoted

$$\mathcal{L}(\Omega) = \left\{ \text{measurable } f : \int_\Omega |f(x)|\, dx < \infty \right\}.$$

Theorem 1.37. *If f is Riemann integrable on a compact set $K \subset \mathbb{R}^d$, then $f \in \mathcal{L}(K)$ and the Riemann and Lebesgue integrals agree.*

Certain properties of the Lebesgue integral are clear from its definition.

Proposition 1.38. *Assume that all functions and sets appearing below are measurable.*

(a) *If $|f|$ is bounded on Ω and $\mu(\Omega) < \infty$, then $f \in \mathcal{L}(\Omega)$.*

(b) *If $a \le f \le b$ on Ω and $\mu(\Omega) < \infty$, then*

$$a\mu(\Omega) \le \int_\Omega f(x)\, dx \le b\mu(\Omega).$$

(c) *If $f \le g$ on Ω, then*

$$\int_\Omega f(x)\, dx \le \int_\Omega g(x)\, dx.$$

(d) *If $f, g \in \mathcal{L}(\Omega)$, then $f + g \in \mathcal{L}(\Omega)$ and*

$$\int_\Omega (f+g)(x)\, dx = \int_\Omega f(x)\, dx + \int_\Omega g(x)\, dx.$$

(e) *If $f \in \mathcal{L}(\Omega)$ and $c \in \mathbb{R}$ (or $\mathbb{C}$), then*

$$\int_\Omega cf(x)\, dx = c \int_\Omega f(x)\, dx.$$

(f) *If $f \in \mathcal{L}(\Omega)$, then $|f| \in \mathcal{L}(\Omega)$ and*

$$\left| \int_\Omega f(x)\, dx \right| \le \int_\Omega |f(x)|\, dx.$$

(g) *If $f \in \mathcal{L}(\Omega)$ and $A \subset \Omega$ is measurable, then $f \in \mathcal{L}(A)$. If also $f \geq 0$, then*

$$0 \leq \int_A f(x)\, dx \leq \int_\Omega f(x)\, dx.$$

(h) *If $\mu(\Omega) = 0$, then*

$$\int_\Omega f(x)\, dx = 0.$$

(i) *If $f \in \mathcal{L}(\Omega)$ and $\Omega = A \cup B$ where $A \cap B = \emptyset$, then*

$$\int_\Omega f(x)\, dx = \int_A f(x)\, dx + \int_B f(x)\, dx.$$

Part (i) has a natural and useful generalization.

Theorem 1.39. *Suppose that $\Omega \subset \mathbb{R}^d$ is measurable, $f \in \mathcal{L}(\Omega)$, $A \subset \Omega$, $A_n \in \mathcal{M}$ for $n = 1, 2, \ldots$, $A_i \cap A_j = \emptyset$ for $i \neq j$, and $A = \bigcup_{n=1}^\infty A_n$. Then, it transpires that*

$$\int_A f(x)\, dx = \sum_{n=1}^\infty \int_{A_n} f(x)\, dx. \tag{1.1}$$

Moreover, if $f \geq 0$, the function $\lambda : \mathcal{M} \to \mathbb{R}$ given by

$$\lambda(A) = \int_A f(x)\, dx$$

is a positive measure.

Proof. That λ is a positive measure follows from (1.1), which provides the countable additivity. If (1.1) is valid when $f \geq 0$, it will follow for any real or complex-valued function via the decomposition $f = f_1 + if_2 = f_1^+ - f_1^- + i(f_2^+ - f_2^-)$, where $f_i^\pm \geq 0$.

For a characteristic function $\mathcal{X}_E$, E measurable, (1.1) holds since μ is countably additive:

$$\int_A \mathcal{X}_E(x)\, dx = \mu(A \cap E) = \sum_{n=1}^\infty \mu(A_n \cap E) = \sum_{n=1}^\infty \int_{A_n} \mathcal{X}_E(x)\, dx.$$

Because of (d) and (e) in Proposition 1.38, (1.1) also holds for any simple function.

If $f \geq 0$ and s is a simple function such that $0 \leq s \leq f$, then

$$\int_A s(x)\, dx = \sum_{n=1}^\infty \int_{A_n} s(x)\, dx \leq \sum_{n=1}^\infty \int_{A_n} f(x)\, dx.$$

Thus,

$$\int_A f(x)\, dx = \sup_{s \leq f} \int_A s(x)\, dx \leq \sum_{n=1}^\infty \int_{A_n} f(x)\, dx.$$

However, by iterating Proposition 1.38(i), it follows that

$$\sum_{k=1}^{n} \int_{A_k} f(x)\, dx = \int_{\cup_{k=1}^{n} A_k} f(x)\, dx \le \int_{A} f(x)\, dx$$

for any n. The last two inequalities imply (1.1) for f. □

From Proposition 1.38(h),(i), it is clear that if A and B are measurable sets and $\mu(A \smallsetminus B) = \mu(B \smallsetminus A) = 0$, then

$$\int_{A} f(x)\, dx = \int_{B} f(x)\, dx$$

for any integrable f. Moreover, if f and g are integrable and $f(x) = g(x)$ for all $x \in A \smallsetminus C$ where $\mu(C) = 0$, then

$$\int_{A} f(x)\, dx = \int_{A} g(x)\, dx.$$

Thus, sets of measure zero are negligible as far as integration is concerned.

If a property P holds for every $x \in E \smallsetminus A$ where $\mu(A) = 0$, then we say that P holds for *almost every* $x \in E$, or that P holds *almost everywhere* on E. A common practice, followed here, is to abbreviate "almost every" or "almost everywhere" as "a.e.". The abbreviation "a.a." for "almost all" is also found in the literature.

Proposition 1.40. *If* $f \in \mathcal{L}(\Omega)$*, where* Ω *is measurable, and if*

$$\int_{A} f(x)\, dx = 0$$

for every measurable $A \subset \Omega$*, then* $f = 0$ *a.e. on* Ω*.*

Proof. Suppose not. Decompose f as $f = f_1 + i f_2 = f_1^{+} - f_1^{-} + i(f_2^{+} - f_2^{-})$. At least one of $f_1^{\pm}$, $f_2^{\pm}$ is not zero a.e. Let g denote one such component of f. Thus $g \ge 0$ and g is not zero a.e. on Ω. However, $\int_{A} g(x)\, dx = 0$ for every measurable $A \subset \Omega$. Let

$$A_n = \{x \in \Omega : g(x) > 1/n\}.$$

Then $\mu(A_n) = 0$ for all n and $A_0 = \bigcup_{n=1}^{\infty} A_n = \{x \in \Omega : g(x) > 0\}$. But $\mu(A_0) = \mu(\bigcup_{n=1}^{\infty} A_n) \le \sum_{n=1}^{\infty} \mu(A_n) = 0$, contradicting the fact that g is not zero a.e. □

The following interesting fact, while not needed here, is illuminating. It shows that Riemann integration is restricted to a much narrower class of functions than is Lebesgue integration.

Proposition 1.41. *If f is bounded on a compact interval $[a, b] \subset \mathbb{R}$, then f is Riemann integrable on $[a, b]$ if and only if f is continuous at all but a countable number of points in the interval.*

The Lebesgue integral is absolutely continuous in the following sense.

Theorem 1.42. *If $f \in \mathcal{L}(\Omega)$, then $\int_A |f| \, dx \to 0$ as $\mu(A) \to 0$, where $A \subset \Omega$ is measurable. That is, given $\epsilon > 0$, there is $\delta > 0$ such that*

$$\int_A |f(x)| \, dx \leq \epsilon$$

whenever $\mu(A) < \delta$.

Proof. Given $\epsilon > 0$, there is a simple function $s(x)$ such that

$$\int_A |f(x) - s(x)| \, dx \leq \epsilon/2,$$

by the definition of the Lebesgue integral. Moreover, the proof of the existence of $s(x)$ in Theorem 1.35 reveals that we can choose $s(x)$ in such a way that

$$|s(x)| \leq M(\epsilon)$$

for some constant $M(\epsilon)$ depending only on f and ϵ. Then, if $A \subset \Omega$ is measurable,

$$\int_A |s(x)| \, dx \leq \mu(A) M(\epsilon),$$

so if $\mu(A) < \delta \equiv \epsilon/2M(\epsilon)$, then

$$\int_A |f(x)| \, dx \leq \int_A |f(x) - s(x)| \, dx + \int_A |s(x)| \, dx \leq \epsilon/2 + \epsilon/2 = \epsilon. \qquad \square$$

Theorem 1.35 states that a measurable function f can be approximated by a sequence of simple functions. We can go further, and approximate by a sequence of continuous functions, at least when the function being approximated is well-behaved near infinity. Let $C_0(\Omega)$ be the set of continuous functions with *compact support*, i.e., continuous functions that vanish outside a bounded set.

Theorem 1.43 (Lusin's theorem). *Suppose that a measurable function f on Ω is such that $f(x) = 0$ for $x \notin A$, where A has finite measure. Given $\epsilon > 0$, there is $g \in C_0(\Omega)$ such that the measure of the set where f and g differ is less than ϵ. Moreover, it may be arranged that*

$$\sup_{x \in \Omega} |g(x)| \leq \sup_{x \in \Omega} |f(x)|.$$

A proof can be found in [20], for example. The following lemma is easily demonstrated (and left to the reader), but it turns out to be quite useful.

Lemma 1.44 (Chebyshev's inequality). *If $f \geq 0$ and $\Omega \subset \mathbb{R}^d$ are both measurable, then*

$$\mu(\{x \in \Omega : f(x) > \alpha\}) \leq \frac{1}{\alpha} \int_\Omega f(x)\,dx$$

for any $\alpha > 0$.

The overview of Lebesgue measure and integration theory is concluded with statements of the three basic convergence theorems, Fubini's theorem on integration over product spaces, and the fundamental theorem of calculus, each without proof. Especially the convergence theorems set Lebesgue integration apart from Riemann integration.

Theorem 1.45 (monotone convergence theorem of Lebesgue). *If $\Omega \subset \mathbb{R}^d$ is measurable and $\{f_n\}_{n=1}^\infty$ is a sequence of measurable functions satisfying $0 \leq f_1(x) \leq f_2(x) \leq \cdots$ for a.e. $x \in \Omega$, then*

$$\lim_{n \to \infty} \int_\Omega f_n(x)\,dx = \int_\Omega \left(\lim_{n \to \infty} f_n(x) \right) dx.$$

Theorem 1.46 (Fatou's lemma). *If $\Omega \subset \mathbb{R}^d$ is measurable and $\{f_n\}_{n=1}^\infty$ is a sequence of nonnegative, measurable functions, then*

$$\int_\Omega \left(\liminf_{n \to \infty} f_n(x) \right) dx \leq \liminf_{n \to \infty} \int_\Omega f_n(x)\,dx.$$

Theorem 1.47 (dominated convergence theorem of Lebesgue). *Let $\Omega \subset \mathbb{R}^d$ be measurable and $\{f_n\}_{n=1}^\infty$ be a sequence of measurable functions that converge pointwise for a.e. $x \in \Omega$. If there is a function $g \in \mathcal{L}(\Omega)$ such that*

$$|f_n(x)| \leq g(x) \quad \text{for every } n \text{ and a.e. } x \in \Omega,$$

then

$$\lim_{n \to \infty} \int_\Omega f_n(x)\,dx = \int_\Omega \left(\lim_{n \to \infty} f_n(x) \right) dx.$$

Theorem 1.48 (Fubini's theorem). *Let f be measurable on $\mathbb{R}^{n+m}$. If at least one of the integrals*

$$I_1 = \int_{\mathbb{R}^{n+m}} f(x,y)\,dx\,dy,$$

$$I_2 = \int_{\mathbb{R}^m} \left(\int_{\mathbb{R}^n} f(x,y)\,dx \right) dy,$$

$$I_3 = \int_{\mathbb{R}^n} \left(\int_{\mathbb{R}^m} f(x,y)\,dy \right) dx$$

exists in the Lebesgue sense (i.e., when f is replaced by $|f|$) and is finite, then each exists and $I_1 = I_2 = I_3$.

Note that in Fubini's theorem, the claim is that the following are equivalent:

(i) $f \in \mathcal{L}(\mathbb{R}^{n+m})$,

(ii) $f(\cdot, y) \in \mathcal{L}(\mathbb{R}^n)$ for a.e. $y \in \mathbb{R}^m$ and $\int_{\mathbb{R}^n} f(x, \cdot) \, dx \in \mathcal{L}(\mathbb{R}^m)$,

(iii) $f(x, \cdot) \in \mathcal{L}(\mathbb{R}^m)$ for a.e. $x \in \mathbb{R}^n$ and $\int_{\mathbb{R}^m} f(\cdot, y) \, dy \in \mathcal{L}(\mathbb{R}^n)$,

and the three full integrals agree. Among other things, f being measurable on $\mathbb{R}^{n+m}$ implies that $f(\cdot, y)$ is measurable for a.e. $y \in \mathbb{R}^m$ and $f(x, \cdot)$ is measurable for a.e. $x \in \mathbb{R}^n$. It is worth noting, however, that it is not the case that $f(x, \cdot)$ or $f(\cdot, y)$ is necessarily measurable for *every* x or y.

Theorem 1.49 (fundamental theorem of calculus). *If $f \in \mathcal{L}([a, b])$ and*

$$F(x) = \int_a^x f(t) \, dt,$$

then $F'(x) = f(x)$ for a.e. $x \in [a, b]$. Conversely, if F is differentiable everywhere (not a.e.!) on $[a, b]$ and $F' \in \mathcal{L}([a, b])$, then

$$F(x) - F(a) = \int_a^x F'(t) \, dt$$

for any $x \in [a, b]$.

1.3 HOLOMORPHIC FUNCTIONS AND COMPLEX CONTOUR INTEGRATION

In this section, $\Omega \subset \mathbb{C}$ will denote an open set in the complex plane. A *holomorphic* or *analytic* function $f : \Omega \to \mathbb{C}$ is a C^∞-function that, in a neighborhood of each point of $z \in \Omega$, agrees with its Taylor series expansion at that point. This turns out to be equivalent to the function possessing a complex derivative at each point of Ω.

Definition. A function $f : \Omega \to \mathbb{C}$ is *differentiable* or *complex differentiable* at $z_0 \in \Omega$ in the complex sense if the limit

$$\lim_{z \to z_0} \frac{f(z) - f(z_0)}{z - z_0}$$

exists. In that case, denote the limit by $f'(z_0)$ and call it the *derivative* of f at z_0. The function f is called *holomorphic (or analytic)* in Ω if $f'(z)$ exists for all $z \in \Omega$.

Clearly the set of holomorphic functions on Ω is a linear vector space (i.e., differentiation is a linear operator). It is easy to show that the product rule holds for complex differentiation; for f and g holomorphic on Ω, the product

fg is holomorphic on Ω and $(fg)' = fg' + f'g$. Moreover, the chain rule holds as well: if f is holomorphic in Ω and g is holomorphic in $f(\Omega)$, then the composition $g \circ f$ is holomorphic and

$$(g \circ f)'(z) = g'(f(z)) f'(z).$$

Before proceeding, a few facts about power series are recalled. For $\{c_n\}_{n=0}^{\infty} \subset \mathbb{C}$, the *power series*

$$\sum_{n=0}^{\infty} c_n (z - z_0)^n$$

has a *radius of convergence* $\rho \in [0, \infty]$ such that the series converges absolutely for $|z - z_0| < \rho$ and diverges for $|z - z_0| > \rho$, with indeterminate convergence on $|z - z_0| = \rho$. Moreover, the convergence is uniform on compact subsets within $|z - z_0| < \rho$, and therefore many operations such as differentiation, addition, and multiplication can be performed on power series term-by-term. The root test, for example, states that

$$\frac{1}{\rho} = \limsup_{n \to \infty} |c_n|^{1/n}.$$

We say that f is *representable by a power series* in Ω if for every ball $B_r(z_0) \subset \Omega$, f agrees with a power series in that ball. Thus, while f may not have a single power series that represents it throughout Ω, for each $z_0 \in \Omega$, it has such a representation valid in the ball $B_r(z_0)$ where

$$r = \text{dist}(z_0, \partial\Omega) = \inf\{|z - z_0| : z \in \partial\Omega\}.$$

It can be shown that if f is holomorphic, it is in fact infinitely differentiable and agrees locally with its Taylor series (see, e.g., [19], [20]).

Theorem 1.50. *If $f : \Omega \to \mathbb{C}$ is holomorphic, then f is infinitely differentiable and representable by a power series in Ω. Conversely, a convergent power series is holomorphic in its disc of convergence. Moreover, for $z_0 \in \Omega$, f agrees with its Taylor series about z_0 in the ball $B_r(z_0)$, where $r = \text{dist}(z_0, \partial\Omega)$; that is, for $z \in B_r(z_0)$,*

$$f(z) = \sum_{n=0}^{\infty} \frac{f^{(n)}(z_0)}{n!} (z - z_0)^n.$$

Holomorphic functions have a tightly restricted structure. In fact, if a holomorphic function is known on a small set with a limit point, such as a line segment, then it is known everywhere in its domain of definition. This is expressed as follows.

Theorem 1.51. *If f and g are holomorphic on Ω and agree on a set of points which has a limit point in Ω, then f and g agree everywhere in Ω.*

Most elementary functions are holomorphic, since they agree with their Taylor series on at least a large portion of $\mathbb{C}$. Examples that are *entire*, meaning that we can take $\Omega = \mathbb{C}$, include complex polynomials,

$$\sin z = z - \frac{1}{3!}z^3 + \frac{1}{5!}z^5 - \cdots,$$

$$\cos z = 1 - \frac{1}{2!}z^2 + \frac{1}{4!}z^4 - \cdots,$$

$$e^z = e^{x+iy} = e^x(\cos y + i \sin y) \quad \text{(Euler's formula)}$$

$$= 1 + z + \frac{1}{2!}z^2 + \frac{1}{3!}z^3 + \cdots.$$

In fact, these functions are defined on the real axis and extended uniquely by the proceeding theorem into $\mathbb{C}$ as holomorphic functions. Examples of holomorphic functions that are not entire are rational functions (ratios of polynomials), $\log z$, and $\sqrt{z}$. Nonholomorphic functions include, for example, $|z|$ and $\bar{z}$, the complex conjugate of z.

A *curve* is a continuous mapping $\gamma(t)$ taking a compact interval $[a, b]$ to $\mathbb{C}$. We say that the curve is *parametrized* by $t \in [a, b]$. Its *initial point* $\gamma(a)$ and *end point (or terminal point)* $\gamma(b)$ may agree, in which case the curve is *closed*. A *simple closed curve* is a closed curve for which only the initial and end points agree, i.e., a closed curve that does not cross itself and therefore does not wind around any point more than once.

A *contour* or *path* is a piecewise continuously differentiable curve. The range of a contour is a subset of $\mathbb{C}$ on which we can define integration:

$$\int_\gamma f(z)\, dz = \int_a^b f(\gamma(t))\, \gamma'(t)\, dt.$$

While the integrand is complex, and so has real and imaginary parts, the integration itself reduces to two real Lebesgue integrals on a compact real interval, and so is well-defined. This definition is independent of the parametrization; that is, if τ mapping $[\hat{a}, \hat{b}]$ one-to-one and onto $[a, b]$ is differentiable and positively oriented (i.e., $\tau(\hat{a}) = a$), and if $\hat{\gamma}(t) = \gamma(\tau(t))$, then $\hat{\gamma} : [\hat{a}, \hat{b}] \to \mathbb{C}$ and

$$\int_{\hat{\gamma}} f(z)\, dz = \int_{\hat{a}}^{\hat{b}} f(\hat{\gamma}(t))\, \hat{\gamma}'(t)\, dt = \int_{\hat{a}}^{\hat{b}} f(\gamma(\tau(t)))\, \gamma'(\tau(t))\, \tau'(t)\, dt$$

$$= \int_a^b f(\gamma(t))\, \gamma'(t)\, dt = \int_\gamma f(z)\, dz.$$

Note that if the direction of traversal is reversed, e.g., by taking $\tau(t) = a+b-t$, then

$$\int_{\hat{\gamma}} f(z)\, dz = -\int_\gamma f(z)\, dz.$$

Example. Let S_r denote the circle in $\mathbb{C}$ of radius $r > 0$ about the origin. A natural parametrization of the circle, traced in the counter-clockwise direction, is $\gamma_{\text{circle}}(t) = x(t) + iy(t) = r\cos t + ir\sin t$, $t \in [0, 2\pi]$, but it is easier to use polar coordinates, *viz.*

$$\gamma_{\text{circle}}(t) = re^{it} \quad \text{so that} \quad \gamma'_{\text{circle}}(t) = ire^{it} = i\gamma_{\text{circle}}(t).$$

Thus for any integrable f, it appears that

$$\int_{\gamma_{\text{circle}}} f(z)\,dz = ir \int_0^{2\pi} f(re^{it})\,e^{it}\,dt,$$

from which follows the result

$$\int_{\gamma_{\text{circle}}} z^n\,dz = ir^{n+1} \int_0^{2\pi} e^{i(n+1)t}\,dt = 0 \quad \text{for } n \neq -1 \text{ an integer.} \tag{1.2}$$

The case $n = -1$ is calculated directly, leading to the important result

$$\int_{\gamma_{\text{circle}}} \frac{dz}{z} = 2\pi i. \tag{1.3}$$

The reader should note that $1/z$ is holomorphic except at the origin, since for $z_0 \neq 0$,

$$\lim_{z \to z_0} \frac{1/z - 1/z_0}{z - z_0} = \lim_{z \to z_0} \frac{z_0 - z}{zz_0(z - z_0)} = \lim_{z \to z_0} \frac{-1}{zz_0} = -\frac{1}{z_0^2}.$$

Similarly z^n is holomorphic, except at $z = 0$ when $n < 0$.

It is a remarkable fact that the integral of a holomorphic function between two points is independent of the contour joining the two points, provided that the function remains holomorphic on the region between the two contours. This fact is expressed more elegantly as follows.

Theorem 1.52 (Cauchy's theorem). *If f is holomorphic in Ω and γ is a simple closed contour with its interior contained entirely in Ω, then*

$$\oint_\gamma f(z)\,dz = 0.$$

In the notation above, the circle is added to the integral sign merely to emphasize that the contour is a closed curve.

Example. Compute the integral $\int_0^\infty e^{-x^2} \cos(2ax)\,dx$ for $a \geq 0$. First compute the integral when $a = 0$, without using Cauchy's theorem. To this end, note that

$$\left[\int_{-\infty}^\infty e^{-x^2}\,dx \right]^2 = \left[\int_{-\infty}^\infty e^{-x^2}\,dx \right]\left[\int_{-\infty}^\infty e^{-y^2}\,dy \right] = \iint_{\mathbb{R}^2} e^{-(x^2+y^2)}\,dx\,dy.$$

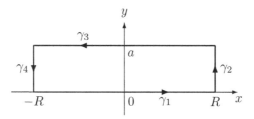

Figure 1.1 A simple closed rectangular contour, for which $R \to \infty$.

A change of variables to polar coordinates gives

$$\left(\int_{-\infty}^{\infty} e^{-x^2} \, dx \right)^2 = \int_0^{2\pi} \int_0^{\infty} e^{-r^2} r \, dr \, d\theta = 2\pi \left[-\frac{1}{2} e^{-r^2} \right]_0^{\infty} = \pi.$$

Since the integrand is an even function,

$$\int_0^{\infty} e^{-x^2} \, dx = \frac{\sqrt{\pi}}{2}.$$

As depicted in Figure 1.1, for $a > 0$, let $R > 0$ and define the simple closed rectangular contour piecewise by

$$\begin{aligned}
\gamma_1(t) &= t & \text{for } -R \leq t \leq R, \\
\gamma_2(t) &= R + it & \text{for } 0 \leq t \leq a, \\
\gamma_3(t) &= -t + ia & \text{for } -R \leq t \leq R, \\
\gamma_4(t) &= -R - it & \text{for } -a \leq t \leq 0.
\end{aligned}$$

The contour is oriented consistently in the counter-clockwise direction. It could be written in terms of a single function γ on a single interval through a re-parametrization, but this would not change our further ruminations. Notice that the integrand $e^{-x^2} \cos(2ax)$ is the real part of the entire function $e^{-z^2} e^{2azi} = e^{-(z-ia)^2 - a^2}$ when $z = x \in \mathbb{R}$. By Cauchy's theorem, it follows that

$$\begin{aligned}
0 = &\int_{-R}^{R} e^{-t^2} e^{2ati} \, dt + i \int_0^a e^{-(R+i(t-a))^2 - a^2} \, dt \\
&- \int_{-R}^{R} e^{-t^2 - a^2} \, dt - i \int_{-a}^0 e^{-(R+i(t+a))^2 - a^2} \, dt.
\end{aligned}$$

Note that the second and fourth integrals vanish as $R \to \infty$, since

$$\left| i \int_0^a e^{-(R+i(t-a))^2 - a^2} \, dt \right| + \left| i \int_{-a}^0 e^{-(R+i(t+a))^2 - a^2} \, dt \right| \leq 2ae^{-R^2} \longrightarrow 0.$$

The third integral has already been computed to have the property

$$\lim_{R \to \infty} \int_{-R}^{R} e^{-t^2 - a^2} \, dt = \sqrt{\pi} e^{-a^2},$$

which is real. It is therefore concluded that

$$\int_{0}^{\infty} e^{-x^2} \cos(2ax) \, dx = \frac{\sqrt{\pi}}{2} e^{-a^2}.$$

Often a function is holomorphic except at certain isolated points. This is given precision in the following definition.

Definition. If $a \in \Omega$ and f is holomorphic on $\Omega \setminus \{a\}$, then f has an *isolated singularity* at a. If f can be defined at a so that f is holomorphic on Ω, then the singularity is called *removable*. If there are constants $c_n \in \mathbb{C}$, $n = -m, \ldots, -1$ with $c_{-m} \neq 0$ such that

$$f(z) - \sum_{n=1}^{m} \frac{c_{-n}}{(z-a)^n}$$

has a removable singularity at a, then the singularity of f at a is called a *pole of order* m. Otherwise, the singularity is called *essential*.

Example. Clearly $1/z$ has a singularity at $z = 0$, which is a pole of order 1, and $(z^2 - 1)/(z - 1)$ has a removable singularity at $z = 1$ since the function equals $z + 1$. On the other hand, the function

$$e^{1/z} = \sum_{n=0}^{\infty} \frac{1}{n! z^n}$$

has an essential singularity at $z = 0$.

Notice that if f has a pole of order m, then it can be expressed as a *Laurent series* with only finitely many negative terms, which is to say

$$f(z) = \sum_{n=-m}^{\infty} c_n (z-a)^n.$$

This follows since

$$f(z) - \sum_{n=-m}^{-1} \frac{c_{-n}}{(z-a)^n}$$

is holomorphic near a and so has a Taylor expansion there. If γ_{circle} is the circle of radius $r > 0$ about a taken in the counter-clockwise direction, then (1.2) and (1.3) imply that

$$\int_{\gamma_{\text{circle}}} f(z) \, dz = 2\pi i c_{-1}.$$

This simple fact has a useful generalization.

Definition. A function is *meromorphic* in Ω if there is a set $A \subset \Omega$ with no limit point in Ω such that f is holomorphic on $\Omega \setminus A$ and f has either a removable singularity or a pole at each point $a \in A$. At $a \in A$, the *residue* of f is the coefficient c_{-1} in its Laurent series, denoted $\mathrm{Res}_a(f)$.

For a pole of order m, the residue can be computed as

$$\lim_{z \to a} \frac{d^{m-1}}{dz^{m-1}} \frac{f(z)(z-a)^m}{(m-1)!}.$$

The order of a pole may be determined by computing these limits successively for $k = 1, 2, \ldots$, until a finite limit is obtained. If a finite limit is never obtained, the singularity is essential.

Theorem 1.53 (residue theorem). *Let f be meromorphic in Ω and let $A \subset \Omega$ be the set of poles of f in Ω. Let γ be a simple closed contour whose interior lies in Ω. If γ lies in $\Omega \setminus A$ and encloses poles $B \subset A$, then*

$$\oint_\gamma f(z)\, dz = \sum_{a \in B} 2\pi i \mathrm{Res}_a(f).$$

Example. Let $f(z) = 1/(e^z - 1)$, which has singularities whenever $e^z = 1$, i.e., when $z = 2\pi n i$ for some integer n. At such a point,

$$\lim_{z \to 2\pi n i} \frac{(z - 2\pi n i)}{e^z - 1} = 1,$$

by L'Hôpital's rule, so the poles are all of order 1 and their residues are all 1. Thus f is meromorphic in $\mathbb{C}$. If, for example, γ is the circle of radius 10 oriented counter-clockwise, then γ encloses three poles and so, by the residue theorem,

$$\oint_\gamma \frac{1}{e^z - 1} dz = 6\pi i.$$

Sometimes, a pole of a function f lies on a contour γ along which we wish to compute the integral. In that case, it is often effective to consider the integral on contours that avoid the poles, and take a limit as these contours approach γ.

Example. Compute the integral $\int_0^\infty \frac{\sin x}{x}\, dx$. In fact, the integrand has no singularity at $x = 0$ since $\sin(x)$ vanishes appropriately at $x = 0$. However, the integral itself is half the imaginary part of $\int_{-\infty}^\infty \frac{e^{ix}}{x} dx$, and the function $\frac{e^{iz}}{z}$ does have a pole at $z = 0$. It is holomorphic at all other points. The pole at the origin is of order 1 and its residue, $\mathrm{Res}_0(e^{iz}/z) = 1$. To avoid the pole,

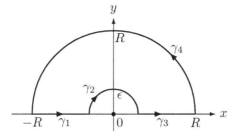

Figure 1.2 A half-annular closed contour, for which $R \to \infty$ and $\epsilon \to 0$.

integrate along the half-annular closed contour γ depicted in Figure 1.2 and defined piecewise for $R > \epsilon > 0$ by

$$
\begin{aligned}
\gamma_1(t) &= t && \text{for } -R \le t \le -\epsilon, \\
\gamma_2(t) &= \epsilon e^{-it} && \text{for } -\pi \le t \le 0, \\
\gamma_3(t) &= t && \text{for } \epsilon \le t \le R, \\
\gamma_4(t) &= R e^{it} && \text{for } 0 \le t \le \pi.
\end{aligned}
$$

This contour encloses no poles, so by Cauchy's theorem,

$$
0 = \int_\gamma \frac{e^{iz}}{z}\,dz = \int_{-R}^{-\epsilon} \frac{e^{it}}{t}\,dt - i \int_{-\pi}^{0} \exp(i\epsilon e^{-it})\,dt
$$
$$
+ \int_\epsilon^R \frac{e^{it}}{t}\,dt + i \int_0^\pi \exp(iR e^{it})\,dt.
$$

Now, Lebesgue's dominated convergence theorem implies that, as $R \to \infty$,

$$
\left| i \int_0^\pi \exp(iR e^{it})\,dt \right| = \int_0^\pi \exp(-R \sin t)\,dt \longrightarrow 0
$$

and, as $\epsilon \to 0$,

$$
i \int_{-\pi}^{0} \exp(i\epsilon e^{-it})\,dt = i \int_{-\pi}^{0} \exp(\epsilon(\sin t + i \cos t))\,dt \longrightarrow i\pi.
$$

In consequence, we see that

$$
\int_0^\infty \frac{\sin x}{x}\,dx = \frac{\pi}{2}.
$$

The reader can verify that if the pole was included by replacing γ_2 by $\gamma_2(t) = \epsilon e^{it}$ for $\pi \le t \le 2\pi$, the same final result obtains by virtue of the residue theorem.

1.4 EXERCISES

1. Show that each of the following defines a topology $\mathcal{T}$ on X, where X is any nonempty set.

   (a) $\mathcal{T} = \{\emptyset, X\}$. This is called the *trivial* topology on X.

   (b) The topology $\mathcal{T}$ for which $\mathcal{T}_B = \{\{x\} : x \in X\}$ is a base. This is called the *discrete* topology on X.

   (c) Let $\mathcal{T}$ consist of $\emptyset$ and all subsets of X with finite complements. If X is finite, what topology is this?

2. Let $X = \{a, b\}$ and $\mathcal{T} = \{\emptyset, \{a\}, X\}$. Show directly that there is no metric $d : X \times X \to \mathbb{R}$ that is compatible with the topology. Thus not every topological space is *metrizable*.

3. Prove that if $A \subset X$, then ∂A is closed,
$$\bar{A} = A^\circ \cup \partial A \quad \text{and} \quad A^\circ \cap \partial A = \emptyset.$$

   Moreover,
$$\partial A = \partial A^c = \{x \in X \; : \; \text{every open } E \text{ containing } x$$
$$\text{intersects both } A \text{ and } A^c\}.$$

4. Prove that if $(X, \mathcal{T})$ is Hausdorff, then every set consisting of a single point is closed. Moreover, limits of sequences are unique.

5. Prove that a set $A \subset X$ is open if and only if, given $x \in A$, there is an open E such that $x \in E \subset A$.

6. Prove that a mapping of X into Y is continuous if and only if the inverse image of every closed set is closed.

7. Prove that if f is continuous and $\lim_{n \to \infty} x_n = x$, then $\lim_{n \to \infty} f(x_n) = f(x)$.

8. Suppose that $f(x) = y$. Let $\mathcal{B}_x$ be a base at $x \in X$, and $\mathcal{C}_y$ a base at $y \in Y$. Prove that f is continuous at x if and only if for each $C \in \mathcal{C}_y$ there is a $B \in \mathcal{B}_x$ such that $B \subset f^{-1}(C)$.

9. Show that every metric space is Hausdorff.

10. Suppose that $f : X \to \mathbb{R}$. Characterize all topologies $\mathcal{T}$ on X that make f continuous. Which is the weakest? Which is the strongest?

11. Construct an infinite open cover of $(0, 1]$ that has no finite subcover. Find a sequence in $(0, 1]$ that does not have a convergent subsequence.

12. Prove that the continuous image of a compact set is compact.

13. Prove that a one-to-one continuous map of a compact space X onto a Hausdorff space Y is necessarily a homeomorphism.

14. Prove that if $f : X \to \mathbb{R}$ is continuous and X is compact, then f takes on its maximum and minimum values.

15. Show that the Borel sets $\mathcal{B}$ is the collection of all sets that can be constructed by a countable number of basic set operations, starting from open sets. The basic set operations consist of taking unions, intersections, or complements.

16. Prove each of the following.

    (a) If $f : \mathbb{R}^d \to \mathbb{R}$ is measurable and $g : \mathbb{R} \to \mathbb{R}$ is continuous, then $g \circ f$ is measurable.

    (b) If $\Omega \subset \mathbb{R}^d$ is measurable and $f : \Omega \to \mathbb{R}$ is continuous, than f is measurable.

17. Let $x \in \mathbb{R}^d$ be fixed. Define δ_x for any $A \subset \mathbb{R}^d$ by

$$\delta_x(A) = \begin{cases} 1 & \text{if } x \in A, \\ 0 & \text{if } x \notin A. \end{cases}$$

    Show that δ_x is a measure on the Borel sets $\mathcal{B}$. This measure is called the *Dirac* or *point measure* at x.

18. The divergence or Gauss theorem from advanced calculus says that if $\Omega \subset \mathbb{R}^d$ has a smooth boundary and $\mathbf{v} \in (C^1(\bar{\Omega}))^d$ is a vector-valued function, then

$$\int_\Omega \nabla \cdot \mathbf{v}(x) \, dx = \int_{\partial\Omega} \mathbf{v}(x) \cdot \nu(x) \, ds(x),$$

    where $\nu(x)$ is the outward pointing unit normal vector to Ω for any $x \in \partial\Omega$, and $ds(x)$ is the surface differential (i.e., measure) on $\partial\Omega$. Note that here dx is a d-dimensional measure, and ds is a $(d-1)$-dimensional measure.

    (a) Interpret the formula when $d = 1$ in terms of the Dirac measure from Problem 17.

    (b) Show that for $\phi \in C^1(\bar{\Omega})$,

$$\nabla \cdot (\phi\mathbf{v}) = \nabla\phi \cdot \mathbf{v} + \phi\nabla \cdot \mathbf{v}.$$

    (c) Let $\phi \in C^1(\bar{\Omega})$ and apply the divergence theorem to the vector $\phi\mathbf{v}$ in place of $\mathbf{v}$. We call this new formula *integration by parts*. Show that it reduces to ordinary integration by parts when $d = 1$.

19. Prove that if $f \in \mathcal{L}(\Omega)$ and $g : \Omega \to \mathbb{R}$, where g and Ω are measurable and g is bounded, then $fg \in \mathcal{L}(\Omega)$.

20. Construct an example of a sequence of nonnegative measurable functions from $\mathbb{R}$ to $\mathbb{R}$ that shows strict inequality can result in Fatou's lemma.

21. Let

$$f_n(x) = \begin{cases} \dfrac{1}{n}, & |x| \le n, \\ 0, & |x| > n. \end{cases}$$

Show that $f_n(x) \to 0$ uniformly on $\mathbb{R}$, but

$$\int_{-\infty}^{\infty} f_n(x)\,dx = 2.$$

Comment on the applicability of the dominated convergence theorem.

22. Let

$$f(x, y) = \begin{cases} 1, & 0 \le x - y \le 1, \\ -1, & 0 \le y - x \le 1, \\ 0, & \text{otherwise.} \end{cases}$$

Show that

$$\int_0^\infty \left(\int_0^\infty f(x, y)\,dx \right) dy \neq \int_0^\infty \left(\int_0^\infty f(x, y)\,dy \right) dx.$$

Comment on the applicability of Fubini's theorem.

23. Suppose that f is integrable on $[a, b]$, and define

$$F(x) = \int_a^x f(t)\,dt.$$

Prove that F is continuous on $[a, b]$. (In fact, $F' = f$ a.e., but it is a little more involved to prove this.)

24. Let γ be a contour traversing the circle of radius 2 counter-clockwise. Evaluate:

(a) $\displaystyle \oint_\gamma \frac{dz}{z(z^2 - 1)}$;

(b) $\displaystyle \oint_\gamma \frac{e^z}{1 - \cos z}\,dz.$

25. Evaluate the following integrals:

(a) $\displaystyle \int_{-\infty}^{\infty} \frac{\sin x}{x(1 + x^2)}\,dx$;

(b) $\displaystyle \int_0^\infty \frac{dx}{1 + x^4}$;

(c) $\displaystyle \int_{-\infty}^{\infty} \frac{\cos x}{1 + x + x^2}\,dx$;

(d) $\displaystyle \int_{-\infty}^{\infty} \frac{\sin(2\pi x)}{(1 - x^2)^2}\,dx.$

CHAPTER 2

Normed Linear Spaces and Banach Spaces

Functional analysis grew out of the late 19th/early 20th century study of differential and integral equations arising in physics, quantum mechanics, and the calculus of variations. Especially the work of the Swedish mathematician Erik Fredholm on integral equations had a great influence in the first decade of the 20th century. His work prompted David Hilbert to run a seminar on integral equations and led to the emergence of what we now call Hilbert spaces (Chapter 3). Functional analysis emerged as a subject in its own right between World War I and World War II under the influence of Jacques Hadamard and especially the Polish school under the leadership of Stefan Banach.

For the first sixty or seventy years of its history, functional analysis was a major topic within mathematics, attracting a large following among both pure and applied mathematicians. Lately, the pure end of the subject has become the purview of a more restricted coterie who are concerned with very difficult and often quite subtle issues. On the other hand, the applications of the basic theory and even some of its finer elucidations have grown steadily, to the point where one can no longer intelligently read papers in, for example, much of numerical analysis, partial differential equations, parts of stochastic analysis, new thrusts in data science, etc. without a working knowledge of functional analysis.

The aim of this chapter is to expound the elements of the subject with an eye especially for aspects that lend themselves to applications. We begin with a formal development, as this is both intellectually attractive and efficient.

2.1 BASIC CONCEPTS AND DEFINITIONS

Vector spaces are fundamental in science and engineering. They encapsulate notions of scaling (scalar multiplication) and translation (vector addition). The formal definition is recalled here.

DOI: 10.1201/9781003492139-2

Definition. A *vector space* or *linear space* over a field $\mathbb{F}$ is a nonempty set X of objects, called *vectors*, together with two algebraic operations. The first, called *addition*, associates to any two vectors $x, y \in X$ their *sum* $x + y \in X$ in such a way that

(a) $x + y = y + x$ for all $x, y \in X$;

(b) $(x + y) + z = x + (y + z)$ for all $x, y, z \in X$;

(c) there is a *zero* vector, denoted 0, such that $x + 0 = x$ for all $x \in X$; and

(d) to each $x \in X$, there is a vector called the *negative* of x and denoted $-x$, such that $x + (-x) = 0$.

The other operation is called *scalar multiplication*, and it associates to any vector $x \in X$ and scalar $\lambda \in \mathbb{F}$ the vector $\lambda x \in X$ in such a way that

(e) $\lambda(x + y) = \lambda x + \lambda y$ for all $x, y \in X$ and $\lambda \in \mathbb{F}$;

(f) $(\lambda + \mu)x = \lambda x + \mu x$ for all $x \in X$ and $\lambda, \mu \in \mathbb{F}$;

(g) $(\lambda\mu)x = \lambda(\mu x)$ for all $x \in X$ and $\lambda, \mu \in \mathbb{F}$; and

(h) $1x = x$ for all $x \in X$.

Throughout, the field $\mathbb{F}$, sometimes called the ground field, from which scalars are drawn will always be either the real numbers $\mathbb{R}$ or the complex number field $\mathbb{C}$.

Vector spaces become very powerful tools when they are augmented with a notion of the size or *norm* of a vector.

Definition. Let X be a vector space over the real numbers $\mathbb{R}$ or the complex numbers $\mathbb{C}$. We say X is a *normed linear space* (NLS for short) if there is a mapping

$$\| \cdot \| : X \to \mathbb{R}^+ = [0, \infty),$$

called the *norm* on X, satisfying the following set of rules which apply to all $x, y \in X$ and $\lambda \in \mathbb{R}$ or $\mathbb{C}$ (the ground field):

(a) $\|x\| = 0$ if and only if $x = 0$,

(b) $\|\lambda x\| = |\lambda| \, \|x\|$,

(c) $\|x + y\| \leq \|x\| + \|y\|$ (triangle inequality).

In situations where more than one NLS is under consideration, it is often convenient to write $\| \cdot \|_X$ for the norm on the space X to indicate which norm is intended.

An NLS X is *finite-dimensional* if it is finite-dimensional as a vector space, which is to say there is a finite collection $\{x_n\}_{n=1}^N \subset X$ such that any $x \in X$ can be written as a linear combination of the $\{x_n\}_{n=1}^N$, viz.

$$x = \lambda_1 x_1 + \lambda_2 x_2 + \cdots + \lambda_N x_N,$$

where the λ_i are scalars (members of the ground field $\mathbb{R}$ or $\mathbb{C}$). Otherwise, X is called *infinite-dimensional*. Interest here is mainly in infinite-dimensional spaces.

Remark. In a good deal of the theory developed here, it will not matter for the outcome whether the NLSs are real or complex vector spaces. When this point is moot, we will write $\mathbb{F}$ rather than $\mathbb{R}$ or $\mathbb{C}$. The reader should understand when the symbol $\mathbb{F}$ appears that it stands for either $\mathbb{R}$ or for $\mathbb{C}$, and the discussion at that juncture holds for both.

Examples.

(1) For $d \geq 1$, consider $\mathbb{F}^d$ with the usual Euclidean length of a vector $x = (x_1, \ldots, x_d)$ denoted $|x| = \left(\sum_{n=1}^d |x_n|^2\right)^{1/2}$. If we define, for $x \in \mathbb{F}^d$, $\|x\| = |x|$, then $(\mathbb{F}^d, \|\cdot\|)$ is a finite-dimensional NLS of dimension d over $\mathbb{F}$.

(2) Let a and b be real numbers, $a < b$, with $a = -\infty$ or $b = +\infty$ allowed as possible values. The collection of bounded, continuous functions on $[a, b]$ taking values in $\mathbb{F}$ is denoted

$$C([a,b]) = C([a,b]; \mathbb{F})$$
$$= \left\{ f : [a,b] \to \mathbb{F} \mid f \text{ is continuous and } \sup_{x \in [a,b]} |f(x)| < \infty \right\}.$$

Impose a vector space structure by pointwise multiplication and addition; that is, for $x \in [a,b]$ and $\lambda \in \mathbb{F}$, define

$$(f+g)(x) = f(x) + g(x) \quad \text{and} \quad (\lambda f)(x) = \lambda f(x).$$

For $f \in C([a,b])$, let

$$\|f\|_{C([a,b])} = \sup_{x \in [a,b]} |f(x)|.$$

This is easily shown to define a norm. Thus, $\left(C([a,b]), \|\cdot\|_{C([a,b])}\right)$ becomes an NLS, which is also infinite-dimensional. (This latter fact follows upon consideration of the collection $\{\sin(nx)\}_{n=1}^\infty$, for example.)

(3) We can impose a different norm on the space $C([a,b])$ when a and b are finite, defined by

$$\|f\|_{L^1([a,b])} = \int_a^b |f(x)|\, dx.$$

One verifies straightforwardly that $\left(C([a,b]), \|\cdot\|_{L^1([a,b])}\right)$ is also an NLS, but *different* from $\left(C([a,b]), \|\cdot\|_{C([a,b])}\right)$. These two NLSs have the same set of objects and the same vector space structure, but different norms; they measure sizes differently.

Further examples arise as subspaces of NLSs. This fact follows directly from the definitions, and is stated formally below.

Proposition 2.1. *If $(X, \|\cdot\|)$ is an NLS and $V \subset X$ is a linear subspace, then $(V, \|\cdot\|)$ is an NLS.*

Let $(X, \|\cdot\|)$ be an NLS. Then X is a *metric space* if a metric d is defined on X by
$$d(x, y) = \|x - y\|;$$
d is called the *induced metric*. To see that d is a metric, just note that for $x, y, z \in X$,

$$0 = d(x, y) = \|x - y\| \iff x - y = 0 \iff x = y,$$
$$d(x, y) = \|x - y\| = \|-(y - x)\| = |-1| \, \|y - x\| = d(y, x),$$
$$d(x, y) = \|x - y\| = \|x - z + z - y\| \le \|x - z\| + \|z - y\| = d(x, z) + d(z, y).$$

Consequently, the concepts of elementary topology are available in any NLS. In particular, we may talk about the norm topology of open sets and closed sets in any NLS.

A set $U \subset X$ is *open* if for each $x \in U$, there is an $r > 0$ (depending on x in general) such that

$$B_r(x) = \{y \in X : d(y, x) < r\} \subset U.$$

The set $B_r(x)$ is referred to as the (open) ball of radius r about x. A set $F \subset X$ is closed if $F^c = X \smallsetminus F = \{y \in X : y \notin F\}$ is open. As with any metric space, F is closed if and only if it is *sequentially closed*. That is, F is closed means that whenever $\{x_n\}_{n=1}^{\infty} \subset F$ and $x_n \to x$ as $n \to \infty$ for the metric, then it must be the case that $x \in F$.

Proposition 2.2. *In an NLS X, the operations of addition, $+ : X \times X \to X$, scalar multiplication, $\cdot : \mathbb{F} \times X \to X$, and taking the norm, $\|\cdot\| : X \to \mathbb{R}$, are continuous.*

Proof. Let $\{x_m\}_{m=1}^{\infty}$ and $\{y_n\}_{n=1}^{\infty}$ be sequences in X converging to $x, y \in X$, respectively. Then

$$\|(x_m + y_n) - (x + y)\| = \|(x_m - x) + (y_n - y)\| \le \|x_m - x\| + \|y_n - y\| \to 0.$$

We leave the proof for scalar multiplication to the reader, which requires the fact that a convergent sequence of scalars is bounded.

Continuity of the norm follows from what is sometimes called the *reverse triangle inequality*. This states that if x and y lie in an NLS X, then

$$\left| \|x\| - \|y\| \right| \leq \|x - y\|. \tag{2.1}$$

The latter inequality follows since

$$\|x\| \leq \|x - y\| + \|y\|$$

from the triangle inequality. This inequality and its analog with the roles of x and y reversed gives (2.1). If $x_n \to x$ in an NSL X, apply the reverse triangle inequality to x and $y = x_n$ to conclude that $\|x_n\| \to \|x\|$. □

Recall that a sequence $\{x_n\}_{n=1}^{\infty}$ in a metric space (X, d) is called a *Cauchy sequence* if

$$\lim_{n,m \to \infty} d(x_n, x_m) = 0.$$

Equivalently, given $\epsilon > 0$, there is an $N = N(\epsilon)$ such that if $n, m \geq N$, then

$$d(x_n, x_m) \leq \epsilon.$$

A metric space is called *complete* if every Cauchy sequence converges to a point in X. An NLS $(X, \| \cdot \|)$ that is complete as a metric space is called a *Banach space* after the Polish mathematician Stefan Banach, who was one of the principal pioneers of functional analysis.

A set $M \subset X$ of an NLS X is said to be *bounded* if there is an $R > 0$ such that $M \subset \overline{B_R(0)}$, which is to say

$$\|x\| \leq R < \infty \quad \text{for all } x \in M.$$

Proposition 2.3. *Every Cauchy sequence in an NLS is bounded.*

Proof. Let $\{x_n\}_{n=1}^{\infty}$ be a Cauchy sequence in an NLS. By definition, there is an N such that if $n \geq N$, then

$$\|x_n - x_N\| \leq 1,$$

say. By the reverse triangle inequality, this means

$$\|x_n\| - \|x_N\| \leq 1 \quad \text{or} \quad \|x_n\| \leq \|x_N\| + 1,$$

for $n \geq N$. The initial segment, $\{x_1, x_2, \ldots, x_{N-1}\}$ of the sequence is bounded since it is finite, say $\|x_n\| \leq K$ for $1 \leq n \leq N - 1$. It follows that

$$\|x_n\| \leq \max\{K, \|x_N\| + 1\} = R,$$

for all n. □

Examples.

(1) The spaces $\mathbb{R}^d$ and $\mathbb{C}^d$ are complete as we learn in advanced calculus or elementary analysis.

(2) For a and b in $[-\infty, \infty]$, $a < b$, the space $\left(C([a,b]), \|\cdot\|_{C([a,b])}\right)$ is complete, since the uniform limit of continuous functions is continuous. (The details of the full proof are left for the reader to provide in one of the exercises at the end of the chapter.)

(3) The space $\left(C([a,b]), \|\cdot\|_{L^1([a,b])}\right)$, a and b finite, is *not* complete. To see this, suppose that $a = -1$ and $b = 1$ (we can translate and scale if this is not the case) and define for $n = 1, 2, 3, \ldots,$

$$f_n(x) = \begin{cases} 1 & \text{if } x \leq 0, \\ 1 - nx & \text{if } 0 < x < 1/n, \\ 0 & \text{if } x \geq 1/n. \end{cases}$$

Each $f_n \in C([-1,1])$, and this is a Cauchy sequence for the given norm since

$$\int_{-1}^{1} |f_n(x) - f_m(x)| \, dx \leq \int_{0}^{1} (|f_n(x)| + |f_m(x)|) \, dx = \frac{1}{2n} + \frac{1}{2m}$$

can be made as small as we like for n and m large enough (note that the sequence is *not* Cauchy using the norm $\|\cdot\|_{C([-1,1])}$!). However, f_n does not converge to a function in $C([-1,1])$, since it must converge to 1 for $x < 0$ and to 0 for $x > 0$.

Unless otherwise specified, we use the norm $\|\cdot\| = \|\cdot\|_{C([a,b])}$ on $C([a,b])$ which renders it a Banach space.

If X is a linear space over $\mathbb{F}$ and d is a metric on X induced from a norm on X, then for all $x, y, a \in X$ and $\lambda \in \mathbb{F}$,

$$d(x + a, y + a) = d(x, y) \quad \text{and} \quad d(\lambda x, \lambda y) = |\lambda| d(x, y). \tag{2.2}$$

Suppose now that X is a linear space over $\mathbb{F}$ and d is a metric on X satisfying (2.2). Is it necessarily the case that there is a norm $\|\cdot\|$ on X such that $d(x, y) = \|x - y\|$? This question is left for the reader to ponder.

If X is a vector space and $\|\cdot\|_1$ and $\|\cdot\|_2$ are two norms on X, they are said to be *equivalent norms* if there exist constants $c, d > 0$ such that

$$c\|x\|_1 \leq \|x\|_2 \leq d\|x\|_1 \tag{2.3}$$

for all $x \in X$. Equivalent norms do not measure size in the same way, but they agree about when something is "small" or "large." In particular, a sequence converges for the $\|\cdot\|_1$-norm if and only if it converges for the $\|\cdot\|_2$-norm.

As will be seen later, on a finite-dimensional NLS any pair of norms is equivalent, whereas this is *not* the case in infinite-dimensional spaces. For example, if $f \in C([0,1])$, then

$$\|f\|_{L^1([0,1])} \leq \|f\|_{C([0,1])},$$

but the opposite bound is lacking. That this is so follows by contemplating the sequence

$$f_n(x) = \begin{cases} n^2 x & \text{if } x \leq 1/n, \\ 2n - n^2 x & \text{if } 1/n < x < 2/n, \\ 0 & \text{if } x \geq 2/n, \end{cases}$$

for which

$$\|f_n\|_{C([0,1])} = n \quad \text{but} \quad \|f_n\|_{L^1([0,1])} = 1.$$

If $\|\cdot\|_1$ and $\|\cdot\|_2$ are two equivalent norms on an NLS X as in (2.3), then the collections O_1 and O_2 of open sets induced by these two norms are the same. To see this, let $B_r^i(x)$ be the ball about $x \in X$ of radius r measured using norm $\|\cdot\|_i$. Then,

$$B_{r/d}^1(x) \subset B_r^2(x) \subset B_{r/c}^1(x)$$

where c and d are as in (2.3). Hence, these open balls are nested, from which one sees at once that topologically, $(X, \|\cdot\|_1)$ and $(X, \|\cdot\|_2)$ are indistinguishable or, what is the same, the identity map $I : (X, \|\cdot\|_1) \to (X, \|\cdot\|_2)$ is a homeomorphism.

Convexity is another important property that arises in a vector space context.

Definition. A set C in a linear space X over $\mathbb{F}$ is *convex* if whenever $x, y \in C$, then

$$tx + (1-t)y \in C$$

whenever $0 \leq t \leq 1$.

In spite of the name, balls in an NLS need not possess a "rounded" shape. For examples, consider the unit balls in the NLSs $(\mathbb{R}^2, \|\cdot\|_{\ell^p})$ for $1 \leq p \leq \infty$, where the norm is

$$\|(x,y)\|_{\ell^p} = (x^p + y^p)^{1/p}, \quad 1 \leq p < \infty, \quad \text{and} \quad \|(x,y)\|_{\ell^\infty} = \max\{|x|, |y|\}.$$

Some of these unit balls B_1^p are depicted in Figure 2.1. While they may not be "round," balls are always convex.

Proposition 2.4. *Suppose $(X, \|\cdot\|)$ is an NLS and $r > 0$. For any $x \in X$, $B_r(x)$ is convex.*

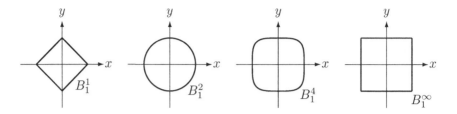

Figure 2.1 Shown are the boundaries of the unit balls $B_1^p = \{(x,y) \in \mathbb{R}^2 : \|(x,y)\|_{\ell^p} < 1\}$ in the NLSs $(\mathbb{R}^2, \|\cdot\|_{\ell^p})$ for $p = 1, 2, 4, \infty$.

Proof. Let $y, z \in B_r(x)$ and $t \in [0,1]$ and compute as follows:

$$\begin{aligned}
\|ty + (1-t)z - x\| &= \|t(y-x) + (1-t)(z-x)\| \\
&\leq \|t(y-x)\| + \|(1-t)(z-x)\| \\
&= |t|\,\|y-x\| + |1-t|\,\|z-x\| \\
&< tr + (1-t)r = r.
\end{aligned}$$

Thus, $B_r(x)$ is convex. □

2.2 LINEAR MAPS AND THE DUAL SPACE

One reason vector spaces are so important and ubiquitous is that they are the natural domain of definition for linear maps or operators, and the latter pervade mathematics and its applications. Remember, a *linear operator* is one that commutes with addition and scalar multiplication, so that

$$\begin{aligned}
T(x+y) &= T(x) + T(y), \\
T(\lambda x) &= \lambda T(x),
\end{aligned}$$

for $x, y \in X$, $\lambda \in \mathbb{F}$. For linear maps, one often writes Tx for $T(x)$, leaving out the parentheses. The scaling property requires that $T(0) = 0$ for every linear map.

An operator $T : X \to Y$, X and Y NLSs, is *bounded* if it takes bounded sets to bounded sets. (Note that this does *not* require that the entire image of T be bounded!)

Proposition 2.5. *If X and Y are NLSs and $T : X \to Y$ is linear, then T is bounded if and only if there is $C > 0$ such that*

$$\|Tx\|_Y \leq C\|x\|_X \quad \text{for all } x \in X.$$

Proof. The result follows from scaling considerations. Suppose first that T is bounded. Corresponding to the bounded set $M = B_1(0)$, there is an $R > 0$ such that

$$\|Ty\|_Y \leq R \quad \text{for all } y \in M.$$

Now let $x \in X$ be given. If $x = 0$, the conclusion holds trivially. Otherwise, let $y = x/(2\|x\|_X) \in M$. Then

$$\|Tx\|_Y = \left\|T(2\|x\|_X y)\right\|_Y = \left\|2\|x\|_X Ty\right\|_Y = 2\|x\|_X \|Ty\|_Y \leq 2R\|x\|_X,$$

which is the desired conclusion with $C = 2R$.

Conversely, suppose that there is $C > 0$ such that

$$\|Tx\|_Y \leq C\|x\|_X \quad \text{for all } x \in X.$$

Let $M \subset B_R(0)$ be bounded and fix $x \in M$. Then

$$\|Tx\|_Y \leq C\|x\|_X \leq CR < \infty,$$

so T takes bounded sets to bounded sets. □

On the one hand, linear maps are the natural (i.e., structure preserving) maps on a vector space. In fact, two vector spaces X and Y are said to be *isomorphic* if there is a one-to-one linear map T mapping X onto Y. In this case, X and Y have elements that are in one-to-one correspondence, and addition and scalar multiplication agree whether performed in X or in Y, since the inverse map T^{-1} is also linear (as the reader can verify). When two spaces X and Y share the same structure, we sometimes denote this state of affairs by $X \cong Y$. The map T that establishes X and Y to be isomorphic is called an *isomorphism*.

On the other hand, the natural mappings between topological spaces, and metric spaces in particular, are the continuous maps. If (X, d) and (Y, ρ) are two metric spaces and $f : X \to Y$ is a function, then f is *continuous* if for any $x \in X$ and $\epsilon > 0$, there exists a $\delta = \delta(x, \epsilon) > 0$ such that

$$d(x, y) \leq \delta \quad \text{implies} \quad \rho(f(x), f(y)) \leq \epsilon.$$

For NLSs, this becomes

$$\|x - y\| \leq \delta \quad \text{implies} \quad \|f(x) - f(y)\| \leq \epsilon.$$

If $(X, \|\cdot\|_X)$ and $(Y, \|\cdot\|_Y)$ are NLSs, then they are simultaneously linear spaces and metric spaces. Thus one might expect the collection

$$B(X, Y) = \{T : X \to Y \mid T \text{ is linear and continuous}\}$$

of mappings that are consistent with both the algebraic and metric structures of the underlying NLS to be of considerable interest. Continuous linear mappings between NLSs are variously referred to as *bounded operators* or *bounded linear operators* or *continuous linear operators*.

Example. If $X = \mathbb{R}^{d_1}$ and $Y = \mathbb{R}^{d_2}$, then $B(\mathbb{R}^{d_1}, \mathbb{R}^{d_2})$ may be identified with the set of (real) $d_2 \times d_1$ matrices under matrix-vector multiplication. These maps are easily seen to be continuous linear operators. In fact, in finite dimensions, every linear map is continuous. This is *not* the case in infinite dimensions.

Proposition 2.6. *Let X and Y be NLSs and $T : X \to Y$ a linear map. The following are equivalent:*

(i) *T is continuous;*

(ii) *T is continuous at some point;*

(iii) *T is bounded.*

Proof. ((i) $\implies$ (ii)) Trivial.

((ii) $\implies$ (iii)) Suppose T is continuous at $x_0 \in X$. Then there is a $\delta = \delta(x_0, 1) > 0$ such that

$$\|x - x_0\|_X \le \delta \quad \text{implies} \quad \|Tx - Tx_0\|_Y \le 1. \tag{2.4}$$

But, by linearity $Tx - Tx_0 = T(x - x_0)$. Thus, (2.4) is equivalent to the condition

$$\|y\|_X \le \delta \quad \text{implies} \quad \|Ty\|_Y \le 1.$$

It follows readily that if $x \in X$, $x \ne 0$, then

$$\|Tx\|_Y = \left\| \frac{\|x\|_X}{\delta} T\left(\frac{\delta}{\|x\|_X} x \right) \right\|_Y = \frac{1}{\delta} \left\| T\left(\frac{\delta}{\|x\|_X} x \right) \right\|_Y \|x\|_X \le \frac{1}{\delta} \|x\|_X,$$

since

$$\left\| \frac{\delta}{\|x\|_X} x \right\|_X = \delta.$$

((iii) $\implies$ (i)) It is supposed that T is linear and bounded, so there is a $C > 0$ such that

$$\|Tx\|_Y \le C \|x\|_X \quad \text{for all } x \in X.$$

Let $\epsilon > 0$ be given and let $\delta = \epsilon/C$. Suppose $\|x - x_0\|_X \le \delta$. Then

$$\|Tx - Tx_0\|_Y = \|T(x - x_0)\|_Y \le C \|x - x_0\|_X \le \epsilon.$$

Therefore T is continuous at x_0, and x_0 was an arbitrary point in X. □

Let X, Y be NLSs and let $T \in B(X, Y)$ be a continuous linear operator from X to Y. We know therefore that T is bounded on any bounded set of X, so the quantity

$$\|T\| = \|T\|_{B(X,Y)} = \sup_{x \in B_1(0)} \|Tx\|_Y \tag{2.5}$$

is finite. As the notation suggests, the mapping $\| \cdot \|_{B(X,Y)} : B(X,Y) \to [0, \infty)$ is expected to be a norm. There are several things to check.

We begin by observing that $B(X, Y)$ is a vector space in its own right if we define $S + T$ and λS by

$$(S + T)(x) = Sx + Tx \quad \text{and} \quad (\lambda S)(x) = \lambda Sx$$

for all $x \in X$ and $\lambda \in \mathbb{F}$. With this definition, it is easy to check that $S + T, \lambda S \in B(X,Y)$, and in particular, that

$$(S + T)(\lambda x + y) = \lambda(S + T)(x) + (S + T)(y)$$

for all $x, y \in X$ and $\lambda \in \mathbb{F}$. Moreover, the eight rules of algebra in the definition of a vector space hold, since they hold in X and Y.

Proposition 2.7. *Let X and Y be NLSs. The formula* (2.5) *defines a norm on $B(X,Y)$. Moreover, if $T \in B(X,Y)$, then*

$$\|T\| = \sup_{\|x\|_X \leq 1} \|Tx\|_Y = \sup_{x \in B_1(0)} \|Tx\|_Y$$

$$= \sup_{\|x\|_X = 1} \|Tx\|_Y = \sup_{x \in S_1(0)} \|Tx\|_Y = \sup_{x \neq 0} \frac{\|Tx\|_Y}{\|x\|_X}. \tag{2.6}$$

If Y is a Banach space, then so is $B(X,Y)$ with this norm.

The notation $S_r(x) = \{x \in X : \|x\| = r\}$ for the sphere of radius $r > 0$ about x in X has been introduced here.

Proof. Attention is turned first to checking that $\| \cdot \|_{B(X,Y)}$ is a norm. If T is the zero map, then clearly $\|T\| = 0$. On the other hand, if $\|T\| = 0$, then T vanishes on the unit ball. For any $x \in X$, a suitable choice of $r > 0$ has $\|rx\| < 1$. Hence, $0 = T(rx) = rTx$ implies $Tx = 0$. Hence T vanishes everywhere.

Plainly, by definition of scalar multiplication,

$$\|\lambda T\| = \sup_{x \in B_1(0)} \|(\lambda T)(x)\|_Y = |\lambda| \sup_{x \in B_1(0)} \|Tx\|_Y = |\lambda| \|T\|.$$

The triangle inequality is just as simple, *viz.*

$$\|T + S\| = \sup_{x \in B_1(0)} \|(T + S)(x)\|_Y = \sup_{x \in B_1(0)} \|Tx + Sx\|_Y$$

$$\leq \sup_{x \in B_1(0)} \{\|Tx\|_Y + \|Sx\|_Y\}$$

$$\leq \sup_{x \in B_1(0)} \|Tx\|_Y + \sup_{x \in B_1(0)} \|Sx\|_Y = \|T\| + \|S\|.$$

Thus $(B(X,Y), \| \cdot \|_{B(X,Y)})$ is indeed an NLS.

The alternative formulas for the norm expressed in (2.6) are straightforward to deduce. Notice that the last formula makes it obvious that for all $x \in X$ and $T \in B(X,Y)$,

$$\|Tx\|_Y \leq \|T\|_{B(X,Y)}\|x\|_X, \tag{2.7}$$

an inequality that will find frequent use.

The more interesting fact is that $B(X,Y)$ is complete even if it is only presumed that Y is complete. This simple result has far-reaching consequences. To establish this point, suppose $\{T_n\}_{n=1}^{\infty}$ is a Cauchy sequence in $B(X,Y)$. We must show it converges in $B(X,Y)$. Let $x \in X$ and consider the sequence $\{T_n x\}_{n=1}^{\infty}$ in Y. Because of (2.7), it follows that

$$\|T_n x - T_m x\|_Y \leq \|T_n - T_m\|_{B(X,Y)} \|x\|_X,$$

and thus $\{T_n x\}_{n=1}^{\infty}$ is seen to be Cauchy in Y. As Y is a Banach space, $\{T_n x\}_{n=1}^{\infty}$ must converge to some element of Y that depends upon x of course; call this element Tx. There is thus established a correspondence

$$x \longmapsto Tx$$

between X and Y. We claim it is a continuous linear correspondence, whence $T \in B(X,Y)$. It is further asserted that $T_n \to T$ in $B(X,Y)$.

First note that

$$T(x+y) = \lim_{n\to\infty} T_n(x+y) = \lim_{n\to\infty} (T_n x + T_n y)$$
$$= \lim_{n\to\infty} T_n x + \lim_{n\to\infty} T_n y = Tx + Ty.$$

Similarly, $T(\lambda x) = \lambda Tx$ for $x \in X$ and $\lambda \in \mathbb{F}$. Thus T is a linear map. To see that T is a bounded map, first note that $\{T_n\}_{n=1}^{\infty}$, being Cauchy, must be a bounded sequence (Proposition 2.3). From this, it follows at once that T is a bounded operator; for if $x \in X$, then

$$\|Tx\|_Y = \lim_{n\to\infty} \|T_n x\|_Y \leq \limsup_{n\to\infty} \|T_n\|_{B(X,Y)} \|x\|_X \leq M\|x\|,$$

where M is the bound on $\{\|T_n\|_{B(X,Y)}\}_{n=1}^{\infty}$.

The last point to check is that $T_n \to T$ in $B(X,Y)$. Let $x \in B_1(0)$ in X and observe that

$$\|Tx - T_n x\|_Y = \lim_{m\to\infty} \|T_m x - T_n x\|_Y = \lim_{m\to\infty} \|(T_m - T_n)x\|$$
$$\leq \limsup_{m\to\infty} \|T_m - T_n\|_{B(X,Y)} \|x\|_X \leq \epsilon(n),$$

for some function $\epsilon(n)$. Since x was an arbitrary element in $B_1(0)$, this means

$$\|T - T_n\|_{B(X,Y)} \leq \epsilon(n),$$

and because $\{T_k\}_{k=1}^{\infty}$ is Cauchy, $\epsilon(n) \to 0$ as $n \to \infty$. ☐

The structure of bounded linear maps $T : X \to Y$ on infinite-dimensional spaces can be quite complex. It will turn out to be fruitful at the beginning to study the simpler case, albeit still quite complex, of a map $T : X \to \mathbb{F}$ for which the range has a single real or complex dimension, depending on whether $\mathbb{F} = \mathbb{R}$ or $\mathbb{F} = \mathbb{C}$. Maps of a vector space into the ground field are called *functionals*.

Definition. Let X be an NLS over $\mathbb{F}$. The *dual space* X^* of X is the Banach space $B(X, \mathbb{F})$. The elements of X^* are called *bounded linear functionals* on X.

The dual space is complete because $\mathbb{R}$ and $\mathbb{C}$ are complete.

2.3 SOME IMPORTANT EXAMPLES

2.3.1 Finite-dimensional spaces

Let X be a vector space of dimension $d < \infty$ over $\mathbb{F}$, and let $\{e_n\}_{n=1}^d \subset X$ be a *basis*, which is to say that for any $x \in X$, there are unique $x_n \in \mathbb{F}$ such that

$$x = \sum_{n=1}^d x_n\, e_n.$$

Define a map $T : X \to \mathbb{F}^d$ by $T(x) = (x_1, \ldots, x_d)$, which gives a one-to-one correspondence between X and $\mathbb{F}^d$. This map, called the *coordinate mapping*, is easily seen to be linear, so we have a vector space isomorphism between the spaces. Consequently, X and $\mathbb{F}^d$ have the same vector space structure. That is, given a dimension d, there is only one vector space structure of the given dimension over the ground field $\mathbb{F}$.

For $1 \le p \le \infty$, define the map $\| \cdot \|_{\ell^p} : \mathbb{F}^d \to [0, \infty)$ by

$$\|x\|_{\ell^p} = \begin{cases} \left(\displaystyle\sum_{n=1}^d |x_n|^p \right)^{1/p} & \text{for } p < \infty, \\ \max_{n=1,\ldots,d} |x_n| & \text{for } p = \infty. \end{cases}$$

When $p = 2$, it is common practice to write the ℓ^2-norm of $x \in \mathbb{F}^d$ as $|x|_{\ell^2}$ or simply $|x|$, since in that case it is the ordinary *Euclidean norm*. Moreover, when no confusion would arise, it is common practice to write $\| \cdot \|_p$ in lieu of $\| \cdot \|_{\ell^p}$. It is easily verified that $\|x\|_{\ell^1}$ and $\|x\|_{\ell^\infty}$ are norms on $\mathbb{F}^d$. If $1 < p < \infty$ the zero and scaling properties of the ℓ^p-norm on $\mathbb{F}^d$ are straightforward. A few facts are needed to establish the triangle inequality. Once these are established, one immediately deduces that $\|T(\cdot)\|_{\ell^p}$ is also a norm on X, where T is the coordinate map introduced earlier.

The following classical inequality will yield information leading to a proof of the triangle inequality for $(X, \| \cdot \|_{\ell^p})$.

Lemma 2.8 (Young's inequality for products). *Let $1 < p < \infty$ and let q denote the conjugate exponent to p defined by*

$$\frac{1}{p} + \frac{1}{q} = 1.$$

If a and b are nonnegative real numbers, then

$$ab \le \frac{a^p}{p} + \frac{b^q}{q}, \tag{2.8}$$

with equality if and only if $a^p = b^q$. Moreover, for any $\epsilon > 0$, then there is $C = C(p, \epsilon) > 0$ such that

$$ab \leq \epsilon a^p + C b^q.$$

Proof. The result holds for $a = 0$ or $b = 0$, so suppose, say, $b > 0$. The function $u : [0, \infty) \to \mathbb{R}$ given by

$$u(t) = \frac{t^p}{p} + \frac{1}{q} - t$$

has minimum value 0, attained only when $t = 1$. Apply this fact to the value $t = ab^{-q/p}$ to obtain the main result. Replace ab by $[(\epsilon p)^{1/p}a][(\epsilon p)^{-1/p}b]$ and apply (2.8) to obtain the final result. □

This lemma leads immediately to the next result, known as Hölder's inequality. When $p = 2$, the inequality is also called the Cauchy-Schwarz inequality, or more correctly, the Cauchy-Bunyakovsky-Schwarz inequality.

Theorem 2.9 (Hölder's inequality). *Let $1 \leq p \leq \infty$ and let q denote the conjugate exponent (i.e., $1/p + 1/q = 1$, with the convention that $q = \infty$ if $p = 1$ and $q = 1$ if $p = \infty$). If $x, y \in \mathbb{F}^d$, then*

$$\sum_{n=1}^{d} |x_n y_n| \leq \|x\|_{\ell^p} \|y\|_{\ell^q}.$$

Proof. The result is trivial if either p or q is infinity, or if either $x = 0$ or $y = 0$. Otherwise, simply apply (2.8) to $a = |x_n|/\|x\|_{\ell^p}$ and $b = |y_n|/\|y\|_{\ell^q}$ and sum on n to see that

$$\sum_{n=1}^{d} \frac{|x_n|}{\|x\|_{\ell^p}} \frac{|y_n|}{\|y\|_{\ell^q}} \leq \sum_{n=1}^{d} \frac{|x_n|^p}{p\|x\|_{\ell^p}^p} + \sum_{n=1}^{d} \frac{|y_n|^q}{q\|y\|_{\ell^q}^q} = \frac{1}{p} + \frac{1}{q} = 1.$$

The conclusion follows. □

Attention is returned to showing that each $\|\cdot\|_{\ell^p}$ is in fact a norm on $\mathbb{F}^d$. It remains only to show the triangle inequality holds for $1 < p < \infty$; as mentioned, the triangle inequality for $p = 1$ or $p = \infty$ is trivial. For $x, y \in \mathbb{F}^d$, simply apply Hölder's inequality twice, *viz.*

$$\|x + y\|_{\ell^p}^p = \sum_{n=1}^{d} |x_n + y_n|^p \leq \sum_{n=1}^{d} |x_n + y_n|^{p-1}(|x_n| + |y_n|)$$

$$\leq \left(\sum_{n=1}^{d} |x_n + y_n|^{(p-1)q} \right)^{1/q} (\|x\|_{\ell^p} + \|y\|_{\ell^p}).$$

Since $(p - 1)q = p$ and $1/q = 1 - 1/p$, the triangle inequality

$$\|x + y\|_{\ell^p} \leq \|x\|_{\ell^p} + \|y\|_{\ell^p}$$

follows.

Proposition 2.10. *Let $1 \le p \le \infty$. For any $x \in \mathbb{F}^d$,*

$$\|x\|_{\ell^\infty} \le \|x\|_{\ell^p} \le d^{1/p}\|x\|_{\ell^\infty},$$

with equality possible.

This result is trivial, and shows that all the ℓ^p-norms $\|x\|_{\ell^p}$, $1 \le p \le \infty$, are equivalent on $\mathbb{F}^d$. The constants 1 and $d^{1/p}$ in the last inequality are sharp, meaning that they cannot be improved.

A fundamental difference between finite-dimensional spaces and infinite-dimensional NLSs is that in finite dimensions, a closed and bounded set is always compact, but such a statement turns out to be untrue in infinite dimensions. This is closely related to another fundamental difference, namely that in finite dimensions, all norms are equivalent, and so there is in fact only one norm topology.

Proposition 2.11. *Let X be a finite-dimensional NLS. All norms on X are equivalent. Moreover, a subset of X is compact if and only if it is closed and bounded.*

Proof. Let d be the dimension of X, and let $\{e_n\}_{n=1}^d$ be a basis. We defined earlier the coordinate mapping $T : X \to \mathbb{F}^d$. Let $\|\cdot\|$ denote any norm of X, and let

$$\|x\|_1 = \|T(x)\|_{\ell^1}$$

be a second norm. We will show that these two norms are equivalent, which then implies that any pair of norms on X are equivalent.

The space $(X, \|\cdot\|_1)$ is essentially $\mathbb{F}^d$ equipped with the norm $\|\cdot\|_{\ell^1}$. In fact, by definition of the norm $\|\cdot\|_1$, the coordinate map $T : (X, \|\cdot\|_1) \to \mathbb{F}^d$ is bounded, i.e., continuous, as is its inverse, which is also a linear function. Thus, $(X, \|\cdot\|_1)$ and $\mathbb{F}^d$ are homeomorphic as topological spaces, and also isomorphic as vector spaces. The Heine-Borel theorem states that every closed and bounded subset of $\mathbb{F}^d$ is compact, so the same is true of $(X, \|\cdot\|_1)$. In particular, $S_1^1 = \{x \in X : \|x\|_1 = 1\}$ is compact.

For $x \in X$,

$$\|x\| = \left\| \sum_{n=1}^d x_n e_n \right\| \le \sum_{n=1}^d |x_n| \|e_n\| \le C\|x\|_1,$$

where $C = \max_n \|e_n\| < \infty$. This is one of the two bounds needed to show that the norms are equivalent, and it implies that $B_r^1 = \{x \in X : \|x\|_1 < r\} \subset B_{Cr} = \{x \in X : \|x\| < Cr\}$. It is concluded at once that the topology on X generated by $\|\cdot\|_1$ is *stronger* than that generated by $\|\cdot\|$ (that is, every open set of $(X, \|\cdot\|)$ is an open set of $(X, \|\cdot\|_1)$). Consequently, the function $\|\cdot\| : (X, \|\cdot\|_1) \to \mathbb{R}$ is continuous since the inverse image of every open set in $\mathbb{R}$ is open in $(X, \|\cdot\|)$, and thus also open in $(X, \|\cdot\|_1)$.

Now, let
$$a = \inf_{x \in S_1^1} \|x\|.$$

Since S_1^1 is compact and $\| \cdot \|$ is continuous, the function must take on its minimal value. That is, there is some $x_1 \in S_1^1$ such that $a = \|x_1\|$, which is then strictly positive since $x_1 \neq 0$. Thus, for any $x \in X$, scaling implies that

$$\|x\| \geq a\|x\|_1,$$

which is the other bound needed to show that the two norms are equivalent. Finally, the compactness result now holds with respect to any norm on X since there is in fact only one norm topology on X. $\qquad\square$

Within the course of the above proof, it was established that, for a given dimension d, there is only one d-dimensional NLS structure. Infinite-dimensional spaces are more interesting.

Corollary 2.12. *Every NLS of dimension $d < \infty$ is isomorphic and homeomorphic to $\mathbb{F}^d$, and so is a Banach space.*

We leave the proof of the following corollary for the exercises at the end of the chapter.

Corollary 2.13. *If X and Y are NLSs, X finite-dimensional, and $T : X \to Y$ linear, then T is bounded. The dual space $X^* = B(X, \mathbb{F})$ is isomorphic and homeomorphic to $\mathbb{F}^d$.*

2.3.2 The spaces ℓ^p

Let p lie in the range $[1, \infty)$. Define the vector spaces and norms

$$\ell^p = \left\{ x = \{x_n\}_{n=1}^{\infty} : x_n \in \mathbb{F} \text{ and } \|x\|_{\ell^p} = \left(\sum_{n=1}^{\infty} |x_n|^p \right)^{1/p} < \infty \right\},$$

and, for $p = \infty$,

$$\ell^{\infty} = \left\{ x = \{x_n\}_{n=1}^{\infty} : x_n \in \mathbb{F} \text{ and } \|x\|_{\ell^{\infty}} = \sup_n |x_n| < \infty \right\}.$$

These spaces are NLSs over $\mathbb{F}$, and they are sometimes called "little-ℓ^p" spaces. Indeed, it is straightforward to verify that they are vector spaces and $\| \cdot \|_{\ell^p}$ can be shown to be a norm using the techniques of the previous section. In fact, an infinite-dimensional version of Hölder's inequality holds.

Theorem 2.14 (Hölder's inequality in ℓ^p). *Let $1 \leq p \leq \infty$ and let q denote the conjugate exponent (i.e., $1/p + 1/q = 1$). If $x \in \ell^p$ and $y \in \ell^q$, then*

$$\sum_{n=1}^{\infty} |x_n y_n| \leq \|x\|_{\ell^p} \|y\|_{\ell^q}.$$

This follows as in the finite-dimensional case. The details of the proof are left to the reader. It is not difficult to show also that

$$\ell^p \subset \ell^q \quad \text{whenever } p \leq q.$$

An algebraic *basis,* or *Hamel basis,* for a vector space V is a subset B for which every element of V can be expressed uniquely as a finite linear combination of the vectors in B. Such a collection of vectors is not generally useful for infinite-dimensional Banach spaces, since a Hamel basis for such spaces is very large. If $1 \leq p < \infty$, ℓ^p is *countably* infinite-dimensional, in the sense that it has a type of basis consisting of the collection $\{e_n\}_{n=1}^{\infty}$, where $e_{n,i} = 0$ for $i \neq n$ and $e_{n,n} = 1$. Such a basis is tied to the topology of the space, and it is referred to as a *Schauder basis* to distinguish it from the very much larger, purely algebraic basis.

Definition. Let X be a Banach space. A sequence $\{\varphi_n\}_{n=1}^{\infty}$ is a *Schauder basis* of X if every $x \in X$ has is a unique representation in the form

$$x = \sum_{j=1}^{\infty} x_j \varphi_j,$$

and the latter infinite series converges to x in the norm on X. That is, the partial sums

$$\sum_{j=1}^{N} x_j \varphi_j \longrightarrow x \quad \text{as } N \to \infty.$$

As just mentioned, the spaces ℓ^p all have a Schauder basis if $1 \leq p < \infty$. However, the NLS ℓ^∞ does *not* have a Schauder basis.

Let $c_0 \subset \ell^\infty$ be the linear subspace defined as

$$c_0 = \left\{ \{x_n\}_{n=1}^{\infty} : \lim_{n \to \infty} x_n = 0 \right\}.$$

This is an NLS using the norm $\| \cdot \|_{\ell^\infty}$. Another interesting subspace is

$$f_0 = \left\{ \{x_n\}_{n=1}^{\infty} : x_n = 0 \text{ except for a finite number of values of } n \right\}.$$

These normed linear spaces are related to each other; indeed, if $1 \leq p < \infty$,

$$f_0 \subset \ell^p \subset c_0 \subset \ell^\infty.$$

On f_0, it is easy to construct a linear functional that is *not* continuous. Consider, for example,

$$T(x) = \sum_{n=1}^{\infty} n \, x_n.$$

This is well-defined and linear on f_0. However, for each $n > 0$, $T(e_n) = n$, but $\|e_n\|_{\ell^\infty} = 1$. Thus it cannot be the case that $|T(x)| \le C\|x\|_{\ell^\infty}$ for some fixed constant C.

The spaces ℓ^p, $1 \le p \le \infty$ are complete, though this requires a proof that is left for the exercises. However, if we take the vector space ℓ^1 and equip it with the ℓ^∞-norm, this is an NLS, but *not* a Banach space. To see this, first note that ℓ^1 is a linear subspace of ℓ^∞. Indeed, if $x = (x_1, x_2, \ldots) \in \ell^1$, then

$$\|x\|_{\ell^\infty} = \sup_{i \ge 1} |x_i| \le \sum_{j=1}^{\infty} |x_j| = \|x\|_{\ell^1} < \infty.$$

Hence ℓ^1 with the ℓ^∞-norm is an NLS. To see it is not complete, consider the sequence $\{y_k\}_{k=1}^{\infty} \subset \ell^1$ defined to be

$$y_k = (y_{k,1}, y_{k,2}, \ldots) = \left(1, \frac{1}{2}, \frac{1}{3}, \frac{1}{4}, \ldots, \frac{1}{k}, 0, 0, \ldots\right),$$

$k = 1, 2, 3, \ldots$. Then $\{y_k\}_{k=1}^{\infty}$ is Cauchy in the ℓ^∞-norm since

$$\|y_k - y_m\|_{\ell^\infty} \le \frac{1}{m+1}$$

provided $k \ge m$. If ℓ^1 were complete in the ℓ^∞-norm, then $\{y_k\}_{k=1}^{\infty}$ would converge to some element $z \in \ell^1$. Thus we would have that

$$\|y_k - z\|_{\ell^\infty} \longrightarrow 0$$

as $k \to \infty$. But, for $j \ge 1$,

$$|y_{k,j} - z_j| \le \|y_k - z\|_{\ell^\infty}$$

where $y_{k,j}$ and z_j are the jth components of y_k and z, respectively. In consequence, it is clear that $z_j = 1/j$ for all $j \ge 1$. However, the element

$$z = \left(1, \frac{1}{2}, \frac{1}{3}, \frac{1}{4}, \ldots, \frac{1}{k}, \frac{1}{k+1}, \ldots\right)$$

does not lie in ℓ^1, a contradiction. An outcome of this example is the conclusion that an infinite-dimensional vector space may have multiple distinct NLS structures imposed on it.

Lemma 2.15. *If $p < 1$, then $\|\cdot\|_{\ell^p}$ is not a norm on ℓ^p.*

To prove this, show the unit ball $B_1(0)$ is not convex, which would contradict Proposition 2.4. It is also easy to see directly that the triangle inequality does not always hold on ℓ^p, $p < 1$. The details are left to the reader.

The Hölder inequality immediately implies the existence of many continuous linear functionals. Indeed, let $1 \leq p \leq \infty$ and let q be conjugate to p. Then any $y \in \ell^q$ can be viewed as a function $Y : \ell^p \to \mathbb{F}$, $viz.$

$$Y(x) = \sum_{n=1}^{\infty} x_n \, y_n \quad \text{for all } x \in \ell^p. \tag{2.9}$$

This is clearly a linear functional and it is bounded since

$$|Y(x)| \leq (\|y\|_{\ell^q}) \|x\|_{\ell^p}.$$

An amusing exercise is to show that the norm of Y considered as a linear operator has the value

$$\|Y\| = \|y\|_{\ell^q}.$$

The converse is also true if $1 \leq p < \infty$. For any continuous linear functional Y on ℓ^p, there is a $y \in \ell^q$ such that (2.9) holds. Thus, the dual of ℓ^p can be identified as ℓ^q under the action defined by (2.9). The details are saved for the exercises, but it should be clear that one must define $y_n = Y(e_n)$, where the e_n are the elements of the standard basis $\{e_i\}_{i=1}^{\infty}$. The main problem is to show that $y \in \ell^q$. It is also true that $c_0^* = \ell^1$, but the dual of ℓ^∞ is larger than ℓ^1.

2.3.3 The Lebesgue spaces $L^p(\Omega)$

Let $\Omega \subset \mathbb{R}^d$ be measurable and let $0 < p < \infty$. Denote by $L^p(\Omega)$ the class of all measurable functions $f : \Omega \to \mathbb{F}$ such that

$$\int_{\Omega} |f(x)|^p \, dx < \infty. \tag{2.10}$$

An interesting point arises here. Suppose f and g lie in $L^p(\Omega)$ and that $f(x) = g(x)$ for a.e. $x \in \Omega$. Then, as far as integration is concerned, one really cannot distinguish f from g. For example, if $A \subset \Omega$ is measurable, then

$$\int_A |f|^p \, dx = \int_A |g|^p \, dx.$$

Thus within the space $L^p(\Omega)$, f and g are equivalent. This is formalized by modifying the definition of the elements of $L^p(\Omega)$. We declare two measurable functions that are equal a.e. to be equivalent, and define the equivalence classes

$$[f] = \{g : \Omega \to \mathbb{F} \mid g = f \text{ a.e. on } \Omega\}.$$

The elements of $L^p(\Omega)$ are then defined to be the equivalence classes $[f]$ such that one representative function g, say, of the class satisfies (2.10) (and hence all functions in the class will satisfy this condition). However, for convenience, we continue to speak of and denote elements of $L^p(\Omega)$ as "functions" which may be modified on a set of measure zero without consequence. For example, $f = 0$ in $L^p(\Omega)$ means only that $f = 0$ a.e. in Ω.

The integral in (2.10) arises frequently. Its pth root is denoted

$$\|f\|_{L^p(\Omega)} = \|f\|_p = \left(\int_\Omega |f(x)|^p \, dx \right)^{1/p}$$

and called the $L^p(\Omega)$-*norm*. It is shown presently to be a norm when $p \geq 1$.

A function $f(x)$ is normally said to be *bounded* on Ω by $R \geq 0$ if $|f(x)| \leq R$ for all $x \in \Omega$. The notion of boundedness is modified for measurable functions.

Definition. A measurable function $f : \Omega \to \mathbb{C}$ is *essentially bounded* on Ω by $R \geq 0$ if $|f(x)| \leq R$ for a.e. $x \in \Omega$. The infimum of such values R is called the *essential supremum* of $|f|$ on Ω, and denoted ess $\sup_{x \in \Omega} |f(x)|$.

For $p = \infty$, define $\|f\|_{L^\infty(\Omega)} = \|f\|_\infty = $ ess $\sup_{x \in \Omega} |f(x)|$. Then for all p with $0 < p \leq \infty$,

$$L^p(\Omega) = \{f : \|f\|_p < \infty\},$$

again with the proviso that the elements of $L^p(\Omega)$ are equivalence classes and not pointwise defined functions.

Proposition 2.16. *If $0 < p \leq \infty$, then $L^p(\Omega)$ is a vector space and $\|f\|_p = 0$ if and only if $f = 0$ a.e. in Ω. Moreover, for all $\alpha \in \mathbb{F}$, $\|\alpha f\|_p = |\alpha| \|f\|_p$.*

Proof. We first show that $L^p(\Omega)$ is closed under addition. For $p < \infty$, $f, g \in L^p(\Omega)$, and $x \in \Omega$,

$$|f(x) + g(x)|^p \leq \big(|f(x)| + |g(x)|\big)^p \leq 2^p \big(|f(x)|^p + |g(x)|^p\big).$$

Integrating and extracting the pth root of the result yields $\|f+g\|_p \leq 2(\|f\|_p^p + \|g\|_p^p)^{1/p} < \infty$. The case $p = \infty$ is clear.

For scalar multiplication, note that for $\alpha \in \mathbb{F}$,

$$\|\alpha f\|_p = |\alpha| \, \|f\|_p,$$

so $f \in L^p(\Omega)$ implies $\alpha f \in L^p(\Omega)$. The remark that $\|f\|_p = 0$ implies $f = 0$ a.e. is also clear. □

These spaces are interrelated in a number of ways.

Theorem 2.17 (Hölder's inequality in L^p). *Let $1 \leq p \leq \infty$ and let q denote the conjugate exponent defined by*

$$\frac{1}{p} + \frac{1}{q} = 1 \quad (q = \infty \text{ if } p = 1, \ q = 1 \text{ if } p = \infty).$$

If $f \in L^p(\Omega)$ and $g \in L^q(\Omega)$, then $fg \in L^1(\Omega)$ and

$$\|fg\|_1 \leq \|f\|_p \|g\|_q.$$

If $1 < p < \infty$, equality occurs if and only if $|f(x)|^p$ and $|g(x)|^q$ are proportional a.e. in Ω.

Proof. The result is clear if $p = 1$ or $p = \infty$. Suppose $1 < p < \infty$. Recall that Young's inequality (2.8) asserts that for $a, b > 0$,

$$ab \leq \frac{a^p}{p} + \frac{b^q}{q},$$

with equality if and only if $a^p/b^q = 1$. If $\|f\|_p = 0$ or $\|g\|_q = 0$, then $fg = 0$ a.e. on Ω and the result follows. Otherwise choose $a = |f(x)|/\|f\|_p$ and $b = |g(x)|/\|g\|_q$ in the last inequality. Integrating the resulting inequality over Ω secures the result. □

The proof that $L^p(\Omega)$ is an NLS is completed by showing the triangle inequality holds. In the context of the L^p-spaces, the triangle inequality is called *Minkowski's inequality*.

Theorem 2.18 (Minkowski's inequality). *If $1 \leq p \leq \infty$ and f and g are measurable, then*

$$\|f + g\|_p \leq \|f\|_p + \|g\|_p.$$

Proof. If f or $g \notin L^p(\Omega)$, the result is clear, since then the right-hand side is infinite. The result is also clear for $p = 1$ or $p = \infty$, so suppose $1 < p < \infty$ and $f, g \in L^p(\Omega)$. Then

$$\|f + g\|_p^p = \int_\Omega |f(x) + g(x)|^p \, dx \leq \int_\Omega |f(x) + g(x)|^{p-1} \big(|f(x)| + |g(x)|\big) \, dx$$

$$\leq \left(\int_\Omega |f(x) + g(x)|^{(p-1)q} \, dx \right)^{1/q} \big(\|f\|_p + \|g\|_p\big),$$

by two applications of Hölder's inequality, where $1/p + 1/q = 1$. Since $(p-1)q = p$ and $1/q = (p-1)/p$, it transpires that

$$\|f + g\|_p^p \leq \|f + g\|_p^{p-1} \big(\|f\|_p + \|g\|_p\big).$$

The integral on the left is finite, so we can cancel terms (unless $\|f + g\|_p = 0$, in which case there is nothing to prove) and the result follows. □

Proposition 2.19. *Suppose $\Omega \subset \mathbb{R}^d$ has finite measure ($\mu(\Omega) < \infty$) and $1 \leq p \leq q \leq \infty$. If $f \in L^q(\Omega)$, then $f \in L^p(\Omega)$ and*

$$\|f\|_p \leq \big(\mu(\Omega)\big)^{1/p - 1/q} \|f\|_q.$$

If $f \in L^\infty(\Omega)$, then

$$\lim_{p \to \infty} \|f\|_p = \|f\|_\infty.$$

If $f \in L^p(\Omega)$ for $1 \leq p < \infty$ and there is $R > 0$ such that

$$\|f\|_p \leq R,$$

for all such p, then $f \in L^\infty(\Omega)$ and $\|f\|_\infty \leq R$.

Finding a proof of these facts is an exercise; the latter two results are not completely trivial.

Proposition 2.20. *Suppose that $1 \leq p \leq \infty$ and $\Omega \subset \mathbb{R}^d$ is measurable. Then $L^p(\Omega)$ is complete, and hence a Banach space.*

Proof. Let $\{f_n\}_{n=1}^{\infty}$ be a Cauchy sequence in $L^p(\Omega)$. First note that f_n can be expressed as the collapsing sum

$$f_n = f_1 + (f_2 - f_1) + \cdots + (f_n - f_{n-1}).$$

Select a subsequence of $\{f_n\}_{n=1}^{\infty}$ such that

$$\|f_{n_{j+1}} - f_{n_j}\|_p \leq 2^{-j}, \quad j = 1, 2, \ldots.$$

Define the monotone increasing sequence of positive functions

$$F_m(x) = |f_{n_1}(x)| + \sum_{j=1}^{m} |f_{n_{j+1}}(x) - f_{n_j}(x)|.$$

Because the $\{F_n\}$ are increasing, the limit $F(x) = \lim_{m \to \infty} F_m(x)$ always exists, though $F(x)$ may be $+\infty$ for some points. On the other hand,

$$\|F_m\|_p \leq \|f_{n_1}\|_p + \sum_{j=1}^{m} 2^{-j} \leq \|f_{n_1}\|_p + 1.$$

We claim that $F \in L^p(\Omega)$, and, in particular, that $F(x) < \infty$ for a.e. $x \in \Omega$. When $p < \infty$, Lebesgue's monotone convergence theorem (Theorem 1.45) implies that

$$\int_{\Omega} |F(x)|^p \, dx = \int_{\Omega} \lim_{m \to \infty} |F_m(x)|^p \, dx = \lim_{m \to \infty} \|F_m\|_p^p \leq (\|f_{n_1}\|_p + 1)^p < \infty.$$

When $p = \infty$, we let A_m be a set of measure zero such that

$$|F_m(x)| \leq \|F_m\|_\infty \leq \|f_{n_1}\|_\infty + 1 \quad \text{for } x \notin A_m,$$

and let A be the (countable) union of the A_m, which continues to have measure zero. Thus, provided that $x \notin A$, our bound holds for every m and therefore also for the limit function F.

Now, the collapsing sequence

$$f_{n_{j+1}}(x) = f_{n_1}(x) + \big(f_{n_2}(x) - f_{n_1}(x)\big) + \cdots + \big(f_{n_{j+1}}(x) - f_{n_j}(x)\big)$$

converges absolutely, pointwise for a.e. $x \in \Omega$, to some measurable function $f(x)$. Because

$$|f_{n_j}(x)| \leq F(x),$$

it follows immediately that $f \in L^p(\Omega)$ for $p = \infty$. For $p < \infty$, Lebesgue's dominated convergence theorem (Theorem 1.47) implies the same conclusion since $F^p \in \mathcal{L}(\Omega) = L^1(\Omega)$ provides an integrable bounding function for $|f|^p$.

Finally, it is asserted that $\|f_{n_j} - f\|_p \to 0$. When $p < \infty$, the dominated convergence theorem applies again with the bounding function $F + |f|$, since

$$|f_{n_j}(x) - f(x)| \leq F(x) + |f(x)| \in L^p(\Omega).$$

If $p = \infty$, let B_{n_j,n_k} be a set of measure zero such that

$$|f_{n_j}(x) - f_{n_k}(x)| \leq \|f_{n_j} - f_{n_k}\|_\infty \quad \text{for } x \notin B_{n_j,n_k},$$

and let B be the union of A and the B_{n_j,n_k}, which continues to have measure zero. The right-hand side of the last inequality can be made as small as we please, say smaller than a given $\epsilon > 0$, provided n_j and n_k are large enough. Taking the limit $n_k \to \infty$, there obtains

$$|f_{n_j}(x) - f(x)| \leq \epsilon \quad \text{for } x \notin B.$$

As $\epsilon > 0$ was arbitrary, the desired convergence follows.

It remains to consider the entire sequence and not just a subsequence. Given $\epsilon > 0$, we can choose $N > 0$ such that for $n, n_j > N$, $\|f_n - f_{n_j}\|_p \leq \epsilon/2$ and also $\|f_{n_j} - f\|_p \leq \epsilon/2$. Thus

$$\|f_n - f\|_p \leq \|f_n - f_{n_j}\|_p + \|f_{n_j} - f\|_p \leq \epsilon.$$

Thus the entire sequence converges to f in $L^p(\Omega)$. □

The proof of the last result has as a consequence an important and useful result.

Corollary 2.21. *Suppose that $1 \leq p < \infty$ and $\Omega \subset \mathbb{R}^d$ is measurable. If $\{f_n\}_{n=1}^\infty \subset L^p(\Omega)$ converges to $f \in L^p(\Omega)$ as $n \to \infty$, then there is a subsequence $\{f_{n_j}\}_{j=1}^\infty$ such that $f_{n_j}(x) \to f(x)$ as $j \to \infty$ pointwise for a.e. $x \in \Omega$.*

The reader can verify that the result holds when $p = \infty$; however, a direct proof shows that restriction to a subsequence is unnecessary.

Just as for the ℓ^p–spaces, a consequence of Hölder's inequality is the existence of a large class of linear functionals defined on $L^p(\Omega)$. Let $1 \leq p \leq \infty$ and let q be the conjugate exponent. For $g \in L^q(\Omega)$, define $T_g : L^p(\Omega) \to \mathbb{F}$ for $f \in L^p(\Omega)$ by

$$T_g(f) = \int_\Omega f(x)\, g(x)\, dx.$$

The Hölder inequality shows that this is well-defined (i.e., finite), and further that

$$|T_g(f)| \leq \|g\|_q \|f\|_p.$$

So the mapping T_g is bounded, and it is easy to verify it is linear; that is, $T_g \in L^p(\Omega)^*$. Moreover, it is also the case that

$$\|T_g\|_{L^p(\Omega)^*} = \|g\|_q,$$

a point left for the reader to verify.

It is natural to ask if these are all the continuous linear functionals on $L^p(\Omega)$. The answer is "yes" when $1 \leq p < \infty$, but "no" when $p = \infty$. To present the details of the proof here would take us far afield of functional analysis. (We refer to [20] for a proof, which uses the Radon-Nikodym theorem, an important result not otherwise needed in our development.)

Proposition 2.22. *Suppose that $1 \leq p < \infty$ and $\Omega \subset \mathbb{R}^d$ is measurable. Then*

$$(L^p(\Omega))^* = \{T_g : g \in L^q(\Omega)\}.$$

Moreover, $\|T_g\|_{(L^p(\Omega))^*} = \|g\|_q$.

Functions in $L^p(\Omega)$ can be quite complicated. Fortunately, at least for $p < \infty$, each is arbitrarily close to a simple function $s(x)$ with *compact support*, which is to say, $s(x) = 0$ outside some ball.

Proposition 2.23. *For $1 \leq p < \infty$, the set S of all measurable simple functions with compact support is dense in $L^p(\Omega)$.*

Proof. It is clear that $S \subset L^p(\Omega)$. For measurable $f : \Omega \to \mathbb{R}$, $f \geq 0$, a sequence of measurable simple functions has been constructed in the proof of Theorem 1.35. Now suppose also that $f \in L^p(\Omega)$, and note that these constructed simple functions may be taken to have compact support for each n (just restrict the support sets to the ball of radius n). For a.e. $x \in \Omega$, $0 \leq s_n(x) \leq f(x)$, $s_n(x) \to f(x)$ as $n \to \infty$, and $|f - s_n|^p \leq f^p \in \mathcal{L}(\Omega)$, so the dominated convergence theorem (Theorem 1.47) implies that $\|f - s_n\|_p \to 0$ as $n \to \infty$. The case of a general f follows from the positive case. □

Because S is dense in $L^p(\Omega)$, many questions about L^p-functions can be reduced to questions about simple functions. However, sometimes it is preferable to reduce to the set $C_0(\Omega)$ of continuous functions with compact support.

Proposition 2.24. *For $1 \leq p < \infty$, $C_0(\Omega)$ is dense in $L^p(\Omega)$.*

Proof. Given $f \in L^p(\Omega)$ and $\epsilon > 0$, there is a measurable simple function s such that

$$\|f - s\|_p \leq \epsilon.$$

By Lusin's theorem (Theorem 1.43), given $\eta > 0$, there is $g \in C_0(\Omega)$ such that g and s agree except on a set A of measure less than η, and $|g| \leq \|s\|_\infty$. Thus

$$\|f - g\|_p \leq \|f - s\|_p + \|s - g\|_p \leq \epsilon + \left\{ \int_A |s - g|^p \, dx \right\}^{1/p} \leq \epsilon + 2\|s\|_\infty \eta^{1/p},$$

which can be made as small as desired by choosing ϵ small, fixing s, and then choosing η small. □

2.4 HAHN-BANACH THEOREMS

Attention is now turned to the three principal results in the elementary theory of Banach spaces. These theorems will find frequent use in many parts of the course. They are the Hahn-Banach theorem, the open mapping theorem, and the uniform boundedness principle. A fourth theorem, the Banach-Alaoglu theorem, is also presented, as it too is of fundamental importance in applied mathematics.

The Hahn-Banach theorems enable us to extend linear functionals defined on a subspace to the entire space. The theory begins with the case when the underlying field $\mathbb{F} = \mathbb{R}$ is real, and the first crucial lemma enables extension by a single dimension. The main theorem then follows from this result and an involved induction argument. The corresponding result over $\mathbb{C}$ follows as a corollary to an important observation relating complex and real linear functionals. In the case of an NLS, we can even extend the functional continuously. But first, a definition.

Definition. Let X be a vector space over $\mathbb{F}$. We say that $p : X \to [0, \infty)$ is *sublinear* if, for any $x, y \in X$ and $\lambda \geq 0$, it satisfies

$$p(\lambda x) = \lambda\, p(x) \qquad \text{(positive homogeneous)},$$
$$p(x + y) \leq p(x) + p(y) \quad \text{(triangle inequality)}.$$

If p also satisfies, for any $\lambda \in \mathbb{F}$,

$$p(\lambda x) = |\lambda|\, p(x),$$

then p is said to be a *seminorm*.

Thus, a sublinear function p is a seminorm if and only if it satisfies the stronger homogeneity property $p(\lambda x) = |\lambda| p(x)$ for any $\lambda \in \mathbb{F}$, and a seminorm p is a norm if and only if $p(x) = 0$ implies that $x = 0$.

Lemma 2.25. *Let X be a vector space over $\mathbb{R}$ and let $Y \subset X$ be a linear subspace such that $Y \neq X$. Let p be sublinear on X and $f : Y \to \mathbb{R}$ be a linear map that respects the inequality*

$$f(y) < p(y) \tag{2.11}$$

for all $y \in Y$. For a given $x_0 \in X \smallsetminus Y$, let

$$\tilde{Y} = \mathrm{span}\{Y, x_0\} = Y + \mathbb{R}\, x_0 = \{y + \lambda x_0 : y \in Y,\ \lambda \in \mathbb{R}\}.$$

Then there exists a linear map $\tilde{f} : \tilde{Y} \to \mathbb{R}$ such that

$$\tilde{f}|_Y = f \quad \text{and} \quad -p(-x) \leq \tilde{f}(x) \leq p(x) \tag{2.12}$$

for all $x \in \tilde{Y}$.

Proof. It is only necessary to find an $\tilde{f}$ such that $\tilde{f}(x) \leq p(x)$ for all $x \in \tilde{Y}$, since then $-\tilde{f}(x) = \tilde{f}(-x) \leq p(-x)$.

Suppose there was such an $\tilde{f}$. What would it have to look like? Let $\tilde{y} = y + \lambda x_0 \in \tilde{Y}$. Then, by linearity,

$$\tilde{f}(\tilde{y}) = \tilde{f}(y) + \lambda \tilde{f}(x_0) = f(y) + \lambda \alpha, \tag{2.13}$$

where $\alpha = \tilde{f}(x_0)$ is some real number. Therefore, such an $\tilde{f}$, were it to exist, is completely determined by α. Conversely, a choice of α determines a well-defined linear mapping. Indeed, if

$$\tilde{y} = y + \lambda x_0 = y' + \lambda' x_0,$$

then

$$y - y' = (\lambda' - \lambda)x_0.$$

The left-hand side lies in Y, while the right-hand side can lie in Y only if $\lambda' - \lambda = 0$. Thus $\lambda = \lambda'$ and then $y = y'$. Hence the representation of x in the form $y + \lambda x_0$ is unique and so a choice of $\tilde{f}(x_0) = \alpha$ determines a unique linear mapping by using the formula (2.13) as its definition.

It remains to be seen whether it is possible to choose α so that (2.12) holds. This amounts to asking that for all $y \in Y$ and $\lambda \in \mathbb{R}$,

$$f(y) + \lambda \alpha = \tilde{f}(y + \lambda x_0) \leq p(y + \lambda x_0). \tag{2.14}$$

Now, (2.14) is true for $\lambda = 0$ by the hypothesis (2.11). If $\lambda \neq 0$, write $y = -\lambda x$, or $x = -y/\lambda$ (our intention is to remove λ by *rescaling*). Then, (2.14) becomes

$$-\lambda(f(x) - \alpha) \leq p(-\lambda(x - x_0)).$$

So, when $\lambda < 0$, it must be the case that

$$f(x) - \alpha \leq p(x - x_0),$$

whilst when $\lambda > 0$,

$$-(f(x) - \alpha) \leq p(-(x - x_0)),$$

for all $x \in Y$. The latter two inequalities are identical to the two-sided inequality

$$-p(x_0 - x) \leq f(x) - \alpha \leq p(x - x_0),$$

or

$$f(x) - p(x - x_0) \leq \alpha \leq f(x) + p(x_0 - x). \tag{2.15}$$

Thus, any choice of α that respects (2.15) for all $x \in Y$ leads via (2.13) to a linear map $\tilde{f}$ with the desired property. Is there such an α? Let

$$a = \sup_{x \in Y}\{f(x) - p(x - x_0)\}$$

and

$$b = \inf_{x \in Y} \{f(x) + p(x_0 - x)\}.$$

If it is demonstrated that $a \le b$, then there certainly is such an α and any choice in the nonempty interval $[a, b]$ will do. But, a calculation shows that for $x, y \in Y$,

$$f(x) - f(y) = f(x - y) \le p(x - y) \le p(x - x_0) + p(x_0 - y),$$

on account of (2.11) and the triangle inequality. In consequence, it appears that

$$f(x) - p(x - x_0) \le f(y) + p(x_0 - y),$$

and this holds for any $x, y \in Y$. Fixing y, we see that

$$\sup_{x \in Y} \{f(x) - p(x - x_0)\} \le f(y) + p(x_0 - y).$$

As this is valid for every $y \in Y$, it must be the case that

$$a = \sup_{x \in Y} \{f(x) - p(x - x_0)\} \le \inf_{y \in Y} \{f(y) + p(x_0 - y)\} = b.$$

The lemma is thereby established. □

The program now is to successively extend f to all of X, one dimension at a time. This can be done trivially if X is a finite-dimensional extension of Y. If X is a countable vector space extension of Y, ordinary induction would suffice to establish the result. However, this situation rarely comes up in applications. Consequently, we are led to consider the general case of a possibly uncountable vector space basis, and this requires the use of what is known as *transfinite induction*.

This process is explained now.

Definition. For a set S, an *ordering*, denoted by $\preceq$, is a binary relation such that

(a) $x \preceq x$ for every $x \in S$ (reflexivity);

(b) If $x \preceq y$ and $y \preceq x$, then $x = y$ (antisymmetry);

(c) If $x \preceq y$ and $y \preceq z$, then $x \preceq z$ (transitivity).

A set S is *partially ordered* if S has an ordering that may apply only to certain pairs of elements of S. That is, there may be x and y in S such that neither $x \preceq y$ nor $y \preceq x$ holds. In that case, x and y are said to be *incomparable;* otherwise they are *comparable*. A *totally ordered set* or *chain* C is a partially ordered set such that every two elements in C are comparable.

Lemma 2.26 (Zorn's lemma). *Suppose S is a nonempty, partially ordered set. Suppose also that every chain $C \subset S$ has an* upper bound; *that is, for any chain C there is some $u \in S$ such that*

$$x \preceq u \quad \text{for all } x \in C.$$

Then S has at least one maximal element, *which is to say, there is some $m \in S$ such that for any $x \in S$,*

$$m \preceq x \quad \Longrightarrow \quad m = x.$$

This lemma follows from the *axiom of choice*, which states in one of its versions that given any set S and any collection of its subsets, we can choose a single element from each subset. Equivalently, an arbitrary Cartesian product of nonempty sets is itself nonempty. The axiom of choice is usually taken as an underlying tenet of set theory. It seems a very natural assertion, but it does not follow from the other axioms of set theory. It came as a surprise that Zorn's lemma implies the axiom of choice, and is therefore equivalent to it. Since the proof of this equivalence would take us far afield from functional analysis, we plan to accept it as an axiom of set theory and proceed. In the interest of being historically correct, we point out that the Polish mathematician Kuratowski put forward a version of this lemma more than a decade before Zorn, and suggested it be added to the axioms of set theory. In Eastern Europe, this axiom goes by the name Kuratowski-Zorn lemma.

Theorem 2.27 (Hahn-Banach theorem for real vector spaces). *Suppose that X is a vector space over $\mathbb{R}$, Y is a linear subspace, and p is sublinear on X. If f is a linear functional on Y such that*

$$f(y) \leq p(y)$$

for all $y \in Y$, then there is a linear functional F on X such that

$$F|_Y = f$$

(which is to say, F is a linear extension of f) and

$$-p(-x) \leq F(x) \leq p(x)$$

for all $x \in X$.

Proof. Let S be the set of all linear extensions g of f, defined on a vector space $D(g) \subset X$, and satisfying the property $g(x) \leq p(x)$ for all $x \in D(g)$. Since $f \in S$, S is not empty. Define a partial ordering on S by $g \preceq h$ means that h is a linear extension of g satisfying the bound. More precisely, $g \preceq h$ means that $D(g) \subset D(h)$, $h(x) \leq p(x)$ for all $x \in D(h)$ and $g(x) = h(x)$ for all $x \in D(g)$.

For any chain $\mathcal{C} \subset \mathcal{S}$, let

$$D = \bigcup_{g \in \mathcal{C}} D(g).$$

This is easily seen to be a vector space since $\mathcal{C}$ is a chain. For $x \in D$, define $g_{\mathcal{C}} : D \to \mathbb{R}$ by

$$g_{\mathcal{C}}(x) = g(x)$$

for any $g \in \mathcal{C}$ such that $x \in D(g)$. The mapping $g_{\mathcal{C}}$ is well-defined since $\mathcal{C}$ is a chain. It is linear and $D(g_{\mathcal{C}}) = D$. Moreover, for any $x \in D$, $g_{\mathcal{C}}(x) = g(x) \leq p(x)$ for some $g \in \mathcal{C}$. Hence, $g_{\mathcal{C}}$ provides an upper bound for $\mathcal{C}$ in $\mathcal{S}$.

An application of Zorn's lemma shows that S has at least one maximal element F. By definition, F is a linear extension of f satisfying $F(x) \leq p(x)$ for all $x \in D(F)$. It remains to show that $D(F) = X$. If not, there is some nonzero $x \in X \smallsetminus D(F)$, and by Lemma 2.25, there is a one-dimensional extension $\tilde{F}$ of F defined on $D(F) + \mathbb{R}x$, contradicting the maximality of F. Thus F is a linear extension of f with the advertised properties. □

Theorem 2.28 (Hahn-Banach theorem for general vector spaces). *Suppose that X is a vector space over $\mathbb{F}$ ($\mathbb{R}$ or $\mathbb{C}$), Y is a linear subspace, and p is a seminorm on X. If f is a linear functional on Y such that*

$$|f(y)| \leq p(y)$$

for all $y \in Y$, then there is a linear functional F on X such that

$$F|_Y = f$$

(i.e., F is a linear extension of f) and

$$|F(x)| \leq p(x)$$

for all $x \in X$.

Proof. When the ground field is real, the result is contained in Theorem 2.27, so assume that $\mathbb{F} = \mathbb{C}$. Write f in terms of its real and imaginary parts, $f = g + ih$, where g and h are real-valued. Clearly $g(y + z) = g(y) + g(z)$ and $h(y + z) = h(y) + h(z)$. If $\lambda \in \mathbb{R}$, then

$$f(\lambda x) = g(\lambda x) + ih(\lambda x)$$
$$\|$$
$$\lambda f(x) = \lambda g(x) + i\lambda h(x).$$

Taking real and imaginary parts in these relations and combining with the fact that g and h commute with addition shows them both to be real linear

(linear when Y is considered as a vector space over $\mathbb{R}$.) Moreover, g and h are intimately related. To see this, remark that for $x \in Y$,

$$f(ix) = if(x) = ig(x) - h(x) = -h(x) + ig(x)$$
$$\|$$
$$g(ix) + ih(ix).$$

Taking the real part of this relation leads to the formula

$$g(ix) = -h(x),$$

so that, in fact,

$$f(x) = g(x) - ig(ix). \tag{2.16}$$

Since g is the real part of f, clearly for $y \in Y$,

$$|g(y)| \le |f(y)| \le p(y) \tag{2.17}$$

by assumption. Thus g is a real-linear map defined on Y, considered as a vector subspace of X over $\mathbb{R}$. Because of (2.17), g satisfies the hypotheses of Theorem 2.27, so there is an extension G of g such that G is an $\mathbb{R}$-linear map of X into $\mathbb{R}$ for which

$$|G(x)| \le p(x)$$

for all $x \in X$. Informed by (2.16), define F in terms of G by

$$F(x) = G(x) - iG(ix).$$

Thus, $F : X \to \mathbb{C}$ and F is $\mathbb{R}$-linear, since G is.

It is to be shown that F is a $\mathbb{C}$-linear extension of f to X such that, for all $x \in X$,

$$|F(x)| \le p(x). \tag{2.18}$$

Since F is $\mathbb{R}$-linear, it automatically commutes with vector addition, so we need only check it commutes with scalar multiplication. As remarked above, it suffices for this to show $F(ix) = iF(x)$. This is true since

$$F(ix) = G(ix) - iG(-x) = G(ix) + iG(x) = i\big(G(x) - iG(ix)\big) = iF(x).$$

Inequality (2.18) holds for the following reason. Let $x \in X$ and write $F(x) = re^{i\theta}$ for some $r \ge 0$. Then, $\mathbb{C}$-linearity implies that

$$r = |F(x)| = e^{-i\theta}F(x) = F(e^{-i\theta}x) = G(e^{-i\theta}x) \le p(e^{-i\theta}x) = p(x),$$

since $F(e^{-i\theta}x)$ is real. The proof is complete. $\qquad\square$

Corollary 2.29 (Hahn-Banach theorem for normed linear spaces). *Let X be an NLS over $\mathbb{F}$ ($\mathbb{R}$ or $\mathbb{C}$) and let Y be a linear subspace. Let $f \in Y^*$ be a continuous linear functional on Y. Then there is an $F \in X^*$ such that*

$$F|_Y = f$$

and

$$\|F\|_{X^*} = \|f\|_{Y^*}.$$

Proof. Simply apply the Hahn-Banach theorem to f, using the seminorm

$$p(x) = \|f\|_{Y^*} \|x\|_X.$$

The details are left to the reader. □

Remark. The strength of this result lies especially in being able to make the extension without increasing the norm.

2.5 APPLICATIONS OF THE HAHN-BANACH THEOREM

The Hahn-Banach theorem has many interesting and useful applications, a few of which are now developed.

Corollary 2.30. *Let X be an NLS, and fix $x_0 \neq 0$ in X. Then there is an $f \in X^*$ such that*

$$\|f\|_{X^*} = 1 \quad and \quad f(x_0) = \|x_0\|.$$

Proof. Let $Z = \mathbb{F}x_0 = \text{span}\{x_0\}$. Define h on Z by

$$h(\lambda x_0) = \lambda \|x_0\|.$$

Then $h : Z \to \mathbb{F}$ is linear and has norm equal to 1 on Z, since for $x \in Z$, say $x = \lambda x_0$,

$$|h(x)| = |h(\lambda x_0)| = \big|\lambda \|x_0\|\big| = \|\lambda x_0\| = \|x\|.$$

By the Hahn-Banach theorem, there exists $f \in X^*$ such that $f|_Z = h$ and $\|f\| = \|h\| = 1$. □

Corollary 2.31. *Let X be an NLS, and fix $x_0 \in X$. There exists an $f \in X^*$ such that*

$$f(x_0) = \|f\|_{X^*} \|x_0\|.$$

The proof is similar to the proof of the last corollary.

Proposition 2.32. *Let X be an NLS and $x \in X$. It follows that*

$$\|x\| = \sup_{\substack{f \in X^* \\ f \neq 0}} \frac{|f(x)|}{\|f\|_{X^*}} = \sup_{\substack{f \in X^* \\ \|f\| = 1}} |f(x)|.$$

Proof. In any event, it is always the case that for any nonzero $f \in X^*$,

$$\frac{|f(x)|}{\|f\|_{X^*}} \leq \frac{\|f\|_{X^*} \|x\|_X}{\|f\|_{X^*}} = \|x\|,$$

and consequently

$$\sup_{\substack{f \in X^* \\ f \neq 0}} \frac{|f(x)|}{\|f\|_{X^*}} \leq \|x\|.$$

On the other hand, by Corollary 2.31, there is an $\tilde{f} \in X^*$ such that $\tilde{f}(x) = \|\tilde{f}\| \|x\|$. It follows that

$$\sup_{\substack{f \in X^* \\ f \neq 0}} \frac{|f(x)|}{\|f\|_{X^*}} \geq \frac{|\tilde{f}(x)|}{\|\tilde{f}\|_{X^*}} = \|x\|. \qquad \square$$

Proposition 2.33. *Let X be an NLS. Then X^* separates points in X. That is, if $x, y \in X$ are such that $f(x) = f(y)$ for every $f \in X^*$, then $x = y$.*

Proof. Let $x_1, x_2 \in X$, with $x_1 \neq x_2$. Then $x_2 - x_1 \neq 0$, so by Corollary 2.30, there is an $f \in X^*$ such that

$$f(x_2 - x_1) \neq 0.$$

Since f is linear, this means

$$f(x_2) \neq f(x_1),$$

which is the desired conclusion. $\qquad \square$

Corollary 2.34. *Let X be an NLS and $x \in X$ such that $f(x) = 0$ for all $f \in X^*$. Then $x = 0$.*

Proof. This follows from either of the last two results. $\qquad \square$

Lemma 2.35 (Mazur separation lemma 1). *Let X be an NLS, Y a linear subspace of X, and $w \in X \setminus Y$. Suppose*

$$d = \text{dist}(w, Y) = \inf_{y \in Y} \|w - y\|_X > 0.$$

Then there exists $f \in X^*$ such that $\|f\|_{X^*} \leq 1$,

$$f(w) \neq 0 \quad and \quad f(y) = 0 \quad for \; all \; y \in Y.$$

Moreover, one may define f so that $f(w) = d > 0$.

Proof. As before, any element $x \in Z = Y + \mathbb{F}w$ has a unique representation in the form $x = y + \lambda w$, where $y \in Y$ and $\lambda \in \mathbb{F}$. Define $g : Z \to \mathbb{F}$ by

$$g(y + \lambda w) = \lambda d.$$

It is straightforward to ascertain that g is $\mathbb{F}$-linear. We show that $\|g\|_{Z^*} \leq 1$ as follows. Let $x \in Z$, $x = y + \lambda w \neq 0$. If $\lambda = 0$, $x \in Y$ and so $|g(x)| = 0 \leq 1$, whereas if $\lambda \neq 0$, then

$$\left| g\left(\frac{y + \lambda w}{\|y + \lambda w\|} \right) \right| = \frac{|\lambda|}{\|y + \lambda w\|} d = \frac{1}{\|\frac{1}{\lambda}y + w\|} d.$$

Since $-\frac{1}{\lambda}y \in Y$, it follows that

$$\left\| \frac{1}{\lambda}y + w \right\| \geq d.$$

In consequence, we have

$$\left| g\left(\frac{y + \lambda w}{\|y + \lambda w\|} \right) \right| \leq \frac{d}{d} = 1.$$

Use the Hahn-Banach theorem to extend g to an $f \in X^*$ without increasing its norm. The functional f meets the requirements in view. $\quad\square$

Remark. The hypothesis that $d > 0$ is necessary. To see this fact, consider within ℓ^∞

$$x = (1, 0, 0, \dots) \quad \text{and} \quad x_n = (1, \underbrace{1/n, \dots, 1/n}_{n \text{ times}}, 0, 0, \dots),$$

for which $x \notin Y = \operatorname{span}\{x_1, x_2, \dots\}$ but $x_n \to x$.

Definition. A topological space is *separable* if it contains a countable, dense subset.

Nontrivial separable NLSs, although uncountable, are in some ways nearly countable. Given any point in a separable space, it can be approximated arbitrarily well by an element of the countable, dense subset. Separable spaces arise frequently in applied mathematics. Spaces with a Schauder basis are separable. The converse is not true, however.

Examples.

(1) The rational numbers $\mathbb{Q}$ are countable, and dense in $\mathbb{R}$. Also, the countable set $\mathbb{Q}+i\mathbb{Q}$ is dense in $\mathbb{C}$. Thus $\mathbb{F}^d$ is separable for any finite dimension d. If $1 \leq p < \infty$, the collection of unit vectors $\{e_n\}_{n=1}^\infty$ with $e_{n,m} = 0$ for $m \neq n$ and $e_{n,n} = 1$ is a countable subset of ℓ^p. A countable dense subset is obtained by taking finite linear combinations of the $\{e_n\}_{n=1}^\infty$ with rational coefficients (i.e., coefficients in $\mathbb{Q}$ or $\mathbb{Q} + i\mathbb{Q}$). The details are left to the reader.

(2) If $\Omega \subset \mathbb{R}^d$ is measurable and $1 \leq p < \infty$, then $L^p(\Omega)$ is separable. This follows from Proposition 2.23, which allows us to approximate an $f \in L^p(\Omega)$ by a simple function with compact support. These in turn can be approximated (in the $L^p(\Omega)$-norm) by simple functions with range in $\mathbb{Q}$ or $\mathbb{Q}+i\mathbb{Q}$, depending on the base field $\mathbb{F}$, whose characteristic functions are based on rectangles having edges in $\mathbb{Q}^d$. That is, the countable set of rational simple functions on rational rectangles is dense in $L^p(\Omega)$, so $L^p(\Omega)$ is separable. Note this argument fails if $p = \infty$, and indeed, $L^\infty(\Omega)$ is not separable.

Proposition 2.36. *Let X be an NLS and X^* its dual. If X^* is separable, then so is X.*

Proof. Let $\{f_n\}_{n=1}^{\infty}$ be a countable dense subset of X^*. Let $\{x_n\}_{n=1}^{\infty} \subset X$, be such that

$$\|x_n\| = 1 \quad \text{and} \quad |f_n(x_n)| \geq \tfrac{1}{2}\|f_n\|, \quad n = 1, 2, \ldots.$$

Such elements $\{x_n\}_{n=1}^{\infty}$ exist by definition of the norm on X^*. Let $\mathcal{D}$ be the countable set

$$\mathcal{D} = \big\{\text{all finite linear combinations of } \{x_n\}_{n=1}^{\infty} \text{ with rational coefficients}\big\}.$$

We claim that $\mathcal{D}$ is dense in X. If $\mathcal{D}$ is *not* dense in X, then there is an element $w \in X \setminus \overline{\mathcal{D}}$. The point w is at positive distance from $\overline{\mathcal{D}}$, for if not, there is a sequence $\{z_n\}_{n=1}^{\infty} \subset \overline{\mathcal{D}}$ such that $z_n \to w$. As $\overline{\mathcal{D}}$ is closed, this means $w \in \overline{\mathcal{D}}$ and that contradicts the choice of w. Since $\overline{\mathcal{D}}$ is a linear subspace of X, Lemma 2.35 implies there is an $f \in X^*$ such that

$$f|_{\overline{\mathcal{D}}} = 0 \quad \text{and} \quad f(w) = d = \inf_{z \in \overline{\mathcal{D}}} \|z - w\|_X > 0.$$

Since $f \in X^*$, there is a subsequence $\{f_{n_k}\}_{k=1}^{\infty} \subset X^*$ such that $f_{n_k} \xrightarrow{X^*} f$, by density. In consequence,

$$\|f - f_{n_k}\|_{X^*} \geq |(f - f_{n_k})(x_{n_k})| = |f_{n_k}(x_{n_k})| \geq \tfrac{1}{2}\|f_{n_k}\|_{X^*}.$$

Hence, $\|f_{n_k}\|_{X^*} \to 0$ as $k \to \infty$, and this means $f = 0$, a contradiction since $f(w) = d > 0$. $\qquad\square$

The Hahn-Banach theorem can also be used to distinguish sets that are not necessarily subspaces, as long as the linear geometry is respected. The next two lemmas consider convex sets.

Lemma 2.37 (Mazur separation lemma 2). *Let X be an NLS and C a nonempty, closed, convex subset of X such that $\lambda x \in C$ whenever $x \in C$ and $|\lambda| \leq 1$ (such a set C is called* balanced*). For any $w \in X \setminus C$, there exists an $f \in X^*$ such that $|f(x)| \leq 1$ for all $x \in C$ and $f(w) > 1$.*

Proof. Since C is closed and $w \notin C$, there is an open ball $B \subset X$ centered about the origin such that $\text{dist}(B + w, C) = \inf_{b \in B, c \in C} \|w + b - c\| = d > 0$.

Define the *Minkowski functional* $p : X \to [0, \infty)$ based on the balanced set $C + B$ by

$$p(x) = \inf\left\{t > 0 : \frac{x}{t} \in C + B\right\}.$$

Since $0 \in C$, $p(x)$ is indeed finite for every $x \in X$ (because for t large enough, every point can be contracted at least into the ball $0 + B$). Moreover, $p(x) \leq 1$ for $x \in C$ and likewise $p(w) \geq 1$. In fact, $p(w) > 1$, for if $t > 1$ is such that $\frac{1}{t}w \in C + B$, then $\frac{1}{t}w + b = c$ for some $b \in B$, $c \in C$. Hence $0 < d \le \|w + b - c\| = \frac{t-1}{t}\|w\|$ which can be made small enough to elicit a contradiction if t is near enough to 1.

The function $p : X \to \mathbb{R}^+$ is a seminorm. To see this, let $x \in X$, $\lambda \in \mathbb{F}$, and $t > 0$. The condition $\lambda x / t \in C + B$ is equivalent to $|\lambda| x / t \in (|\lambda|/\lambda)(C + B) = C + B$, since C and B are balanced. Thus

$$p(\lambda x) = p(|\lambda| x) = |\lambda| p(x).$$

Now take $x, y \in X$ and choose any $r > 0$, $s > 0$ such that $x/r \in C + B$ and $y/s \in C + B$. It then transpires that the convex combination

$$\frac{r}{s+r}\frac{x}{r} + \frac{s}{s+r}\frac{y}{s} = \frac{x+y}{s+r} \in C + B,$$

and so it is concluded that

$$p(x + y) \le p(x) + p(y).$$

Now, let $Y = \mathbb{F}w$ and define on Y the linear functional

$$f(\lambda w) = \lambda p(w),$$

so $f(w) = p(w) > 1$. Then, it follows that

$$|f(\lambda w)| = |\lambda| p(w) = p(\lambda w),$$

so the Hahn-Banach theorem breeds a linear extension with the property that

$$|f(x)| \le p(x),$$

which is to say, $|f(x)| \le 1$ for $x \in C \subset C + B$, as required. Finally, f is bounded on B, so it is continuous. □

Not all convex sets are balanced. The following lemma replaces the Mazur separation result for general convex sets.

Lemma 2.38 (separating hyperplane theorem). *Let A and B be disjoint, nonempty, convex sets in an NLS X.*

(a) *If A is open, then there is an $f \in X^*$ and a $\gamma \in \mathbb{R}$ such that*

$$\mathrm{Re} f(x) \le \gamma \le \mathrm{Re} f(y) \quad \forall x \in A,\ y \in B.$$

(b) *If both A and B are open, then there is an $f \in X^*$ and a $\gamma \in \mathbb{R}$ such that*

$$\mathrm{Re} f(x) < \gamma < \mathrm{Re} f(y) \quad \forall x \in A,\ y \in B.$$

(c) *If A is compact and B is closed, then there is an $f \in X^*$ and a $\gamma \in \mathbb{R}$ such that*

$$\mathrm{Re} f(x) < \gamma < \mathrm{Re} f(y) \quad \forall x \in A,\ y \in B.$$

Remark. Notice that when the ground field $\mathbb{F} = \mathbb{C}$, it is the real part of f that separates A and B, not f itself.

Proof. It is sufficient to prove the result when the ground field $\mathbb{F} = \mathbb{R}$. For, if $\mathbb{F} = \mathbb{C}$, first view X as a real Banach space and infer existence of a continuous, real-linear functional g satisfying the separation result. Then, construct $f \in X^*$ by using formula (2.16), *viz.*

$$f(x) = g(x) - ig(ix).$$

Thus attention is restricted to the case $\mathbb{F} = \mathbb{R}$.

For (a), fix $-w \in A - B = \{x - y : x \in A, y \in B\}$ and let

$$C = A - B + w,$$

an open, convex neighborhood of 0 in X. Then, $w \notin C$ since A and B are disjoint. Define the subspace $Y = \mathbb{R}w$ and the linear functional $g : Y \to \mathbb{R}$ by

$$g(tw) = t.$$

Now let $p : X \to [0, \infty)$ be the Minkowski functional based on C, to wit,

$$p(x) = \inf\left\{t > 0 : \frac{1}{t}x \in C\right\}.$$

We saw in the previous proof that p is sublinear. However, it is not necessarily a seminorm, since C may not be balanced, but it does satisfy the triangle inequality and positive homogeneity. Since $w \notin C$, $p(w) \geq 1$ and so $g(y) \leq p(y)$ for $y \in Y$. Use the Hahn-Banach theorem for real-linear functionals (Theorem 2.27) to extend g to a linear mapping on all of X which is still bounded by p. Now $g \leq 1$ on C, so also $g \geq -1$ on $-C$, and therefore $|g| \leq 1$ on $C \cap (-C)$, which is a neighborhood of 0. Thus g is bounded, and so continuous.

If $a \in A$ and $b \in B$, then $a - b + w \in C$, so

$$1 \geq g(a - b + w) = g(a) - g(b) + g(w) = g(a) - g(b) + 1,$$

which implies that $g(a) \leq g(b)$, and the result follows with $\gamma = \sup_{a \in A} g(a)$.

For (b), use the previous construction. It is left to the reader to show that $g(A)$ is an open subset of $\mathbb{R}$ since g is bounded and linear and A is open. Now both $g(A)$ and $g(B)$ are open subsets of $\mathbb{R}$ that can intersect at most at one point, so they must in fact be disjoint.

For (c), consider $S \equiv B - A$. Since A is compact, S must be closed. For suppose there are points $x_n \in S$, $n = 1, 2, 3, \ldots$, such that $x_n = b_n - a_n$ with $b_n \in B$ and $a_n \in A$ and $x_n \to x$ in X. Since A is compact, there is a subsequence (still denoted by a_n for convenience), such that $a_n \to a \in A$. But then $b_n = x_n + a_n \to x + a = b \in B$, since B is closed. But this implies that $x \in S$, and the claim follows.

Since $0 \notin S$, there is some open convex set $U \subset X$ containing 0 such that $U \cap S$ is empty. Let $A' = A + \frac{1}{2}U$ and $B' = B - \frac{1}{2}U$. Then A' and B' are disjoint, convex, open sets, and so (b) gives a separating functional g for A' and B'. As $A \subset A'$ and $B \subset B'$, g separates A and B as well. □

2.6   THE OPEN MAPPING THEOREM

The second of the three major principles of elementary functional analysis is the open mapping theorem (or equivalently the closed graph theorem). The third is the principle of uniform boundedness (also called the Banach-Steinhaus theorem). Both of these rely on the following theorem of Baire.

Theorem 2.39 (Baire category theorem). *Let X be a complete metric space. Then the intersection of any countable collection of dense open sets in X is itself dense in X.*

Proof. Let $\{V_j\}_{j=1}^{\infty}$ be a countable collection of dense open sets. Let W be any nonempty open set in X. It is required to show that if $V = \bigcap_{j=1}^{\infty} V_j$, then $V \cap W \neq \emptyset$.

Since V_1 is dense, $V_1 \cap W$ is a nonempty open set. Thus there is an $x_1 \in W$ and, without loss of generality, an r_1 with $0 < r_1 < 1$ such that

$$\overline{B_{r_1}(x_1)} \subset V_1 \cap W.$$

Similarly, V_2 is open and dense. Hence, there is an x_2 and an r_2 with $0 < r_2 < 1/2$ such that
$$\overline{B_{r_2}(x_2)} \subset V_2 \cap B_{r_1}(x_1).$$

Inductively, there is determined x_n and r_n with $0 < r_n < 1/n$ such that

$$\overline{B_{r_n}(x_n)} \subset V_n \cap B_{r_{n-1}}(x_{n-1}), \quad n = 2, 3, 4, \ldots.$$

Consider the sequence $\{x_n\}_{n=1}^{\infty}$ just generated. If $i, j \geq n$, then by construction
$$x_i, x_j \in B_{r_n}(x_n) \quad \Longrightarrow \quad d(x_i, x_j) \leq \frac{2}{n}.$$

This shows that $\{x_n\}_{n=1}^{\infty}$ is a Cauchy sequence. As X is complete, there is an x for which $x_n \to x$ as $n \to \infty$. Because $x_i \subset \overline{B_{r_n}(x_n)} \subset V_n$ for $i > n$, it follows that $x \in \overline{B_{r_n}(x_n)} \subset V_n$, $n = 1, 2, \ldots$. Clearly, since $x \in \overline{B_{r_1}(x_1)} \subset W$, $x \in W$ also. Hence

$$x \in W \cap \bigcap_{n=1}^{\infty} V_n = W \cap V$$

and the proof is complete. □

Corollary 2.40. *The intersection of countably many dense open subsets of a complete metric space is nonempty.*

Definition. A subset A of a topological space is called *nowhere dense* if the interior of its closure is empty, in symbols, $(\bar{A})^\circ = \emptyset$. A set is called *first category* if it is a countable union of nowhere dense sets. Otherwise, it is called *second category*.

Corollary 2.41. *A complete metric space is second category, which is to say that it is* not *the countable union of nowhere dense sets.*

Proof. If $X = \bigcup_{j=1}^{\infty} M_j$ where each M_j is nowhere dense, then $X = \bigcup_{j=1}^{\infty} \bar{M}_j$, so by De Morgan's law,

$$\emptyset = \bigcap_{j=1}^{\infty} \bar{M}_j^c.$$

But, for each j, $\bar{M}_j^c$ is open and it is dense since, by Proposition 1.7,

$$\overline{\bar{M}_j^c} = \left((\bar{M}_j)^\circ\right)^c = \emptyset^c = X.$$

This contradicts Baire's theorem. □

Theorem 2.42 (open mapping theorem). *Let X and Y be Banach spaces and let $T : X \to Y$ be a bounded linear surjection. Then T is an open mapping, i.e., T maps open sets to open sets.*

Proof. It is required to demonstrate that if U is open in X, then $T(U)$ is open in Y. If $y \in T(U)$, we must show $T(U)$ contains an open set about y. Suppose it is known that there is an $r > 0$ for which $T(B_1(0)) \supset B_r(0)$. Let $x \in U$ be such that $Tx = y$ and let $t > 0$ be such that $B_t(x) \subset U$. Then, it transpires that

$$\begin{aligned}
T(U) \supset T(B_t(x)) &= T(tB_1(0) + x) \\
&= tT(B_1(0)) + Tx \supset tB_r(0) + y = B_{rt}(y).
\end{aligned}$$

As $rt > 0$, y would be an interior point of $T(U)$ and the result would then be established. Thus, attention is concentrated on showing that $T(U) \supset B_r(0)$ for some $r > 0$ when $U = B_1(0) = B_1$.

Since T is onto,

$$Y = \bigcup_{k=1}^{\infty} T(kB_1).$$

Since Y is a complete metric space, at least one of the sets $T(kB_1)$, $k = 1, 2, \ldots$, is not nowhere dense. Hence there is a nonempty open set W_1 such that

$$W_1 \subset \overline{T(kB_1)} \quad \text{for some } k \geq 1.$$

Multiplying this inclusion by $1/2k$ yields a nonempty open set $W = \frac{1}{2k}W_1$ included in $\overline{T(\frac{1}{2}B_1)}$. Hence there is a $y_0 \in Y$ and an $r > 0$ such that

$$B_r(y_0) \subset W \subset \overline{T(\tfrac{1}{2}B_1)}.$$

But then, it must be the case that

$$B_r(0) = B_r(y_0) - y_0 \subset B_r(y_0) - B_r(y_0) \subset \overline{T(\tfrac{1}{2}B_1)} - \overline{T(\tfrac{1}{2}B_1)} \subset \overline{T(B_1)}. \quad (2.19)$$

The latter inclusion is very nearly the desired conclusion. It suffices to remove the closure operation on the right-hand side to complete the proof. Note that since multiplication by a nonzero constant is a homeomorphism, (2.19) implies that for any $s > 0$,

$$B_{rs}(0) \subset \overline{T(sB_1)} = \overline{T(B_s(0))}. \quad (2.20)$$

We will show that $B_r(0) \subset T(2B_1)$, so then $B_{r/2}(0) \subset T(B_1)$.

Fix $y \in B_r(0) \subset \overline{T(B_1)}$. Then by Proposition 1.16 there exists $x_1 \in B_1$ such that

$$\|y - Tx_1\|_Y < \tfrac{1}{2}r.$$

For $z = y - Tx_1 \subset B_{r/2}(0) \subset \overline{T(B_{1/2}(0))}$, there exists $x_2 \in B_{1/2}(0)$ such that

$$\|y - Tx_1 - Tx_2\|_Y < 2^{-2}r.$$

We proceed by mathematical induction. Let $n \geq 1$ and suppose $x_1, x_2, \ldots, x_n$ have been chosen so that $\|x_j\| < 2^{-j+1}$ for $j = 1, 2, 3, \ldots, n$, and

$$\|y - Tx_1 - Tx_2 - \cdots - Tx_n\|_Y < 2^{-n}r. \quad (2.21)$$

Let $z = y - (Tx_1 + \cdots + Tx_n)$, so $z \in B_{2^{-n}r}(0) \subset \overline{T(2^{-n}B_1)}$, because of (2.20). We conclude that there is an $x_{n+1} \in 2^{-n}B_1$, which is to say

$$\|x_{n+1}\| < 2^{-n}, \quad (2.22)$$

such that x_{n+1} is as close to z as we like, e.g.,

$$\|z - Tx_{n+1}\| < 2^{-(n+1)}r.$$

Thus the induction proceeds and (2.21) and (2.22) are inferred to hold for all $n \geq 1$.

It is clear from (2.22) that $\sum_{j=1}^{n} x_j = s_n$ is a Cauchy sequence. Hence, there is an $x \in X$ such that $s_n \to x$ as $n \to \infty$. Clearly,

$$\|x\| \leq \sum_{j=1}^{\infty} \|x_j\| < \sum_{n=1}^{\infty} 2^{-n+1} = 2.$$

By continuity of T, $Ts_n \to Tx$ as $n \to \infty$. By (2.21), $Ts_n \to y$ as $n \to \infty$. Hence $Tx = y$. It is concluded that

$$T(2B_1) \supset B_r(0),$$

or, what is the same,

$$T(B_1) \supset B_{r/2}(0),$$

and the result is established. □

Corollary 2.43. *Let X, Y be Banach spaces and T a bounded, linear surjection that is also an injection. Then T^{-1} is a continuous linear operator.*

Proof. This follows since $(T^{-1})^{-1} = T$ is open, hence T^{-1} is continuous. We have discussed earlier that T^{-1} must be linear. □

The foregoing results allow us to posit conditions implying when two Banach spaces have the same structure. They have the same vector space (i.e., linear) structure if there is a linear bijection between them. They have the same topological structure if this map, and its inverse, is continuous. By the open mapping theorem, the latter requirement is reduced to merely asking that the bijection be bounded. Of course, the norm structure may differ unless the bijection is an isometry (see Section 2.8), but the topologies will agree.

Definition. Two Banach spaces X and Y are said to be *isomorphic as Banach spaces* if there exists a bounded linear bijection $T : X \to Y$. Moreover, if T is an isometry (i.e., $\|Tx\| = \|x\|$ for every $x \in X$), then X and Y are said to be *isometrically isomorphic*.

Example. If $X_1, \ldots, X_m$ are NLSs over $\mathbb{F}$, we can define their (external) *direct sum*

$$Y = X_1 \times \cdots \times X_m = \{(x_1, \ldots, x_m) : x_i \in X_i, 1 \le i \le m\}.$$

This is a vector space if the operations are defined componentwise. It is an NLS with the norm defined to be any of the equivalent functions

$$\|(x_1, \ldots, x_m)\|_Y = \left(\sum_{i=1}^{m} \|x_i\|_{X_i}^p \right)^{1/p} = \left\| \left(\|x_1\|_{X_1}, \ldots, \|x_m\|_{X_m} \right) \right\|_{\ell^p},$$

where $p \in [1, \infty]$ (modified in the usual way if $p = \infty$). It is not hard to show that Y is a Banach space if and only if X_i is a Banach space for each $1 \le i \le m$. As an alternative, we may have an NLS X decomposed into closed subspaces $X_1, \ldots, X_m$ so that $X = X_1 + \cdots + X_m$ and $X_i \cap \text{span}\{X_1, \ldots, X_{i-1}, X_{i+1}, \ldots, X_m\} = \{0\}$ for all i. In this case, given $x \in X$, there is a unique decomposition $x = x_1 + \cdots + x_m$, where $x_i \in X_i$ for $1 \le i \le m$, and we define the (internal) *direct sum*

$$X = X_1 \oplus \cdots \oplus X_m = \{x_1 + \cdots + x_m : x_i \in X_i, 1 \le i \le m\}.$$

We can view X as the Cartesian product Y by associating $x = x_1 + \cdots + x_m \in X$ with $y = (x_1, \ldots, x_m) \in Y$. In other words, the map $S : Y \to X$ is defined by $S(x_1, \ldots, x_m) = x_1 + \ldots + x_m$, and it maps Y one-to-one and onto X. The mapping S is bounded because

$$\|S(x_1, \ldots, x_m)\|_X = \|x_1 + \ldots + x_m\|_X \le \sum_{i=1}^{m} \|x_i\|_X \le C\|(x_1, \ldots, x_m)\|_Y.$$

So when X is a Banach space, one concludes that X and Y are isomorphic as Banach spaces.

A result closely related to the open mapping theorem is the closed graph theorem. If X, Y are sets, $D \subset X$, and $f : D \to Y$ a function defined on the subset D, the *graph* of f is the set

$$\mathrm{graph}(f) = \{(x, y) \in X \times Y : x \in D \text{ and } y = f(x)\}.$$

It is a subset of the Cartesian product $X \times Y$.

Proposition 2.44. *Let X be a topological space, Y a Hausdorff space, and $f : X \to Y$ continuous. Then $\mathrm{graph}(f)$ is closed in $X \times Y$.*

Proof. Let $U = X \times Y \smallsetminus \mathrm{graph}(f)$. We show that U is open. Fix $(x_0, y_0) \in U$, so that $y_0 \neq f(x_0)$. Because Y is Hausdorff, there exist open sets V and W with $y_0 \in V$, $f(x_0) \in W$, and $V \cap W = \emptyset$. Since f is continuous, $f^{-1}(W)$ is open in X. Thus, the open set $f^{-1}(W) \times V$ lies in U. $\qquad\square$

The reader is left to ponder the question: is the last result true if the hypothesis that Y is Hausdorff is omitted?

As will be seen in a moment, $f : X \to Y$ and $\mathrm{graph}(f)$ closed in $X \times Y$ does not imply f is continuous. However, in special circumstances, the reverse conclusion is warranted.

Definition. Let X and Y be NLSs and let D be a linear subspace of X. Suppose $T : D \to Y$ is linear. Then T is a *closed operator* if $\mathrm{graph}(T)$ is a closed subset of $X \times Y$.

Since both X and Y are metric spaces, $\mathrm{graph}(T)$ being closed means exactly that if $\{x_n\}_{n=1}^{\infty} \subset D$ with

$$x_n \xrightarrow{X} x \quad \text{and} \quad T x_n \xrightarrow{Y} y,$$

then it follows that $x \in D$ and $y = Tx$.

Theorem 2.45 (closed graph theorem). *Let X and Y be Banach spaces and $T : X \to Y$ linear. Then T is continuous (i.e., bounded) if and only if T is closed.*

Proof. On account of Proposition 2.44, T being continuous implies that $\mathrm{graph}(T)$ is closed since a Banach space is Hausdorff.

For the converse, suppose $\mathrm{graph}(T)$ to be closed. Then $\mathrm{graph}(T)$ is a closed linear subspace of the Banach space $X \times Y$. Hence $\mathrm{graph}(T)$ is a Banach space in its own right with the *graph norm*

$$\|(x, Tx)\| = \|x\|_X + \|Tx\|_Y.$$

Consider the continuous projections Π_1 and Π_2 on $X \times Y$ given by

$$\Pi_1(x, y) = x \quad \text{and} \quad \Pi_2(x, y) = y.$$

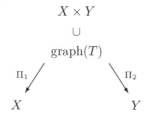

Figure 2.2 An illustration of the proof that if T is a closed operator, then $T = \Pi_2 \circ \Pi_1^{-1} : X \to Y$ is continuous.

Restrict these to the subspace graph(T). The mapping Π_1 is a one-to-one, continuous linear map of the Banach space graph(T) onto X. By the open mapping theorem,

$$\Pi_1^{-1} : X \to \text{graph}(T)$$

is continuous. But then

$$T = \Pi_2 \circ \Pi_1^{-1} : X \to Y$$

is continuous since it is the composition of continuous maps. The proof of the converse is illustrated in Figure 2.2. □

Corollary 2.46. *Let X and Y be Banach spaces and D a linear subspace of X. Let $T : D \to Y$ be a closed linear operator on D. Then T is bounded on D if and only if D is a closed subspace of X.*

Proof. If D is closed, it is a Banach space, so the closed graph theorem applied to $T : D \to Y$ shows T to be continuous.

Conversely, suppose T is bounded as a map from D to Y. Let $\{x_n\}_{n=1}^{\infty} \subset D$ and suppose $x_n \to x$ in X. Since T is bounded, it follows that $\{Tx_n\}_{n=1}^{\infty}$ is a Cauchy sequence, for

$$\|Tx_n - Tx_m\| \leq \|T\| \, \|x_n - x_m\| \to 0$$

as $n, m \to \infty$. Since Y is complete, there is a $y \in Y$ such that $Tx_n \to y$. But since T is closed, we infer that $x \in D$ and $y = Tx$. In particular, D has all its limit points, so D is closed. □

Example. Closed does not imply continuous in general, even for linear operators. Take $X = C([0,1])$ with the max norm. Let $Tf = f'$ for $f \in D = C^1([0,1])$. Consider T as a mapping of D into X. Note that $D \subset X$ is *not* a closed subspace. In fact, $\bar{D} = X$, so T is defined on a dense subspace of X.

T *is not bounded*. Let $f_n(t) = t^n$. Then $\|f_n\| = 1$ for all n, but $Tf_n = nt^{n-1}$ so $\|Tf_n\| = n$.

T is closed. Let $\{f_n\}_{n=1}^{\infty} \subset D$ and suppose $f_n \xrightarrow{X} f$ and $f_n' \xrightarrow{X} g$. Then, by the fundamental theorem of calculus, for any $t \in [0,1]$,

$$f_n(t) = f_n(0) + \int_0^t f_n'(s)\, ds$$

for $n = 1, 2, \ldots$. Taking the limit of this equation as $n \to \infty$ yields

$$f(t) = f(0) + \int_0^t g(s)\, ds,$$

so $g = f'$, by another application of the fundamental theorem of calculus.

2.7 THE UNIFORM BOUNDEDNESS PRINCIPLE

The third basic result in Banach space theory is the Banach-Steinhaus theorem, also known as the principle of uniform boundedness.

Theorem 2.47 (uniform boundedness principle, Banach-Steinhaus theorem). *Let X be a Banach space, Y an NLS and $\{T_\alpha\}_{\alpha \in \mathcal{I}} \subset B(X,Y)$ a collection of bounded linear operators from X to Y. Then one of the following two conclusions must obtain; either*

(a) *there is a constant M such that for all $\alpha \in \mathcal{I}$,*

$$\|T_\alpha\|_{B(X,Y)} \leq M,$$

which is to say, the T_α are uniformly bounded, or

(b) *there is an $x \in X$ such that*

$$\sup_{\alpha \in \mathcal{I}} \|T_\alpha x\| = +\infty,$$

i.e., $\{T_\alpha\}_{\alpha \in \mathcal{I}}$ is not pointwise bounded, meaning that there is at least one fixed $x \in X$ for which the set $\{T_\alpha x\}_{\alpha \in \mathcal{I}}$ is unbounded.

Proof. For each $\alpha \in \mathcal{I}$, the map φ_α defined by

$$\varphi_\alpha(x) = \|T_\alpha x\|$$

is continuous from X to $[0, \infty)$, since it is the composition of two continuous maps. Thus the sets

$$\{x : \|T_\alpha x\| > n\} = \varphi_\alpha^{-1}((n, \infty))$$

are open, and consequently,

$$V_n = \bigcup_{\alpha \in \mathcal{I}} \varphi_\alpha^{-1}((n, \infty))$$

is a union of open sets, so is itself open.

Each V_n is either dense in X or it is not. If, for some N, V_N is *not* dense in X, then there is an $r > 0$ and an $x_0 \in X$ such that

$$B_r(x_0) \cap V_N = \emptyset.$$

Therefore, if $x \in B_r(x_0)$, then for all $\alpha \in \mathcal{I}$,

$$\|T_\alpha(x)\| \leq N.$$

Hence, if $\|z\| < r$, then for all $\alpha \in \mathcal{I}$,

$$\|T_\alpha(z)\| \leq \|T_\alpha(z + x_0)\| + \|T_\alpha(x_0)\| \leq 2N.$$

In consequence, we have

$$\|T_\alpha\| \leq \frac{2N}{r},$$

and so condition (a) holds.

On the other hand, if all the V_n are dense, then they are all dense and open. By Baire's theorem,

$$\bigcap_{n=1}^{\infty} V_n$$

is nonempty, so let $x \in \bigcap_{n=1}^{\infty} V_n$. This means that $x \in V_n$ for all $n = 1, 2, 3, \ldots$, and so there is $\alpha_n \in \mathcal{I}$ such that $\|T_{\alpha_n} x\| > n$, and (b) follows. ☐

Because the above, standard proof of the uniform boundedness principle uses Baire's theorem, it is not viewed as completely elementary. Over the years, more elementary proofs have appeared in the literature. One of these is presented now[1].

Lemma 2.48. *Let X, Y be NLSs and $T \in B(X, Y)$. For any $x \in X$ and $r > 0$,*

$$\sup_{z \in B_r(x)} \|Tz\| \geq \|T\| \, r.$$

Proof. For any $z \in X$, we have that

$$\max\{\|T(x+z)\|, \|T(x-z)\|\} \geq \tfrac{1}{2}\{\|T(x+z)\| + \|T(x-z)\|\}$$
$$\geq \tfrac{1}{2}\|T(x+z) - T(x-z)\| = \|Tz\|.$$

Taking the supremum over all $z \in B_r(0)$ gives the result. ☐

Alternate proof of Theorem 2.47. Suppose that (a) does not hold, so that

$$\sup_{\alpha \in \mathcal{I}} \|T_\alpha\| = +\infty.$$

[1] A. D. Sokal, *A Really Simple Elementary Proof of the Uniform Boundedness Theorem*, American Mathematical Monthly, **118** (5), 2011, pp. 2-7. The references in this short paper point to some of the earlier elementary proofs.

Then for each $n = 1, 2, \ldots, n$, there is an index $\alpha_n \in \mathcal{I}$ so that $\|T_{\alpha_n}\| \geq 8^n$. A sequence $\{x_n\}_{n=1}^\infty \subset X$ is constructed inductively by setting $x_0 = 0$ and choosing, by Lemma 2.48, $x_n \in B_{4^{-n}}(x_{n-1})$ such that

$$\|T_{\alpha_n} x_n\| \geq \tfrac{2}{3} 4^{-n} \|T_{\alpha_n}\|.$$

By construction, the sequence is Cauchy, since for $m > n \geq N$,

$$\begin{aligned}
\|x_m - x_n\| &\leq \|x_m - x_{m-1}\| + \cdots + \|x_{n+1} - x_n\| \\
&< 4^{-m} + \cdots + 4^{-n-1} \\
&= 4^{-n-1} \frac{1 - 4^{-m+n}}{1 - 4^{-1}} \longrightarrow 0
\end{aligned}$$

as $N \to \infty$. So the sequence has a limit $x \in X$, and the calculation above shows that $\|x - x_n\| \leq \tfrac{1}{3} 4^{-n}$ (just take $m \to \infty$). Combining these results shows that

$$\begin{aligned}
\|T_{\alpha_n} x\| &\geq -\|T_{\alpha_n}(x - x_n)\| + \|T_{\alpha_n} x_n\| \\
&\geq -\|T_{\alpha_n}\| \, \|x - x_n\| + \tfrac{2}{3} 4^{-n} \|T_{\alpha_n}\| \\
&\geq \left(-\tfrac{1}{3} + \tfrac{2}{3}\right) 4^{-n} 8^n \\
&= \tfrac{1}{3} 2^n \longrightarrow +\infty.
\end{aligned}$$

This is conclusion (b), and so the theorem is established. $\qquad\square$

We will need to wait till later in the chapter for a chance to see the uniform boundedness principle in action.

2.8   THE EMBEDDING OF X INTO ITS DOUBLE DUAL X^{**}

Let X be an NLS and X^* its dual space. Since X^* is a Banach space, it in turn has a dual space written X^{**} which is sometimes referred to as the *double dual* of X. There is a natural construction whereby X may be viewed as a subspace of X^{**} that is described now.

For any $x \in X$, define $E_x \in X^{**}$ as follows: if $f \in X^*$, then

$$E_x(f) = f(x).$$

We call E_x the *evaluation map* at $x \in X$. First, let us check that this is an element of X^{**}, which is to say, E_x is a bounded linear map on X^*. For $f, g \in X^*$ and $\lambda \in \mathbb{F}$, compute as follows:

$$E_x(f + g) = (f + g)(x) = f(x) + g(x) = E_x(f) + E_x(g)$$

and

$$E_x(\lambda f) = (\lambda f)(x) = \lambda f(x) = \lambda E_x(f).$$

Thus E_x is a linear map from X^* into $\mathbb{F}$ for each fixed x. It is bounded since, using Proposition 2.32,

$$\|E_x\|_{X^{**}} = \sup_{\substack{f \in X^* \\ f \neq 0}} \frac{|E_x(f)|}{\|f\|_{X^*}} = \sup_{\substack{f \in X^* \\ f \neq 0}} \frac{|f(x)|}{\|f\|_{X^*}} = \|x\|.$$

Thus, not only is E_x bounded, but its norm in X^{**} is the same as the norm of x in X. Thus we may view X as the linear subspace $\tilde{X} = \{E_x \in X^{**} : x \in X\}$ of X^{**}, and in this guise, X is faithfully represented in X^{**}. That is to say, X and $\tilde{X}$ are isometrically isomorphic by the map $x \mapsto E_x$.

Definition. Let (M, d) and (N, ρ) be two metric spaces and $f : M \to N$. The function f is called an *isometry* if f preserves distances, which is to say

$$\rho(f(x), f(y)) = d(x, y).$$

The spaces M and N are called *isometric* if there is a surjective isometry $f : M \to N$.

Note that an isometry is automatically an injective map. If it is also surjective, it is a one-to-one correspondence that preserves open balls,

$$f(B_r(x)) = B_r(f(x)),$$

and so is also a homeomorphism. Metric spaces that are isometric are indistinguishable as metric spaces. If the metric spaces are NLSs $(X, \|\cdot\|_X)$ and $(Y, \|\cdot\|_Y)$ and $T : X \to Y$ is a *linear* isometry, then $T(X)$ and X are indistinguishable as NLSs. Of course, in the context of linear maps on an NLS, T being an isometry means simply that $\|x\|_X = \|T(x)\|_Y$ for all $x \in X$.

In this terminology, the correspondence $F : X \to X^{**}$ given by

$$F(x) = E_x$$

is an isometry. Indeed, F is itself linear and injective by Proposition 2.33. Thus, X is isomorphic as a vector space, homeomorphic as a topological space and isometric as a metric space to the linear subspace $F(X) \subset X^{**}$. In other words, when X is complete, X and $F(X)$ are isometrically isomorphic as Banach spaces. We identify X with $F(X)$ via the map F; that is, we identify x and E_x, and speak of X as being a subspace of its double dual.

An NLS X is called *reflexive* if F is surjective, i.e., $F(X) = X^{**}$. A reflexive space is necessarily complete, i.e., a Banach space. We leave it to the exercises to show that if X is reflexive, then so is X^*. In general, the isometric inclusions

$$X \subset X^{**} \subset X^{****} \subset \cdots$$

and

$$X^* \subset X^{***} \subset X^{*****} \subset \cdots$$

are strict. That is, equality need not hold. However, if X is reflective, both these chains collapse to X and X^*, respectively.

Example. For $1 \leq p < \infty$, $(\ell^p)^* = \ell^q$, where q is the conjugate exponent for p, and consequently ℓ^p is reflexive for $1 < p < \infty$. For the Lebesgue space $L^p(\Omega)$, $\Omega \subset \mathbb{R}^d$ open, it is true that $(L^p(\Omega))^* = L^q(\Omega)$ when $1 \leq p < \infty$; however, $(L^\infty(\Omega))^* \neq L^1(\Omega)$, so $L^p(\Omega)$ is reflexive only for $1 < p < \infty$.

2.9 COMPACTNESS AND WEAK CONVERGENCE IN AN NLS

In a metric space, an infinite sequence within a compact set always has a convergent subsequence. This is oftentimes a very salutary intermediate conclusion. It is therefore useful to characterize compact sets. Compact sets in an infinite-dimensional NLS tend to be quite small. Fortunately, NLSs can be endowed with weaker topologies, and in that setting, there are larger compact sets and associated weaker notions of sequential convergence than the one induced by the norm. Some naturally weaker topologies play an interesting and helpful role in numerical analysis and in the theory of partial differential equations.

2.9.1 The norm or strong topology

We begin our study of compactness by noting that the Heine-Borel theorem is *not* true in infinite dimensions. That is, a closed and norm-bounded subset of an infinite-dimensional NLS need not be compact. Indeed, if a subset is closed, bounded and has a nonempty interior, it *cannot* be compact, as we now show. The following technical result will aid us in this endeavor.

Lemma 2.49. *Let X be an NLS, Y a closed subspace, and Z a subspace containing Y. If $Z \neq Y$ and θ is given with $0 < \theta < 1$, then there is some $z \in Z$ such that $\|z\| = 1$ and*

$$\mathrm{dist}(z, Y) \geq \theta.$$

Proof. Let $z_0 \in Z \setminus Y$, and define

$$d = \mathrm{dist}(z_0, Y) = \inf_{y \in Y} \|z_0 - y\|.$$

Since Y is closed, $d > 0$, so we can find $y_0 \in Y$ such that

$$\frac{d}{\theta} \geq \|z_0 - y_0\| \geq d.$$

Define z by

$$z = \frac{z_0 - y_0}{\|z_0 - y_0\|} \in Z.$$

Then, for $y \in Y$,

$$\|z - y\| = \frac{\left\| z_0 - y_0 - y\|z_0 - y_0\| \right\|}{\|z_0 - y_0\|} = \frac{\|z_0 - y_1\|}{\|z_0 - y_0\|} \geq \|z_0 - y_1\| \frac{\theta}{d} \geq \theta,$$

since $y_1 = y_0 + y\|z_0 - y_0\| \in Y$, Y being a subspace. □

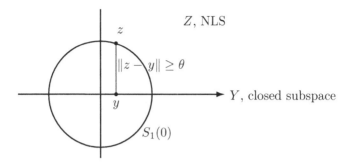

Figure 2.3 An illustration of a point z on the unit sphere far from a subspace Y.

Remark. In a general NLS, there is no notion of angle between two vectors. The last theorem can be viewed as asserting the existence of a point that is nearly 'orthogonal' to a given, closed subspace, although such a point need not be unique, even up to some symmetry conditions. The result is illustrated in Figure 2.3, where the unit sphere is shown as round, but it could have been depicted as square-like (see Figure 2.1).

Corollary 2.50. *If X is an infinite-dimensional NLS and M is a closed bounded set with nonempty interior, then M is not compact.*

Proof. It is enough to show the result for the closed unit ball. Let $x_1 \in X$ have norm 1. Suppose we have managed to define a sequence $x_1, x_2, \ldots, x_n$ such that each x_j has norm 1 and

$$\|x_i - x_j\| \geq 1/2$$

for $i, j \leq n$ and $i \neq j$. We then continue inductively as follows. Let $Y = \operatorname{span}\{x_1, x_2, \ldots, x_n\}$ and choose any $x \in X \smallsetminus Y$, which is nonempty since X has infinite dimension. With $Z = \operatorname{span}\{Y, x\}$ and $\theta = 1/2$, the previous lemma gives $x_{n+1} \in Z$ of size 1 and such that

$$\operatorname{dist}(x_{n+1}, Y) \geq 1/2.$$

Thus we have inductively constructed an infinite sequence of points that are at least a distance 1/2 apart from each other. This collection $\{x_j\}_{j=1}^{\infty}$ clearly has no convergent subsequence, and so we conclude that the closed unit ball cannot be compact. □

2.9.2 The weak and weak-∗ topologies

Here is a weaker notion of convergence than norm convergence.

Definition. Let X be an NLS and $\{x_n\}_{n=1}^\infty$ a sequence in X. We say that $\{x_n\}_{n=1}^\infty$ *converges weakly* to $x \in X$ if

$$f(x_n) \to f(x) \quad \text{as } n \to \infty$$

for all $f \in X^*$. We write $x_n \rightharpoonup x$ or $x_n \xrightarrow{w} x$ for weak convergence. Let $\{f_n\}_{n=1}^\infty$ be a sequence in X^* and $f \in X^*$. We say that f_n *converges weak-** to $f \in X^*$ if

$$f_n(x) \to f(x) \quad \text{as } n \to \infty$$

for each $x \in X$. We write $f_n \xrightarrow{w^*} f$ to indicate weak-* convergence.

Proposition 2.51. *Let X be an NLS and $\{x_n\}_{n=1}^\infty$ a sequence from X. If $x_n \to x$ as $n \to \infty$, then $x_n \rightharpoonup x$.*

Example. The converse to the above result does not hold. For example, consider the Banach space ℓ^p, $1 < p < \infty$, and the elements e_n, where $e_{n,i} = 0$ if $i \neq n$ and $e_{n,n} = 1$. Clearly e_n is not Cauchy, since $\|e_n - e_m\|_{\ell^p} = 2^{1/p}$ whenever $n \neq m$. However, it is the case that $e_n \rightharpoonup 0$. Indeed, fix an $f \in (\ell^p)^* = \ell^q$, where $1/p + 1/q = 1$. Then $f(e_n) = f_n \to 0$, since $\sum_{n=1}^\infty |f_n|^q < \infty$ implies that the nth term f_n in the series tends to zero.

Proposition 2.52. *Let X be an NLS, $\{x_n\}_{n=1}^\infty$ a sequence from X, and $\{f_n\}_{n=1}^\infty$ a sequence from X^*. If $\{x_n\}_{n=1}^\infty$ converges weakly, then its weak limit is unique and $\{\|x_n\|_X\}_{n=1}^\infty$ is bounded. If $\{f_n\}_{n=1}^\infty \subset X^*$ converges weak-*, then its weak-* limit is unique. If in addition X is a Banach space, then $\{\|f_n\|_{X^*}\}_{n=1}^\infty$ is bounded.*

Proof. Suppose $x_n \rightharpoonup x$ and $x_n \rightharpoonup y$. That means that for any $f \in X^*$,

$$f(x_n) \longrightarrow f(x)$$
$$\downarrow$$
$$f(y)$$

as $n \to \infty$. Consequently $f(x) = f(y)$ for all $f \in X^*$, which means $x = y$ by the Hahn-Banach theorem.

Fix an $f \in X^*$. Then the sequence $\{f(x_n)\}_{n=1}^\infty$ converges, and so is bounded in $\mathbb{F}$, say

$$|f(x_n)| \leq C_f \quad \text{for all } n.$$

View x_n as the evaluation map $E_{x_n} \in X^{**}$. In this context, the last condition amounts to

$$|E_{x_n}(f)| \leq C_f$$

for all n. Thus we have a collection of continuous linear maps $\{E_{x_n}\}_{n=1}^\infty$ in $X^{**} = B(X^*, \mathbb{F})$ which are bounded at each point of their domain X^*. By the uniform boundedness principle, which can be applied since X^* is a Banach space, we must have

$$\sup_n \|E_{x_n}\|_{X^{**}} \leq C.$$

But, by the Hahn-Banach theorem,

$$\|E_{x_n}\|_{X^{**}} = \|x_n\|_X.$$

Drawing the advertised conclusions for weak-* convergence is left to the reader.

□

Proposition 2.53. *Let X be an NLS and $\{x_n\}_{n=1}^{\infty} \subset X$. If $x_n \rightharpoonup x$, then $\|x\| \leq \liminf\limits_{n \to \infty} \|x_n\|$.*

This result is left as an exercise; it follows from the Hahn-Banach theorem.

New topologies may be defined on both X and X^* that are consistent with the notions of weak and weak-* convergence.

Definition. Suppose X is an NLS with dual X^*. The *weak topology* on X is the smallest topology on X such that each $f \in X^*$ is continuous. The *weak-* topology* on X^* is the smallest topology on X^* making continuous each evaluation map $E_x : X^* \to \mathbb{F}$, $x \in X$ (defined by $E_x(f) = f(x)$).

It is not difficult to describe a base for these topologies. A basic open set containing zero in the weak topology of X is of the form

$$U = \{x \in X : |f_i(x)| < \epsilon_i, \; i = 1, \ldots, n\} = \bigcap_{i=1}^{n} f_i^{-1}(B_{\epsilon_i}(0))$$

for some n, $\epsilon_i > 0$, and $f_i \in X^*$, where $B_{\epsilon_i}(0) = \{z \in \mathbb{F} : |z| < \epsilon_i\}$. Similarly for the weak-* topology of X^*, a basic open set containing zero is of the form

$$V = \{f \in X^* : |f(x_i)| < \epsilon_i, \; i = 1, \ldots, n\} = \bigcap_{i=1}^{n} E_{x_i}^{-1}(B_{\epsilon_i}(0))$$

for some n, $\epsilon_i > 0$, and $x_i \in X$. The rest of the topology is given by translations and unions of these. If X is infinite-dimensional, these topologies are not compatible with any metric, so some care is warranted. The weak and weak-* limit processes arise from these topologies as we show now.

Proposition 2.54. *Suppose X is an NLS with dual X^*. Let $x \in X$ and $\{x_n\}_{n=1}^{\infty} \subset X$. Then x_n converges to x in the weak topology if and only if $x_n \rightharpoonup x$ (i.e., $f(x_n) \to f(x)$ in $\mathbb{F}$ for every $f \in X^*$). Moreover, if $f \in X^*$ and $\{f_n\}_{n=1}^{\infty} \subset X^*$, then f_n converges to f in the weak-* topology if and only if $f_n \xrightarrow{w^*} f$ (i.e., $f_n(x) \to f(x)$ in $\mathbb{F}$ for every $x \in X$).*

Proof. If x_n converges to x in the weak topology, then, since $f \in X^*$ is continuous in the weak topology (by definition), $f(x_n) \to f(x)$. That is $x_n \rightharpoonup x$. Conversely, suppose $f(x_n) \to f(x)$ for all $f \in X^*$. Let U be a basic open set containing x. Then, U has the form

$$U = x + \{y \in X : |f_i(y)| < \epsilon_i, \; i = 1, \ldots, m\}$$

for some m, $\epsilon_i > 0$, and $f_i \in X^*$. Now there is some $N > 0$ such that

$$|f_i(x_n) - f_i(x)| = |f_i(x_n - x)| < \epsilon_i$$

for all $n \geq N$ and $i = 1, \ldots, m$, since $f_i(x_n) \to f_i(x)$, so $x_n = x + (x_n - x) \in U$. That is, x_n converges to x in the weak topology. Similar reasoning gives the result for weak-* convergence. $\qquad\square$

Remark. It is interesting to note that weak sequential convergence can be used independently to generate the weak topology. Define a set U in an NLS X to be open if whenever $x_n \rightharpoonup x$ for some $x \in U$, then there exists an N such that $x_n \in U$ for all $n \geq N$. It is an exercise to show that this collection of sets $\mathcal{O}$ comprises a topology and that all the weakly open sets are contained in $\mathcal{O}$.

By the Hahn-Banach theorem (e.g., Proposition 2.33), the weak and weak-* topologies are Hausdorff. The weak topology on X is weaker than the *strong* or *norm* topology (for which more than just the linear functionals are continuous). For if U is a basic weakly open set and $x \in U$, it is straightforward to check that $B_r(x) \subset U$ for $r = r(x)$ small enough.

On X^*, there are three topologies, the weak-* topology (weakest topology for which the evaluation maps $\{E_x : x \in X\} \subset X^{**}$ are continuous), the weak topology (weakest for which all elements of X^{**} are continuous maps), and the strong or norm topology. The weak-* topology is weaker than the weak topology, which is weaker than the strong topology. Of course, if X is reflexive, the weak-* and weak topologies agree.

It is easier for a sequence to converge in weaker topologies, as then there are fewer open sets to consider. In an infinite-dimensional NLS, the unit ball is never a compact set as we know. Hence, open covers of the unit ball need not have finite sub-covers and sequences drawn from the unit ball need not have convergent subsequences. However, if the open sets in a cover are restricted to being weakly open sets, we might hope that there is a finite subcover. This is in fact the case in X^*.

Theorem 2.55 (Banach-Alaoglu theorem). *Suppose X is an NLS with dual X^* and let $B_1^* = \{f \in X^* : \|f\| \leq 1\}$ be the closed unit ball in X^*. Then B_1^* is compact in the weak-* topology.*

By a scaling argument, we can immediately generalize the theorem to show that a closed ball of any radius $r > 0$ is weak-* compact.

Proof. For each $x \in X$, let

$$B_x = \{\lambda \in \mathbb{F} : |\lambda| \leq \|x\|\}.$$

Each B_x is closed and bounded in $\mathbb{F}$, and so is compact. By Tychonoff's theorem,

$$C = \underset{x \in X}{\times} B_x$$

is also compact. An element of C can be viewed as a function $g : X \to \mathbb{F}$ satisfying $|g(x)| \leq \|x\|$. In this way, B_1^* is the subset of C consisting of the linear functionals. The product topology on C is the weakest one making all coordinate projection maps $g \mapsto g(x)$ continuous. As these maps are the evaluation maps, the inherited topology on B_1^* is precisely the weak-$*$ topology.

Since C is compact, we can complete the proof by showing that B_1^* is closed in C. Since X^* is not a metric space when endowed with the weak-$*$ topology, we must show that any accumulation point g of B_1^* is in B_1^*, i.e., linear with norm at most one. Fix $x, y \in X$ and $\lambda \in \mathbb{F}$. Since g is an accumulation point, every neighborhood of the form

$$U = g + \{h \in C : |h(x_i)| < \epsilon_i, \ i = 1, \ldots, m\}$$

intersects B_1^*. Given $\epsilon > 0$, there is a neighborhood with $m = 4$ containing $f \in B_1^*$ such that

$$f = g + h,$$

where

$$|h(x)| < \frac{\epsilon}{3 \max(1, |\lambda|)}, \quad |h(y)| < \frac{\epsilon}{3}, \quad |h(x+y)| < \frac{\epsilon}{3}, \quad \text{and} \quad |h(\lambda x)| < \frac{2\epsilon}{3}.$$

Thus, since f is linear,

$$|g(x+y) - g(x) - g(y)| = |h(x+y) - h(x) - h(y)| \leq \epsilon$$

and

$$|g(\lambda x) - \lambda g(x)| = |h(\lambda x) - \lambda h(x)| \leq \epsilon.$$

As ϵ is arbitrary, g is linear. Moreover,

$$|g(x)| = |f(x) - h(x)| \leq |f(x)| + \frac{\epsilon}{3} \leq \|x\| + \frac{\epsilon}{3},$$

so also $|g(x)| \leq \|x\|$. That is, $g \in B_1^*$, so B_1^* is closed. $\qquad\square$

What does compactness say about sequences? If the space is metrizable (i.e., there is a metric that gives the same topology), a sequence in a compact subset has a convergent subsequence (see Proposition 1.28). An infinite-dimensional NLS equipped with its weak topology is not metrizable, so it is not clear that a bounded sequence must have a weakly convergent subsequence. However, with an extra hypothesis, such a conclusion may be drawn.

Theorem 2.56. *Let X be a separable NLS and $K \subset X^*$. If K is weak-$*$ compact, then K is metrizable in the weak-$*$ topology, and hence K is weak-$*$ sequentially compact (i.e., given any sequence $\{f_n\}_{n=1}^{\infty} \subset K$, there exists a subsequence $\{f_{n_i}\}_{i=1}^{\infty}$ that converges weak-$*$ in K).*

Proof. Recall that separability means that we can find a dense subset $D = \{x_n\}_{n=1}^{\infty} \subset X$. The evaluation maps $E_n : X^* \to \mathbb{F}$, defined by $E_n(x^*) = x^*(x_n)$, are weak-* continuous by definition. If $E_n(x^*) = E_n(y^*)$ for each n, then x^* and y^* are two continuous functions that agree on the dense set D, and so they must agree everywhere. That is, the set $\{E_n\}_{n=1}^{\infty}$ is a countable set of continuous functions that separates points on X^*.

Now let $C_n = \sup_{x^* \in K} |E_n(x^*)|$. This is finite since K is compact and E_n is continuous. Define $f_n = E_n/C_n$ if $C_n > 0$ and $f_n = 0$ otherwise. Then $|f_n| \leq 1$ on K, and

$$d(x^*, y^*) = \sum_{n=1}^{\infty} 2^{-n} |f_n(x^*) - f_n(y^*)|$$

is a metric on K because the collection $\{f_n\}$ separates points.

The compact set K may be endowed with the weak-* topology τ or with the topology τ_d induced by the metric just defined. The claim is that these are the same. First, let's check that $\tau_d \subset \tau$. For $N \geq 1$, let

$$d_N(x^*, y^*) = \sum_{n=1}^{N} 2^{-n} |f_n(x^*) - f_n(y^*)|,$$

and consider the function $d_N : x^* \mapsto d_N(x^*, y^*)$ which, for a fixed y^*, associates to x^* the nonnegative real number $d_N(x^*, y^*)$. Since the sum defining d_N is finite, it is clear that $d_N(\cdot, y^*)$ is continuous for the weak-* topology. But, $d_N(\cdot, y^*)$ converges uniformly on K to $d(\cdot, y^*)$, so it is concluded that $d(\cdot, y^*)$ is also τ-continuous. Finally, any metric ball $B_r(y^*) = \{x^* \in K : d(x^*, y^*) < r\}$ is the inverse image under $d(\cdot, y^*)$ of the open set $(-\infty, r)$, and so must be τ-open.

To show the opposite inclusion $\tau \subset \tau_d$, let $A \in \tau$. Then $A^c \subset K$ is τ-closed and thus τ-compact (Proposition 1.25). But $\tau_d \subset \tau$ implies that A^c is also τ_d-compact since any τ_d-open cover of A^c is also a τ-open cover and so has a finite subcover. Proposition 1.25 now implies that A^c is τ_d-closed, therefore $A \in \tau_d$. Thus K is metrizable. Reference to Proposition 1.28 completes the proof. □

Corollary 2.57. *If X is a separable NLS, $\{f_n\}_{n=1}^{\infty} \subset X^*$, and there is some $R > 0$ such that $\|f_n\| \leq R$ for all n, then there is a subsequence $\{f_{n_i}\}_{i=1}^{\infty}$ that converges weak-* in X^*.*

Theorem 2.58 (generalized Heine-Borel theorem 1). *Suppose X is a separable Banach space with dual X^* and $K \subset X^*$. Then the following are equivalent: (i) K is weak-* compact, (ii) K is weak-* closed and bounded, and (iii) K is weak-* sequentially compact.*

Proof. That (i) implies (iii) is contained in Theorem 2.56. Any weak-* closed and bounded set K is weak-* compact, as it sits in a large closed ball, which is weak-∗ compact by the Banach-Alaoglu theorem (Theorem 2.55), so (ii) implies (i).

It remains to show that (iii) implies (ii). Suppose K is weak-∗ sequentially compact. First, if K is not bounded, there is a sequence $\{f_n\}_{n=1}^{\infty} \subset K$ such that $\|f_n\|_{X^*} \geq n$. This sequence has no weak-∗ convergent subsequence, however, since every weak-∗ convergent sequence is bounded. This contradiction forces the conclusion that K is bounded.

Let $\{x_n\}_{n=1}^{\infty}$ be a countable dense subset of X and let f be any accumulation point of K. Then, for each $n = 1, 2, \ldots$, there must be a point $f_n \in K$ which belongs to the weak-∗ open set

$$V_n = f + \left\{ g \in X^* : |g(x_j)| < \frac{1}{n}, \text{ for } j = 1, \ldots, n \right\}$$

containing f. The sequence $\{f_n(x_j)\}_{n=1}^{\infty}$ clearly converges to $f(x_j)$ for each $j = 1, 2, \ldots$. By density and the fact that K is norm-bounded, the same is true of $\{f_n(x)\}_{n=1}^{\infty}$ for any $x \in X$. Thus the sequence converges weak-∗ to f and by sequential compactness, f must lie in K. The proof is complete. □

Since the weak and weak-∗ topologies agree in reflexive spaces, the last two results can be restated in terms of weak convergence when the context is a separable, reflexive space.

Corollary 2.59. *If a Banach space X is separable and reflexive and $\{x_n\}_{n=1}^{\infty} \subset X$ is a bounded sequence, then there is a subsequence $\{x_{n_i}\}_{i=1}^{\infty}$ that converges weakly in X.*

Example. If $1 < p < \infty$ and $\Omega \subset \mathbb{R}^d$ is measurable, then $L^p(\Omega)$ is separable and reflexive, so a bounded sequence $\{f_n\}_{n=1}^{\infty}$ always has a weakly convergent subsequence. That is, if $\|f_n\|_p \leq M$ for some $M > 0$, then there is an $f \in L^p(\Omega)$ and a subsequence $\{f_{n_k}\}_{k=1}^{\infty}$ such that $f_{n_k} \rightharpoonup f$ in $L^p(\Omega)$.

Theorem 2.60 (generalized Heine-Borel theorem 2). *Suppose X is a separable, reflexive Banach space and $K \subset X$. Then the following are equivalent: (i) K is weakly compact, (ii) K is weakly closed and bounded, and (iii) K is weakly sequentially compact.*

Remark. Corollary 2.59 is true even without the assumption that X is separable. It follows from the Eberlein-Šmulian theorem (see, e.g., [28]) that in fact a Banach space is reflexive if and only if every bounded sequence has a weakly convergent subsequence.

This section closes with a few interesting results relating weak and strong convergence.

Proposition 2.61. *Let X and Y be NLSs, $T \in B(X, Y)$, and $\{x_n\}_{n=1}^{\infty} \subset X$. If $x_n \rightharpoonup x$, then $Tx_n \rightharpoonup Tx$ in Y.*

That is, a bounded (i.e., strongly continuous) linear operator between NLSs is weakly continuous. It is an exercise to provide a proof of this result.

Theorem 2.62 (Banach-Saks theorem). *Suppose that X is an NLS and $\{x_n\}_{n=1}^{\infty}$ is a sequence in X that converges weakly to $x \in X$. Then for every $n \geq 1$, there are constants $\alpha_j^n \geq 0$ with $\sum_{j=1}^{n} \alpha_j^n = 1$, such that the sequence $y_n = \sum_{j=1}^{n} \alpha_j^n x_j$ converges strongly to x as $n \to \infty$.*

That is to say, whenever $x_n \rightharpoonup x$, there is a sequence $\{y_n\}_{n=1}^{\infty}$ of finite, convex, linear combinations of the x_n such that $y_n \to x$ as $n \to \infty$.

Proof. Consider the convex set

$$M = \left\{ \sum_{j=1}^{n} \alpha_j^n x_j : n \geq 1, \ \alpha_j^n \geq 0, \ \text{and} \ \sum_{j=1}^{n} \alpha_j^n = 1 \right\}.$$

The conclusion of the theorem is that $x \in \bar{M}$, the (norm) closure of M. Suppose this is not the case. Then the separating hyperplane theorem (Theorem 2.38) applied to the closed, convex set $\bar{M}$ and the compact set $\{x\}$ implies that there is a continuous linear functional f and a number γ such that $\text{Re} f(x_n) > \gamma$ but $\text{Re} f(x) < \gamma$. Thus $\liminf_{n \to \infty} \text{Re} f(x_n) \geq \gamma$, so $f(x_n) \not\to f(x)$, a contradiction. It follows that $x \in \bar{M}$, as required. $\qquad \square$

Corollary 2.63. *Suppose that X is an NLS and $K \subset X$ is nonempty, convex, and closed. If $\{x_n\}_{n=1}^{\infty} \subset K$ converges weakly to x as $n \to \infty$, then $x \in K$.*

2.10 THE DUAL OF AN OPERATOR

Suppose X and Y are NLSs and $T \in B(X, Y)$. The operator T induces an operator

$$T^* : Y^* \to X^*,$$

called the *dual, conjugate* or *adjoint* of T, as follows. Let $g \in Y^*$ and define $T^* g : X^* \to \mathbb{F}$ by the formula

$$(T^* g)(x) = g(Tx)$$

for $x \in X$. It is clear that $T^* g \in X^*$ since $T^* g = g \circ T$ is a composition of continuous linear maps,

$$X \xrightarrow{\ T\ } Y \xrightarrow{\ g\ } \mathbb{F} \ ,$$

$$T^* g$$

and so is itself continuous and linear. Moreover, if $g \in Y^*$, $x \in X$, then

$$\begin{aligned} |T^* g(x)| = |g(Tx)| &\leq \|g\|_{Y^*} \|Tx\|_Y \\ &\leq \|g\|_{Y^*} \|T\|_{B(X,Y)} \|x\|_X \\ &= \left(\|g\|_{Y^*} \|T\|_{B(X,Y)} \right) \|x\|_X. \end{aligned}$$

Hence, not only is T^*g bounded, but

$$\|T^*g\|_{X^*} \leq \|T\|_{B(X,Y)}\|g\|_{Y^*}. \tag{2.23}$$

Thus we have defined a map $T^* : Y^* \to X^*$. In fact, T^* is itself a bounded linear map, which is to say $T^* \in B(Y^*, X^*)$. For linearity, it is necessary to show that for $g, h \in Y^*$, $\lambda \in \mathbb{F}$,

$$T^*(g + h) = T^*g + T^*h \quad \text{and} \quad T^*(\lambda g) = \lambda T^*g. \tag{2.24}$$

Let $x \in X$ and evaluate both sides of these potential equalities at x, *viz.*

$$T^*(g + h)(x) = (g + h)(Tx) = g(Tx) + h(Tx)$$
$$= T^*(g)(x) + T^*(h)(x) = (T^*g + T^*h)(x)$$

and

$$T^*(\lambda g)(x) = (\lambda g)(Tx) = \lambda g(Tx) = \lambda T^*g(x).$$

As $x \in X$, was arbitrary, it follows that the formulas (2.24) are valid. Thus T^* is linear. The fact that T^* is bounded follows from (2.23), and, moreover,

$$\|T^*\|_{B(Y^*, X^*)} \leq \|T\|_{B(X,Y)}. \tag{2.25}$$

In fact, equality always holds in (2.25). To see this, first note that if $T = 0$ is the zero operator, then $T^* = 0$ also and so their norms certainly agree. If $T \neq 0$, then $\|T\|_{B(X,Y)} > 0$. Let $\epsilon > 0$ be given and let $x_0 \in X$, $\|x_0\|_X = 1$ be such that

$$\|Tx_0\|_Y \geq \|T\|_{B(X,Y)} - \epsilon.$$

Let $g_0 \in Y^*$ be such that $\|g_0\|_{Y^*} = 1$ and

$$g_0(Tx_0) = \|Tx_0\|.$$

Such a g_0 exists by one of the corollaries of the Hahn-Banach theorem. Then, it transpires that

$$\|T^*\|_{B(Y^*, X^*)} \geq \|T^*g_0\|_{X^*} = \sup_{\|x\|_X=1} |T^*g_0(x)|$$
$$\geq |T^*g_0(x_0)| = |g_0(Tx_0)| = \|Tx_0\|_Y$$
$$\geq \|T\|_{B(X,Y)} - \epsilon.$$

In consequence of these ruminations, it is seen that

$$\|T^*\|_{B(Y^*, X^*)} \geq \|T\|_{B(X,Y)} - \epsilon,$$

and $\epsilon > 0$ was arbitrary. Hence,

$$\|T^*\|_{B(Y^*, X^*)} \geq \|T\|_{B(X,Y)}$$

and, along with (2.25), this establishes the result

$$\|T^*\|_{B(Y^*,X^*)} = \|T\|_{B(X,Y)}.$$

The 'star' map

$$* : B(X,Y) \longrightarrow B(Y^*,X^*)$$

that sends T to T^* has many simple properties of its own, which are left for the reader to verify.

Proposition 2.64. *Let X, Y, and Z be NLSs, $S, T \in B(X,Y)$, $R \in B(Y,Z)$ and $\lambda, \mu \in \mathbb{F}$. Then*

(a) $\|T^*\|_{B(Y^*,X^*)} = \|T\|_{B(X,Y)}$ (* *is norm preserving*),

(b) $(\lambda T + \mu S)^* = \lambda T^* + \mu S^*$ (* *is a linear map*),

(c) $(RS)^* = S^* R^*$,

(d) $(I_X)^* = I_{X^*}$,

where $I_X \in B(X,X)$ is the identity mapping of X to itself.

Examples.

(1) If $X = \mathbb{R}^d$, then a linear operator $T : X \to X$ may be represented by a $d \times d$ matrix M_T once a basis $\{e_n\}_{n=1}^d \subset X$ is given. In fact, then, $M_T = [T(e_1) \cdots T(e_d)]$. The dual operator T^* also has a matrix representation M_{T^*} in any given basis for X^*. Given the basis $\{e_n\}_{n=1}^d \subset X$, the *dual basis* $\{e_n^*\}_{n=1}^d \subset X^*$ is defined by the requirement $e_n^*(e_m) = 0$ if $n \neq m$ and $e_n^*(e_n) = 1$. If T^* is represented in the dual basis, then $M_{T^*} = M_T^T$, the transpose of M_T.

(2) Here is a less elementary, but related example. Let $1 < p < \infty$ and, for $f \in L^p(0,1)$ and $x \in (0,1)$, set

$$Tf(x) = \int_0^1 K(x,y)f(y)\, dy,$$

where K is, say, a bounded measurable function. It is easily determined that T is a bounded linear map of $L^p(0,1)$ into itself. We have seen that the dual space of $L^p(0,1)$ may be realized concretely as $L^q(0,1)$, where $1/p + 1/q = 1$. That is, for $g \in L^q(0,1)$, $\Lambda_g \in (L^p(0,1))^*$ is given by

$$\Lambda_g(f) = \int_0^1 f(x)\, g(x)\, dx,$$

where $f \in L^p(0,1)$. To understand T^*, compute its action on a $g \in L^q(0,1)$ and evaluate the result at an $f \in L^p(0,1)$, *viz.*

$$(T^*\Lambda_g)(f) = \Lambda_g(Tf) = \int_0^1 g(x)Tf(x)\,dx$$

$$= \int_0^1 g(x)\int_0^1 K(x,y)f(y)\,dy\,dx$$

$$= \int_0^1 f(y)\int_0^1 K(x,y)g(x)\,dx\,dy.$$

Thus, it is determined that for $y \in (0,1)$,

$$T^*(g)(y) = \int_0^1 K(x,y)g(x)\,dx.$$

Lemma 2.65. *Let X, Y be NLSs and $T \in B(X,Y)$. Then $T^{**} : X^{**} \to Y^{**}$ is a bounded linear extension of T. If X is reflexive, then $T = T^{**}$.*

Proof. Let $x \in X$ and $g \in Y^*$. Realize x as the evaluation map $E_x \in X^{**}$. Then, by definition,

$$(T^{**}E_x)(g) = E_x(T^*g) = T^*g(x) = g(Tx) = E_{Tx}(g),$$

and so

$$T^{**}E_x = E_{Tx}.$$

Thus $T^{**}|_X = T$. If $X = X^{**}$, then this means $T = T^{**}$. □

Lemma 2.66. *Let X and Y be Banach spaces and $T \in B(X,Y)$. Then T has a bounded inverse defined on all of Y if and only if T^* has a bounded inverse defined on all of X^*. When either exists, then*

$$(T^{-1})^* = (T^*)^{-1}.$$

Proof. If $S = T^{-1} \in B(Y,X)$, then

$$S^*T^* = (TS)^* = (I_Y)^* = I_{Y^*}.$$

This shows that T^* is one-to-one. The other way around,

$$T^*S^* = (ST)^* = (I_X)^* = I_{X^*},$$

shows T^* is onto. Moreover, S^* is the inverse of T^*, and of course S^* is bounded since it is the dual of a bounded map.

Conversely, if $T^* \in B(Y^*, X^*)$ has a bounded inverse, then applying the preceding argument, it is ascertained that $(T^{**})^{-1} \in B(Y^{**}, X^{**})$ is one-to-one and onto. But,

$$T^{**}|_X = T,$$

so T must be one-to-one. We claim that T maps X onto Y. If so, the open mapping theorem implies that the inverse is bounded, and we are done. Since

T^{**} has a continuous inverse, it is a homeomorphism, and so T^{**} takes a closed set to a closed set. Since X is a Banach space, it is closed in X^{**}, so $T^{**}(X)$ is closed in Y^{**}, which is to say that $T(X)$ is closed in Y^{**}, and hence in Y. Now suppose T is not onto. Then there is a $y \in Y \setminus T(X)$. By the Hahn-Banach theorem, since $T(X)$ is closed, there is a $y^* \in Y^*$ such that

$$y^*\big|_{T(X)} = 0, \quad \text{but} \quad y^*(y) \neq 0.$$

But then, for all $x \in X$,

$$T^*y^*(x) = y^*(Tx) = 0,$$

whence $T^*y^* = 0$, and therefore $y^* = 0$, since T^* is one-to-one. But y^* is not the zero element in Y^* since $y^*(y) \neq 0$. This contradiction leaves only the possibility that T is surjective, and the lemma is established. □

2.11 EXERCISES

1. Suppose that X is a vector space.

   (a) If $A, B \subset X$ are convex, show that $A + B$ and $A \cap B$ are convex. What about $A \cup B$ and $A \setminus B$?

   (b) Show that $2A \subset A + A$. When is it true that $2A = A + A$?

2. Let (X, d) be a metric space. Show that, for all $x, y, z, w \in X$:

   (a) $|d(x, y) - d(z, w)| \leq d(x, z) + d(y, w)$,

   (b) $|d(x, y) - d(y, z)| \leq d(x, z)$.

3. Let (X, d) be a metric space and $A \subset X$ a nonempty subset. Define the *diameter* of A to be
$$\text{diam}(A) = \sup_{x,y \in A} d(x, y),$$
   and say that A is *bounded* if this number is finite. By convention, the empty set has diameter 0 and is bounded. Show that if $A \subset B$, then $\text{diam}(A) \leq \text{diam}(B)$, and thereby show that any subset of a bounded set is bounded.

4. Let (X, d) be a metric space.

   (a) Show that
$$\rho(x, y) = \min\{1, d(x, y)\}$$
   is also a metric on X.

   (b) Show that $U \subset X$ is open in (X, d) if and only if U is open in (X, ρ).

   (c) Draw the same conclusions for
$$\sigma(x, y) = \frac{d(x, y)}{1 + d(x, y)}.$$

5. Let X be an NLS, x_0 a fixed vector in X and $\alpha \neq 0$ a fixed scalar. Show that the mappings $x \mapsto x + x_0$ and $x \mapsto \alpha x$ are homeomorphisms of X onto itself.

6. Show that if X is an NLS, then X is homeomorphic to $B_r(0)$ for fixed r. $\left[\text{Hint: consider the mapping } x \mapsto \dfrac{xr}{1 + \|x\|}. \right]$

7. Prove that a subset A of a metric space (X, d) is bounded if and only if every countable subset of A is bounded.

8. Let A and B be NLSs that are also linear subspaces of a vector space X. Prove that $A \cap B$ is an NLS with the norm $\| \cdot \|_{A \cap B} = \| \cdot \|_A + \| \cdot \|_B$. If $1 \leq p < q < \infty$ and $\Omega \subset \mathbb{R}^d$ is measurable, prove that $L^p(\Omega) \cap L^q(\Omega)$ is a Banach space.

9. Let X be a vector space. Define the *convex hull* of $A \subset X$ to be

$$\mathrm{co}(A) = \left\{ x \in X : x = \sum_{i=1}^{n} t_i y_i \text{ for some } n \geq 1, t_i \in [0, 1], \right.$$
$$\left. \sum_{i=1}^{n} t_i = 1, \text{ and } y_i \in A \right\}.$$

(a) Prove that the convex hull is convex, and that it is the intersection of all convex subsets of X containing A.

(b) If X is a normed linear space, prove that the convex hull of an open set is open.

(c) If X is a normed linear space, is the convex hull of a closed set always closed? [Hint: look in $\mathbb{R}^2$ for a counterexample.]

(d) Prove that if X is a normed linear space, then the convex hull of a bounded set is bounded.

10. Finite-dimensional matrices. Let $M^{n \times m}$ be the set of $n \times m$ matrices with real-valued coefficients a_{ij}, for $1 \leq i \leq n$ and $1 \leq j \leq m$.

(a) For every $A \in M^{n \times m}$, define

$$\|A\| = \max_{\substack{x \in \mathbb{R}^m \\ x \neq 0}} \frac{|Ax|}{|x|}.$$

Show that $\left(M^{n \times m}, \|\cdot\| \right)$ is an NLS. (Recall that $|\cdot| = \|\cdot\|_{\ell^2}$ is the Euclidean norm on $\mathbb{F}^d$.)

(b) Each $A \in M^{n \times m}$ defines a linear map of $\mathbb{R}^m$ into $\mathbb{R}^n$. Show that when $m = n$,

$$\|A\| = \max_{|x|=|y|=1} y^T A x,$$

where y^T is the transpose of y.

(c) Show that each $A \in M^{n \times m}$ is continuous.

11. Prove Corollary 2.13.

12. Show that $\left(C([a, b]), \|\cdot\|_{C([a,b])}\right)$, the set of real-valued bounded continuous functions on the interval $[a, b]$ with the sup-norm (L^∞-norm), is a Banach space.

13. Consider the space $(\ell^p, \|\cdot\|_p)$.

(a) Prove that $\|\cdot\|_p$ is a norm for $1 \leq p \leq \infty$.

(b) Prove that for $1 \leq p \leq \infty$, ℓ^p is a Banach space. [Hint: use that $\mathbb{R}$ is complete.]

(c) Show that $\|\cdot\|_p$ is *not* a norm for $0 < p < 1$. [Hint: first show the result in $\mathbb{R}^2$.]

(d) In Euclidean space $\mathbb{R}^2$, what does the "unit ball" look like for $p < 1$?

14. Show that an infinite-dimensional Banach space X must have an uncountably infinite Hamel basis as follows: consider any finite-dimensional linear subspace $Y \subset X$. Why is Y closed in X (and so *not* dense)? Show that Y has empty interior in X, and use the Baire category theorem to conclude the result.

15. If an infinite-dimensional vector space X is also an NLS and contains a sequence $\{e_n\}_{n=1}^{\infty}$ with the property that for every $x \in X$ there is a unique sequence of scalars $\{a_n\}_{n=1}^{\infty}$ such that

$$\|x - (a_1 e_1 + \cdots + a_n e_n)\| \to 0 \quad \text{as } n \to \infty,$$

then $\{e_n\}_{n=1}^{\infty}$ is called a *Schauder basis* for X, and we have the *expansion* of x

$$x = \sum_{n=1}^{\infty} a_n e_n.$$

(a) Find a Schauder basis for ℓ^p, $1 \leq p < \infty$.

(b) Show that if an NLS has a Schauder basis, then it is separable. (The converse is *not* true.)

16. If $f \in L^p(\Omega)$ show that

$$\|f\|_p = \sup \left| \int_\Omega f(x)\, g(x)\, dx \right| = \sup \int_\Omega |f(x)\, g(x)|\, dx,$$

where the supremum is taken over all $g \in L^q(\Omega)$ such that $\|g\|_q \leq 1$, $1 \leq p, q \leq \infty$, and $1/p + 1/q = 1$.

17. Suppose that $\Omega \subset \mathbb{R}^d$ is measurable with finite measure and $1 \leq p \leq q \leq \infty$.

(a) Prove that if $f \in L^q(\Omega)$, then $f \in L^p(\Omega)$ and

$$\|f\|_p \leq (\mu(\Omega))^{1/p-1/q}\|f\|_q.$$

(b) Prove that if $f \in L^\infty(\Omega)$, then

$$\lim_{p \to \infty} \|f\|_p = \|f\|_\infty.$$

(c) Prove that if $f \in L^p(\Omega)$ for all p with $1 \leq p < \infty$ and there is $R > 0$ such that $\|f\|_p \leq R$, then $f \in L^\infty(\Omega)$ and $\|f\|_\infty \leq R$.

18. Let $1 \leq p < \infty$ and define, for each $y \in \mathbb{R}^d$, the *translation operator* $\tau_y : L^p(\mathbb{R}^d) \to L^p(\mathbb{R}^d)$ by

$$\tau_y(f)(x) = f(x - y).$$

(a) Verify that $\tau_y(f) \in L^p(\mathbb{R}^d)$ and that τ_y is bounded and linear. What is the norm of τ_y?

(b) Show that as $y \to z$, $\|\tau_y f - \tau_z f\|_{L^p} \to 0$. [Hint: the set of continuous functions with compact support is dense in $L^p(\mathbb{R}^d)$ for $p < \infty$.]

19. Let p be a sublinear functional on a vector space X. For $r > 0$, show that the set $\{x \in X : p(x) \leq r\}$ is convex.

20. Let Y be a subspace of a vector space X. The *coset* of an element $x \in X$ with respect to Y is denoted by $x + Y$ and is defined to be the set

$$x + Y = \{z \in X : z = x + y \text{ for some } y \in Y\}.$$

(a) Show that the distinct cosets form a partition of X. Show that under algebraic operations defined by

$$(x_1 + Y) + (x_2 + Y) = (x_1 + x_2) + Y \quad \text{and} \quad \lambda(x + Y) = \lambda x + Y,$$

for any $x_1, x_2, x \in X$ and λ in the ground field, these cosets form a vector space. This space is called the *quotient space* of X by (or *modulo*) Y, and it is denoted X/Y.

(b) Let Y be a closed subspace of an NLS X. Show that for $\hat{x} \in X/Y$, the mapping $\hat{x} \mapsto \|\hat{x}\|_{X/Y}$ where

$$\|\hat{x}\|_{X/Y} = \inf_{x \in \hat{x}} \|x\|_X$$

defines a norm on X/Y.

21. Let X and Y be Banach spaces and $T \in B(X,Y)$. Let $N = N(T) = \ker(T) = \{x \in X : Tx = 0\}$ and $\hat{X} = X/N$ be the quotient space from the previous exercise.

(a) Show that the map $\hat{T} : \hat{X} \to Y$ given by $\hat{T}\hat{x} = Tx$ is a well-defined linear operator that is also bounded and injective. In fact, show $\|\hat{T}\| = \|T\|$.

(b) Show that if T is surjective, then $\hat{T} : \hat{X} \to Y$ is invertible with a bounded inverse (and so $\hat{X}$ and Y are isomorphic as Banach spaces). Furthermore, given y_n in the image of T (so there are x_n such that $y_n = Tx_n$), if $y_n \to y$ for some $y \in Y$, show that x_n may not converge if $N \neq \{0\}$, but $\hat{x}_n$ must converge. [Hint: use the open mapping theorem. For this exercise, you may presuppose that $\hat{X}$ is a Banach space.]

22. If X and Y are NLSs, then the *product space* $X \times Y$ is also an NLS with any of the norms

$$\|(x,y)\|_{X \times Y} = \max\{\|x\|_X, \|y\|_Y\}$$

or, for any $1 \leq p < \infty$,

$$\|(x,y)\|_{X \times Y} = \left(\|x\|_X^p + \|y\|_Y^p\right)^{1/p}.$$

(a) Why are these norms equivalent?

(b) If X and Y are Banach spaces, prove that $X \times Y$ is a Banach space.

23. Let X be an NLS and M a nonempty subset. The *annihilator* M^a of M is defined to be the set of all bounded linear functionals $f \in X^*$ such that f restricted to M is zero. Show that M^a is a closed subspace of X^*. What are X^a and $\{0\}^a$?

24. Let $a : [0, \infty) \to [0, \infty)$ be a continuous bijection such that $a(0) = 0$, and let b denote its inverse. Define A and B by

$$A(t) = \int_0^t a(s)\, ds \quad \text{and} \quad B(t) = \int_0^t b(s)\, ds,$$

which are convex functions. For $s \geq 0$, assume the growth condition

$$A(st) \leq k(s)\, A(t) \quad \text{for all } t \geq 0,$$

where k is continuous and $k(s) \to 0$ as $s \to 0$. Let

$$L_A(\mathbb{R}) = \left\{ \text{equivalence classes of } u : \mathbb{R} \to \mathbb{F} \;\middle|\; \int_{\mathbb{R}} A(|u(x)|)\, dx < \infty \right\}$$

and define $\| \cdot \|_A$ by

$$\|u\|_A = \inf\left\{ r > 0 : \int_{\mathbb{R}} A\left(\frac{|u(x)|}{r}\right) dx \leq 1 \right\}.$$

This is an example of an *Orlicz space*.

(a) Show that $L_A(\mathbb{R})$ is a vector space.

(b) Show that $\|\cdot\|_A$ is a norm on $L_A(\mathbb{R})$ (be sure to show that it is finite).

(c) Show that

$$\left| \int_{\mathbb{R}} u(x)\, v(x)\, dx \right| \leq 2\|u\|_A \|v\|_B.$$

[Hint: use the elementary fact that $st \leq A(s) + B(t)$.]

25. Fix $y \in C([a,b])$ and scalars $\alpha, \beta \in \mathbb{F}$. Show that the functionals

$$f_1(x) = \int_a^b x(t)\, y(t)\, dt \quad \text{and} \quad f_2(x) = \alpha x(a) + \beta x(b),$$

for $x \in C([a,b])$, are linear and bounded.

26. Find the norm of the linear functional f defined on $C([-1,1])$ by

$$f(x) = \int_{-1}^0 x(t)\, dt - \int_0^1 x(t)\, dt.$$

27. Let $T : C([0,1]) \to C([0,1])$ be defined by

$$y(t) = \int_0^t x(s)\, ds.$$

Find the range or image $R(T)$ of T, and show that T is invertible on its range, $T^{-1} : R(T) \to C([0,1])$. Is T^{-1} linear and bounded?

28. The space $C^1([a,b])$ is the NLS of all continuously differentiable functions defined on $[a,b]$ with the norm

$$\|x\| = \sup_{t \in [a,b]} |x(t)| + \sup_{t \in [a,b]} |x'(t)|.$$

(a) Show that $\|\cdot\|$ is indeed a norm.

(b) Show that $f(x) = x'\big((a+b)/2\big)$ defines a continuous linear functional on $C^1([a,b])$.

(c) Show that f defined above is *not* bounded on the subspace of $C([a,b])$ consisting of all continuously differentiable functions with the norm inherited from $C([a,b])$.

29. Define the operator T by the formula

$$T(f)(x) = \int_a^b K(x,y)\, f(y)\, dy.$$

Suppose that $K \in L^q([a, b] \times [a, b])$, where q lies in the range $1 \le q \le \infty$. Determine the values of p for which T is necessarily a bounded linear operator from $L^p(a, b)$ to $L^q(a, b)$. In particular, if a and b are both finite, show that $K \in L^\infty([a, b] \times [a, b])$ implies T to be bounded on all the L^p-spaces.

30. Suppose that X, Y, and Z are Banach spaces and that $T : X \times Y \to Z$ is bilinear. By T being *bilinear*, we mean that it is linear in each of its arguments separately; that is, $T(x, y)$ is linear in $x \in X$ for each fixed $y \in Y$, and also linear in $y \in Y$ for each fixed $x \in X$.

(a) If T is continuous, prove that there is a constant $M < \infty$ such that

$$\|T(x, y)\| \le M\|x\|\,\|y\| \quad \text{for all } x \in X, y \in Y.$$

In this case, we say that T is bounded. Is completeness needed here?

(b) Prove that T is continuous if and only if it is continuous at the origin $(0, 0)$.

(c) Prove that T is continuous if and only if it is bounded.

31. Suppose that X is a Banach space, M and N are linear subspaces, and that $X = M \oplus N$ is the direct sum of M and N, which means that

$$X = M + N = \{m + n : m \in M, n \in N\}$$

and $M \cap N = \{0\}$ is the trivial linear subspace consisting only of the zero element. Let P denote the projection of X onto M. That is, if $x = m + n$, then

$$P(x) = m.$$

Show that P is well-defined and linear. Prove that P is bounded if and only if both M and N are closed.

32. Let X be a Banach space with closed linear subspaces Y and Z such that $Y \cap Z = \{0\}$ and $X = Y + Z$ (i.e., $X = Y \oplus Z$). For $f \in Z^*$, show that $F : X \to \mathbb{F}$ is a well-defined linear extension of f, where

$$F(y + z) = f(z) \quad \forall y \in Y, \ z \in Z,$$

and that F is bounded on X.

33. Suppose X is a vector space. The *algebraic dual* of X is the vector space of all linear functionals on X. Suppose also that X is an NLS. Show that X has finite dimension if and only if the algebraic dual and the dual space X^* coincide.

34. Let $U = B_r(0) = \{x : \|x\| < r\}$ be an open ball about 0 in a real normed linear space, and let $y \notin \bar{U}$. Show that there is a bounded linear functional f that separates U from y. (That is, U and y lie in opposite

half spaces determined by f, which is to say there is an α such that U lies in $\{x : f(x) < \alpha\}$ and $f(y) > \alpha$.)

35. Prove that $L^2([0,1])$ is of the first category in $L^1([0,1])$. [Hint: show that $A_k = \{f \in L^1([0,1]) : \|f\|_{L^2} \le k\}$ is closed in L^1 but has empty interior.]

36. For each $\alpha \in \mathbb{R}$, let E_α be the set of all continuous functions f on $[-1,1]$ such that $f(0) = \alpha$. Show that the E_α are convex, and that each is dense in $L^2([-1,1])$.

37. For $1 \le p < \infty$, prove that the dual of ℓ^p is isomorphic and isometric to ℓ^q, where $1/p + 1/q = 1$, via the mapping $F : \ell^q \to (\ell^p)^*$ defined by

$$F(y)(x) = \sum_{n=1}^\infty y_n x_n \quad \text{for } x \in \ell^p, y \in \ell^q.$$

Showing that this map is well-defined is part of the exercise.

38. Prove that if X is a normed linear space and $B = B_1(0)$ is the unit ball, then X is infinite-dimensional if and only if B contains an infinite collection of non-overlapping balls of radius $1/4$.

39. Show that every finite-dimensional vector space is reflexive. In fact, show that the dual of $\mathbb{F}^d$ is isomorphic to $\mathbb{F}^d$.

40. If a Banach space X is reflexive, show that X^* is also reflexive. A much more difficult question: is the converse true? Give a proof. [Hint: first show that a closed subspace of a reflexive space is reflexive.]

41. Let $y = (y_1, y_2, y_3, \ldots) \in \mathbb{C}^\infty$ be a vector of complex numbers such that $\sum_{i=1}^\infty y_i x_i$ converges for every $x = (x_1, x_2, x_3, \ldots) \in c_0$, where $c_0 = \{x \in \mathbb{C}^\infty : x_i \to 0 \text{ as } i \to \infty\}$. Prove that

$$\sum_{i=1}^\infty |y_i| < \infty.$$

What does this result say about c_0^*? [Hint: c_0 is complete, so consider applying the uniform boundedness principle.]

42. Let f_0 be the space of sequences of complex numbers $x = \{x_i\}_{i=1}^\infty$, only finitely many of whose terms are nonzero. Define a norm on f_0 by $\|x\| = \sup_i |x_i|$. Let $T : f_0 \to f_0$ be defined by

$$y = Tx = \left(x_1, \frac{1}{2}x_2, \frac{1}{3}x_3, \ldots \right).$$

Show that T is a bounded linear map, but that T^{-1} is unbounded. Why does this *not* contradict the open mapping theorem?

43. Give an example of a function that is closed but *not* continuous.

44. Let X be a Banach space, Y an NLS, and $\{T_n\}_{n=1}^\infty \subset B(X,Y)$ be such that $\{T_n x\}_{n=1}^\infty$ is a Cauchy sequence in Y for each $x \in X$. Show that $\{\|T_n\|\}_{n=1}^\infty$ is bounded. If, in addition, Y is a Banach space, show that if we define T by $T_n x \to Tx$, then $T \in B(X,Y)$.

45. Show that $C([0,1])$ is not reflexive.

46. Let X be an NLS. If $\{x_n\}_{n=1}^\infty \subset X$ converges strongly to $x \in X$ ($x \to x$), show that it also converges weakly to x ($x_n \rightharpoonup x$), as $n \to \infty$. If $\{f_n\}_{n=1}^\infty \subset X^*$ converges weakly to f ($f_n \rightharpoonup f$), show that it also converges weak-$*$ to f ($f_n \xrightarrow{w^*} f$), as $n \to \infty$.

47. Prove that a bounded linear operator between NLSs is weakly sequentially continuous, i.e., prove Proposition 2.61. Is a weakly sequentially continuous linear operator necessarily bounded?

48. Let X be an NLS, $\{x_n\}_{n=1}^\infty \subset X$, and $\{f_n\}_{n=1}^\infty \subset X^*$. If $f_n \to f \in X^*$ and $x_n \rightharpoonup x \in X$, prove that $f_n(x_n) \to f(x)$.

49. Consider $X = C([a,b])$, the continuous functions defined on the real interval $[a,b]$ with the maximum norm. Let $\{f_n\}_{n=1}^\infty$ be a sequence in X and suppose that $f_n \rightharpoonup f$. Prove that $\{f_n\}_{n=1}^\infty$ is pointwise convergent. That is,

$$f_n(x) \to f(x) \quad \text{for all } x \in [a,b].$$

Prove that a weakly convergent sequence in $C^1([a,b])$, where a and b are finite, is norm-convergent in $C([a,b])$. A much more difficult question: is this still true when $[a,b]$ is replaced by $\mathbb{R}$?

50. Prove Corollary 2.63. Let X be a normed linear space and Y a closed linear subspace. Show that Y is weakly sequentially closed.

51. Let X be a normed linear space. We say that a sequence $\{x_n\}_{n=1}^\infty \subset X$ is *weakly Cauchy* if $\{Tx_n\}_{n=1}^\infty$ is Cauchy for all $T \in X^*$, and we say that X is *weakly complete* if each weak Cauchy sequence converges weakly. If X is reflexive, prove that X is weakly complete.

52. Let X be a Banach space and $T \in X^* = B(X, \mathbb{F})$. Identify the range of $T^* \in B(\mathbb{F}, X^*)$.

53. Show that if X is a Banach space, but not necessarily separable, $K \subset X^*$, and K is weak-$*$ compact, then K is weak-$*$ closed and bounded. [Hint: first show that, for any $x \in X$, the set $E_x(K) \subset \mathbb{F}$ is compact, and then invoke the uniform boundedness principle on K.]

54. Let X and Y be NLSs and let $T \in B(X,Y)$ be such that T^* is surjective.

(a) Show that T is injective.

(b) Suppose that T is also surjective. Show that T^{-1} is weakly sequentially continuous.

55. Let X be a vector space and let W be a vector space of linear functionals on X. Suppose that W separates points of X, meaning that for any $x, y \in X$, $x \neq y$, there exists $w \in W$ such that $w(x) \neq w(y)$. Let X be endowed with the smallest topology such that each $w \in W$ is continuous (we call this the W-weak topology of X).

(a) Describe a W-weak open set about 0.

(b) Let T_i, $i = 1, 2, \ldots, n$, and L be linear functionals on X with null spaces $N(T_i)$ and $N(L)$. Suppose that $K = \bigcap_{i=1}^{n} N(T_i) \subset N(L)$. Prove that L is a linear combination of the T_i. [Hint: consider $\pi : X \to \mathbb{F}^n$ defined by $\pi(x) = (T_1(x), \ldots, T_n(x))$ and relate L to π.]

(c) Prove that if L is a W-weakly continuous linear functional on X, then $L \in W$. [Hint: consider the inverse image of $B_1(0) \subset \mathbb{F}$, which must contain a W-weak open set about 0.]

(d) If X is an NLS, characterize the set of weak-$*$ continuous linear functionals on X^*.

56. Let $a = (a_1, a_2, \ldots) \in \ell^{3/2}$ and let $A : \ell^3 \to \ell^1$ be the bounded linear operator defined for $x = (x_1, x_2, \ldots) \in \ell^3$ by $A(x) = y$, where $y_n = a_n x_n$. Find $A^*(z)$ for $z \in \ell^\infty$.

CHAPTER 3

Hilbert Spaces

The norm of a normed linear space provides a notion of size for the elements of the space. While linear spaces equipped with norms in which they are complete are mathematically interesting and often practically useful, they lack an important part of the geometric structure that we are used to from our experience with ordinary Euclidean space. In this chapter, a notion of "angle" between two vectors in an NLS will be added. In particular, a notion of orthogonality will be introduced through the device known as an inner-product.

The material in this chapter emanates from the work of David Hilbert, a German mathematician who was active in the late 19th and early 20th centuries. Especially his work on integral equations in the first decade of the 20th, stemming from Fredholm's work on integral equations with the help of Erhard Schmidt (his student) and Frigyes Riesz, led to the work that figures here. The term 'Hilbert space' was coined by John von Neumann, who was briefly an assistant to Hilbert. It should be noted that Hilbert made many, many significant contributions to mathematics. His work in what was to become known as functional analysis is just a small part of his total corpus.

3.1 BASIC PROPERTIES OF INNER-PRODUCTS

Definition. An *inner-product* on a vector space H is a map $(\cdot, \cdot) : H \times H \to \mathbb{F}$ satisfying the following properties:

(a) The map $(\cdot, \cdot)$ is linear in its first argument; that is, for $\alpha, \beta \in \mathbb{F}$ and $x, y, z \in H$,
$$(\alpha x + \beta y, z) = \alpha(x, z) + \beta(y, z).$$

(b) The map $(\cdot, \cdot)$ is *conjugate symmetric* (or just *symmetric* if $\mathbb{F} = \mathbb{R}$), meaning that for $x, y \in H$,
$$(x, y) = \overline{(y, x)}.$$

DOI: 10.1201/9781003492139-3

(c) The map $(\cdot, \cdot)$ is *positive definite*, meaning that for any $x \in H$, $(x, x) \geq 0$ and $(x, x) = 0$ if and only if $x = 0$.

If H has such an inner-product, then H is called an *inner-product space* (IPS) or a *pre-Hilbert space*. Any map satisfying (a) and (b) is said to be a *Hermitian form* (or *bilinear*, if $\mathbb{F} = \mathbb{R}$). An inner-product on a vector space H is usually denoted $(\cdot, \cdot)_H$ or $\langle \cdot, \cdot \rangle_H$. If the space H is understood from context, then the subscript is dropped and its inner-product denoted simply by $(\cdot, \cdot)$ or $\langle \cdot, \cdot \rangle$.

Proposition 3.1. *If $(\cdot, \cdot)$ is Hermitian on H, then for $\alpha, \beta \in \mathbb{F}$ and x, y, $z \in H$,*

$$(x, \alpha y + \beta z) = \overline{\alpha}(x, y) + \overline{\beta}(x, z). \tag{3.1}$$

Any mapping $(\cdot, \cdot) : H \times H \to \mathbb{F}$ satisfying (3.1) is said to be *conjugate linear* in its second argument. Moreover, if $(\cdot, \cdot)$ is also linear in its first argument, the form is said to be *sesquilinear*.

Examples.

(1) $\mathbb{F}^d$ (i.e., $\mathbb{C}^d$ or $\mathbb{R}^d$) is an IPS with the inner-product

$$(x, y) = x \cdot \bar{y} = \sum_{i=1}^{d} x_i \bar{y}_i \quad \forall x, y \in \mathbb{F}^d.$$

(2) Similarly ℓ^2 is an IPS with

$$(x, y) = \sum_{i=1}^{\infty} x_i \bar{y}_i \quad \forall x, y \in \ell^2.$$

The Hölder inequality shows that this quantity is finite.

(3) For any measurable set $\Omega \subset \mathbb{R}^d$, $L^2(\Omega)$ has the inner-product

$$(f, g) = \int_{\Omega} f(x) \overline{g(x)} \, dx \quad \forall f, g \in L^2(\Omega).$$

Definition. If $(H, (\cdot, \cdot))$ is an IPS, define the map $\| \cdot \| : H \to \mathbb{R}$ by

$$\|x\| = (x, x)^{1/2}$$

for any $x \in H$. This map is called the *induced norm*.

As the notation suggests, $\| \cdot \|$ is a norm on H. This is shown in the next two results.

Lemma 3.2 (Cauchy-Bunyakovsky-Schwarz inequality, Cauchy-Schwarz inequality). *If $(H, (\cdot, \cdot))$ is an IPS with induced norm $\|\cdot\|$, then for any $x, y \in H$,*

$$|(x, y)| \leq \|x\| \, \|y\|,$$

with equality holding if and only if x or y is a multiple of the other.

Proof. Let $x, y \in H$. If $y = 0$, there is nothing to prove, so assume $y \neq 0$. Then for any $\lambda \in \mathbb{F}$,

$$
\begin{aligned}
0 \leq \|x - \lambda y\|^2 &= (x - \lambda y, x - \lambda y) \\
&= (x, x) - \overline{\lambda}(x, y) - \lambda(y, x) + |\lambda|^2(y, y) \\
&= \|x\|^2 - \left(\overline{\lambda(y, x)} + \lambda(y, x) \right) + |\lambda|^2 \|y\|^2 \\
&= \|x\|^2 - 2 \operatorname{Real}\{\lambda(y, x)\} + |\lambda|^2 \|y\|^2.
\end{aligned}
$$

Choose λ to be

$$
\lambda = \frac{(x, y)}{\|y\|^2}.
$$

This is well-defined since $y \neq 0$, and so

$$
0 \leq \|x\|^2 - 2 \operatorname{Real}\left\{ \frac{(x, y)(y, x)}{\|y\|^2} \right\} + \frac{|(x, y)|^2}{\|y\|^4} \|y\|^2 = \|x\|^2 - \frac{|(x, y)|^2}{\|y\|^2},
$$

since $(x, y)(y, x) = |(x, y)|^2$ is real. A rearrangement gives the result, with equality only if $x - \lambda y = 0$. $\qquad \square$

Corollary 3.3. *The induced norm is indeed a norm, and thus an IPS is an NLS.*

Proof. For $\alpha \in \mathbb{F}$ and $x \in H$, $\|x\| \geq 0$, $\|\alpha x\| = (\alpha x, \alpha x)^{1/2} = |\alpha|(x, x)^{1/2} = |\alpha| \|x\|$ and $\|x\| = 0$ if and only if $(x, x) = 0$ if and only if $x = 0$.

It remains only to demonstrate the triangle inequality. For $x, y \in H$,

$$
\begin{aligned}
\|x + y\|^2 &= (x + y, x + y) \\
&= \|x\|^2 + 2 \operatorname{Real}(x, y) + \|y\|^2 \\
&\leq \|x\|^2 + 2|(x, y)| + \|y\|^2 \\
&\leq \|x\|^2 + 2\|x\| \|y\| + \|y\|^2 \\
&= (\|x\| + \|y\|)^2. \qquad \square
\end{aligned}
$$

It is an interesting and occasionally useful fact that in an IPS, the norm determines the inner-product. For any x, y in an IPS, calculate thusly:

$$
\|x + y\|^2 = \|x\|^2 + \|y\|^2 + (x, y) + (y, x) = \|x\|^2 + \|y\|^2 + 2 \operatorname{Real}(x, y).
$$

If $\mathbb{F} = \mathbb{C}$ (the argument simplifies if $\mathbb{F} = \mathbb{R}$), then

$$
\begin{aligned}
\|x + iy\|^2 &= \|x\|^2 + \|y\|^2 + (x, iy) + (iy, x) \\
&= \|x\|^2 + \|y\|^2 - i(x, y) + i(y, x) = \|x\|^2 + \|y\|^2 + 2 \operatorname{Imag}(x, y),
\end{aligned}
$$

and the following result emerges.

Proposition 3.4. *For any* x, y *in an IPS,*

$$(x,y) = \frac{1}{2}\{\|x+y\|^2 - \|x\|^2 - \|y\|^2\} + \frac{i}{2}\{\|x+iy\|^2 - \|x\|^2 - \|y\|^2\},$$

where the imaginary term on the right is omitted if $\mathbb{F} = \mathbb{R}$.

An inner-product along with the Cauchy-Bunyakovsky-Schwarz inequality can be used to provide a notion of angle. If the underlying field is real, then naturally the angle θ between x and y is defined by

$$\cos\theta = \frac{(x,y)}{\|x\|\,\|y\|},$$

just as in Euclidean space. This makes sense because of the Cauchy-Bunyakovsky-Schwartz inequality, but of course has the usual multi-valued aspect associated with the function $\cos^{-1}(\theta) = \arccos(\theta)$. The case where $\theta = \pi/2$ is of the most importance, and generalizes to complex IPSs.

Definition. If $(H, (\cdot, \cdot))$ is an IPS, $x, y \in H$ and $(x, y) = 0$, then we say that x and y are *orthogonal*, and denote this state of affairs symbolically by $x \perp y$. In a similar way, when $A, B \subset H$, x and B are orthogonal $(x \perp B)$ if $(x, y) = 0$ for all $y \in B$ and A and B are orthogonal $(A \perp B)$ if $(x, y) = 0$ for all $x \in A$ and $y \in B$.

Proposition 3.5 (parallelogram and Pythagorean laws). *If* $x, y \in H$, *an IPS, then*

$$\|x+y\|^2 + \|x-y\|^2 = 2(\|x\|^2 + \|y\|^2) \quad (parallelogram\ law).$$

Moreover, if $x \perp y$, *then*

$$\|x+y\|^2 = \|x\|^2 + \|y\|^2 \quad (Pythagorean\ law),$$

and the converse holds when $\mathbb{F} = \mathbb{R}$.

The proof is left as an exercise for the reader. The first law expresses the geometry of a parallelogram in $\mathbb{R}^2$, generalized to an arbitrary IPS. A consequence of the parallelogram law is that not all norms are generated by an inner-product, as there are norms that do not obey this rule.

Lemma 3.6. *If* $(H, \langle \cdot, \cdot \rangle)$ *is an IPS, then* $\langle \cdot, \cdot \rangle : H \times H \to \mathbb{F}$ *is continuous.*

Proof. Since $H \times H$ is a metric space, it is enough to show sequential continuity. So suppose that the points $(x_n, y_n) \to (x, y)$ in $H \times H$, or what is the same, $x_n \to x$ and $y_n \to y$ in H. The following sequence of inequalities comes to our rescue:

$$|\langle x_n, y_n \rangle - \langle x, y \rangle| = |\langle x_n, y_n \rangle - \langle x_n, y \rangle + \langle x_n, y \rangle - \langle x, y \rangle|$$
$$\leq |\langle x_n, y_n \rangle - \langle x_n, y \rangle| + |\langle x_n, y \rangle - \langle x, y \rangle|$$
$$= |\langle x_n, y_n - y \rangle| + |\langle x_n - x, y \rangle|$$
$$\leq \|x_n\|\,\|y_n - y\| + \|x_n - x\|\,\|y\|.$$

Since $x_n \to x$, the real sequence $\{\|x_n\|\}_{n=1}^{\infty}$ is bounded and the advertised conclusion follows. □

Corollary 3.7. *If $\lambda_n \to \lambda$ and $\mu_n \to \mu$ in $\mathbb{F}$ and $x_n \to x$ and $y_n \to y$ in H, then*

$$\langle \lambda_n x_n, \mu_n y_n \rangle \to \langle \lambda x, \mu y \rangle.$$

Proof. Just note that $\lambda_n x_n \to \lambda x$ and $\mu_n y_n \to \mu y$. □

Definition. A complete IPS H is called a *Hilbert space*.

Hilbert spaces are thus Banach spaces with additional structure.

3.2 BEST APPROXIMATION AND ORTHOGONAL PROJECTION

The following is an important geometric concept that obtains in an IPS.

Theorem 3.8 (best approximation theorem). *Suppose $(H, (\cdot, \cdot))$ is an IPS and $M \subset H$ is nonempty, convex, and complete (e.g., closed if H is a Hilbert space). If $x \in H$, then there is a unique $y = y(x) \in M$ such that*

$$\operatorname{dist}(x, M) \equiv \inf_{z \in M} \|x - z\| = \|x - y\|.$$

We call y the best approximation *of or* closest point *to x from M.*

Proof. Let

$$\delta = \operatorname{dist}(x, M) = \inf_{z \in M} \|x - z\|.$$

Then there is a sequence $\{y_n\}_{n=1}^{\infty} \subset M$ such that as $n \to \infty$,

$$\|x - y_n\| \equiv \delta_n \to \delta.$$

We claim that $\{y_n\}_{n=1}^{\infty}$ is Cauchy. By the parallelogram law,

$$
\begin{aligned}
\|y_n - y_m\|^2 &= \|(y_n - x) + (x - y_m)\|^2 \\
&= 2 \left(\|y_n - x\|^2 + \|x - y_m\|^2 \right) - \|y_n + y_m - 2x\|^2 \\
&= 2(\delta_n^2 + \delta_m^2) - 4 \left\| \frac{y_n + y_m}{2} - x \right\|^2 \\
&\leq 2(\delta_n^2 + \delta_m^2) - 4\delta^2,
\end{aligned}
$$

since by convexity $(y_n + y_m)/2 \in M$. Thus as $n, m \to \infty$, $\|y_n - y_m\| \to 0$. By completeness, $y_n \to y$ for some $y \in M$. Since $\|\cdot\|$ is continuous, $\|x - y\| = \delta$.

To see that y is unique, suppose that for some $z \in M$, $\|x - z\| = \delta$. Then using the parallelogram law again reveals that

$$
\begin{aligned}
\|y - z\|^2 &= \|(y - x) + (x - z)\|^2 \\
&= 2 \left(\|y - x\|^2 + \|x - z\|^2 \right) - \|y + z - 2x\|^2 \\
&= 4\delta^2 - 4 \left\| \frac{y + z}{2} - x \right\|^2 \\
&\leq 4\delta^2 - 4\delta^2 = 0.
\end{aligned}
$$

Thus $y = z$. □

Corollary 3.9. *Suppose* $(H, (\cdot, \cdot))$ *is an IPS and* M *is a complete linear subspace. If* $x \in H$ *and* $y \in M$ *is the best approximation to* x *in* M, *then*

$$x - y \perp M.$$

Proof. Let $m \in M$, $m \neq 0$. Since y is the best approximation of x in M, for any $\lambda \in \mathbb{F}$,

$$\|x - y\|^2 \leq \|x - y + \lambda m\|^2 = \|x - y\|^2 + \overline{\lambda}(x - y, m) + \lambda(m, x - y) + |\lambda|^2 \|m\|^2.$$

If λ is taken to be $\lambda = -(x - y, m)/\|m\|^2$, then the last inequality implies

$$0 \leq -\overline{\lambda}\lambda\|m\|^2 - \lambda\overline{\lambda}\|m\|^2 + |\lambda|^2\|m\|^2 = -|\lambda|^2\|m\|^2,$$

so $\lambda = 0$, which means

$$(x - y, m) = 0.$$

As m was an arbitrary element of M, it follows that $x - y \perp M$. $\qquad\square$

Definition. Given an IPS H and $M \subset H$,

$$M^{\perp} = \{x \in H : (x, m) = 0, \ \forall m \in M\}.$$

The space $M^{\perp}$ is called "M-perp." When M is a subspace of H, $M^{\perp}$ is also called the *orthogonal complement* of M.

Proposition 3.10. *Suppose* H *is an IPS and* $M \subset H$. *Then* $M^{\perp}$ *is a closed linear subspace of* H, $M \perp M^{\perp}$, *and* $M \cap M^{\perp}$ *is either* $\{0\}$ *or* $\emptyset$.

Proof. Closure follows easily from the continuity of the inner-product. The rest of the assertion is also straightforward to demonstrate. $\qquad\square$

If X is an NLS and $P : X \to X$ satisfies $P^2 = P$, then P is called an *idempotent operator*. If P is linear and idempotent, then it is called a *projection operator*. General properties of a projection operator appear in the following theorem, the proof of which is left to the exercises at the end of the chapter. The results are illustrated in Figure 3.1.

Theorem 3.11. *Suppose that* X *is an NLS and* $P : X \to X$ *is a projection operator mapping onto* $M \subset X$ *with kernel or null space* N. *Then* P *is the identity on* M. *Let* $Q : X \to X$ *be defined by* $Q = I - P$. *Then* Q *is a projection operator, maps onto* N, *has null space* M, *and* $QP = PQ = 0$. *Moreover,* X *is the (internal) direct sum of the linear subspaces* M *and* N *(denoted* $X = M \oplus N$ *and meaning* $X = M + N$ *and* $M \cap N = \{0\}$). *If* P *is bounded and* $M \neq \{0\}$, *then* $\|P\| \geq 1$.

In an IPS, the best approximation defines a projection operator. The best approximation operator is an *orthogonal projection* in light of the next result.

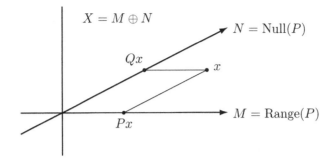

$X = M \oplus N$

$N = \mathrm{Null}(P)$

Qx

x

$M = \mathrm{Range}(P)$

Px

Figure 3.1 An illustration of projection operators P and $Q = I - P$ on $X = M \oplus N$.

Theorem 3.12. *Suppose $(H, (\cdot, \cdot))$ is an IPS and $M \subset H$ is a complete linear subspace. Then there are two unique, bounded, surjective projection operators*

$$P : H \to M \quad and \quad P^\perp : H \to M^\perp$$

defined for any $x \in H$ by

(a) *$Px = z$ where $\|x - z\| = \inf_{y \in M} \|x - y\|$ (Px is the best approximation to x in M),*

(b) *$x = Px + P^\perp x$ $(P^\perp = I - P)$.*

The operator P, applied to $x \in H$, is characterized by

(c) *$y = Px$ if and only if $y \in M$ satisfies $(x - y, m) = 0$ for all $m \in M$, i.e., $x - y \in M^\perp$.*

Moreover,

(d) *$(Px, P^\perp x) = 0$ for all $x \in H$,*

(e) *$\|x\|^2 = \|Px\|^2 + \|P^\perp x\|^2$,*

(f) *$\|P\| = 1$ unless $M = \{0\}$, and $\|P^\perp\| = 1$ unless $M = H$.*

Proof. By the best approximation theorem, (a) defines the map P uniquely. The proof of the best approximation theorem shows that if $x \in M$, then $Px = x$. Consequently, P is an idempotent operator mapping onto M. Then (b) defines $P^\perp : H \to H$ uniquely. If $x \in H$, then for $m \in M$,

$$(P^\perp x, m) = (x - Px, m) = 0,$$

by Corollary 3.9. It follows that $(P^\perp x, Px) = 0$ for all $x \in H$ (which is (d)), the image of $P^\perp$ is contained in $M^\perp$, and $Px = 0$ for any $x \in M^\perp$. To see that $P^\perp$ maps onto $M^\perp$, consider $x \in M^\perp$ and note that $x = Px + (I-P)x = P^\perp x$.

To show that P and $P^\perp$ are linear, let $\alpha, \beta \in \mathbb{F}$ and $x, y \in H$. Then by (b),

$$\alpha x + \beta y = P(\alpha x + \beta y) + P^\perp(\alpha x + \beta y),$$

and

$$\alpha x + \beta y = \alpha(Px + P^\perp x) + \beta(Py + P^\perp y)$$
$$= \alpha Px + \beta Py + \alpha P^\perp x + \beta P^\perp y.$$

Thus, it is deduced that

$$\alpha Px + \beta Py - P(\alpha x + \beta y) = P^\perp(\alpha x + \beta y) - \alpha P^\perp x - \beta P^\perp y.$$

Since M and $M^\perp$ are vector spaces, the left side above is in M and the right side is in $M^\perp$. So, both sides are in $M \cap M^\perp = \{0\}$, whence

$$P(\alpha x + \beta y) = \alpha Px + \beta Py,$$
$$P^\perp(\alpha x + \beta y) = \alpha P^\perp x + \beta P^\perp y,$$

which is to say, P and $P^\perp$ are linear. We conclude that P is a projection, and hence also $P^\perp$ by Theorem 3.11.

For (c), Corollary 3.9 gives the forward implication. For the converse, note that for all $x \in H$ and $y, m \in M$,

$$\|x - (y - m)\|^2 = ((x - y) + m, (x - y) + m) = \|x - y\|^2 + \|m\|^2,$$

which is minimized for $m = 0$. Thus $y = Px$ is the best approximation.

By (d), the property (e) is just the Pythagorean law from Proposition 3.5. Now (e) implies that

$$\|Px\|^2 = \|x\|^2 - \|P^\perp x\|^2 \le \|x\|^2,$$

so $\|P\| \le 1$. But $\|P\| \ge 1$ for any nonzero projection, by Theorem 3.11, so $\|P\| = 1$ in this case. Similar remarks apply to $P^\perp$. We conclude that P and $P^\perp$ are bounded and (f) holds. □

As a corollary to the previous two theorems, we have the following additional properties.

Corollary 3.13. *The projection operators P and $P^\perp$ also have the properties*

(g) $PP^\perp = P^\perp P = 0$, $P^2 = P$, and $(P^\perp)^2 = P^\perp$,

(h) $x \in M$ if and only if $P^\perp x = 0$ if and only if $x = Px$,

(i) $x \in M^\perp$ if and only if $Px = 0$ if and only if $x = P^\perp x$.

Moreover, H is decomposed into the (orthogonal) direct sum of M and $M^\perp$, i.e., $H = M \oplus M^\perp$.

Corollary 3.14. *Let $(H, (\cdot, \cdot))$ be a Hilbert space and $M \subset H$ a closed linear subspace. Let P_M be the best approximation of H in M. Then, it follows that $P_M^\perp$ is the best approximation of H in $M^\perp$, i.e., $P_M^\perp = P_{M^\perp}$.*

Proof. There are unique operators P_M and $(P_M)^\perp$ satisfying the conclusions of the last theorem. Now $M^\perp$ is closed, so we can also apply the last theorem to the subspace $M^\perp$ to obtain unique orthogonal projections $P_{M^\perp}$ and $(P_{M^\perp})^\perp$. It is not difficult to conclude that $P_{M^\perp} = (P_M)^\perp$ using (c) of Theorem 3.12, which is to say, $P_M^\perp$ is the best approximation of H in $M^\perp$. □

3.3 DUALITY IN HILBERT SPACES

The discussion is now turned to consideration of the dual H^* of a Hilbert space $(H, (\cdot, \cdot))$. First observe that if $y \in H$, then the functional L_y defined by

$$L_y(x) = (x, y)$$

is linear in x and bounded because of the Cauchy-Bunyakovsky-Schwarz inequality. In fact,

$$|L_y(x)| \leq \|y\| \, \|x\|,$$

so $\|L_y\| \leq \|y\|$. But $|L_y(y/\|y\|)| = \|y\|$, so in fact

$$\|L_y\| = \|y\|.$$

It is concluded that $L_y \in H^*$, and, as y is arbitrary,

$$\{L_y\}_{y \in H} \subset H^*.$$

It will transpire that the collection $\{L_y\}_{y \in H}$ exhausts H^*. By identifying L_y with y, H can then be seen to be its own dual.

Theorem 3.15 (Riesz representation theorem). *If $(H, (\cdot, \cdot))$ is a Hilbert space and $L \in H^*$, then there is a unique $y \in H$ such that*

$$Lx = L_y x = (x, y) \quad \forall x \in H.$$

Moreover, $\|L\|_{H^*} = \|y\|_H$.

Proof. The question of uniqueness is easily settled. If

$$Lx = (x, y_1) = (x, y_2) \quad \forall x \in H,$$

then

$$(x, y_1 - y_2) = 0 \quad \forall x \in H.$$

Substitute $x = y_1 - y_2$ to conclude $y_1 = y_2$.

If $L \equiv 0$ (i.e., $Lx = 0$ for all $x \in H$), then take $y = 0$. Suppose then that $L \not\equiv 0$. Let

$$M = N(L) \equiv \ker(L) \equiv \{x \in H : Lx = 0\}.$$

Since M is the inverse image of the closed set $\{0\}$ under the continuous function L, M is closed. The null space $N(L)$ of a linear map is always a linear subspace of the domain. We conclude that M is a closed, hence complete, linear subspace of the Hilbert space H. Since L vanishes on M, if $y \in H$ is to exist with the desired properties, then it must lie in $M^\perp$.

Because $L \not\equiv 0$, infer that $M \neq H$ and $M^\perp \neq \{0\}$ by Theorem 3.12. Although it is perhaps not obvious, we will in effect show that the dimension of $M^\perp$ is one. Take *any* $z \in M^\perp \setminus \{0\}$, normalized so $\|z\| = 1$. For $x \in H$, define

$$u = (Lx)z - (Lz)x.$$

Then, of course,

$$Lu = (Lx)(Lz) - (Lz)(Lx) = 0,$$

so $u \in M$ and therefore $u \perp z$. This means that

$$0 = (u, z) = \big((Lx)z - (Lz)x, z\big) = Lx(z, z) - Lz(x, z) = Lx - Lz(x, z),$$

or, what is the same,

$$Lx = Lz(x, z) = (x, (\overline{Lz})z).$$

Thus, $L = L_y$ with $y = (\overline{Lz})z$.

Finally, we saw already that $\|L\| = \|L_y\| = \|y\|$. □

Define a map $R : H \to H^*$, called the *Riesz map*, by

$$Rx = L_x \quad \forall x \in H.$$

The Riesz representation theorem may be interpreted as saying that R is well-defined and comprises a one-to-one, surjective isometry. As suggested earlier, the dual H^* is identified with H via R. Given $x \in H$, there is a unique $Rx = L_x \in H^*$ and $\|Rx\|_{H^*} = \|x\|_H$, and conversely given $L \in H^*$, there is a unique $x = R^{-1}L \in H$ such that $L = L_x$ and $\|R^{-1}L\|_H = \|L\|_{H^*}$. The Riesz map R is linear when $\mathbb{F} = \mathbb{R}$, but it is conjugate linear when $\mathbb{F} = \mathbb{C}$. That is to say,

$$R(x + y) = Rx + Ry \quad \forall x, y \in H,$$
$$R(\lambda x) = \overline{\lambda} Rx \quad \forall x \in H, \lambda \in \mathbb{F}.$$

If X and Y are Hilbert spaces and $T : X \to Y$ is a bounded linear map, then the dual operator T^* maps Y^* to X^*. Since Y^* and X^* can be identified with Y and X, respectively, we may view T^* as taking Y to X. To be precise, let R_Y and R_X be the Riesz maps for Y and X, respectively. Then the map $T' = R_X^{-1} T^* R_Y : Y \to X$ is a well-defined bounded linear map for which $\|T'\| = \|T^*\|$. This operator has the property that for all $x \in X$ and $y \in Y$,

$$(x, T'y)_X = (x, R_X^{-1} T^* R_Y y)_X = T^* R_Y y(x) = R_Y y(Tx) = (Tx, y)_Y.$$

We call T' the *Hilbert adjoint* of T. Frequently, this operator is used in place of T^*, and it is simply called the dual operator, when no confusion would arise. The domain of the operator determines which one of T^* and T' is intended.

Proposition 3.16. *Let X and Y be Hilbert spaces and $T \in B(X, Y)$. Then (the Hilbert adjoint) $T^* \in B(Y, X)$, $T = T^{**}$ and*

$$(Tx, y)_Y = (x, T^*y)_X \quad and \quad (T^*y, x)_X = (y, Tx)_Y \quad \forall\, x \in X,\ y \in Y.$$

Proof. Since the usual adjoint is bounded and linear and the Riesz map and its inverse are bounded and conjugate linear, the Hilbert adjoint is bounded and linear. We verified above the formula $(Tx, y)_Y = (x, T^*y)_X$ for all $x \in X$, $y \in Y$. The other formula follows upon taking complex conjugates. Iterating, we have $(y, Tx)_Y = (T^*y, x)_X = (y, T^{**}x)_Y$ for all $y \in Y$, implying that $Tx = T^{**}x$ for all $x \in X$, i.e., $T = T^{**}$. $\square$

3.4 ORTHONORMAL SUBSETS AND BASES

In finite dimensions, a vector space V over $\mathbb{F}$ is isomorphic to $\mathbb{F}^d$ for some natural number d and the isomorphism can be described by way of an orthogonal basis for V. Similar results hold for infinite-dimensional Hilbert spaces.

Definition. Suppose H is an IPS and $\mathcal{I}$ is some index set. A set $A = \{x_\alpha\}_{\alpha \in \mathcal{I}} \subset H$ is said to be *orthogonal* if $x_\alpha \neq 0$ for all $\alpha \in \mathcal{I}$ and

$$x_\alpha \perp x_\beta \quad (\text{i.e., } (x_\alpha, x_\beta) = 0)$$

for all $\alpha, \beta \in \mathcal{I}$, $\alpha \neq \beta$. If it is also the case that $\|x_\alpha\| = 1$ for all $\alpha \in \mathcal{I}$, then A is said to be *orthonormal* (frequently abbreviated as ON).

Definition. If $A \subset X$, a vector space, then A is *linearly independent* if every finite subset of A is linearly independent. That is, every collection $\{x_i\}_{i=1}^n \subset A$ must satisfy the property that if there are scalars $c_i \in \mathbb{F}$ with

$$\sum_{i=1}^n c_i x_i = 0 \tag{3.2}$$

then necessarily $c_i = 0$ for $1 \leq i \leq n$.

Proposition 3.17. *If a subset A of a Hilbert space H is orthogonal, then A is linearly independent.*

Proof. If $\{x_i\}_{i=1}^n \subset A$ and $c_i \in \mathbb{F}$ satisfy (3.2), then for $1 \leq j \leq n$,

$$0 = \left(\sum_{i=1}^n c_i x_i, x_j \right) = \sum_{i=1}^n c_i(x_i, x_j) = c_j \|x_j\|^2.$$

As $x_j \neq 0$, necessarily each $c_j = 0$. $\square$

Suppose $\{x_1, \ldots, x_n\}$ are linearly independent elements of a Hilbert space H and let

$$M = \operatorname{span}\{x_1, \ldots, x_n\}.$$

Then, M is closed in H as it is finite-dimensional. Let's compute the orthogonal projection $P_M x$ of $x \in H$ onto M. As $P_M x \in M$, there are constants $c_1, \ldots, c_n \in \mathbb{F}$ such that $P_M x = \sum_{j=1}^{n} c_j x_j$ and of course $P_M x - x \perp M$ by Corollary 3.9. The latter is equivalent to the condition

$$(P_M x, x_i) = (x, x_i)$$

for $1 \leq i \leq n$. But,

$$(P_M x, x_i) = \sum_{j=1}^{n} c_j (x_j, x_i),$$

so defining

$$a_{ij} = (x_i, x_j) \quad \text{and} \quad b_i = (x, x_i)$$

for $1 \leq i, j \leq n$, we have that the $n \times n$ matrix $A = (a_{ij})$ and n-vectors $b = (b_i)$ and $c = (c_j)$ satisfy

$$Ac = b.$$

We already know that a unique solution c exists to this equation, so A is invertible and the solution c can be found as $A^{-1}b$, thereby providing $P_M x$.

Theorem 3.18. *Suppose H is a Hilbert space and $\{u_1, \ldots, u_n\} \subset H$ is ON. Let $x \in H$. Then the orthogonal projection of x onto $M = \operatorname{span}\{u_1, \ldots, u_n\}$ is given by*

$$P_M x = \sum_{i=1}^{n} (x, u_i) u_i.$$

Moreover,

$$\|P_M x\|^2 = \sum_{i=1}^{n} |(x, u_i)|^2 \leq \|x\|^2.$$

Proof. In this case, the matrix $A = ((u_i, u_j)) = I$, the $n \times n$ identity matrix, so the coefficients c are the values $b = ((x, u_i))$. The final remark follows from the fact that $\|P_M x\| \leq \|x\|$ and the calculation

$$\|P_M x\|^2 = \sum_{i=1}^{n} |(x, u_i)|^2,$$

left to the reader. $\qquad\qquad\qquad\qquad\qquad\qquad\qquad\qquad\qquad\qquad\qquad$ □

This result can be extended to larger ON sets. To do so, we need to note a few facts about infinite series. Let $\mathcal{I}$ be any index set (possibly uncountable!)

and $\{x_\alpha\}_{\alpha \in \mathcal{I}}$ a collection of nonnegative real numbers. Define the sum of $\{x_\alpha\}_{\alpha \in \mathcal{I}}$ by

$$\sum_{\alpha \in \mathcal{I}} x_\alpha = \sup_{\substack{\mathcal{J} \subset \mathcal{I} \\ \mathcal{J} \text{ finite}}} \sum_{\beta \in \mathcal{J}} x_\beta.$$

If $\mathcal{I} = \mathbb{N} = \{0, 1, 2, \ldots\}$ is countable, this agrees with the usual definition

$$\sum_{\alpha=0}^{\infty} x_\alpha = \lim_{n \to \infty} \sum_{\alpha=0}^{n} x_\alpha.$$

The reader will have no trouble showing that if

$$\sum_{\alpha \in \mathcal{I}} x_\alpha < \infty,$$

then at most countably many x_α are nonzero. If $\{x_\alpha\}_{\alpha \in \mathcal{I}}$ is a collection of real numbers, we need to separate the sum into positive and negative parts, viz. $x_\alpha = x_\alpha^+ - x_\alpha^-$, where $x_\alpha^\pm \geq 0$. Then just define

$$\sum_{\alpha \in \mathcal{I}} x_\alpha = \sup_{\substack{\mathcal{J} \subset \mathcal{I} \\ \mathcal{J} \text{ finite}}} \sum_{\alpha \in \mathcal{J}} x_\alpha^+ - \sup_{\substack{\mathcal{J} \subset \mathcal{I} \\ \mathcal{J} \text{ finite}}} \sum_{\alpha \in \mathcal{J}} x_\alpha^-,$$

provided at least one supremum is finite. For complex numbers, separate real and imaginary parts before applying the definition above, although now we would require all four sums to be finite.

Theorem 3.19 (Bessel's inequality). *Let H be a Hilbert space and $\{u_\alpha\}_{\alpha \in \mathcal{I}} \subset H$ an ON set. For $x \in H$,*

$$\sum_{\alpha \in \mathcal{I}} |(x, u_\alpha)|^2 \leq \|x\|^2.$$

Proof. By the previous theorem, for any finite $\mathcal{J} \subset \mathcal{I}$,

$$\sum_{\alpha \in \mathcal{J}} |(x, u_\alpha)|^2 \leq \|x\|^2,$$

so the same is true of the supremum. □

Corollary 3.20. *At most countably many of the (x, u_α) are nonzero.*

In a sense to be made precise below in the Riesz-Fischer theorem (Theorem 3.23), $x \in H$ can be associated to its coefficients $\{(x, u_\alpha)\}_{\alpha \in \mathcal{I}}$, where $\mathcal{I}$ is some index set. The subtlety is that $\mathcal{I}$ may be uncountable. We define a space of coefficients next.

Definition. Let $\mathcal{I}$ be a set. Denote by $\ell^2(\mathcal{I})$ the set

$$\ell^2(\mathcal{I}) = \left\{ f : \mathcal{I} \to \mathbb{F} \;\Big|\; \sum_{\alpha \in \mathcal{I}} |f(\alpha)|^2 < \infty \right\}.$$

If $\mathcal{I} = \mathbb{N}$, we have the usual space ℓ^2, which is a Hilbert space. In general, we have an inner-product on $\ell^2(\mathcal{I})$ given by

$$(f, g) = \sum_{\alpha \in \mathcal{I}} f(\alpha) \, \overline{g(\alpha)},$$

as the reader can verify. Moreover, $\ell^2(\mathcal{I})$ is complete in the associated norm topology.

Lemma 3.21. *Let H be a Hilbert space and $\{u_\alpha\}_{\alpha \in \mathcal{I}}$ any ON set in H. Define the mapping $F : H \to \ell^2(\mathcal{I})$ by $F(x) = f_x$, where*

$$f_x(\alpha) = x_\alpha \equiv (x, u_\alpha) \quad \text{for } \alpha \in \mathcal{I}.$$

Then F is a surjective, bounded linear map.

The mapping F defined above is called the *Riesz-Fischer map*.

Proof. Denoting the map f_x by $\{x_\alpha\}_{\alpha \in \mathcal{I}}$ (think of it as a long vector), the mapping F is linear since for $\lambda \in \mathbb{F}$ and $x, y \in H$,

$$\begin{aligned}
F(\lambda x + y) &= \{(\lambda x + y)_\alpha\}_{\alpha \in \mathcal{I}} = \{(\lambda x + y, u_\alpha)\}_{\alpha \in \mathcal{I}} \\
&= \{\lambda(x, u_\alpha) + (y, u_\alpha)\}_{\alpha \in \mathcal{I}} \\
&= \lambda\{(x, u_\alpha)\}_{\alpha \in \mathcal{I}} + \{(y, u_\alpha)\}_{\alpha \in \mathcal{I}} \\
&= \lambda F(x) + F(y).
\end{aligned}$$

That F maps into $\ell^2(\mathcal{I})$ follows from Bessel's inequality, *viz.*

$$\|F(x)\|^2_{\ell^2(\mathcal{I})} = \sum_{\alpha \in \mathcal{I}} |x_\alpha|^2 \leq \|x\|^2_H.$$

The same remark reveals F to be bounded and that

$$\|F\|_{B(H, \ell^2(\mathcal{I}))} \leq 1.$$

The interesting point is that F is surjective. Let $f \in \ell^2(\mathcal{I})$ and let $n \geq 1$. Define

$$\mathcal{I}_n = \left\{ \alpha \in \mathcal{I} : |f(\alpha)| > \frac{1}{n} \right\}$$

and let $|\mathcal{I}_n|$ denote the number of α in $\mathcal{I}_n$. Notice that

$$|\mathcal{I}_n| = \sum_{\alpha \in \mathcal{I}_n} 1 < \sum_{\alpha \in \mathcal{I}_n} \left(n \, |f(\alpha)| \right)^2 \leq n^2 \|f\|^2_{\ell^2(\mathcal{I})},$$

so $\mathcal{I}_n$ is finite for all n. Let $\mathcal{J} = \bigcup_{n=1}^\infty \mathcal{I}_n$. Then $\mathcal{J}$ is countable and if $\beta \notin \mathcal{J}$, then $f(\beta) = 0$. In H, define x_n by

$$x_n = \sum_{\alpha \in \mathcal{I}_n} f(\alpha) u_\alpha.$$

Since $\mathcal{I}_n$ is a finite set, x_n is a well-defined element of H. We expect that $\{x_n\}_{n=1}^{\infty}$ is Cauchy in H. To see this, let $n > m \geq 1$ and compute

$$\|x_n - x_m\|^2 = \left\| \sum_{\alpha \in \mathcal{I}_n \setminus \mathcal{I}_m} f(\alpha) u_\alpha \right\|^2 = \sum_{\alpha \in \mathcal{I}_n \setminus \mathcal{I}_m} |f(\alpha)|^2 \leq \sum_{\alpha \in \mathcal{I} \setminus \mathcal{I}_m} |f(\alpha)|^2,$$

and the latter is the tail of an absolutely convergent series, so can be made small by taking m large enough. Since H is a Hilbert space, there is an $x \in H$ such that $x_n \to x$. As F is continuous, $F(x_n) \to F(x)$. We claim that $F(x) = f$. By continuity of the inner-product, for $\alpha \in \mathcal{I}$,

$$F(x)(\alpha) = (x, u_\alpha) = \lim_{n \to \infty} (x_n, u_\alpha)$$

$$= \lim_{n \to \infty} \sum_{\beta \in \mathcal{I}_n} f(\beta)(u_\beta, u_\alpha) = \begin{cases} f(\alpha), & \alpha \in \mathcal{J}, \\ 0 = f(\alpha), & \alpha \notin \mathcal{J}. \end{cases} \qquad \square$$

The Riesz-Fischer map is most useful if the ON set is as large as possible.

Theorem 3.22. *Let H be a Hilbert space. The following are equivalent conditions on an ON set $\{u_\alpha\}_{\alpha \in \mathcal{I}} \subset H$:*

(i) *$\{u_\alpha\}_{\alpha \in \mathcal{I}}$ is a maximal ON set (also called an ON basis for H);*

(ii) *$\mathrm{span}\{u_\alpha : \alpha \in \mathcal{I}\}$ is dense in H;*

(iii) *$\|x\|_H^2 = \sum_{\alpha \in \mathcal{I}} |(x, u_\alpha)|^2$ for all $x \in H$ (called the Parseval identity);*

(iv) *$(x, y) = \sum_{\alpha \in \mathcal{I}} (x, u_\alpha)\overline{(y, u_\alpha)}$ for all $x, y \in H$.*

Proof. (i) $\implies$ (ii). Let $M = \overline{\mathrm{span}\{u_\alpha\}}$. Then M is a closed linear subspace of H. If M is not all of H, $M^\perp \neq \{0\}$ since $H = M + M^\perp$. Let $x \in M^\perp$, $x \neq 0$, $\|x\| = 1$. Then the set $\{u_\alpha : \alpha \in \mathcal{I}\} \cup \{x\}$ is an ON set, so $\{u_\alpha\}_{\alpha \in \mathcal{I}}$ is not maximal, a contradiction.

(ii) $\implies$ (iii). We are assuming $M = H$ in the notation of the last paragraph. Let $x \in H$. Because of Bessel's inequality,

$$\|x\|^2 \geq \sum_{\alpha \in \mathcal{I}} |x_\alpha|^2,$$

where $x_\alpha = (x, u_\alpha)$ for $\alpha \in \mathcal{I}$. Let $\epsilon > 0$ be given. Since $\mathrm{span}\{u_\alpha : \alpha \in \mathcal{I}\}$ is dense, there is a finite set $\alpha_1, \dots, \alpha_N$ and constants $c_1, \dots, c_N$ such that

$$\left\| x - \sum_{i=1}^{N} c_i u_{\alpha_i} \right\| \leq \epsilon.$$

By the Best Approximation analysis, on the other hand,

$$\left\| x - \sum_{i=1}^{N} x_{\alpha_i} u_{\alpha_i} \right\| \le \left\| x - \sum_{i=1}^{N} c_i u_{\alpha_i} \right\|.$$

It follows from orthonormality of $\{u_\alpha\}_{\alpha \in \mathcal{I}}$ that

$$\epsilon^2 \ge \left\| x - \sum_{i=1}^{N} x_{\alpha_i} u_{\alpha_i} \right\|^2 = \|x\|^2 - \sum_{i=1}^{N} |x_{\alpha_i}|^2 \ge \|x\|^2 - \sum_{\alpha \in \mathcal{I}} |x_\alpha|^2.$$

In consequence,

$$\|x\|^2 \le \sum_{\alpha \in \mathcal{I}} |x_\alpha|^2 + \epsilon,$$

and $\epsilon > 0$ was arbitrary. Thus equality holds in Bessel's inequality.

(iii) $\implies$ (iv). This follows because, in a Hilbert space, the norm determines the inner-product (Proposition 3.4). Because of (iii), if $x, y \in H$, then

$$\begin{aligned}
(x, y) &= \frac{1}{2}\{\|x+y\|^2 - \|x\|^2 - \|y\|^2\} + \frac{i}{2}\{\|x+iy\|^2 - \|x\|^2 - \|y\|^2\} \\
&= \frac{1}{2}\sum_{\alpha \in \mathcal{I}}\{|x_\alpha + y_\alpha|^2 - |x_\alpha|^2 - |y_\alpha|^2 + i(|x_\alpha + iy_\alpha|^2 - |x_\alpha|^2 - |y_\alpha|^2)\} \\
&= \frac{1}{2}\sum_{\alpha \in \mathcal{I}}\{x_\alpha \bar{y}_\alpha + \bar{x}_\alpha y_\alpha + i(-ix_\alpha \bar{y}_\alpha + i\bar{x}_\alpha y_\alpha)\} \\
&= \sum_{\alpha \in \mathcal{I}} x_\alpha \bar{y}_\alpha,
\end{aligned}$$

which is the desired result. This calculation works if $\mathbb{F} = \mathbb{C}$ or $\mathbb{R}$, but it is a little simpler over $\mathbb{R}$.

(iv) $\implies$ (i). If $\{u_\alpha\}_{\alpha \in \mathcal{I}}$ is not a maximal ON set, let $u \in H$, $u \perp u_\alpha$ for all $\alpha \in \mathcal{I}$, and $\|u\| = 1$. Then, because of (iv),

$$1 = \|u\|^2 = \sum_{\alpha \in \mathcal{I}} |(u, u_\alpha)|^2 = 0,$$

a contradiction. □

Theorem 3.23 (Riesz-Fischer theorem). *If H is an infinite-dimensional Hilbert space, $\{u_\alpha\}_{\alpha \in \mathcal{I}} \subset H$ is a maximal ON set, and $x \in H$, then there are $\alpha_i \in \mathcal{I}$ for $i = 1, 2, \ldots$ such that*

$$x = \sum_{i=1}^{\infty} (x, u_{\alpha_i}) u_{\alpha_i}.$$

At this point in the development, the few remaining details needed to prove this theorem are left for the reader to provide (as an exercise at the end of the chapter). The expansion above is denoted by

$$x = \sum_{\alpha \in \mathcal{I}} (x, u_\alpha) u_\alpha.$$

A maximal ON set is, indeed, a type of basis for the Hilbert space. The numbers $\{(x, u_\alpha)\}_{\alpha \in \mathcal{I}}$ are often called the *Fourier coefficients* of x in the ON basis $\{u_\alpha\}_{\alpha \in \mathcal{I}}$.

Corollary 3.24. *If H is a Hilbert space containing $\{u_\alpha\}_{\alpha \in \mathcal{I}}$, a maximal ON set, then the Riesz-Fischer map $F : H \to \ell^2(\mathcal{I})$ is a Hilbert space isomorphism, i.e., a map that establishes that the two spaces are isomorphic as Hilbert spaces.*

That F is injective follows, for a linear map, from $Fx = 0$ implying that $x = 0$, which follows from (iii) of Theorem 3.22. To clarify the terminology, the statement that F is a Hilbert space isomorphism means that it preserves the linear structure and topology. It then also preserves the inner-product by (iv) of Theorem 3.22, so it is automatically an isometric isomorphism with equivalent inner-products.

Does every Hilbert space have a maximal ON set, i.e., an ON basis? Indeed it does, as follows from the next theorem.

Theorem 3.25. *Let H be a Hilbert space and $\{u_\alpha\}_{\alpha \in \mathcal{I}}$ any ON set in H. Then $\{u_\alpha\}_{\alpha \in \mathcal{I}} \subset \{u_\beta\}_{\beta \in \mathcal{J}}$, where the latter is ON and maximal, and thus an ON basis for H.*

Proof. The general result follows from transfinite induction, e.g., by using Zorn's lemma (Lemma 2.26). The result is proved here assuming that H is separable, and making use of a Gram-Schmidt orthogonalization argument.

Let $\{\tilde{x}_j\}_{j=1}^\infty$ be dense in H and

$$M = \overline{\text{span}\{u_\alpha\}_{\alpha \in \mathcal{I}}}.$$

Define

$$\hat{x}_j = \tilde{x}_j - P_M \tilde{x}_j \in M^\perp,$$

where P_M is orthogonal projection onto M. Then the span of

$$\{u_\alpha\}_{\alpha \in \mathcal{I}} \cup \{\hat{x}_j\}_{j=1}^\infty$$

is dense in H. Define successively for $j = 1, 2, \ldots$ (with $x_1 = \hat{x}_1$)

$$N_j = \overline{\text{span}\{x_1, \ldots, x_j\}},$$
$$x_{j+1} = \hat{x}_{j+1} - P_{N_j} \hat{x}_{j+1} \in N_j^\perp.$$

Then the span of

$$\{u_\alpha\}_{\alpha \in \mathcal{I}} \cup \{x_j\}_{j=1}^\infty$$

is dense in H and any two elements are orthogonal. Remove any zero vectors and normalize to complete the proof by the equivalence of (i) and (ii) in Theorem 3.22. □

Corollary 3.26. *Every Hilbert space H is isometrically isomorphic to $\ell^2(\mathcal{I})$ for some $\mathcal{I}$. Moreover, H is infinite-dimensional and separable if and only if H is isomorphic to $\ell^2(\mathbb{N})$.*

Since $\ell^2(\mathbb{N})$ has a Schauder basis, so too does every separable Hilbert space.

Orthogonality in a Hilbert space is nicely illustrated in the context of Fourier series. If $f : \mathbb{R} \to \mathbb{C}$ is periodic of period T, then $g : \mathbb{R} \to \mathbb{C}$ defined by $g(x) = f(\lambda x)$ for some $\lambda \neq 0$ is periodic of period T/λ. So when considering periodic functions, it is enough to restrict to the case $T = 2\pi$.

Let

$$L^2_{\text{per}}(-\pi, \pi) = \{ f : \mathbb{R} \to \mathbb{C} \mid f \in L^2([-\pi, \pi)) \text{ and } f(x + 2n\pi) = f(x)$$
$$\text{for a.e. } x \in [-\pi, \pi) \text{ and each integer } n \}.$$

With the inner-product

$$(f, g) = \frac{1}{2\pi} \int_{-\pi}^{\pi} f(x) \overline{g(x)} \, dx,$$

$L^2_{\text{per}}(-\pi, \pi)$ is a Hilbert space (it is left to the reader to verify these assertions). The set

$$\{e^{inx}\}_{n=-\infty}^{\infty} \subset L^2_{\text{per}}(-\pi, \pi)$$

is ON, as can be readily verified.

Theorem 3.27. *The set $\text{span}\{e^{inx}\}_{n=-\infty}^{\infty}$ is dense in $L^2_{\text{per}}(-\pi, \pi)$, and it is thus a maximal ON set (an ON basis) for $L^2_{\text{per}}(-\pi, \pi)$.*

Proof. First remark that $C_{\text{per}}([-\pi, \pi])$, the continuous functions defined on $(-\infty, \infty)$ that are periodic, are dense in $L^2_{\text{per}}(-\pi, \pi)$. In fact, $C_0((-\pi, \pi))$ is dense in $L^2(-\pi, \pi)$ (Proposition 2.24), so its periodic extension is dense in $L^2_{\text{per}}(-\pi, \pi)$. Thus it is enough to show that a continuous and periodic function f of period 2π is the limit of functions in $\text{span}\{e^{inx}\}_{n=-\infty}^{\infty}$.

For any integer $m \geq 0$, let

$$k_m(x) = c_m \left(\frac{1 + \cos x}{2} \right)^m \geq 0$$

where c_m is defined so that

$$\frac{1}{2\pi} \int_{-\pi}^{\pi} k_m(x) \, dx = 1. \tag{3.3}$$

As $m \to \infty$, $k_m(x)$ is concentrated about $x = 0$ but maintains total integral 2π (in more advanced terms, $k_m/2\pi \to \delta_0$, the Dirac distribution to be defined in Chapter 5). By Euler's formula,

$$k_m(x) = c_m \left[\frac{2 + e^{ix} + e^{-ix}}{4} \right]^m \in \text{span } \{e^{inx}\}_{n=-m}^m,$$

and so, for some $\lambda_n \in \mathbb{C}$,

$$f_m(x) \equiv \frac{1}{2\pi} \int_{-\pi}^{\pi} k_m(x-y) f(y)\, dy$$

$$= \frac{1}{2\pi} \int_{-\pi}^{\pi} \sum_{n=-m}^m \lambda_n e^{in(x-y)} f(y)\, dy$$

$$= \sum_{n=-m}^m \left(\frac{\lambda_n}{2\pi} \int_{-\pi}^{\pi} e^{-iny} f(y)\, dy \right) e^{inx} \in \text{span } \{e^{inx}\}_{n=-m}^m.$$

The claim is that in fact $f_m \to f$ uniformly (i.e., in the L^∞ norm), so also in L^2, and this fact will complete the proof. By periodicity,

$$f_m(x) = \frac{1}{2\pi} \int_{-\pi}^{\pi} f(x-y)\, k_m(y)\, dy,$$

and, by (3.3),

$$f(x) = \frac{1}{2\pi} \int_{-\pi}^{\pi} f(x)\, k_m(y)\, dy.$$

Thus, for any $\delta \in (0, \pi)$,

$$|f_m(x) - f(x)| = \frac{1}{2\pi} \left| \int_{-\pi}^{\pi} (f(x-y) - f(x))\, k_m(y)\, dy \right|$$

$$\leq \frac{1}{2\pi} \int_{-\pi}^{\pi} |f(x-y) - f(x)|\, k_m(y)\, dy$$

$$= \frac{1}{2\pi} \int_{\delta < |y| \leq \pi} |f(x-y) - f(x)|\, k_m(y)\, dy \qquad (3.4)$$

$$+ \frac{1}{2\pi} \int_{|y| \leq \delta} |f(x-y) - f(x)|\, k_m(y)\, dy.$$

Since f is continuous on $[-\pi, \pi]$, it is uniformly continuous there. Thus, given $\epsilon > 0$ there is a $\delta > 0$, which can be taken to lie in $(0, \pi)$, such that

$$|f(x-y) - f(x)| < \epsilon/2 \quad \text{for all } |y| \leq \delta.$$

For such a choice of δ, the last term on the right-hand side of (3.4) is bounded by $\epsilon/2$. As for the other term on the right-hand side of (3.4), (3.3) implies that

$$
\begin{aligned}
1 &= \frac{c_m}{\pi} \int_0^\pi \left(\frac{1 + \cos x}{2} \right)^m dx \\
&\geq \frac{c_m}{\pi} \int_0^\pi \left(\frac{1 + \cos x}{2} \right)^m \sin x \, dx \\
&= \frac{2c_m}{(m+1)\pi},
\end{aligned}
$$

which means that

$$
c_m \leq \frac{(m+1)\pi}{2}.
$$

As f is continuous on $[-\pi, \pi]$, there is an $M \geq 0$ such that $|f(x)| \leq M$ for $x \in [-\pi, \pi]$. For $|y| \geq \delta$,

$$
k_m(y) \leq \frac{(m+1)\pi}{2} \left(\frac{1 + \cos \delta}{2} \right)^m < \frac{\epsilon}{4M}
$$

for m sufficiently large. Gathering these ruminations together, it follows that

$$
|f_m(x) - f(x)| \leq \frac{1}{2\pi} \int_{-\pi}^\pi 2M \frac{\epsilon}{4M} \, dy + \frac{\epsilon}{2} = \epsilon,
$$

for all sufficiently large m, so completing the demonstration that $f_m \to f$ uniformly. □

Thus, given $f \in L^2_{\text{per}}(-\pi, \pi)$, we have the *Fourier series* representation

$$
f(x) = \sum_{n=-\infty}^{\infty} (f, e^{-in(\cdot)}) e^{-inx} = \sum_{n=-\infty}^{\infty} \left(\frac{1}{2\pi} \int_{-\pi}^\pi f(y) \, e^{iny} \, dy \right) e^{-inx},
$$

where convergence of the infinite series is in the sense of the Hilbert space norm.

3.5 WEAK CONVERGENCE IN A HILBERT SPACE

Because of the Riesz representation theorem, a sequence $\{x_n\}_{n=1}^\infty$ from a Hilbert space H converges weakly to x if and only if

$$(x_n, y) \longrightarrow (x, y) \tag{3.5}$$

for all $y \in H$.

Lemma 3.28. *If $\{e_\alpha\}_{\alpha \in \mathcal{I}}$ is an ON basis for the Hilbert space H and x_n and x lie in H, then $x_n \xrightarrow{w} x$ if and only if $\|x_n\|$ is bounded and the Fourier coefficients*

$$(x_n, e_\alpha) \longrightarrow (x, e_\alpha) \quad \text{as } n \to \infty \tag{3.6}$$

for all $\alpha \in \mathcal{I}$.

Proof. Weakly convergent sequences are bounded, and clearly (3.5) implies (3.6). On the other hand suppose (3.6) is valid. Since $\{e_\alpha\}_{\alpha\in\mathcal{I}}$ is an ON basis, we know from the Riesz-Fischer theorem that $\mathrm{span}\{e_\alpha\}_{\alpha\in\mathcal{I}}$ is dense in H. Let $\epsilon > 0$ and $y \in H$ be given, and let $\{c_\alpha\}_{\alpha\in\mathcal{I}}$ be a collection of constants such that $c_\alpha = 0$ for all but a finite number of α and so that $z = \sum_{\alpha\in\mathcal{I}} c_\alpha e_\alpha \in \mathrm{span}\{e_\alpha\}_{\alpha\in\mathcal{I}}$ satisfies

$$\|y - z\| < \epsilon.$$

Because of (3.6),

$$(x_n, z) \longrightarrow (x, z) \quad \text{as } n \to \infty$$

since z is a finite linear combination of the e_α's. But then,

$$\limsup_{n\to\infty} |(x_n - x, y)| \leq \limsup_{n\to\infty} |(x_n - x, y - z)| + \limsup_{n\to\infty} |(x_n - x, z)|$$

$$= \limsup_{n\to\infty} |(x_n - x, y - z)|$$

$$\leq \left(\sup_{n\geq 1} \|x_n\| + \|x\| \right) \|y - z\|$$

$$\leq C\epsilon,$$

because $\{\|x_n\|\}_{n=1}^\infty$ is bounded. Thus,

$$\lim_{n\to\infty} (x_n, y) = (x, y),$$

as required. □

Since $H^* = H$, the Banach-Alaoglu theorem (Theorem 2.55) has the following further conclusion in the Hilbert space context.

Lemma 3.29. *If H is a separable Hilbert space and $\{x_n\}_{n=1}^\infty \subset H$ is a bounded sequence, then there exists a subsequence x_{n_j} converging weakly to some $x \in H$.*

Although we have only proved this result assuming H is separable, it holds in general by the Eberlein-Šmulian theorem (see Corollary 2.59 and the remark after Theorem 2.60).

Example. Consider $L_{\mathrm{per}}^2(-\pi, \pi)$ and consider the ON basis

$$\{e^{inx}\}_{n=-\infty}^\infty$$

discussed earlier. This sequence converges weakly to zero, for obviously if m is fixed,

$$(e^{inx}, e^{imx}) = 0$$

for $n > m$. However, as $\|e^{inx} - e^{imx}\| = \sqrt{2}$ for $n \neq m$, the sequence is *not* Cauchy in norm, and so has no strong limit.

3.6   EXERCISES

1. Prove the parallelogram law in a Hilbert space.

2. If H is a real Hilbert space, show that if $x, y \in H$ satisfy $\|x\|^2 + \|y\|^2 = \|x + y\|^2$, then $x \perp y$. Show by example that this result can fail for a complex Hilbert space.

3. Let H be a real Hilbert space and let $S \subset H$ be a closed convex subset. For fixed $x \in H$ and $y \in S$, show that $\|x - y\| = \inf_{z \in S} \|x - z\|$ if and only if $(x - y, z - y) \leq 0$ for all $z \in S$. [Hint: for the direct implication, consider $\alpha \in [0, 1]$ and $\alpha y + (1 - \alpha)z$ (and $\alpha \to 1$).]

4. Let H be a Hilbert space and $P : H \to H$ a projection.

   (a) Prove that if $\|P\| = 1$, then P is the orthogonal projection onto some subspace of H.

   (b) Prove that in general if $P \neq 0$, $\|P\| \geq 1$. Show by example that if H has at least two dimensions, then there is a nonorthogonal projection defined on H.

5. Prove Theorem 3.11 regarding projection operators on an NLS.

6. Let H be a Hilbert space and M a closed linear subspace. Let $A : H \to H$ be a bounded linear operator with a bounded inverse.

   (a) For a given fixed $x \in H$, show that there is a unique $y \in M$ such that

   $$\inf_{z \in M} \|A(z - x)\| = \|A(y - x)\|.$$

   This defines an operator $P : H \to M$ by $Px = y$.

   (b) Show that P is a projection onto M.

   (c) Show that

   $$(A^* A(Px - x), y)_H = 0 \quad \forall y \in M.$$

   (d) Show that P is bounded. In fact, show that $\|P\| \leq \|A\| \|A^{-1}\|$.

7. Let X be an NLS and Y a finite-dimensional subspace of X. Let

   $$\text{dist}(x, S) = \inf_{z \in S} \|x - z\|$$

   denote the distance from $x \in X$ to $S \subset X$. Fix $x_0 \in X$.

   (a) Show that $\text{dist}(x_0, Y) = \text{dist}(x_0, \mathcal{B})$ where $\mathcal{B} = \{y \in Y : \|y\| \leq 2\|x_0\|\}$.

   (b) Show that there is a *best approximation* $y_0 \in \mathcal{B} \subset Y$ to x_0, i.e., such that $\text{dist}(x_0, y_0) = \text{dist}(x_0, Y)$. [Hint: why is $\mathcal{B}$ compact?]

   (c) Show by example that a best approximation may not be unique. [Hint: try $X = (\mathbb{R}^2, \| \cdot \|_{\ell^1})$ and $Y = \text{span}(1, 1)$.]

8. Let H be a Hilbert space.

   (a) If M is a nonempty subset of H, show that the span of M is dense in H if and only if $M^\perp = \{0\}$.

   (b) Let $T : H \to H$ be a bounded linear operator. Let $N = N(T)$ be the null space of T and $R(T)$ be the range or image of T. Let $P : H \to N$ be orthogonal projection onto N. Show that $S = T \circ P^\perp$ is a one-to-one mapping when restricted to $N^\perp$ and that $R(S) = R(T)$.

9. Let H be a Hilbert space and $R : H \to H^*$ the Riesz map.

   (a) Show that R is conjugate linear.

   (b) Show that the map $(\cdot, \cdot)_{H^*} : H^* \times H^* \to \mathbb{F}$ defined by $(L_1, L_2)_{H^*} = (R^{-1}L_2, R^{-1}L_1)_H$ is an inner-product.

10. Let X and Y be Hilbert spaces, and identify $X^* = X$ and $Y^* = Y$. Suppose $T : X \to Y$ is a bounded linear map. Show that the map $S = I + T^*T$ is injective.

11. Suppose H is an IPS and $x, y \in H$ are such that $(x, z)_H = (y, z)_H$ for all $z \in H$. Show that $x = y$.

12. If H is a complex Hilbert space, show that if $T : H \to H$ is a bounded linear operator such that $(Tx, x)_H = 0$ for all $x \in H$, then $T = 0$. Show by example that this result can fail for a real Hilbert space.

13. Let H be a Hilbert space and $U \in B(H, H)$ a *unitary operator*, which means that U is bijective and $U^{-1} = U^*$.

   (a) Show that U is an isometry, and when $H \neq \{0\}$, $\|U\| = 1$.

   (b) If V is another isometry on H, show that UV is an isometry.

   (c) If H is a complex Hilbert space, show that $U \in B(H, H)$ is unitary if and only if it is isometric and surjective.

14. Let H be a Hilbert space.

   (a) If Y is a subspace of H, show that Y is closed if and only if $Y = (Y^\perp)^\perp$.

   (b) If X is any nonempty subset of H, show that $(X^\perp)^\perp$ is the smallest closed subspace of H which contains X.

15. Let H be a Hilbert space and Y a subspace (not necessarily closed).

   (a) Prove that
   $$(Y^\perp)^\perp = \bar{Y} \quad \text{and} \quad Y^\perp = (\bar{Y})^\perp.$$

   (b) If Y is not trivial, show that orthogonal projection P onto $\bar{Y}$ has norm 1 and that
   $$(Px, y) = (x, y)$$
   for all $x \in H$ and $y \in \bar{Y}$.

16. Let H be a Hilbert space and $T \in B(H, H)$, $T \neq 0$. Let $N = \ker(T)$ and suppose M is a closed linear subspace such that $H = N \oplus M$. Let $R = R(T)$ be the image of T, and suppose that R is closed. (Note that we could take $M = N^{\perp}$, but this is not required for the validity of the following conclusions.)

   (a) Show that T is one-to-one on M.

   (b) Show that there is a $\gamma > 0$ such that

$$\gamma \|x\| \leq \|Tx\| \quad \forall x \in M.$$

   (c) Let γ_M^* be the maximal value of γ for which the inequality in (b) holds for all elements of M. Show that $\gamma_M^* \leq \gamma_{N^{\perp}}^*$.

17. Show that if $\mathcal{I}$ is an index set and $\{x_\alpha\}_{\alpha \in \mathcal{I}}$ is a collection of nonnegative real numbers satisfying

$$\sum_{\alpha \in \mathcal{I}} x_\alpha < \infty,$$

   then at most countably many of the x_α are different from zero.

18. Prove that for any index set $\mathcal{I}$, the space $\ell^2(\mathcal{I})$ is a Hilbert space.

19. Prove the Riesz-Fischer theorem (Theorem 3.23).

20. If $\{u_\alpha\}_{\alpha \in \mathcal{I}}$ is a maximal ON set in a Hilbert space $(H, (\cdot, \cdot))$ and $x \in H$, give a precise meaning to the expression

$$x = \sum_{\alpha \in \mathcal{I}} (x, u_\alpha) u_\alpha.$$

21. Prove Theorem 3.25 using Zorn's lemma (Lemma 2.26), *not* assuming that H is separable.

22. Verify that

$$L_{\text{per}}^2(-\pi, \pi) = \left\{ f : \mathbb{R} \to \mathbb{C} \mid f \in L^2([-\pi, \pi)) \text{ and } f(x + 2n\pi) = f(x) \right.$$
$$\left. \text{for a.e. } x \in [-\pi, \pi) \text{ and integer } n \right\}$$

   with the inner-product

$$(f, g) = \frac{1}{2\pi} \int_{-\pi}^{\pi} f(x) \overline{g(x)} \, dx$$

   is a Hilbert space, and that

$$\{e^{inx}\}_{n=-\infty}^{\infty} \subset L_{\text{per}}^2(-\pi, \pi)$$

   is a maximal ON set contained in $L_{\text{per}}^2(-\pi, \pi)$.

23. The Weierstrass approximation theorem states that the set of polynomials $\mathbb{P}$ is dense in $C([0,1])$.

(a) Show that $\mathbb{P}$ is dense in $L^p([0,1])$ for any $1 \le p < \infty$.

(b) The Legendre polynomials $p_0(x) = 1$, $p_1(x)$, ... have the properties:
   (i) p_n has strict degree n for each $n = 0, 1, 2, \ldots$;
   (ii) $\int_0^1 p_n(x) p_m(x) \, dx = 0$ if $n \ne m$, and it is 1 if $n = m$.

Show that they exist and form a maximal ON set for $L^2([0,1])$. [Hint: use induction, and relate p_n to an appropriate orthogonal projection of the monomial x^n onto some subspace.]

(c) Let $f \in L^2([0,1])$ and $n \ge 0$. In terms of the Legendre polynomials, find an explicit expression for the polynomial q of degree n that minimizes $\|f - p\|_{L^2([0,1])}$ over all polynomials p of degree n.

24. Let H be a separable Hilbert space and $\{x_n\}_{n=1}^\infty$ a bounded sequence in H.

(a) Show that $\{x_n\}_{n=1}^\infty$ has a weakly convergent subsequence.

(b) Suppose that $x_n \rightharpoonup x$. Prove that $x_n \to x$ if and only if $\|x_n\| \to \|x\|$.

(c) Show that if $x_n \rightharpoonup x$, then there exists a set of nonnegative constants $\{\{\alpha_i^n\}_{i=1}^n\}_{n=1}^\infty$ such that $\sum_{i=1}^n \alpha_i^n = 1$ and

$$\sum_{i=1}^n \alpha_i^n x_i \equiv y_n \to x \quad \text{(strong convergence)}.$$

25. Let H be a separable Hilbert space, $\{v_i\}_{i=1}^\infty$ a countable orthonormal basis, and $L : H \to H$ the operator such that $Lv_i = \sum_{j=1}^\infty 2^{-(i+j)} v_j$. Show that L takes any bounded set to a precompact set (a set whose closure is compact). Such an operator is simply called *compact*.

26. Let V and W be real Hilbert spaces, $A \in B(V,V)$, and $B \in B(W,V)$. By identifying a space with its dual, consider $B^* \in B(V,W)$. Assume that B^* maps V onto W, that A is self-adjoint (which means $(Av, w) = (v, Aw) \; \forall v, w \in V$) and that there are constants $\alpha > 0$ and $\gamma > 0$ for which

$$(Av, v)_V \ge \alpha \|v\|_V^2 \quad \forall v \in N(B^*) = \ker(B^*),$$

$$\sup_{v \in V} \frac{(B^*v, w)_W}{\|v\|_V} \ge \gamma \|w\|_W \quad \forall w \in W.$$

Assume also that the following problem has a solution: find $u \in V$ and $p \in W$ such that

$$\begin{cases} Au + Bp = f \in V, \\ B^*u \quad\quad\; = g \in W. \end{cases}$$

(a) Show the inequality

$$\|p\|_W \leq \frac{1}{\gamma}\{\|f\|_V + \|A\| \, \|u\|_V\}$$

holds for any such solution.

(b) Prove that the solution (u, p) is unique.

(c) Find $C \geq 0$ such that $\|u\|_V \leq C < \infty$, where C may depend on anything but u and p.

CHAPTER 4

Spectral Theory and Compact Operators

The focus of this chapter is spectral theory, which is concerned with questions of invertibility of an operator. Initially, our theory will be developed for operators in a Banach space. Later, the discussion will be restricted to Hilbert spaces where more detailed information is available. Some of the most complete theory applies to a special type of operator called *compact*, which were introduced already in the problem set for Chapter 3.

The mathematical spectral theory in infinite dimensions, as it appears here, stems from Hilbert's work, with various of his many students, on quadratic forms in infinitely many variables. One can think of it as an infinite-dimensional version of the old principle-axis result for an ellipsoid in finite dimensions. It is amusing to remark that Hilbert himself, who coined the name spectrum, was completely surprised when it later turned out that his mathematical spectrum corresponded to spectrum as measured by physicists in certain real-world situations.

Before continuing, there are two prototypical examples that will guide the conversation. The first is the case of a square $d \times d$ matrix A viewed as a linear transformation

$$A : \mathbb{F}^d \longrightarrow \mathbb{F}^d$$

via matrix multiplication. From elementary linear algebra, we know that the eigenvalues $\lambda \in \mathbb{C}$ and nonzero eigenvectors $x \in \mathbb{C}^d$, for which

$$Ax = \lambda x,$$

play a critical role in understanding the mapping properties of A. For such values of λ, $A - \lambda I$ is not invertible. The matrix A itself is invertible if and only if 0 is *not* an eigenvalue. In some cases, the entire action of A is revealed simply by knowing its eigenstructure. The best case is when A is real and symmetric. Then the eigenvalues are real and the eigenvectors may be taken to be real. Moreover, if there are d distinct eigenvalues λ_i, with corresponding

DOI: 10.1201/9781003492139-4

nonzero eigenvectors x_i, $i = 1, \ldots, d$, then the eigenvectors form an orthogonal basis for $\mathbb{R}^d$. Hence, given $x \in \mathbb{R}^d$, if

$$\alpha_i = \frac{x \cdot x_i}{x_i \cdot x_i},$$

then

$$x = \sum_{i=1}^{d} \alpha_i x_i.$$

This is an important way to generate an orthogonal basis. Moreover, this particular basis is well tailored to the operator A, since

$$Ax = \sum_{i=1}^{d} \alpha_i \lambda_i x_i;$$

that is, for each i, the component of x in the x_i direction is scaled by λ_i under the action of A. Another way of looking at it is that in this basis, the matrix A of the transformation is diagonal. These sorts of facts will be proved in the more general, infinite-dimensional setting of Banach and Hilbert spaces.

The second prototypical example is that of differentiation,

$$D : C^1 \longrightarrow C^0,$$

which is *not* invertible, but nearly so. A family of eigenvalues and *eigenfunctions* is

$$De^{\lambda x} = \lambda e^{\lambda x} \quad \forall \lambda \neq 0.$$

On the surface there is a difference from the first example; the domain and range are not the same in this case. However, they are both separable and infinite-dimensional, so in a certain sense D is like a square matrix.

4.1 DEFINITIONS OF THE RESOLVENT AND SPECTRUM

Even in the finite-dimensional case, complex eigenvalues arise and are needed. So let X be a complex NLS ($\mathbb{F} = \mathbb{C}$) and $D = D(T) \subset X$ a dense linear subspace. Suppose that

$$T : D \longrightarrow X$$

is a linear operator. The domain of T is $D = D(T)$, the range or image is

$$R(T) = T(D) = \{y \in X : y = Tx \text{ for some } x \in D\} \subset X,$$

and the null space or kernel is

$$N(T) = \ker(T) = \{x \in D : Tx = 0\} \subset X.$$

For $\lambda \in \mathbb{C}$, consider the operator

$$T_\lambda = T - \lambda I : D \longrightarrow R_\lambda = R(T_\lambda) = T_\lambda(D(T)) \subset X,$$

where I is the identity operator on X.

If T_λ is injective, then $T_\lambda^{-1} : R_\lambda \to D$ exists, and it is necessarily a linear operator. However, this is not a helpful operator unless it is bounded and defined on at least most of X, i.e., unless $R_\lambda \subset X$ is dense. These are subtle points that arise only in infinite dimensions. In finite dimensions, linear operators are necessarily bounded, and as soon as R_λ is dense, it is all of X.

Definition. If T_λ is injective, maps onto a dense subset of X, and T_λ^{-1} is bounded (i.e., continuous), then λ is said to be in the *resolvent* set of T, denoted $\rho(T) \subset \mathbb{C}$. The mapping T_λ^{-1} is then called the *resolvent operator* of T for the given value λ.

There are three reasons why $\lambda \in \mathbb{C}$ may fail to lie in $\rho(T)$.

Definition. If $\lambda \notin \rho(T)$, then we say that λ lies in the *spectrum* of T, denoted by $\sigma(T) = \mathbb{C} \setminus \rho(T)$, which is subdivided into the *point spectrum* of T,

$$\sigma_p(T) = \{\mu \in \mathbb{C} : T_\mu \text{ is not one-to-one}\},$$

the *continuous spectrum* of T,

$$\sigma_c(T) = \{\mu \in \mathbb{C} : T_\mu \text{ is one-to-one and } R(T_\mu) \text{ is dense in } X,$$
$$\text{but } T_\mu^{-1} \text{ is not bounded}\},$$

and the *residual spectrum* of T,

$$\sigma_r(T) = \{\mu \in \mathbb{C} : T_\mu \text{ is one-to-one and } R(T_\mu) \text{ is not dense in } X\}.$$

The following result is a simple consequence of the definition.

Proposition 4.1. *The point, continuous, and residual spectra are pairwise disjoint and their union is $\sigma(T)$. That is,*

$$\sigma(T) = \mathbb{C} \setminus \rho(T) = \sigma_p(T) \cup \sigma_c(T) \cup \sigma_r(T),$$

and $\sigma_p(T) \cap \sigma_c(T) = \sigma_p(T) \cap \sigma_r(T) = \sigma_c(T) \cap \sigma_r(T) = \emptyset$.

If $\lambda \in \sigma_p(T)$, then
$$N(T_\lambda) \neq \{0\},$$
so there are vectors $x \in X$, $x \neq 0$, such that $T_\lambda x = 0$ or, what is the same, $Tx = \lambda x$.

Definition. The complex numbers in $\sigma_p(T)$ are called *eigenvalues*, and any $x \in X$ such that $x \neq 0$ and
$$Tx = \lambda x$$
is called an *eigenfunction* or *eigenvector* of T corresponding to $\lambda \in \sigma_p(T)$.

Examples.

(1) The linear operator $T : \ell^2 \to \ell^2$ defined by

$$Tx = T(x_1, x_2, \ldots) = (0, x_1, x_2, \ldots)$$

clearly has an inverse, but the range is not dense. Thus $0 \in \sigma_r(T)$.

(2) The linear operator $D : C^1(\mathbb{R}) \to C^0(\mathbb{R})$, given by differentiation, has $\sigma(D) = \sigma_p(D) = \mathbb{C}$ and $\rho(D) = \emptyset$, because of the aforementioned eigenfunctions. This operator is densely defined, as required, since $C^1(\mathbb{R})$ is dense in $C^0(\mathbb{R})$, but it is also unbounded (consider D applied to $f_n(x) = \sin nx$).

4.2 BASIC SPECTRAL THEORY IN BANACH SPACES

Things are somewhat simpler if we consider bounded linear operators $T : X \to X$ defined on the full space, and not just a linear subspace.

Lemma 4.2. *If X is a Banach space, $T \in B(X, X)$, and $\lambda \in \rho(T)$, then T_λ maps onto X.*

Proof. Let $y \in X$. Since T_λ maps onto a dense subspace of X, there is a sequence $\{x_n\}_{n=1}^\infty \subset X$ such that

$$y_n \equiv T_\lambda x_n \longrightarrow y.$$

This sequence is Cauchy, since for any $m, n \in \mathbb{N}$,

$$\|x_m - x_n\| = \|T_\lambda^{-1}(y_m - y_n)\| \leq \|T_\lambda^{-1}\| \, \|y_m - y_n\|,$$

and the right-hand side converges to zero in the limit of large values of the indices. Therefore, there is some $x \in X$ such that $x_n \to x$ and, by continuity, $T_\lambda x_n \to T_\lambda x$. We conclude that $y = T_\lambda x$ and, therefore, that T_λ is a surjection. $\square$

Corollary 4.3. *If X is a Banach space and $T \in B(X, X)$, then $\lambda \in \rho(T)$ if and only if T_λ is invertible on all of X (i.e., T_λ is injective and surjective).*

Proof. Apply the last lemma and, for the converse, the open mapping theorem (Theorem 2.42). $\square$

Lemma 4.4. *Let X be a Banach space and $V \in B(X, X)$ with $\|V\| < 1$. Then $I - V \in B(X, X)$ is one-to-one and onto, hence by the open mapping theorem has a bounded inverse. Moreover,*

$$(I - V)^{-1} = \sum_{n=0}^{\infty} V^n.$$

The latter expression is called the *Neumann series* for $(I - V)^{-1}$.

Proof. Let $N > 0$ be an integer and let

$$S_N = I + V + V^2 + \cdots + V^N = \sum_{n=0}^{N} V^n.$$

Then $S_N \in B(X, X)$ for all N. The sequence $\{S_N\}_{N=1}^{\infty}$ is Cauchy in $B(X, X)$, for if $M > N$, then

$$\|S_M - S_N\|_{B(X,X)} = \left\| \sum_{n=N+1}^{M} V^n \right\|_{B(X,X)} \le \sum_{n=N+1}^{M} \|V^n\|_{B(X,X)}.$$

It is not difficult to show that $\|V^n\| \le \|V\|^n$, so

$$\|S_M - S_N\|_{B(X,X)} \le \sum_{n=N+1}^{M} \|V\|_{B(X,X)}^n,$$

and this tends to zero as $N \to \infty$ since $\mu = \|V\|_{B(X,X)} < 1$ implies $\sum_{k=0}^{\infty} \mu^k < \infty$. Since $B(X, X)$ is a Banach space, there is an $S \in B(X, X)$ such that $S_N \to S$.

It will follow that $(I - V)S = S(I - V) = I$. First, notice that

$$(I - V)S_N = I - V^{N+1} = S_N(I - V). \tag{4.1}$$

On the one hand $V^{N+1} \to 0$ in $B(X, X)$ since

$$\|V^{N+1}\|_{B(X,X)} \le \|V\|_{B(X,X)}^{N+1} \to 0$$

as $N \to \infty$, and on the other hand $S_N \to S$ in $B(X, X)$. It follows readily that $TS_N \to TS$ and $S_N T \to ST$ for any $T \in B(X, X)$. Taking the limit as $N \to \infty$ in (4.1) yields

$$(I - V)S = I = S(I - V).$$

These two relations imply $I - V$ to be onto and one-to-one, respectively. $\square$

Corollary 4.5. *If X is a Banach space, $T \in B(X, X)$, $\lambda \in \mathbb{C}$, and $\|T\|_{B(X,X)} < |\lambda|$, then $\lambda \subset \rho(T)$ and*

$$T_\lambda^{-1} = -\frac{1}{\lambda} \sum_{n=0}^{\infty} \left(\frac{1}{\lambda} T \right)^n.$$

Proof. Simply note that

$$T_\lambda = T - \lambda I = -\lambda \left(I - \frac{1}{\lambda} T \right)$$

and apply the last lemma. $\square$

When X is a Banach space, the set of boundedly invertible operators in $B(X, X)$ is called the *general linear group*, and it is denoted $GL(X, X)$ or sometimes just $GL(X)$.

Corollary 4.6. *If X is a Banach space, then $GL(X)$ is open in $B(X, X)$.*

Proof. Let $A \in B(X, X)$ be such that $A^{-1} \in B(X, X)$. Choose $\epsilon > 0$ such that $\epsilon < 1/\|A^{-1}\|_{B(X,X)}$. For any $B \in B(X, X)$ with $\|B\|_{B(X,X)} < \epsilon$, $A + B$ is invertible. To see this, write

$$A + B = A(I + A^{-1}B)$$

and note that

$$\|A^{-1}B\|_{B(X,X)} \leq \|A^{-1}\|_{B(X,X)} \|B\|_{B(X,X)} < \epsilon \|A^{-1}\|_{B(X,X)} < 1.$$

Hence $I + A^{-1}B$ is boundedly invertible, and thus so is $A(I + A^{-1}B)$ since it is a composition of two invertible operators. □

Corollary 4.7. *Let $T \in B(X, X)$, where X is a Banach space. Then $\rho(T)$ is an open subset of $\mathbb{C}$ and $\sigma(T)$ is compact. Moreover,*

$$|\lambda| \leq \|T\|_{B(X,X)} \quad \text{for all } \lambda \in \sigma(T).$$

Proof. If $\lambda \in \rho(T)$, then $T - \lambda I$ is invertible. Hence $T - \lambda I + B$ is invertible if $\|B\|_{B(X,X)}$ is small enough. In particular,

$$T - \lambda I - \mu I$$

is invertible if $|\mu|$ is small enough. Thus $\lambda \in \rho(T)$ implies $\lambda + \mu \in \rho(T)$ if $|\mu|$ is small enough, and so $\rho(T)$ is open.

Corollary 4.5 shows that if $\lambda > \|T\|$, then $\lambda \in \rho(T)$, so $\sigma(T)$ is bounded and hence compact, since it is closed and bounded in $\mathbb{C}$. □

A cautionary remark is warranted: we have not shown that $\sigma(T) \neq \emptyset$. Operators with an empty spectrum exist. To continue, the discussion is restricted to certain special classes of operators where more can be said.

4.3   COMPACT LINEAR OPERATORS ON A BANACH SPACE

An important class of operators exhibits a compactness property. We will see examples later.

Definition. Suppose X and Y are NLSs. An operator $T : X \to Y$ is a *compact* linear operator if T is linear and if the closure of the image of any bounded set $M \subset X$ is compact, i.e., $\overline{T(M)} \subset Y$ is compact. (A set with compact closure is termed *precompact*.)

Proposition 4.8. *Let X and Y be NLSs. If $T : X \to Y$ is a compact linear operator, then T is bounded, hence continuous.*

Proof. The unit sphere $S_1 = S_1(0) = \{x \in X : \|x\| = 1\}$ in X is bounded, so $\overline{T(S_1)}$ is compact. A compact set in Y is necessarily bounded, so there is some $R > 0$ such that

$$\overline{T(S_1)} \subset B_R(0) \subset Y;$$

that is,

$$\|T\| = \sup_{x \in S_1} \|Tx\| \leq R < \infty,$$

so $T \in B(X, Y)$. □

Compactness gives us convergence of subsequences, as the next two lemmas show. The first is simply a restatement of Proposition 1.28.

Lemma 4.9. *Suppose (X, d) is a metric space. Then X is compact if and only if every sequence $\{x_n\}_{n=1}^{\infty} \subset X$ has a subsequence $\{x_{n_k}\}_{k=1}^{\infty}$ that converges in X.*

Lemma 4.10. *Let X and Y be NLSs and $T : X \to Y$ linear. Then T is compact if and only if T maps every bounded sequence $\{x_n\}_{n=1}^{\infty} \subset X$ onto a sequence $\{Tx_n\}_{n=1}^{\infty} \subset Y$ with a convergent subsequence.*

Proof. If T is compact and $\{x_n\}_{n=1}^{\infty}$ bounded, then the closure in Y of $\{Tx_n\}_{n=1}^{\infty}$ is compact. Since Y is a metric space, the conclusion follows from the previous lemma.

Conversely, suppose every bounded sequence $\{x_n\}_{n=1}^{\infty}$ gives rise to a convergent subsequence within $\{Tx_n\}_{n=1}^{\infty}$. Let $B \subset X$ be bounded and consider $\overline{T(B)}$. This set is compact if every sequence $\{y_n\}_{n=1}^{\infty} \subset \overline{T(B)}$ has a convergent subsequence. Take such a sequence out of $\overline{T(B)}$ and for each y_n, choose $z_n \in T(B)$ such that

$$\|z_n - y_n\| \leq \frac{1}{n}$$

and a corresponding $x_n \in B$ for which $z_n = Tx_n$. Then $\{x_n\}_{n=1}^{\infty}$ is bounded, so by hypothesis, there is a convergent subsequence

$$z_{n_k} = Tx_{n_k} \longrightarrow y \in \overline{T(B)} \quad \text{as } k \to \infty.$$

But then,

$$\|y_{n_k} - y\| \leq \|y_{n_k} - z_{n_k}\| + \|z_{n_k} - y\|$$

$$\leq \frac{1}{n_k} + \|z_{n_k} - y\| \longrightarrow 0 \quad \text{as } k \to \infty.$$

Thus, $y_{n_k} \to y$ and, by the previous lemma, $\overline{T(B)}$ is compact. □

Trivial examples of compact operators abound, as seen in the following two propositions.

Proposition 4.11. *Let X and Y be NLSs and $T : X \to Y$ a linear operator.*

(a) *If X is finite-dimensional, then T is compact.*

(b) *If T is bounded and Y is finite-dimensional, then T is compact.*

(c) *If X is infinite-dimensional, then $I : X \to X$ is not compact.*

Proof. Note that (a) follows from (b) since necessarily if T is linear and $\dim X < \infty$, then T is bounded and $R(T)$ is finite-dimensional. On the other hand, (b) is trivial since closed and bounded sets in finite-dimensional spaces are compact. The fact that the closed unit ball in infinite dimensions is never compact (Corollary 2.50) shows (c) to be valid. ☐

Proposition 4.12. *Let X, Y, and Z be Banach spaces, $S \in B(X, Y)$, and $T \in B(Y, Z)$. If either S or T is compact, then $T \circ S$ is compact.*

We leave the proof of this result to the reader.

The collection of all compact linear operators $T : X \to Y$ is denoted by

$$C(X, Y) \subset B(X, Y).$$

Clearly $C(X, Y)$ is a linear subspace, as a finite linear combination of compact linear operators is compact. When X is a Banach space, Proposition 4.12 shows that the composition of a bounded operator with a compact operator is compact, so $C(X, X)$ is a two-sided ideal in the algebra $B(X, X)$. The following theorem shows that $C(X, Y)$ is closed in $B(X, Y)$ when Y is complete.

Theorem 4.13. *Suppose X is an NLS and Y is a Banach space. Let $\{T_n\}_{n=1}^{\infty} \subset C(X, Y)$ be convergent in norm to $T \in B(X, Y)$,*

$$\|T_n - T\| \longrightarrow 0 \quad \text{as } n \to \infty.$$

Then $T \in C(X, Y)$. That is, $C(X, Y)$ is a closed linear subspace of $B(X, Y)$.

Proof. The proof makes extensive use of Lemma 4.10. Let $\{x_n\}_{n=1}^{\infty} \subset X$ be bounded. Then $\{T_1 x_n\}_{n=1}^{\infty} \subset Y$ has a convergent subsequence. Denote it by $\{T_1 x_{1,n}\}_{n=1}^{\infty}$. The new sequence $\{x_{1,n}\}_{n=1}^{\infty}$ is still bounded, so $\{T_2 x_{1,n}\}_{n=1}^{\infty}$ has a convergent subsequence. Denote it by $\{T_2 x_{2,n}\}_{n=1}^{\infty}$. Continuing, there obtains subsequences of $\{x_n\}_{n=1}^{\infty}$ satisfying

$$\{x_{k,n}\}_{n=1}^{\infty} \supset \{x_{k+1,n}\}_{n=1}^{\infty} \quad \forall\, k$$

and $T_n x_{n,m}$ converges as $m \to \infty$.

Now apply a *Cantor diagonalization argument* by considering the subsequence

$$\{\tilde{x}_n\}_{n=1}^{\infty} \equiv \{x_{n,n}\}_{n=1}^{\infty} \subset X.$$

For each $n \geq 1$, the sequence $\{T_n \tilde{x}_m\}_{m=1}^{\infty}$ converges, since convergence depends only on the tail of the sequence. We claim also that $\{T \tilde{x}_m\}_{m=1}^{\infty}$ is

Cauchy, and therefore T is compact. Let $\epsilon > 0$ be given and find $N \geq 1$ such that

$$\|T_N - T\| < \epsilon.$$

Let M bound $\{\|x_n\|\}_{n=1}^{\infty}$. Then for any n and m,

$$\|T\tilde{x}_n - T\tilde{x}_m\| \leq \|T\tilde{x}_n - T_N\tilde{x}_n\| + \|T_N\tilde{x}_n - T_N\tilde{x}_m\| + \|T_N\tilde{x}_m - T\tilde{x}_m\|$$
$$\leq 2\epsilon M + \|T_N\tilde{x}_n - T_N\tilde{x}_m\|.$$

Since the last term above tends to zero as $n, m \to \infty$, and $\epsilon > 0$ was arbitrary, the conclusion follows. □

Example. Let $X = Y = \ell^2$ and define $T \in B(\ell^2, \ell^2)$ by

$$Tx = T(x_1, x_2, \ldots) = \left(x_1, \frac{1}{2}x_2, \frac{1}{3}x_3, \ldots \right). \tag{4.2}$$

If we let

$$T_n x = \left(x_1, \frac{1}{2}x_2, \ldots, \frac{1}{n}x_n, 0, \ldots \right),$$

then T_n is compact since it is bounded and has finite-dimensional range. Calculating directly shows

$$\|T - T_n\|^2 = \sup_{\|x\|=1} \|T_n x - Tx\|^2 = \sup_{\|x\|=1} \sum_{j=n+1}^{\infty} \frac{1}{j^2} |x_j|^2$$

$$\leq \sup_{\|x\|=1} \frac{1}{(n+1)^2} \sum_{j=n+1}^{\infty} |x_j|^2 = \frac{1}{(n+1)^2},$$

so $T_n \to T$, and thus T is compact.

When X and Y are NLSs, an operator $T \in B(X, Y)$ is called a *completely continuous linear operator* if T takes weakly convergent sequences in X to strongly convergent sequences in Y. In other words, T is sequentially continuous when X is endowed with its weak topology. A useful property of a compact operator $T : X \to Y$ is that it is completely continuous, as we show next.

Theorem 4.14. *Suppose X and Y are NLSs and $T \in C(X, Y)$. If $\{x_n\}_{n=1}^{\infty} \subset X$ is weakly convergent,*

$$x_n \rightharpoonup x,$$

then the entire image sequence converges in norm, viz.

$$Tx_n \to Tx.$$

Proof. By Proposition 2.61, when T is a bounded linear operator, it is weakly (sequentially) continuous, so we conclude that $Tx_n \rightharpoonup Tx$. Since $\{x_n\}_{n=1}^{\infty}$ converges weakly, it is bounded. The compactness of T implies that $\{Tx_n\}_{n=1}^{\infty}$ has a convergent subsequence $\{Tx_{n_j}\}_{j=1}^{\infty}$ with limit $y \in Y$, say. That is, $Tx_{n_j} \to y$ as $j \to \infty$. But then also $Tx_{n_j} \rightharpoonup y$, so $y = Tx$, whence

$$Tx_{n_j} \to Tx.$$

If $y_n = Tx_n$ does not converge to $y = Tx$ for the entire sequence, then for some $\epsilon > 0$, there is a subsequence $y_{n_k} = Tx_{n_k}$ such that $\|y_{n_k} - y\| \geq \epsilon > 0$. Applying the above argument to the subsequence $\{x_{n_k}\}_{k=1}^{\infty}$ leads to a contradiction. □

Proposition 4.15. *Suppose X is an NLS, $T \in C(X,X)$. Then $\sigma_p(T)$ is countable (it could be empty), and, if it is infinite, it must accumulate at 0 and only at 0. In any case, if X is infinite-dimensional, then $0 \in \sigma(T)$.*

This result shows that eigenvalues of T may be ordered so that $|\lambda_1| \geq |\lambda_2| \geq \cdots$, and that $\lambda_n \to 0$ as $n \to \infty$ if there are an infinite number of them.

Proof. We take the last statement first. Suppose that $0 \notin \sigma(T)$. Then T is one-to-one, maps onto a dense subspace of X, and T^{-1} is bounded. Let B be the closed unit ball in X and $\tilde{B} = T^{-1}(B)$. Then $\tilde{B}$ is a bounded set and $T(\tilde{B}) = B$. If X is infinite-dimensional, B is not compact in X, so T cannot be compact (Corollary 2.50 again).

Let $r > 0$ be given. If it can be established that

$$\sigma_p(T) \cap \{\lambda : |\lambda| \geq r\}$$

is finite for any positive r, then the result follows.

Arguing by contradiction, suppose there is an $r > 0$ and a sequence $\{\lambda_n\}_{n=1}^{\infty}$ of distinct eigenvalues of T with $|\lambda_n| \geq r > 0$, for all n. Let $\{x_n\}_{n=1}^{\infty}$ be corresponding eigenvectors, $x_n \neq 0$ of course. The set $\{x_n : n = 1, 2, \ldots\}$ is linearly independent in X, for if

$$\sum_{j=1}^{N} \alpha_j x_j = 0 \qquad (4.3)$$

and N is chosen to be minimal with this property consistent with not all the α_j being zero, then

$$0 = T_{\lambda_N}\left(\sum_{j=1}^{N} \alpha_j x_j\right) = \sum_{j=1}^{N} \alpha_j (\lambda_j - \lambda_N) x_j.$$

Since $\lambda_j - \lambda_N \neq 0$ for $1 \leq j < N$, by the minimality of N, we conclude that $\alpha_j = 0$, $1 \leq j \leq N - 1$. But then $\alpha_N = 0$ since $x_N \neq 0$. We have reached a contradiction unless (4.3) implies $\alpha_j = 0$, $1 \leq j \leq N$.

Define

$$M_n = \text{span}\{x_1, \ldots, x_n\},$$

and note that $M_1 \subsetneq M_2 \subsetneq M_3 \subsetneq \cdots$ are closed subspaces. For any $x \in M_n$, $x = \sum_{j=1}^{n} \alpha_j x_j$ for some $\alpha_j \in \mathbb{F}$. Because $Tx_j = \lambda_j x_j$, $T : M_n \to M_n$ for all n. Moreover, as above, for $x \in M_n$,

$$T_{\lambda_n} x = \sum_{j=1}^{n} \alpha_j (\lambda_j - \lambda_n) x_j = \sum_{j=1}^{n-1} \alpha_j (\lambda_j - \lambda_n) x_j.$$

Thus it transpires that $T_{\lambda_n} : M_n \to M_{n-1}$.

For each $n = 1, 2, \ldots$, Lemma 2.49 yields points $z_n \in M_n \supsetneq M_{n-1}$ such that $\|z_n\| = 1$ and

$$\|z_n - w\| \geq \frac{1}{2} \quad \text{for all } w \in M_{n-1}. \tag{4.4}$$

Let $n > m$ and consider

$$T z_n - T z_m = \lambda_n z_n - \tilde{x},$$

where

$$\tilde{x} = \lambda_n z_n - T z_n + T z_m = -T_{\lambda_n} z_n + T z_m.$$

As above, $T_{\lambda_n} z_n \in M_{n-1}$ and $T z_m \in M_m \subset M_{n-1}$. Thus $\tilde{x} \in M_{n-1}$, and because of (4.4), we adduce that

$$\|T z_n - T z_m\| = \|\lambda_n z_n - \tilde{x}\| = |\lambda_n| \left\| z_n - \frac{1}{|\lambda_n|} \tilde{x} \right\| \geq \frac{1}{2} |\lambda_n| \geq \frac{1}{2} r > 0.$$

Thus $\{T z_n\}_{n=1}^{\infty}$ has no convergent subsequence, contrary to the hypothesis that T is compact and the fact that $\{z_n\}_{n=1}^{\infty}$ is a bounded sequence. □

Example. Let $X = \ell^2$ and consider the compact operator defined in (4.2). It is not hard to verify that $\sigma_p(T) = \{1/n : n \text{ is a positive integer}\}$ and the eigenspaces are $N(T_{1/n}) = \text{span}\{e_n\}$, where e_n is the vector with one in the nth position and zeros elsewhere. Moreover, T maps onto a dense subspace of X. For, given $\epsilon > 0$ and $y = (y_1, y_2, \ldots) \in X$, let n be such that $\sum_{j=n+1}^{\infty} |y_j|^2 \leq \epsilon^2$, take

$$x = (y_1, 2y_2, 3y_3, \ldots, ny_n, 0, \ldots) \in X,$$

and note that $\|y - Tx\| \leq \epsilon$. Thus, in this case, $0 \in \sigma_c(T)$.

Proposition 4.16. *Suppose that X is an NLS and $T \in C(X, X)$. If $\lambda \neq 0$, then $N(T_\lambda)$ is finite-dimensional.*

Proof. If $\lambda \notin \sigma_p(T)$, then $\dim N(T_\lambda) = 0$, so we can assume $\lambda \in \sigma_p(T)$. Let B be the closed unit ball in $N(T_\lambda)$, so that

$$B = \overline{B_1(0)} \cap N(T_\lambda).$$

Let $\{x_n\}_{n=1}^{\infty}$ be any sequence in B. Since B is bounded, there is a subsequence $\{x_{n_k}\}_{k=1}^{\infty}$ such that $\{Tx_{n_k}\}_{k=1}^{\infty}$ converges, say

$$Tx_{n_k} \to z \quad \text{as } k \to \infty.$$

But $Tx_{n_k} = \lambda x_{n_k}$ and $\lambda \neq 0$, so $x_{n_k} \to \frac{1}{\lambda}z = w$, say. As B is closed, $w \in B$. Thus B is sequentially compact, and hence compact. Since $N(T_\lambda)$ is an NLS, its closed unit ball can be compact only if

$$\dim N(T_\lambda) < +\infty. \qquad \square$$

Here is an important lemma. It is stated separately, as it is useful in other contexts.

Lemma 4.17. *Suppose X and Y are Banach spaces and $T \in B(X, Y)$. Suppose that T is bounded below, so that there is some $\gamma > 0$ for which*

$$\|Tx\|_Y \geq \gamma \|x\|_X \quad \forall x \in X. \qquad (4.5)$$

Then T is one-to-one and $R(T)$ is closed in Y.

For such an operator, if also T maps onto Y, then it has an inverse and $\|T^{-1}\| \leq 1/\gamma$.

Proof. That T is one-to-one is clear by linearity. Suppose for $n = 1, 2, \ldots$ that $y_n = Tx_n$ is a sequence in $R(T)$ and that $y_n \to y \in Y$. Then $\{y_n\}_{n=1}^{\infty}$ is Cauchy, therefore by (4.5), so also is $\{x_n\}_{n=1}^{\infty}$. Since X is complete, there is $x \in X$ such that $x_n \to x$. Since T is continuous, $y_n = Tx_n \to Tx = y \in R(T)$, thus $R(T)$ is closed. $\qquad \square$

Proposition 4.18. *Suppose that X is a Banach space and $T \in C(X, X)$. If $\lambda \neq 0$, then $R(T_\lambda)$ is closed. Moreover, $R(T_\lambda^n)$ is closed for any $n = 1, 2, 3, \ldots$.*

Proof. We begin the proof by 'removing' the null space of T_λ. It was seen previously that $N = N(T_\lambda)$ has finite dimension n, say. Let $\{e_1, \ldots, e_n\}$ be a basis for N. For any $x \in N$, we have the expansion

$$x = \sum_{i=1}^{n} \alpha_i(x) \, e_i,$$

where $\alpha_i : N \to \mathbb{F}$ is linear and so lies in N^* for each $i = 1, 2, \ldots, n$. For each i, apply the Hahn-Banach theorem (Theorem 2.28) to obtain a bounded, linear extension to all of X, which we continue to denote by α_i. Consider the closed, linear subspace

$$M = \bigcap_{i=1}^{n} N(\alpha_i).$$

We claim that $X = M \oplus N$. If $x \in M \cap N$, then x is expanded as above since it lies in N, but all the coefficients must equal zero since $x \in M$. Consequently, $M \cap N = \{0\}$. If $x \in X$, define $y = \sum_{i=1}^{n} \alpha_i(x) e_i \in N$ and $z = x - y$. Notice that $\alpha_i(z) = \alpha_i(x) - \alpha_i(y) = 0$ for all i, so $z \in M$ and we conclude $X = M + N$. The claim is established.

Define the operator $S : M \to X$ by $Sx = T_\lambda x$. Clearly S is a bounded linear operator that is injective on M. In fact, it is bounded below: there is a constant $\gamma > 0$ such that

$$\gamma \|x\| \le \|Sx\| \quad \text{for all } x \in M.$$

If not, then there is a sequence $\{x_n\}_{n=1}^{\infty} \subset M$ such that $\|x_n\| = 1$ and $Sx_n \to 0$ as $n \to \infty$. The compactness of T allows us to extract a subsequence such that Tx_{n_k} converges, say to $x \in X$. The convergence of Sx_{n_k} then implies that λx_{n_k} converges to x, which must be in M since M is a closed subspace. Since $\lambda Sx_{n_k} \to Sx = 0$ and S is injective, $x = 0$. However, $\|x_n\| = 1$ implies that $\|x\| = |\lambda|$, which is a contradiction and establishes boundedness below. By Lemma 4.17, $R(S) = R(T_\lambda)$ is closed.

Finally, for any positive integer n, the operator

$$T_\lambda^n = (T - \lambda I)^n = \sum_{k=1}^{n} \frac{n!}{k!\,(n-k)!} (-\lambda)^{n-k} T^k + (-\lambda)^n I$$

has of the form of a compact operator minus a nonzero multiple of the identity. We just proved that such an operator has closed range. □

Theorem 4.19. *Let X be a Banach space and $T \in C(X, X)$. If $\lambda \in \sigma(T)$ and $\lambda \neq 0$, then $\lambda \in \sigma_p(T)$. That is, all nonzero spectral values are eigenvalues.*

Proof. Let $\lambda \in \sigma(T)$ and $\lambda \neq 0$. To draw a contradiction, suppose $\lambda \notin \sigma_p(T)$. Then T_λ is one-to-one but $R(T_\lambda) \neq X$, because otherwise the open mapping theorem (Theorem 2.42) would imply T_λ^{-1} is bounded, so $\lambda \in \rho(T)$.

Consider the nested sequence of closed subspaces

$$X \supsetneq R(T_\lambda) \supset R(T_\lambda^2) \supset \cdots \supset R(T_\lambda^n) \supset \cdots .$$

Here, $R(T_\lambda^0) = R(I) = X$ by convention. This sequence must stabilize for some $n \geq 1$, which is to say

$$R(T_\lambda^n) = R(T_\lambda^{n+1}). \tag{4.6}$$

If not, then use the construction put forward in Proposition 4.15 (see again Lemma 2.49) to produce a sequence $\{x_n\}_{n=0}^{\infty}$, with

$$x_n \in R(T_\lambda^n), \quad \|x_n\| = 1, \quad n = 0, 1, \ldots,$$

having the property

$$\|x_n - x\| \geq \frac{1}{2} \quad \text{for all } x \in R(T_\lambda^{n+1}).$$

As before, if $n > m$, then

$$Tx_m - Tx_n = T_\lambda x_m - T_\lambda x_n + \lambda x_m - \lambda x_n = \lambda x_m - \tilde{x},$$

where

$$\tilde{x} = \lambda x_n + T_\lambda x_n - T_\lambda x_m.$$

But $x_n \in R(T_\lambda^n) \subset R(T_\lambda^{m+1})$, $T_\lambda x_n \in R(T_\lambda^{n+1}) \subset R(T_\lambda^{m+1})$, and $T_\lambda x_m \in R(T_\lambda^{m+1})$. We gather that $\tilde{x} \in R(T_\lambda^{m+1})$, whence by the construction,

$$\|Tx_m - Tx_n\| = \|\lambda x_m - \tilde{x}\| = |\lambda| \left\| x_m - \frac{1}{\lambda}\tilde{x} \right\| \geq \frac{1}{2}|\lambda|.$$

In consequence, $\{x_n\}_{n=1}^\infty$ is a bounded sequence such that $\{Tx_n\}_{n=1}^\infty$ has no convergent subsequence, a contradiction to the compactness of T. Thus there is an $n \geq 1$ for which (4.6) holds.

Let $y \in X \setminus R(T_\lambda)$. Consider $T_\lambda^n y \in R(T_\lambda^n) = R(T_\lambda^{n+1})$. There is an $x \in X$ such that

$$T_\lambda^{n+1} x = T_\lambda^n y,$$

so

$$T_\lambda^n (y - T_\lambda x) = 0.$$

As T_λ is one-to-one, this means

$$y - T_\lambda x = 0,$$

i.e., $y \in R(T_\lambda)$, a contradiction. The proof has come to an end. $\square$

The next theorem summarizes the foregoing analysis of the spectral properties of compact operators.

Theorem 4.20 (spectral theorem for compact operators). *Let X be a Banach space and $T \in C(X, X)$. The spectrum of T consists of at most a countable number of eigenvalues and possibly 0. If $\lambda \in \sigma(T)$ and $\lambda \neq 0$, then the eigenspace $N(T_\lambda)$ is finite-dimensional. If X is infinite-dimensional, then $0 \in \sigma(T)$. If there are infinitely many eigenvalues, they converge to $0 \in \sigma(T)$.*

Corollary 4.21 (Fredholm alternative for compact operators). *Suppose X is a Banach space, $\lambda \in \mathbb{F}$, $\lambda \neq 0$, and $T \in C(X, X)$. Define the operator $A = I - \frac{1}{\lambda}T$, which is a compact perturbation of the identity operator. Let $y \in X$ and consider the problem*

$$Ax = -\frac{1}{\lambda}(T - \lambda I)x = -\frac{1}{\lambda}T_\lambda x = y.$$

Either

(a) *there exists a unique solution $x \in X$ to the equation for every $y \in X$; or*

(b) *if $y \in X$ has a solution, then it has infinitely many solutions.*

Proof. Case (a) corresponds to $\lambda \in \rho(T)$. Otherwise, $\lambda \in \sigma_p(T)$ which we must show is case (b). So for $\lambda \in \sigma_p(T)$, if x is a solution to the equation, then so is $x + z$ for any $z \in N(T_\lambda) \neq \{0\}$, and so there must be infinitely many solutions. $\square$

4.4 BOUNDED SELF-ADJOINT LINEAR OPERATORS ON A HILBERT SPACE

We consider now an operator $T \in B(H, H)$ defined on a Hilbert space H. Because of the Riesz representation theorem (Theorem 3.15), the dual or adjoint operator $T^* : H^* \to H^*$ is also defined on $H \cong H^*$. Let $R : H \to H^*$ be the Riesz map. As explained in Chapter 3, T^* is equivalent to the Hilbert adjoint $T' = R^{-1}T^*R : H \to H$. Henceforth, we simply write T^* for this Hilbert adjoint operator, since the domain of definition makes clear whether one means the usual dual operator or the Hilbert adjoint. With this convention, recall from Proposition 3.16 that $T^* \in B(H, H)$, $T = T^{**}$, and

$$(Tx, y) = (x, T^*y) \quad \text{and} \quad (T^*x, y) = (x, Ty) \quad \forall\, x, y \in H.$$

The spectral knowledge of a special subclass of operators is now investigated. This type of operator arises frequently in applications.

Definition. Suppose that H is a Hilbert space and $T \in B(H, H)$. We say that T is *self-adjoint* or *Hermitian* if $T = T^*$, interpreted as above, i.e., if

$$(x, Ty) = (Tx, y) \quad \forall x, y \in H.$$

Proposition 4.22. *Let H be a Hilbert space and $T \in B(H, H)$.*

(a) *If T is self-adjoint, then*

$$(Tx, x) \in \mathbb{R} \quad \forall x \in H.$$

(b) *If H is a complex Hilbert space, then T is self-adjoint if and only if (Tx, x) is real for all $x \in H$.*

Proof. (a) Compute directly:

$$(Tx, x) = \overline{(x, Tx)} = \overline{(x, T^*x)} = \overline{(Tx, x)},$$

whence $(Tx, x) \in \mathbb{R}$.

(b) By (a), we need only show the converse. This will follow if it can be shown that

$$(Tx, y) = (T^*x, y) \quad \forall\, x, y \in H.$$

Let $\alpha \in \mathbb{C}$ and compute

$$\big(T(x + \alpha y), x + \alpha y\big) = (Tx, x) + |\alpha|^2 (Ty, y) + \alpha(Ty, x) + \bar{\alpha}(Tx, y).$$

By hypothesis, the left-hand side and the first two terms on the right are real, so

$$\alpha(Ty, x) + \bar{\alpha}(Tx, y) = \overline{\bar{\alpha}(T^*x, y)} + \bar{\alpha}(Tx, y) \in \mathbb{R}.$$

If $\alpha = 1$, then the complex parts of (Tx, y) and (T^*x, y) are seen to agree; if $\alpha = i$, the real parts agree.

□

Theorem 4.23 (spectral theorem for self-adjoint operators, part 1). *Let H be a complex Hilbert space and $T \in B(H, H)$ be a self-adjoint operator. Then $\sigma_r(T) = \emptyset$ and*

$$\sigma(T) \subset [r, R] \subset \mathbb{R},$$

where

$$r = \inf_{\|x\|=1} (Tx, x) \quad and \quad R = \sup_{\|x\|=1} (Tx, x).$$

Moreover, $\lambda \in \rho(T)$ if and only if T_λ is bounded below.

Proof. If $\lambda \in \sigma_p(T)$ and $Tx = \lambda x$ for some $x \neq 0$, then

$$\lambda(x, x) = (Tx, x) = (x, Tx) = (x, \lambda x) = \bar{\lambda}(x, x);$$

thus $\lambda = \bar{\lambda}$ is real.

If $\lambda \in \rho(T)$, then T_λ^{-1} is bounded, so

$$\|x\| = \|T_\lambda^{-1} T_\lambda x\| \leq \|T_\lambda^{-1}\| \, \|T_\lambda x\|,$$

showing that T_λ is bounded below since $\|T_\lambda^{-1}\| \neq 0$. Conversely, suppose T_λ is bounded below. By Lemma 4.17, T_λ is one-to-one and $R(T_\lambda)$ is closed. If $R(T_\lambda) \neq H$, there is some nontrivial $x_0 \in R(T_\lambda)^\perp$. Therefore, for all $x \in H$,

$$\begin{aligned} 0 = (T_\lambda x, x_0) &= (Tx, x_0) - \lambda(x, x_0) \\ &= (x, Tx_0) - \lambda(x, x_0) \\ &= (x, T_{\bar{\lambda}} x_0). \end{aligned}$$

Thus $T_{\bar{\lambda}} x_0 = 0$, so $Tx_0 = \bar{\lambda} x_0$ and $\bar{\lambda} \in \sigma_p(T)$. But then $\bar{\lambda} \in \mathbb{R}$ and T_λ is not one-to-one, a contradiction. Thus $R(T_\lambda) = H$ and $\lambda \in \rho(T)$.

To see that the entire spectrum is real, suppose now $\lambda = \alpha + i\beta \in \sigma(T)$, where $\alpha, \beta \in \mathbb{R}$. For any $x \neq 0$ in H,

$$(T_\lambda x, x) = (Tx, x) - \lambda(x, x)$$

and

$$\overline{(T_\lambda x, x)} = (Tx, x) - \bar{\lambda}(x, x),$$

since (Tx, x) is real. Thus,

$$(T_\lambda x, x) - \overline{(T_\lambda x, x)} = -2i\beta(x, x),$$

or

$$|\beta| \, \|x\|^2 = \frac{1}{2} \left| (T_\lambda x, x) - \overline{(T_\lambda x, x)} \right| \leq \|T_\lambda x\| \, \|x\|.$$

As $x \neq 0$ is arbitrary in H, if $\beta \neq 0$, T_λ is bounded below, and thus $\lambda \in \rho(T)$, a contradiction. Therefore, $\sigma(T) \subset \mathbb{R}$.

Now suppose that $\sigma_r(T)$ is not empty and let $\lambda \in \sigma_r(T)$. Then T_λ is invertible on its range, *viz.*

$$T_\lambda^{-1} : R(T_\lambda) \longrightarrow H,$$

but
$$\overline{R(T_\lambda)} \neq H.$$

Let $y \neq 0$ lie in $y \in \overline{R(T_\lambda)}^{\perp}$. Then, for all $x \in H$,
$$0 = (T_\lambda x, y) = (x, T_\lambda y),$$

which implies that $T_\lambda y = 0$, so $\lambda \in \sigma_p(T)$. Since $\sigma_r(T) \cap \sigma_p(T) = \emptyset$, we have our contradiction, and conclude that $\sigma_r(T) = \emptyset$.

To obtain bounds on the spectrum, let $c > 0$ and let $\lambda = R + c > R$, where R is as defined in the statement of the theorem. Let $x \neq 0$ and compute

$$(Tx, x) = \|x\|^2 \left(T\left(\frac{x}{\|x\|}\right), \frac{x}{\|x\|} \right) \leq \|x\|^2 R.$$

On the one hand,

$$-(Tx - \lambda x, x) = -(T_\lambda x, x) \leq \|T_\lambda x\| \, \|x\|,$$

and on the other hand,

$$-(Tx - \lambda x, x) = -(Tx, x) + \lambda\|x\|^2 \geq -\|x\|^2 R + \lambda\|x\|^2 = c\|x\|^2.$$

It is concluded that
$$\|T_\lambda x\| \geq c\|x\|,$$

showing that T_λ is bounded below. Hence, $\lambda \in \rho(T)$.

A similar argument applies in case $\lambda = r - c$ where $c > 0$. Write for $x \neq 0$

$$(Tx, x) = \|x\|^2 \left(T\left(\frac{x}{\|x\|}\right), \frac{x}{\|x\|} \right) \geq \|x\|^2 r.$$

Then

$$(Tx - \lambda x, x) = (T_\lambda x, x) \leq \|T_\lambda x\| \, \|x\|,$$
$$(Tx - \lambda x, x) = (Tx, x) - \lambda\|x\|^2 \geq (r - \lambda)\|x\|^2 = c\|x\|^2,$$

so again, $\lambda \in \rho(T)$. □

The quantity
$$q(x) = \frac{(Tx, x)}{(x, x)}$$

is called the *Rayleigh quotient* of T at x. The result above is that

$$\sigma(T) \subset \left[\inf_{x \neq 0} q(x), \ \sup_{x \neq 0} q(x) \right].$$

The next result shows further importance of the Rayleigh quotient of a self-adjoint operator.

Theorem 4.24 (spectral theorem for self-adjoint operators, part 2). *Let H be a complex Hilbert space and $T \in B(H, H)$ self-adjoint. Then*

$$r = \inf_{\|x\|=1} (Tx, x) \in \sigma(T) \quad and \quad R = \sup_{\|x\|=1} (Tx, x) \in \sigma(T).$$

Moreover,

$$\|T\|_{B(X,X)} = \sup_{\|x\|=1} |(Tx, x)| = \max(|r|, |R|).$$

That is, the minimal real number in $\sigma(T)$ is r, and the maximal number in $\sigma(T)$ is R, the infimal and supremal values of the Rayleigh quotient. These values also give the norm as $\|T\| = \max(|r|, |R|)$.

Proof. Let

$$M = \sup_{\|x\|=1} |(Tx, x)| = \max(|r|, |R|);$$

obviously,

$$M \leq \|T\|.$$

If $T \equiv 0$, the result is trivially true. So let $z \in H$ be such that $Tz \neq 0$ and normalize so that $\|z\| = 1$. Set

$$v = \|Tz\|^{1/2} z \quad and \quad w = \|Tz\|^{-1/2} Tz.$$

Then

$$\|v\|^2 = \|w\|^2 = \|Tz\|$$

and, since T is self-adjoint,

$$\big(T(v + w), v + w\big) - \big(T(v - w), v - w\big) = 2\big[(Tv, w) + (Tw, v)\big] = 4\|Tz\|^2,$$

and

$$\big|\big(T(v + w), v + w\big) - \big(T(v - w), v - w\big)\big|$$
$$\leq \big|\big(T(v + w), v + w\big)\big| + \big|\big(T(v - w), v - w\big)\big|$$
$$\leq M \big(\|v + w\|^2 + \|v - w\|^2\big)$$
$$= 2M \big(\|v\|^2 + \|w\|^2\big)$$
$$= 4M\|Tz\|.$$

It follows that

$$\|Tz\| \leq M,$$

and, taking the supremum over all such z yields

$$\|T\| \leq M.$$

Thus $\|T\| = M$.

Obviously, $\lambda \in \sigma(T)$ if and only if $\lambda - \mu \in \sigma(T_\mu)$, so by making such a translation, we may assume that $0 \leq r \leq R$. Then $\|T\| = R$ and there is a sequence $\{x_n\}_{n=1}^\infty$ such that $\|x_n\| = 1$ and

$$(Tx_n, x_n) \geq R - \frac{1}{n}.$$

But then,

$$\begin{aligned}
\|T_R x_n\|^2 &= \|Tx_n - Rx_n\|^2 \\
&= \|Tx_n\|^2 - 2R(Tx_n, x_n) + R^2 \\
&\leq 2R^2 - 2R\left(R - \frac{1}{n}\right) = \frac{2R}{n} \longrightarrow 0.
\end{aligned}$$

Thus T_R is not bounded below, hence $R \notin \rho(T)$, so $R \in \sigma(T)$. A similar argument shows $r \in \sigma(T)$. □

We know that if $T \in B(H, H)$ is self-adjoint, then $(Tx, x) \in \mathbb{R}$ for all $x \in H$.

Definition. If H is a Hilbert space and $T \in B(H, H)$ satisfies

$$(Tx, x) \geq 0 \quad \forall x \in H,$$

then T is said to be a *positive operator*. We denote this fact by writing $0 \leq T$. If $R, S \in B(H, H)$, then $R \leq S$ means that $0 \leq S - R$.

Proposition 4.25. *Suppose H is a complex Hilbert space and $T \in B(H, H)$. Then T is a positive operator if and only if T is self-adjoint and $\sigma(T) \subset \mathbb{R}^+ = [0, \infty)$.*

Proof. This result is a consequence of Proposition 4.22 and Theorems 4.23 and 4.24. □

Remark. It follows from the last two results that if H is a complex Hilbert space, the binary relation $S \leq T$ is a partial ordering on the algebra $B(H, H)$.

It is easy to show that if T and S are positive operators, then so is $T + S$ and T^n for any integer $n \geq 0$. If $T \geq 0$ has an inverse, then $T^{-1} \geq 0$. If $\alpha \geq 0$, then $\alpha T \geq 0$. Moreover, the set of positive operators is closed in $B(H, H)$. An interesting and useful fact about a positive operator is that it has a positive square root.

Definition. Let H be a Hilbert space and $T \in B(H, H)$ be positive. An operator $S \in B(H, H)$ is said to be a *square root* of T if

$$S^2 = T.$$

If, in addition, S is a positive operator, then it is called a *positive square root* of T and is denoted

$$S = T^{1/2} = \sqrt{T}.$$

Theorem 4.26. *If H is a complex Hilbert space, then every positive operator $T \in B(H, H)$ has a unique positive square root S, and S commutes with T and with any bounded linear operator that commutes with T.*

Proof. As in the proof of Neumann series (Lemma 4.4), we base the argument on the Taylor series of an analytic function (see Theorem 1.50). For $z \in \mathbb{C}$,

$$\sqrt{1 - z} = \sum_{n=0}^{\infty} c_n z^n,$$

which converges absolutely for $|z| < 1$. The coefficients satisfy $c_0 = 1$ and $c_n = n! \frac{d^n}{dz^n}(1 - z)^{1/2}|_{z=0} \leq 0$ for all $n \geq 1$, and so the series converges absolutely for $|z| = 1$. To see this, for $N \geq 1$, compute

$$\sum_{n=0}^{N} |c_n| |z|^n = \sum_{n=0}^{N} |c_n| = 2 - \sum_{n=0}^{N} c_n = 2 - \lim_{x \to 1^-} \sum_{n=0}^{N} c_n x^n$$

$$\leq 2 - \lim_{x \to 1^-} \sqrt{1 - x} = 2,$$

and conclude that $\sum_{n=0}^{\infty} |c_n| = 2 < \infty$. A similar argument applies to the square of the series,

$$1 - z = \left(\sum_{n=0}^{\infty} c_n z^n \right)^2 = \sum_{n=0}^{\infty} d_n z^n,$$

which is also absolutely convergent for $|z| \leq 1$, although clearly in this case $d_0 = 1$, $d_1 = -1$, and $d_n = 0$ for $n \geq 2$.

If $T = 0$, $S = 0$ as well. Otherwise, if $A = T/\|T\|$ has a positive square root B, then $S = \sqrt{\|T\|}\, B$. So it suffices to take T such that $\|T\| \leq 1$. The positivity of T and the spectral theorem for self-adjoint operators (Theorem 4.24) implies that $0 \leq T \leq I$ and $\|I - T\| \leq 1$. Let

$$S_N = \sum_{n=0}^{N} c_n (I - T)^n,$$

which is Cauchy in $B(H, H)$ and converges absolutely to some $S \in B(H, H)$. The terms in the series can be rearranged, so

$$S^2 = \lim_{N \to \infty} S_N^2 = \lim_{N \to \infty} \left(\sum_{n=0}^{N} c_n (I - T)^n \right)^2$$

$$= \lim_{N \to \infty} \sum_{n=0}^{N} d_n (I - T)^n = I - (I - T) = T.$$

Since S_N will commute with any bounded linear operator commuting with T, the same is true of S. For any $n \geq 1$, $(I - T)^n$ is positive and its norm is bounded by 1. If $x \in H$ with $\|x\| = 1$, then

$$(S_N x, x) = \sum_{n=0}^{N} c_n((I - T)^n x, x) \geq 1 + \sum_{n=1}^{N} c_n \geq 1 + \sum_{n=1}^{\infty} c_n = 0,$$

showing that $S_N \geq 0$, and by continuity that also $S \geq 0$.

To show uniqueness, suppose that R is another positive square root. Then $RT = R^3 = TR$, so R commutes with T and thus also with S. Therefore $0 = T - T = R^2 - S^2 = (R - S)(R + S)$. Now $0 = (R - S)(R + S)(R - S) = (R - S)R(R - S) + (R - S)S(R - S)$ is the sum of two positive operators, so they must both vanish, as well as their difference $(R - S)^3 = 0$. Now if A is self-adjoint and $A^2 = 0$, then $A = 0$ as well, since

$$0 = \|A^2\| = \sup_{\|x\|=1} (A^2 x, x) = \sup_{\|x\|=1} (Ax, Ax) = \|A\|^2.$$

We apply this fact to $(R-S)^4$ to conclude that $0 = \|(R-S)^4\| = \|(R-S)^2\|^2 = \|R - S\|^4$, leading to $R = S$. □

Examples.

(1) Let $H = L^2(\Omega)$ for some measurable set $\Omega \subset \mathbb{R}^d$ and $\phi : \Omega \to \mathbb{R}$ a measurable, positive, and bounded function. Then $T : H \to H$ defined by

$$(Tf)(x) = \phi(x)f(x) \quad \forall\, x \in \Omega$$

is a positive operator, with positive square root

$$(Sf)(x) = \sqrt{\phi(x)}\, f(x) \quad \forall\, x \in \Omega.$$

(2) If $T \in B(H, H)$ is any operator, then T^*T is positive.

(3) Let H be a complex Hilbert space and A a bounded linear operator on H. The *absolute value* of the operator A is $|A| = (A^*A)^{1/2}$, which is well-defined and positive.

4.5 COMPACT SELF-ADJOINT LINEAR OPERATORS ON A HILBERT SPACE

On a Hilbert space, we can be very specific about the structure of a self-adjoint, compact operator. For such operators, the spectrum is real, countable, and nonzero values are eigenvalues with finite-dimensional eigenspaces. Moreover, if the number of eigenvalues is infinite, then they converge to 0.

Theorem 4.27 (Hilbert-Schmidt theorem). *Let H be a complex Hilbert space, $T \in C(H, H)$, and $T = T^*$. There is an ON set $\{u_n\}_{n=1}^{N}$ of eigenvectors corresponding to nonzero eigenvalues $\{\lambda_n\}_{n=1}^{N}$ of T, where N may be a finite number or infinity, such that every $x \in H$ has a unique decomposition of the form*

$$x = \sum_{n=1}^{N} \alpha_n u_n + v,$$

where $\alpha_n \in \mathbb{C}$ and $v \in N(T)$.

Proof. We make extensive use of the spectral theorems for compact (Theorem 4.20) and self-adjoint operators (Theorems 4.23–4.24). The construction of the $\{u_n\}_{n=1}^{N}$ takes place inductively, one dimension at a time. If $T = 0$, $N(T) = H$ and there is nothing to prove. Otherwise $T \neq 0$, and there is a nonzero eigenvalue λ_1 of T such that

$$|\lambda_1| = \sup_{\|x\|=1} |(Tx, x)|.$$

Let u_1 be an associated eigenvector, normalized so that $\|u_1\| = 1$. Let $Q_1 = \{u_1\}^{\perp}$. Then Q_1 is a closed linear subspace of H, so Q_1 is a Hilbert space in its own right. Moreover, if $x \in Q_1$, we have by self-adjointness that

$$(Tx, u_1) = (x, Tu_1) = \overline{\lambda_1}(x, u_1) = 0,$$

so $Tx \in Q_1$. Thus $T : Q_1 \to Q_1$ and we may conclude that either $T|_{Q_1} = 0$ or there is a nonzero eigenvalue λ_2 with

$$|\lambda_2| = \sup_{\substack{\|x\|=1 \\ x \in Q_1}} |(Tx, x)| \leq |\lambda_1|.$$

Let u_2 be a normalized eigenvector corresponding to λ_2. Plainly, $u_1 \perp u_2$. Let

$$Q_2 = \{x \in Q_1 : x \perp u_2\} = \{u_1, u_2\}^{\perp},$$

which is a Hilbert space. Arguing inductively, there obtains a sequence of closed linear subspaces $\{Q_n\}$. At the nth stage, we note that if $x \in Q_n = \{u_1, \ldots, u_n\}^{\perp}$, then for $j = 1, \ldots, n$,

$$(Tx, u_j) = (x, Tu_j) = \overline{\lambda}_j(x, u_j) = 0,$$

so $T : Q_n \to Q_n$. Thus either $T|_{Q_n} = 0$ or there is a nonzero eigenvalue λ_{n+1} with

$$|\lambda_{n+1}| = \sup_{\substack{\|x\|=1 \\ x \in Q_n}} |(Tx, x)| \leq |\lambda_n|$$

and an eigenvector u_{n+1} with $\|u_{n+1}\| = 1$ corresponding to λ_{n+1}.

Two possibilities occur. Either we reach a point where $|\lambda_N| > 0$ for some N but $T|_{Q_N} = 0$, or $\lambda_n \neq 0$ for all n (in this case, let $N = \infty$). Let H_1 be the Hilbert space generated by the ON family $\{u_n\}_{n=1}^N$. Every element $x \in H$ is written uniquely in the form

$$x = \sum_{j=1}^{N} (x, u_j) u_j + v$$

for some $v \in H_1^\perp$, since $H = H_1 \oplus H_1^\perp$. It remains to check that $H_1^\perp = N(T)$. Let $v \in H_1^\perp$, $v \neq 0$. Now,

$$H_1^\perp \subset Q_n \quad \text{for all } n = 1, 2, \ldots, N,$$

so it must be the case that

$$\frac{|(Tv, v)|}{\|v\|^2} \leq \sup_{x \in Q_n} \frac{|(Tx, x)|}{\|x\|^2} = |\lambda_{n+1}|.$$

The right-hand side vanishes in the case of finite N, and tends to zero as $n \to \infty$ otherwise, since T is a compact operator. The left-hand side does not depend on n, so it follows that

$$(Tv, v) = 0 \quad \text{for all } v \in H_1^\perp.$$

Thus $T|_{H_1^\perp}$ vanishes, as

$$\left\| T|_{H_1^\perp} \right\| = \sup_{\substack{\|v\|=1 \\ v \in H_1^\perp}} |(Tv, v)| = 0,$$

so $H_1^\perp \subset N(T)$. For $x \in H_1$ and for some scalars β_n,

$$Tx = T\left(\sum_{n=1}^{N} \beta_n u_n \right) = \sum_{n=1}^{N} \beta_n T u_n = \sum_{n=1}^{N} \lambda_n \beta_n u_n \in H_1,$$

so $T : H_1 \to H_1$ is one-to-one since each $\lambda_n \neq 0$. Thus $N(T) \cap H_1 = \{0\}$. If $x \in N(T) \supset H_1^\perp$, it has the unique decomposition $x = h_1 + h_2 \in H_1 \oplus H_1^\perp = H$, and so $0 = Tx = Th_1 + Th_2 = Th_1$ implies that $h_1 = 0$. That is, $H_1^\perp = N(T)$, and the proof is complete. □

Theorem 4.28 (spectral theorem for compact self-adjoint operators). *Let $T \in C(H, H)$ be a compact, self-adjoint operator on a complex Hilbert space H. Then there exists an ON basis $\{v_\alpha\}_{\alpha \in \mathcal{I}}$ for H such that each v_α is an eigenvector for T. Moreover, for every $x \in H$,*

$$Tx = \sum_{\alpha \in \mathcal{I}} \lambda_\alpha (x, v_\alpha) v_\alpha,$$

where λ_α is the (real) eigenvalue corresponding to v_α.

Proof. Let $\{u_n\}_{n=1}^N$ be the ON system constructed in the last theorem (N is finite or ∞). Let H_1 be the closed subspace $\overline{\text{span}\{u_n\}_{n=1}^N}$. Let $\{e_\beta\}_{\beta \in \mathcal{J}}$ be an ON basis for $H_1^\perp = N(T)$. Then

$$\{e_\beta\}_{\beta \in \mathcal{J}} \cup \{u_n\}_{n=1}^N$$

is an ON basis for H. Moreover,

$$T e_\beta = 0 \quad \forall \beta \in \mathcal{J},$$

so the e_β are eigenvectors corresponding to the eigenvalue 0 (if $H_1^\perp \neq \{0\}$).

We know that for $x \in H$, there is $v \in N(T)$ such that

$$\sum_{n=1}^N (x, u_n) u_n + v$$

converges to x in H. Because T is continuous,

$$T\left(\sum_{n=1}^N (x, u_n) u_n + v \right) = \sum_{n=1}^N (x, u_n) T u_n + T v = \sum_{n=1}^N \lambda_n (x, u_n) u_n,$$

since $v \in N(T)$. $\qquad\square$

One can think of the preceding as representing a self-adjoint $T \in C(H, H)$ as an infinite, diagonal matrix with the eigenvalues appearing on the diagonal. It should come as no surprise that if T is also a positive operator, S defined by

$$S x = \sum_{\alpha \in \mathcal{I}} \sqrt{\lambda_\alpha}\, (x, u_\alpha) u_\alpha$$

is the positive square root of T. We leave it to the reader to verify this statement, as well as the implied fact that $S \in C(H, H)$.

Proposition 4.29. *Let $S, T \in C(H, H)$ be compact, self-adjoint operators on a complex Hilbert space H. Suppose $ST = TS$. Then there exists an ON basis $\{v_\alpha\}_{\alpha \in \mathcal{I}}$ for H of common eigenvectors of S and T.*

Proof. Let $\lambda \in \sigma_p(S)$ and let V_λ be the corresponding eigenspace. For any $x \in V_\lambda$,

$$ST x = TS x = T(\lambda x) = \lambda T x \quad \implies \quad T x \in V_\lambda.$$

Therefore $T : V_\lambda \to V_\lambda$. Now T is self-adjoint on V_λ and compact, so it has a complete ON set of T-eigenvectors. This ON set are also eigenvectors for S since everything in V_λ is such. $\qquad\square$

4.6 THE ASCOLI-ARZELÀ THEOREM

The discussion turns to important examples of compact operators called *integral operators*. These are operators of the form

$$(Tf)(x) = \int_\Omega K(x, y)\, f(y)\, dy,$$

where f is drawn from some Hilbert (or Banach) space and K satisfies appropriate hypotheses. A helpful criterion often used to establish that such operators are compact is a general compactness criterion, the Ascoli-Arzelà theorem. This is explained after a basic lemma.

Lemma 4.30. *A compact metric space (M, d) is separable (i.e., it has a countable dense subset).*

Proof. For any integer $n \geq 1$, cover M by balls of radius $1/n$, *viz.*

$$M = \bigcup_{x \in M} B_{1/n}(x).$$

By compactness, there is a finite subcover

$$M = \bigcup_{i=1}^{N_n} B_{1/n}(x_i^n) \tag{4.7}$$

for some $x_i^n \in M$. The set

$$S = \{x_i^n : i = 1, \ldots, N_n;\ n = 1, 2, \ldots\}$$

is a countable union of finite sets, and so countable. We claim that it is dense in M. Let $x \in M$ and $\epsilon > 0$ be given. Choose n large enough that $1/n \leq \epsilon$. For such a value of n, (4.7) implies there is some $x_j^n \in S$ with

$$x \in B_{1/n}(x_j^n);$$

that is, $d(x, x_j^n) < 1/n \leq \epsilon$. Thus S is indeed dense in M. □

Theorem 4.31 (Ascoli-Arzelà theorem). *Let (M, d) be a compact metric space and let*

$$C(M) = C(M; \mathbb{F})$$

denote the Banach space of continuous functions from M to $\mathbb{F}$ with the maximum norm

$$\|f\| = \max_{x \in M} |f(x)|.$$

Let $A \subset C(M)$ be a subset that is bounded and equicontinuous, which is to say, respectively, that for some $R > 0$,

$$\|f\| < R \quad \forall f \in A,$$

(A ⊂ $B_R(0)$), and, given $\epsilon > 0$ there is $\delta > 0$ such that for all $f \in A$,

$$\sup_{x,y \in M,\, d(x,y) < \delta} |f(x) - f(y)| < \epsilon. \qquad (4.8)$$

Then the closure $\bar{A}$ of A is compact in $C(M)$.

Proof. It suffices by Lemma 4.9 to show that an arbitrary sequence $\{f_n\}_{n=1}^\infty \subset \bar{A}$ has a convergent subsequence. For each fixed $x \in M$, $\{f_n(x)\}_{n=1}^\infty$ is bounded in $\mathbb{F}$ and so it has a convergent subsequence. Let $\{x_j\}_{j=1}^\infty$ be a countable dense subset of M. By a diagonalization argument, we can extract a single subsequence $\{f_{n_k}\}_{k=1}^\infty$ such that $\{f_{n_k}(x_j)\}_{k=1}^\infty$ converges for each j, as follows. Let $\{f_{n_k(x_1)}(x_1)\}_{k=1}^\infty$ be convergent, and from the bounded set $\{f_{n_k(x_1)}(x_2)\}_{k=1}^\infty$, select a convergent subsequence $\{f_{n_k(x_2)}(x_2)\}_{k=1}^\infty$. Continuing, there obtains indices

$$\{n_k(x_1)\}_{k=1}^\infty \supset \{n_k(x_2)\}_{k=1}^\infty \supset \cdots$$

such that $\{f_{n_k(x_i)}(x_j)\}_{k=1}^\infty$ converges for all $j \le i$.

We abbreviate the subsequence $\{f_{n_k(x_k)}\}_{k=1}^\infty$ as $\{f_{n_k}\}_{k=1}^\infty$ for ease of reading. This subsequence is convergent in $C(M)$. To see this, it suffices to show that it is uniformly Cauchy. Proceed as follows. First, it may be that for some i, $f_{n_i} \notin A$. Let $\epsilon > 0$ be given and choose $\tilde{f}_{n_i} \in A$ such that

$$\|f_{n_i} - \tilde{f}_{n_i}\| \le \epsilon.$$

Now let $\delta > 0$ correspond to this fixed ϵ so that (4.8) holds. There exists a finite subset $\{\tilde{x}_m\}_{m=1}^N \subset \{x_j\}_{j=1}^\infty$ such that

$$\bigcup_{m=1}^N B_\delta(\tilde{x}_m) \supset M,$$

since M is compact. Fix $x \in M$ and choose $\tilde{x}_\ell$ to be such that

$$d(x, \tilde{x}_\ell) < \delta.$$

For any i, j, by (4.8),

$$\begin{aligned}
|f_{n_i}&(x) - f_{n_j}(x)| \\
&\le |f_{n_i}(x) - \tilde{f}_{n_i}(x)| + |\tilde{f}_{n_i}(x) - \tilde{f}_{n_i}(\tilde{x}_\ell)| + |\tilde{f}_{n_i}(\tilde{x}_\ell) - f_{n_i}(\tilde{x}_\ell)| \\
&\quad + |f_{n_i}(\tilde{x}_\ell) - f_{n_j}(\tilde{x}_\ell)| + |f_{n_j}(\tilde{x}_\ell) - \tilde{f}_{n_j}(\tilde{x}_\ell)| \\
&\quad + |\tilde{f}_{n_j}(\tilde{x}_\ell) - \tilde{f}_{n_j}(x)| + |\tilde{f}_{n_j}(x) - f_{n_j}(x)| \\
&\le 6\epsilon + |f_{n_i}(\tilde{x}_\ell) - f_{n_j}(\tilde{x}_\ell)| \\
&\le 6\epsilon + \max_{1 \le m \le N} |f_{n_i}(\tilde{x}_m) - f_{n_j}(\tilde{x}_m)|.
\end{aligned}$$

As N is fixed, the last term on the right-hand side of this inequality can be made less than ϵ provided i and j are large enough. For such values of i and j, the right-hand side is then less than 7ϵ and this is independent of x. Therefore, the sequence $\{f_{n_k}\}_{k=1}^\infty$ is uniformly Cauchy and the result is established. ☐

Theorem 4.32. *Let* $\Omega \subset \mathbb{R}^d$ *be bounded and open, and* K *continuous on* $\overline{\Omega} \times \overline{\Omega}$. *Let* $X = C(\overline{\Omega})$ *and define* $T : X \to X$ *by*

$$Tf(x) = \int_\Omega K(x, y) f(y) \, dy.$$

(That T *is well-defined follows since both* K *and* f *are continuous, hence bounded and measurable.) Then* T *is compact.*

Proof. Let $\{f_n\}_{n=1}^\infty$ be bounded in $C(\overline{\Omega})$. We must show that $\{Tf_n\}_{n=1}^\infty$ has a convergent subsequence. Since $\overline{\Omega}$ is a compact metric space, the Ascoli-Arzelà theorem implies the result if the image of our sequence is bounded and equicontinuous. The former follows since

$$\|Tf_n\|_{L^\infty(\Omega)} \leq \|f_n\|_{L^\infty(\Omega)} \|K\|_{L^\infty(\Omega \times \Omega)} |\Omega|,$$

where $|\Omega|$, the d-dimensional volume of Ω, is bounded independently of n. For equicontinuity, compute

$$|Tf_n(x) - Tf_n(y)| = \left| \int_\Omega \left(K(x, z) - K(y, z) \right) f_n(z) \, dz \right|$$
$$\leq \|f_n\|_{L^\infty(\Omega)} \sup_{z \in \overline{\Omega}} |K(x, z) - K(y, z)| \, |\Omega|.$$

Since K is uniformly continuous on $\overline{\Omega} \times \overline{\Omega}$, the right-side above can be made uniformly small provided $|x - y|$ is taken small enough, independently of $\{f_n\}_{n=1}^\infty$ save for the bound on the sequence. $\qquad\square$

By an argument based on the density of $C(\overline{\Omega})$ in $L^2(\Omega)$, and the fact that the limit of compact operators is compact, we can extend this result to $L^2(\Omega)$. The details are left to the reader in an exercise at the end of the chapter.

Corollary 4.33. *Let* $\Omega \subset \mathbb{R}^d$ *be bounded and open. Suppose* $K \in L^2(\Omega \times \Omega)$ *and* $T : L^2(\Omega) \to L^2(\Omega)$ *is defined as in the previous theorem. Then* T *is compact. Moreover, if* $K(x, y) = \overline{K(y, x)}$ *for a.e.* $x, y \in \Omega$, *then* T *is also self-adjoint.*

4.7 STURM-LIOUVILLE THEORY

We proceed with a study of spectral theory in the context of ordinary differential equations.

Suppose $I = [a, b] \subset \mathbb{R}$, $a_j \in C^{2-j}(I)$, $j = 0, 1, 2$, and $a_0 > 0$. Consider the differential operator $L : C^2(I) \to C(I)$ defined for $x(t) \in C^2(I)$ by

$$(Lx)(t) = a_0(t)x''(t) + a_1(t)x'(t) + a_2(t)x(t).$$

Note that, assuming only that $a_j \in C^0(I)$, $j = 0, 1, 2$, L is seen to be a bounded linear operator when using, e.g., the L^∞-norm for the function and its derivatives. That is, the norm of $u \in C^n$ is

$$\|u\|_{C^n} = \max_{0 \le k \le n} \left\| \frac{d^k u}{dt^k} \right\|_{L^\infty}.$$

Theorem 4.34 (Picard's theorem). *Given $f \in C(I)$ and $x_0, x_1 \in \mathbb{R}$, there exists a unique solution $x \in C^2(I)$ to the initial value problem (IVP)*

$$\begin{cases} Lx = f, & t \in (a, b], \\ x(a) = x_0, & x'(a) = x_1. \end{cases} \tag{4.9}$$

Consult a text on ordinary differential equations for a proof.

Corollary 4.35. *The null space $N(L)$ is two-dimensional.*

Proof. A two-dimensional basis for $N(L)$ is constructed thusly. Solve (4.9) with $f = 0$, $x_0 = 1$, and $x_1 = 0$. Call this solution $z_0(t)$. Clearly $z_0 \in N(L)$. Now solve for $z_1(t)$ with $f = 0$, $x_0 = 0$, and $x_1 = 1$. Then any $x \in N(L)$ solves (4.9) with $x_0 = x(a)$ and $x_1 = x'(a)$, so

$$x(t) = x(a)z_0(t) + x'(a)z_1(t),$$

by uniqueness. □

As the null space is nontrivial, L^{-1} does not exist. However, with initial conditions imposed, L does have an inverse as will be seen presently. Ignoring these auxiliary conditions for the moment, we study the structure of L within the context of an inner-product space.

Definition. The *formal adjoint* of L is denoted L^* and is defined by

$$\begin{aligned} (L^*x)(t) &= (\bar{a}_0 x)'' - (\bar{a}_1 x)' + \bar{a}_2 x \\ &= \bar{a}_0 x'' + (2\bar{a}_0' - \bar{a}_1)x' + (\bar{a}_0'' - \bar{a}_1' + \bar{a}_2)x \end{aligned}$$

for $x \in C^2(I)$. Here, the stronger assumption that $a_j \in C^{2-j}(I)$, $j = 0, 1, 2$, comes into play so that $L^* : C^2(I) \to C(I)$.

The motivation for this definition comes from the $L^2(I)$ inner-product. If $x, y \in C^2(I)$, then integration by parts shows that

$$\begin{aligned} (Lx, y) &= \int_a^b Lx(t)\, \bar{y}(t)\, dt \\ &= \int_a^b [a_0 x''\bar{y} + a_1 x'\bar{y} + a_2 x\bar{y}]\, dt \\ &= \int_a^b x\, \overline{L^*y}\, dt + [a_0 x'\bar{y} - x(a_0\bar{y})' + a_1 x\bar{y}]_a^b \\ &= (x, L^*y) + \text{ boundary terms.} \end{aligned}$$

Definition. If $L = L^*$, L is said to be *formally self-adjoint*. If a_0, a_1, and a_2 are real-valued functions, we say that L is *real*.

Proposition 4.36. *The real operator* $L = a_0 D^2 + a_1 D + a_2$ *is formally self-adjoint if and only if* $a_0' = a_1$. *In this case,*

$$Lx = (a_0 x')' + a_2 x = \big[D(a_0 D) + a_2\big]x.$$

Proof. As L is a real operator,

$$L^* = a_0 D^2 + (2a_0' - a_1)D + (a_0'' - a_1' + a_2),$$

so $L = L^*$ if and only if

$$a_1 = 2a_0' - a_1,$$
$$a_2 = a_0'' - a_1' + a_2,$$

which is to say,

$$a_1 = a_0' \quad \text{and} \quad a_1' = a_0''.$$

Of course, the first condition implies the second, and in this case,

$$Lx = a_0 D^2 x + a_0' Dx + a_2 x = D(a_0 Dx) + a_2 x. \qquad \square$$

Remark. If $L = a_0 D^2 + a_1 D + a_2$ is real but not formally self-adjoint, it can be rendered so by a small adjustment using an integrating factor. In more detail, let

$$Q(t) = \frac{1}{a_0(t)} P(t) \quad \text{where} \quad P(t) = \exp\left(\int_a^t \frac{a_1(\tau)}{a_0(\tau)} \, d\tau\right) > 0.$$

Then, $P' = a_1 P / a_0$ and a calculation reveals that

$$Lx = f \quad \Longleftrightarrow \quad \tilde{L}x = \tilde{f},$$

where

$$\tilde{L} = QL \quad \text{and} \quad \tilde{f} = Qf.$$

The modified operator $\tilde{L}$ is formally self-adjoint since

$$\tilde{L}x = QLx = Px'' + \frac{a_1}{a_0} Px' + a_2 Qx$$
$$= Px'' + P'x' + a_2 Qx$$
$$= (Px')' + \left(\frac{a_2}{a_0} P\right)x.$$

Examples. Many important examples are posed for $I = (a, b)$, a or b possibly infinite, and $a_j \in C^{2-j}(\bar{I})$, where $a_0 > 0$ on I. Thus, $a_0(a)$ or $a_0(b)$ could vanish. The present discussion excludes this case, though the theory needed for these more general problems is similar to what is outlined here. We list some of the most important examples below.

(1) Legendre: $\qquad$ $Lx = \big((1 - t^2)x'\big)',\quad -1 \leq t \leq 1.$

(2) Chebyshev: $\qquad$ $Lx = (1 - t^2)^{1/2}\big((1 - t^2)^{1/2}x'\big)',\quad -1 \leq t \leq 1.$

(3) Laguerre: $\qquad$ $Lx = e^t\big(te^{-t}x'\big)',\quad 0 < t < \infty.$

(4) Bessel: for $\nu \in \mathbb{R}$, $\qquad$ $Lx = \dfrac{1}{t}(tx')' - \dfrac{\nu^2}{t^2}x,\quad 0 < t < 1.$

(5) Hermite: $\qquad$ $Lx = e^{t^2}\big(e^{-t^2}x'\big)',\quad t \in \mathbb{R}.$

4.7.1 Sturm-Liouville problems and Green's functions

Auxiliary conditions are now brought into play. Rather than the initial conditions used to characterize the null space of L, consideration will be given to one condition at each end of the interval, so-called *boundary conditions*, often abbreviated as simply BCs or BC in the singular.

Definition. Let p, q, and w be real-valued functions on $I = [a, b]$, $a < b$, both finite, with $p \neq 0$ and $w > 0$. Let $\alpha_1, \alpha_2, \beta_1$, and $\beta_2 \in \mathbb{R}$ be such that

$$\alpha_1^2 + \alpha_2^2 \neq 0 \quad \text{and} \quad \beta_1^2 + \beta_2^2 \neq 0.$$

For real-valued f, the problem of finding $x(t) \in C^2(I)$ such that

$$\begin{cases} Ax \equiv \dfrac{1}{w}\Big[(px')' + qx\Big] = f, & t \in (a, b), \\ \alpha_1 x(a) + \alpha_2 x'(a) = 0, \\ \beta_1 x(b) + \beta_2 x'(b) = 0, \end{cases} \tag{4.10}$$

is called a *regular Sturm-Liouville* (regular SL) problem. If f is replaced by λx above, the problem of finding *both* a nontrivial $x(t) \in C^2(I)$ and a $\lambda \in \mathbb{C}$ satisfying the equations is called a regular SL eigenvalue problem. It is the eigenvalue problem for A with the specified BCs.

A regular SL problem is an example of a *boundary value problem*. We remark that if a or b is infinite or p vanishes at a or b, the corresponding BC is lost and the problem is called a *singular* Sturm-Liouville problem.

Example. Let $I = [0, 1]$ and

$$\begin{cases} Ax = -x'' = \lambda x, & t \in (0, 1), \\ x(0) = x(1) = 0. \end{cases} \tag{4.11}$$

Then we need to solve

$$x'' + \lambda x = 0,$$

which as we know, has a two-dimensional solution space that can be expressed as

$$x(t) = B \sin\sqrt{\lambda}\, t + C \cos\sqrt{\lambda}\, t$$

for arbitrary constants B and C. Now the BCs imply that

$$x(0) = C = 0,$$
$$x(1) = B \sin \sqrt{\lambda} = 0.$$

Thus either $B = 0$ or, for some integer n,

$$\sqrt{\lambda} = n\pi;$$

that is, nontrivial solutions are given only for the eigenvalues

$$\lambda_n = n^2 \pi^2,$$

and the corresponding eigenfunctions are

$$x_n(t) = \sin(n\pi t)$$

(or any nonzero multiple).

To analyze a regular SL problem, it is helpful to notice that

$$A : C^2(I) \to C^0(I)$$

has strictly larger range than its domain. However, its inverse (with the BCs) would map $C^0(I)$ to $C^2(I) \subset C^0(I)$. So the inverse might be a bounded linear operator with known spectral properties, which can perhaps be related to spectral properties of A itself. This is the case, and leads us to the classical notion of a Green's function. The Green's function allows us to construct the solution to the boundary value problem (4.10) for any $f \in C^0(I)$.

Definition. A *Green's function* for the regular SL problem (4.10) is a function $G : I \times I \to \mathbb{R}$ such that:

(a) $G \in C^0(I \times I)$ and $G \in C^2(I \times I \setminus D)$, where $D = \{(t,t) : t \in I\}$ is the diagonal in $I \times I$;

(b) for each fixed $s \in I$, $G(\cdot, s)$ satisfies the BCs of the problem;

(c) A applied to the first variable t of $G(t,s)$, denoted $A_t G(t,s)$, vanishes for $(t,s) \in I \times I \setminus D$, viz.

$$A_t G(t,s) \equiv \frac{1}{w} \left[\frac{\partial}{\partial t} \left(p(t) \frac{\partial G}{\partial t}(t,s) \right) + q(t) G(t,s) \right] = 0 \quad \forall t \neq s;$$

(d) $\displaystyle \lim_{s \to t^-} \frac{\partial G}{\partial t}(t,s) - \lim_{s \to t^+} \frac{\partial G}{\partial t}(t,s) = \frac{1}{p(t)}$ for all $t \in (a,b)$.

Example. The conditions defining a Green's function are not immediately intuitive. Here is a very simple example that may help to clarify their origin. Corresponding to (4.11), consider

$$\begin{cases} Ax = -x'' = f, & t \in (0,1), \\ x(0) = x(1) = 0, \end{cases} \tag{4.12}$$

for $f \in C^0(I)$. Let

$$G(t, s) = \begin{cases} (1-t)s, & 0 \le s \le t \le 1, \\ (1-s)t, & 0 \le t \le s \le 1. \end{cases}$$

Then G satisfies (a) and

$$G(0, s) = (1-s) \cdot 0 = 0,$$
$$G(1, s) = (1-(1))s = 0,$$

so (b) holds. Since $w = 1$, $p = -1$, and $q = 0$,

$$A_t G(t, s) = -\frac{\partial^2}{\partial t^2} G(t, s) = 0 \quad \text{for } s \ne t$$

and

$$\lim_{s \to t^-} \frac{\partial G}{\partial t} = -t, \quad \lim_{s \to t^+} \frac{\partial G}{\partial t} = 1 - t,$$

so (c) and (d) also hold. Thus $G(t, s)$ is a Green's function for this SL problem. Moreover, if

$$x(t) = \int_0^1 G(t, s) f(s) \, ds,$$

then $x(0) = x(1) = 0$ and

$$x'(t) = \frac{d}{dt} \left\{ \int_0^t G(t, s) f(s) \, ds + \int_t^1 G(t, s) f(s) \, ds \right\}$$

$$= G(t, t) f(t) + \int_0^t \frac{\partial G}{\partial t}(t, s) f(s) \, ds - G(t, t) f(t) + \int_t^1 \frac{\partial G}{\partial t}(t, s) f(s) \, ds$$

$$= \int_0^1 \frac{\partial G}{\partial t}(t, s) f(s) \, ds,$$

whence

$$x''(t) = \frac{d}{dt} \left\{ \int_0^t \frac{\partial G}{\partial t} f \, ds + \int_t^1 \frac{\partial G}{\partial t} f \, ds \right\}$$

$$= \frac{\partial G}{\partial t}(t, t^-) f(t) + \int_0^t \frac{\partial^2 G}{\partial t^2} f \, ds - \frac{\partial G}{\partial t}(t, t^+) f(t) + \int_t^1 \frac{\partial^2 G}{\partial t^2} f \, ds$$

$$= -f + \int_0^1 \frac{\partial^2 G}{\partial t^2} f \, ds$$

$$= -f(t).$$

Thus, for any continuous right-hand side f, a solution to (4.12) has been constructed via $G(t, s)$. This simple calculation generalizes as is shown next.

Theorem 4.37. *Consider the regular SL system*

$$\begin{cases} Au \equiv \dfrac{1}{w}Lu \equiv \dfrac{1}{w}\left[(pu')' + qu\right] = f, \quad t \in (a, b), \\ \alpha_1 u(a) + \alpha_2 u'(a) = 0, \\ \beta_1 u(b) + \beta_2 u'(b) = 0, \end{cases}$$

on the interval $I = [a, b]$, with $p \in C^1(I)$, $w, q \in C^0(I)$, $p, w > 0$, $\alpha_1^2 + \alpha_2^2 > 0$, and $\beta_1^2 + \beta_2^2 > 0$. Suppose that 0 is not an eigenvalue (so $Au = Lu = 0$ with the BCs implies $u = 0$). Let u_1 and u_2 be any nonzero real solutions of $Au = Lu = 0$ such that for u_1,

$$\alpha_1 u_1(a) + \alpha_2 u_1'(a) = 0,$$

and for u_2,

$$\beta_1 u_2(b) + \beta_2 u_2'(b) = 0.$$

Define $G : I \times I \to \mathbb{R}$ by

$$G(t, s) = \begin{cases} \dfrac{u_2(t)u_1(s)}{pW}, & a \le s \le t \le b, \\ \dfrac{u_1(t)u_2(s)}{pW}, & a \le t \le s \le b, \end{cases}$$

where

$$W(s) = W(s; u_1, u_2) \equiv u_1(s)u_2'(s) - u_1'(s)u_2(s)$$

is the Wronskian *of u_1 and u_2. Then, $p(t)W(t)$ is a nonzero constant and G is a Green's function for the above regular SL problem. Moreover, if $\mathcal{G}$ is any such Green's function and $f \in C^0(I)$, then*

$$u(t) = \int_a^b \mathcal{G}(t, s)\, f(s)\, w(s)\, ds \tag{4.13}$$

is the unique solution of $Au = f$ satisfying the given BCs.

Lemma 4.38 (Abel's theorem). *Let $Lu = (pu')' + qu$ satisfy $p \in C^1(I)$ and $q \in C^0(I)$. For any positive $w \in C^0(I)$ and $\lambda \in \mathbb{C}$, if u_1 and u_2 solve*

$$Lu = \lambda wu \quad (i.e., Au = \lambda u),$$

then

$$p(t)W(t; u_1, u_2)$$

is constant.

Proof. The simple calculation

$$
\begin{aligned}
0 &= \lambda w(u_1 u_2 - u_2 u_1) \\
&= u_1 L u_2 - u_2 L u_1 \\
&= u_1 \left(p u_2'' + p' u_2' + q u_2 \right) - u_2 \left(p u_1'' + p' u_1' + q u_1 \right) \\
&= p \left(u_1 u_2'' - u_2 u_1'' \right) + p' W \\
&= (pW)'
\end{aligned}
$$

establishes the result. ☐

Lemma 4.39. *Suppose* $u, v \in C^1(I)$. *If* $W(t_0; u, v) \neq 0$ *for some* $t_0 \in I$, *then* u *and* v *are linearly independent.*

Proof. Suppose for some scalars α and β,

$$
\alpha u(t) + \beta v(t) = 0.
$$

Then

$$
\alpha u'(t) + \beta v'(t) = 0
$$

also holds. At $t = t_0$, we have a linear system

$$
\begin{bmatrix} u(t_0) & v(t_0) \\ u'(t_0) & v'(t_0) \end{bmatrix} \begin{pmatrix} \alpha \\ \beta \end{pmatrix} = \begin{pmatrix} 0 \\ 0 \end{pmatrix},
$$

which is uniquely solvable since the matrix is invertible because its determinant, $\det W(t_0) \neq 0$. Thus $\alpha = \beta = 0$ and so u and v are linearly independent. ☐

Proof of Theorem 4.37. The existence of u_1 and u_2 follows from Picard's theorem (Theorem 4.34). If we use the standard basis

$$
N(L) = \operatorname{span}\{z_0, z_1\},
$$

for the null space of L, where

$$
\begin{aligned}
z_0(a) &= 1, & z_0'(a) &= 0, \\
z_1(a) &= 0, & z_1'(a) &= 1,
\end{aligned}
$$

then

$$
u_1(t) = -\alpha_2 z_0(t) + \alpha_1 z_1(t) \not\equiv 0.
$$

A similar construction at $t = b$ gives $u_2(t)$.

The product pW is a constant by Abel's theorem (Lemma 4.38). If this constant is zero, then W vanishes identically. Consider the matrix

$$
\begin{bmatrix} u_1(a) & u_1'(a) \\ u_2(a) & u_2'(a) \end{bmatrix},
$$

which is singular since it has zero determinant. This matrix is also nonzero, because if $u_1(a)$ and $u_1'(a)$ both vanished, we would conclude that $u_1 \equiv 0$ by Picard's theorem (neither u_1 nor u_2 are identically zero of course). Thus, the dimension of the null space of the matrix is deduced to be one, and the boundary condition for u_1 identifies this space as $\text{span}\{(\alpha_1, \alpha_2)\}$, i.e., the space of vectors orthogonal to $(\alpha_2, -\alpha_1)$. But $(u_2'(a), -u_2(a))$ is also in the null space, and so $\alpha_1 u_2(a) + \alpha_2 u_2'(a) = 0$. So, u_2 satisfies the boundary conditions at both a and b and so is a nontrivial eigenfunction for the eigenvalue $\lambda = 0$, which is prohibited by assumption. The conclusion is that pW is a nonzero constant, and therefore that $G(t, s)$ is well-defined.

Clearly G is continuous and C^2 when $t \neq s$, since $u_1, u_2 \in C^2(I)$. Moreover, $G(\cdot, s)$ satisfies the BCs by construction, and $A_t G$ is either a function times Au_1 or times Au_2 for $t \neq s$ and hence vanishes there. It remains to show the jump condition on $\partial G/\partial t$ in the definition of Green's function. To this end, just compute as follows:

$$\frac{\partial G}{\partial t}(t, s) = \begin{cases} \dfrac{u_2'(t)u_1(s)}{pW}, & a \leq s \leq t \leq b, \\[2mm] \dfrac{u_1'(t)u_2(s)}{pW}, & a \leq t \leq s \leq b, \end{cases}$$

so

$$\frac{\partial G}{\partial t}(t, t^-) - \frac{\partial G}{\partial t}(t, t^+) = \frac{u_2'(t)u_1(t)}{pW} - \frac{u_1'(t)u_2(t)}{pW} = \frac{1}{p(t)},$$

as required.

If $Au = f$ has a solution, it must be unique since the difference of two such solutions would satisfy the eigenvalue problem with eigenvalue 0, and therefore vanish. Thus it remains only to show that $u(t)$ defined by (4.13) is a solution to $Au = f$. We use only the properties (a)–(d) in the definition of a Green's function. The detailed form of the function G constructed above is not needed.

For $t \in (a, b)$, use the smoothness requirement in (a) to compute $u'(t)$ from (4.13) thusly:

$$u'(t) = \frac{d}{dt}\left[\int_a^t \mathcal{G}(t, s)\, f(s)\, w(s)\, ds + \int_t^b \mathcal{G}(t, s)\, f(s)\, w(s)\, ds \right]$$

$$= \mathcal{G}(t, t)\, f(t)\, w(t) + \int_a^t \frac{\partial \mathcal{G}}{\partial t}(t, s)\, f(s)\, w(s)\, ds$$

$$\quad - \mathcal{G}(t, t)\, f(t)\, w(t) + \int_t^b \frac{\partial \mathcal{G}}{\partial t}(t, s)\, f(s)\, w(s)\, ds$$

$$= \int_a^b \frac{\partial \mathcal{G}}{\partial t}(t, s)\, f(s)\, w(s)\, ds.$$

It is now trivial to see that u satisfies the two BCs because of (b). Applying (d), we see that

$$(p(t)u'(t))' = \frac{d}{dt}\left[\int_a^t p(t)\frac{\partial \mathcal{G}}{\partial t}(t,s)\,f(s)\,w(s)\,ds + \int_t^b p(t)\frac{\partial \mathcal{G}}{\partial t}(t,s)\,f(s)\,w(s)\,ds\right]$$

$$= p(t)\frac{\partial \mathcal{G}}{\partial t}(t,t^-)\,f(t)\,w(t) + \int_a^t \frac{\partial}{\partial t}\left(p(t)\frac{\partial \mathcal{G}}{\partial t}(t,s)\right)f(s)\,w(s)\,ds$$

$$- p(t)\frac{\partial \mathcal{G}}{\partial t}(t,t^+)\,f(t)\,w(t) + \int_t^b \frac{\partial}{\partial t}\left(p(t)\frac{\partial \mathcal{G}}{\partial t}(t,s)\right)f(s)\,w(s)\,ds$$

$$= f(t)\,w(t) + \int_a^b \frac{\partial}{\partial t}\left(p(t)\frac{\partial \mathcal{G}}{\partial t}(t,s)\right)f(s)\,w(s)\,ds.$$

Finally, use (c) to conclude that

$$Au(t) = \frac{1}{w}\left[(pu')' + qu\right]$$

$$= f(t) + \int_a^b A_t\mathcal{G}(t,s)\,f(s)\,w(s)\,ds$$

$$= f(t)$$

as required. $\qquad\square$

4.7.2 Spectral properties of the solution operator

With a Green's function G in hand, define the solution operator

$$T : C^0(I) \to C^0(I)$$

by

$$Tf(t) = \int_a^b G(t,s)\,f(s)\,w(s)\,ds.$$

Endowing $C^0(I)$ with the $L^2(I)$ inner-product, we see that T is a bounded linear operator, since for $f \in C^0(I)$,

$$\|Tf\|_{L^2(I)}^2 \le \int_a^b \left(\int_a^b |G(t,s)\,f(s)\,w(s)|\,ds\right)^2 dt$$

$$\le \int_a^b \left(\int_a^b |G(t,s)|^2\,ds \int_a^b |f(s)|^2\,ds\right)dt\,\|w\|_{L^\infty(I)}^2$$

$$= \|G\|_{L^2(I\times I)}^2\,\|f\|_{L^2(I)}^2\,\|w\|_{L^\infty(I)}^2.$$

For arbitrary $f, g \in C^0(I)$, it is not the case that $(Tf, g) = (f, Tg)$, so T is not self-adjoint on $L^2(I)$ with the usual inner-product. However, T turns out to be self-adjoint if we use an inner-product adapted to it, defined by

$$\langle f, g\rangle_w = \int_a^b f(t)\,\overline{g(t)}\,w(t)\,dt.$$

This induces a norm equivalent to the usual $L^2(I)$-norm, since

$$0 < \min_{s \in I} w(s) \leq w(t) \leq \max_{s \in I} w(s) < \infty$$

for all $t \in I$.

Since $G(s,t) = G(t,s)$ is real, the calculation

$$\langle Tf, g \rangle_w = (Tf, gw) = \int_a^b \int_a^b G(t,s)\, f(s)\, w(s)\, ds\, \overline{g(t)}\, w(t)\, dt$$

$$= \int_a^b f(s) \int_a^b G(s,t)\, \overline{g(t)}\, w(t)\, dt\, w(s)\, ds$$

$$= (f, Tgw) = \langle f, Tg \rangle_w,$$

reveals that T is self-adjoint in $(C^0(I), \langle \cdot, \cdot \rangle_w)$. We would like to apply the Ascoli-Arzelà theorem to show that T is a compact operator. The incompleteness of $C^0(I)$ in the $L^2(I)$-norm is easily rectified since $C^0(I)$ is dense in $L^2(I)$. Simply extend T to $L^2(I)$ by density as follows. Given $f \in L^2(I)$, let $\{f_n\}_{n=1}^\infty \subset C^0(I)$ be such that $f_n \to f$ in $L^2(I)$. Then boundedness implies that $\{Tf_n\}_{n=1}^\infty$ is Cauchy in $L^2(I)$, so it makes sense to define

$$Tf = \lim_{n \to \infty} Tf_n.$$

This extension is a well-defined, bounded linear operator $T : L^2(I) \to L^2(I)$ and it remains self-adjoint. In fact, T is defined by the same integral, $Tf(t) = \int_a^b G(t,s)\, f(s)\, w(s)\, ds$ for $f \in L^2(I)$.

Lemma 4.40. *The operator T is a compact mapping of $L^2(I)$ to $C^0(I)$ and maps $C^0(I)$ to $C^2(I)$.*

Proof. Compactness of T as an operator $T : C^0(I) \to C^0(I)$ and $T : L^2(I) \to L^2(I)$ follows from the Ascoli-Arzelà theorem (Theorem 4.32) and Corollary 4.33, respectively.

Let $f \in L^2(I)$ and compute for $t_1, t_2 \in I$,

$$|Tf(t_1) - Tf(t_2)| = \left| \int_a^b (G(t_1, s) - G(t_2, s)) f(s)\, w(s)\, ds \right|$$

$$\leq \max_{s \in I} |G(t_1, s) - G(t_2, s)|\, \|f\|_{L^2(I)}\, \|w(s)\|_{L^2(I)}.$$

The Green's function $G(s,t)$ is uniformly continuous on $I \times I$, so the right-hand side can be made as small as desired simply by demanding that t_1 and t_2 be sufficiently close, independently of s, so Tf is continous. (Incidentally, if the functions f are drawn from a bounded subset of $L^2(I)$, then the preceding inequality shows the image to be bounded and equicontinuous in $C^0(I)$.) Picard's theorem (Theorem 4.34) implies that T takes $C^0(I)$ to $C^2(I)$. □

Much is known about the spectrum of the compact operator T. These spectral properties need to be related to those of the SL problem.

Proposition 4.41. *If $\lambda = 0$ is not an eigenvalue of the regular SL problem, then $\lambda = 0$ is not an eigenvalue of T either.*

Proof. Suppose $Tf = 0$ for some $f \in L^2(I)$. Then, with $c = (pW)^{-1}$ and u_1, u_2 from Theorem 4.37,

$$
0 = (Tf)'(t) = c\frac{d}{dt}\left[u_2(t) \int_a^t f(s)u_1(s)\,w(s)\,ds + u_1(t) \int_t^b f(s)u_2(s)\,w(s)\,ds \right]
$$
$$
= c\left[u_2' \int_a^t fu_1w\,ds + u_1' \int_t^b fu_2w\,ds \right].
$$

But

$$
0 = Tf(t) = c\left[u_2 \int_a^t fu_1w\,ds + u_1 \int_t^b fu_2w\,ds \right]
$$

also, so since $c \neq 0$ and $W(t; u_1, u_2) \neq 0$, the solution of the linear system composed of the last two equations is trivial; that is, for all $t \in [a, b]$,

$$
\int_a^t fu_1w\,ds = \int_t^b fu_2w\,ds = 0.
$$

It follows by differentiation that

$$
f(t)u_1(t) = f(t)u_2(t) = 0,
$$

so $f = 0$, since u_1 and u_2 cannot both vanish at the same point (because $W \neq 0$). Thus $N(T) = \{0\}$ and $0 \notin \sigma_p(T)$. □

Proposition 4.42. *Suppose $\lambda \neq 0$. Then λ is an eigenvalue of the regular SL problem if and only if $1/\lambda$ is an eigenvalue of T. Moreover, the corresponding eigenspaces coincide.*

Proof. The preceding lemma and Picard's theorem imply that an eigenfunction of T or A must lie in $C^2(I)$. Since T is the solution operator for the SL problem, $Au = f$ with u satisfying the BCs holds if and only if $u = Tf$ holds. Thus, if $f \in C^2(I)$ is an eigenfunction for the SL problem with eigenvalue $\lambda \neq 0$, then f satisfies the BCs and $Af = \lambda f$, so $f = T(\lambda f)$, or $Tf = \frac{1}{\lambda}f$. Conversely, if $f \in C^2(I)$ is an eigenfunction for T with eigenvalue $1/\lambda$, then $Tf = \frac{1}{\lambda}f$, so $A(\frac{1}{\lambda}f) = f$. That is, $Af = \lambda f$ and λf, satisfies the homogeneous BCs. As $\lambda \neq 0$, so does f. □

We know from the general theory that $\dim N(T_\lambda) = \dim N(A_{1/\lambda})$ is finite. More can be said in this concrete situation.

Proposition 4.43. *The nonzero eigenvalues of a regular SL problem are simple.*

Proof. Suppose u and v are eigenvectors for a nonzero eigenvalue λ. The BCs give a nonzero solution to the linear system

$$\begin{bmatrix} u(a) & u'(a) \\ v(a) & v'(a) \end{bmatrix} \begin{pmatrix} \alpha_1 \\ \alpha_2 \end{pmatrix} = \begin{pmatrix} 0 \\ 0 \end{pmatrix},$$

which can only transpire if the matrix is singular, i.e., $W(a) = 0$. Abel's theorem (Lemma 4.38) tells us that pW is constant, and so $W(t) = uv' - u'v = 0$ for all t.

One might be tempted to conclude that u and v are linearly dependent when their Wronskian vanishes. This is incorrect in general (consider, e.g., $u(t) = t^2$ and $v(t) = |t|t$). However, u and v satisfy the differential equation, and so Picard's theorem implies that u is C^2 and that $u(t)$ and $u'(t)$ cannot both vanish at any fixed t (otherwise $u = 0$ for all t and we do not have an eigenvector). The same remarks apply to v. The vanishing of the Wronskian, then, is to say that for each t the nonzero vectors (u, u') and (v, v') are multiples of each other, which implies in turn that there is some nonzero function $c(t)$ such that $u = c(t)v$ and $u' = c(t)v'$. We wish to conclude that $c(t)$ is in fact constant.

The zeros of u agree with the zeros of v; let $S = \{t \in [a, b] : u(t) = v(t) = 0\}$. If S has a limit point t_0, then there exist $t_n \in S$, $t_n \neq t_0$, such that $t_n \to t_0$ as $n \to \infty$. Moreover,

$$u'(t_0) = \lim_{n \to \infty} \frac{u(t_n) - u(t_0)}{t_n - t_0} = 0,$$

which is not possible since also $u(t_0) = 0$. So S has only isolated points. Between them, $c(t) = u(t)/v(t)$ is differentiable, and $c'(t) = W(t)/v(t)^2 = 0$. It is concluded that c is piecewise constant. But L'Hôpital's rule implies that at $t_0 \in S \cap (a, b)$,

$$\lim_{t \to t_0} \frac{u(t)}{v(t)} = \lim_{t \to t_0} \frac{u'(t)}{v'(t)} = \frac{u'(t_0)}{v'(t_0)}$$

exists, so $c(t)$ must be continuous and therefore constant. We conclude that u and v are linearly dependent, whence λ is simple. $\qquad\square$

We summarize what is known about the regular SL problem for A based on the spectral theorem for compact self-adjoint operators as applied to T. The details of the proof are left as an exercise.

Theorem 4.44. *Let $a, b \in \mathbb{R}$, $a < b$, $I = [a, b]$, $p \in C^1(I)$, $p \neq 0$, $q \in C^0(I)$, and $w \in C^0(I)$, $w > 0$. Let*

$$A = \frac{1}{w}\big[DpD + q\big]$$

be a formally self-adjoint regular SL operator with boundary conditions

$$\alpha_1 u(a) + \alpha_2 u'(a) = 0,$$
$$\beta_1 u(b) + \beta_2 u'(b) = 0,$$

for $u \in C^2(I)$, where $\alpha_1^2 + \alpha_2^2 \neq 0$ and $\beta_1^2 + \beta_2^2 \neq 0$, $\alpha_i, \beta_i \in \mathbb{R}$. If 0 is not an eigenvalue of the SL problem, then it has a countable collection of real eigenvalues $\{\lambda_n\}_{n=1}^\infty$ such that

$$|\lambda_n| \to \infty \quad as \ n \to \infty$$

and each eigenspace is one-dimensional. Let $\{u_n\}_{n=1}^\infty$ be the corresponding normalized eigenfunctions. These form an ON basis for $(L^2(I), \langle \cdot, \cdot \rangle_w)$, so if $u \in L^2(I)$,

$$u = \sum_{n=1}^\infty \langle u, u_n \rangle_w u_n$$

and, provided $Au \in L^2(I)$ and u satisfies the boundary conditions,

$$Au = \sum_{n=1}^\infty \lambda_n \langle u, u_n \rangle_w u_n.$$

4.7.3 Some applications

We saw earlier that the regular SL problem

$$\begin{cases} -x'' = \lambda x, \quad t \in (0,1), \\ x(0) = x(1) = 0, \end{cases}$$

has eigenvalues

$$\lambda_n = n^2 \pi^2, \quad n = 1, 2, \ldots$$

and corresponding (normalized) eigenfunctions

$$u_n(t) = \sqrt{2} \sin(n\pi t).$$

Given any $f \in L^2(0,1)$, we have its Fourier *sine* series

$$f(t) = \sum_{n=1}^\infty \left(2 \int_0^1 f(s) \sin(n\pi s) \, ds \right) \sin(n\pi t),$$

where equality holds for a.e. $t \in [0,1]$, i.e., in $L^2(0,1)$. Incidentally, this provides another proof that $L^2(0,1)$ is separable.

By iterating this result, we can decompose any $f \in L^2(I \times I)$, $I = (0,1)$. For a.e. $x \in I$,

$$f(x,y) = \sum_{n=1}^\infty \left(2 \int_0^1 f(x,t) \sin(n\pi t) \, dt \right) \sin(n\pi y)$$

$$= 4 \sum_{n=1}^\infty \int_0^1 \sum_{m=1}^\infty \int_0^1 f(s,t) \sin(m\pi s) \, ds \, \sin(n\pi t) \, dt \, \sin(n\pi y) \sin(m\pi x)$$

$$= \sum_{n=1}^\infty \sum_{m=1}^\infty \left(4 \int_0^1 \int_0^1 f(s,t) \sin(m\pi s) \sin(n\pi t) \, ds \, dt \right) \sin(m\pi x) \sin(n\pi y).$$

So $L^2(I \times I)$ has the ON basis

$$\{\{2 \sin(m\pi x) \sin(n\pi y)\}_{m=1}^\infty\}_{n=1}^\infty,$$

and it is also separable. Continuing, we can find a countable basis for any $L^2(R)$, $R = I^d$, $d = 1, 2, \ldots$. By dilation and translation, we can replace R by any rectangle, and since $L^2(\Omega) \subset L^2(R)$ whenever $\Omega \subset R$ is measurable (if we extend the domain of $f \in L^2(\Omega)$ by defining $f \equiv 0$ on $R \smallsetminus \Omega$), $L^2(\Omega)$ is separable for any bounded Ω, but the construction of an ON basis is not so clear. Finally, for unbounded Ω, we can let $\Omega_N = \{x \in \Omega : |x| < N\}$ and consider $N \to \infty$ to show that $L^2(\Omega)$ is separable.

The regular SL eigenvalue problem

$$\begin{cases} -x'' = \lambda x, & t \in (0,1), \\ x'(0) = x'(1) = 0, \end{cases}$$

has the eigenvalues $\lambda_n = n^2\pi^2$, $n = 0, 1, 2, \ldots$, and eigenfunctions

$$u_0(t) = 1, \quad u_n(t) = \sqrt{2} \cos(n\pi t), \quad n = 1, 2, \ldots.$$

These are used to define *cosine series* expansions, similar to the development above. In this case, 0 is an eigenvalue of A, and the null space consists of the constant functions. The problem

$$\begin{cases} -x'' = \lambda x, & t \in (0,1), \\ x(0) = x(1), \\ x'(0) = x'(1), \end{cases}$$

seeks a *periodic* solution. Although not a regular SL problem as defined above, a similar theory produces a complete set of orthogonal eigenfunctions, which leads us to *Fourier series* with the eigenvalues $\lambda_n = (2n\pi)^2$, $n = 0, 1, 2, \ldots$, and eigenfunctions

$$u_n(t) = \begin{cases} 1, & n = 0, \\ \sqrt{2} \cos(2n\pi t), & n = 1, 2, \ldots, \end{cases}$$

$$v_n(t) = \sqrt{2} \sin(2n\pi t), \quad n = 1, 2, \ldots.$$

Example. Let $\Omega = (0,a) \times (0,b)$, and consider a solution $u(x,y)$ of

$$\begin{cases} -\Delta u = -\dfrac{\partial^2 u}{\partial x^2} - \dfrac{\partial^2 u}{\partial y^2} = f(x,y), & (x,y) \in \Omega, \\ u(x,y) = 0, & (x,y) \in \partial\Omega, \end{cases}$$

where $f \in L^2(\Omega)$. We proceed in a formal manner, computing without attempting to justify the intermediate steps. Only the final result will be justified. The technique of *separation of variables* is used. Suppose $v(x,y) = X(x)Y(y)$ is a solution to the above eigenvalue problem, so that

$$-\Delta v = -X''Y - XY'' = \lambda v = \lambda XY.$$

Then

$$-\frac{X''}{X} = \lambda + \frac{Y''}{Y} = \mu,$$

a constant. The BCs on X and Y are implied by those for v, so

$$X(0) = X(a) = 0,$$
$$Y(0) = Y(b) = 0.$$

Therefore, the function X satisfies a simple SL problem whose eigenvalues and eigenfunctions are

$$\mu = \mu_m = \left(\frac{m\pi}{a}\right)^2, \quad X_m(x) = \sin\left(\frac{m\pi x}{a}\right), \quad m = 1, 2, \ldots.$$

For each m,

$$-Y'' = (\lambda_m - \mu_m)Y$$

has solution

$$\lambda_{m,n} - \mu_m = \left(\frac{n\pi}{b}\right)^2, \quad Y_n(y) = \sin\left(\frac{n\pi y}{b}\right), \quad n = 1, 2, \ldots.$$

That is, for $m, n = 1, 2, \ldots,$

$$\lambda_{m,n} = \left[\left(\frac{m}{a}\right)^2 + \left(\frac{n}{b}\right)^2\right]\pi^2,$$

$$v_{m,n}(x, y) = \sin\frac{m\pi x}{a}\sin\frac{n\pi y}{b}.$$

The collection $\{\{v_{m,n}\}_{m=1}^{\infty}\}_{n=1}^{\infty}$ forms a basis for $L^2((0, a) \times (0, b))$, so f can be rigorously expanded in the form

$$f(x, y) = \sum_{m,n} c_{m,n} v_{m,n}(x, y)$$

where

$$c_{m,n} = \frac{\displaystyle\int_0^b \int_0^a f(x, y) v_{m,n}(x, y)\, dx\, dy}{\displaystyle\int_0^b \int_0^a v_{m,n}^2(x, y)\, dx\, dy}.$$

Forming

$$u(x, y) \equiv \sum_{m,n} \frac{c_{m,n}}{\lambda_{m,n}} v_{m,n}(x, y),$$

one verifies that indeed u is a solution to the problem.

4.8 EXERCISES

1. Prove that eigenfunctions of distinct eigenvalues are linearly independent.

2. Let $X = C([0, 1])$.

   (a) For $g \in X$ fixed, find the spectrum of $T : X \to X$ defined by $Tf = gf$.

   (b) Find an operator $T : X \to X$ with spectrum equal to a given interval $[a, b]$ of the real line.

3. Define $T : \ell^2 \to \ell^2$ by $Tx = y$, where $y_j = \alpha_j x_j$, $j = 1, 2, \ldots$, and with the set $\{\alpha_j\}_{j=1}^{\infty}$ dense in $[0, 1]$. Find $\sigma_p(T)$ and $\sigma(T)$. Show that if $\lambda \in \sigma(T) \setminus \sigma_p(T)$, then T_λ^{-1} is unbounded.

4. Define $T : \ell^\infty \to \ell^\infty$ by $Tx = y$, where $y_j = x_{j+1}$, $j = 1, 2, \ldots$. Show that $\rho(T) = \{\lambda : |\lambda| > 1\}$ and $\sigma(T) = \sigma_p(T) = \{\lambda : |\lambda| \le 1\}$.

5. Let H be a Hilbert space and $T \in B(H, H)$. If $\|T\| \le 1$, show that for all $y \in H$, there is a unique solution to $T^2x - Tx - 12x = y$.

6. If X is a Banach space and T is a bounded linear operator from X to X, show that as $|\lambda| \to \infty$, $\|T_\lambda^{-1}\|_{B(X,X)} \to 0$.

7. Let H be a Hilbert space and $P \in B(H, H)$ a projection.

   (a) Show that P is an orthogonal projection if and only if $P = P^*$.

   (b) If P is an orthogonal projection, find $\sigma_p(P)$, $\sigma_c(P)$, and $\sigma_r(P)$.

8. Let T be a bounded linear operator on a Banach space X such that the sequence $x, Tx, T^2x, \ldots$ is bounded for every $x \in X$. Show that if $\lambda \in \sigma(T)$, then $|\lambda| \le 1$. [Hint: investigate the limit as $N \to \infty$ of $\sum_{n=0}^{N} T^n x / \lambda^n$.]

9. If X, Y are NLSs and $T : X \to Y$ is a bounded linear operator, show that T is compact if and only if the image of the closed unit ball in X is precompact in Y.

10. Prove that if X is a Banach space and $T \in B(X, X)$, then the spectrum of T^2 is the square of the spectrum of T. [Hint: this is an example of the spectral mapping theorem (see, e.g., [12]).]

11. Define $T : \ell^2 \to \ell^2$ by $Tx = y$, where $y_j = \alpha_j x_j$, $j = 1, 2, \ldots$, and $\alpha_j \to 0$ as $j \to \infty$. Show that T is compact.

12. Let $\{x_n\}_{n=1}^{\infty}$ be an orthonormal set in a Hilbert space H. Let $\{a_n\}_{n=1}^{\infty}$ be a sequence of nonnegative numbers and let

$$S = \left\{ x \in H : x = \sum_{n=1}^{\infty} b_n x_n \text{ and } |b_n| \le a_n \text{ for all } n \right\}.$$

Show that S is compact if and only if $\sum_{n=1}^{\infty} a_n^2 < \infty$.

13. Suppose that X is a Banach space and $S, T \in B(X, X)$. Suppose further that T is compact.

   (a) Prove that TS and ST are compact.

   (b) If S is invertible, show that $S + T$ is invertible on all of X provided only that it is injective.

14. Let X be a Banach space, $S, T \in B(X, X)$, and I be the identity map.

   (a) Show by example that $ST = I$ does not imply $TS = I$.

   (b) If T is compact, show that $S(I - T) = I$ if and only if $(I - T)S = I$.

   (c) If $S = (I - T)^{-1}$ exists for some T compact, show that $I - S$ is compact.

15. Let T and S be positive operators on a complex Hilbert space H and let the scalar α be nonnegative.

   (a) Show that $T + S$, T^n for any integer $n \geq 0$, and αT are all positive operators, and if T has an inverse, then T^{-1} is a positive operator.

   (b) Show that the set of positive operators is closed in $B(H, H)$.

   (c) Show by example in the space of 2×2 matrices that TS may not be a positive operator.

16. Let $T \in B(\mathbb{C}^n, \mathbb{C}^m)$ be a (complex) $m \times n$ matrix. Recall that we can factor T into its singular value decomposition, so $T = U\Sigma V^*$, where U and V are unitary matrices ($U^*U = UU^* = I$ and $V^*V = VV^* = I$, $U^* = \bar{U}^T$) and Σ is an $m \times n$ rectangular diagonal matrix with nonnegative real numbers on the diagonal.

   (a) Show that when $m = n$, if $T = V\Sigma V^*$, then T is positive.

   (b) If $m = n$ and $T = V\Sigma V^*$, find $\sqrt{T}$ in terms of V and Σ.

   (c) Find $|T|$ in terms of U, V, and Σ.

   (d) Suppose $S \in B(\mathbb{C}^m, \mathbb{C}^\ell)$. Show that it is possible that $|ST| \neq |S||T|$.

17. Let H be an infinite dimensional Hilbert space over $\mathbb{C}$ and $T : H \to H$ a positive, compact linear operator. Prove that for all $y \in H$, we have that

$$\inf_{x \in H} \|(Tx - x) - y\|_H = 0.$$

[Hint: suppose not and then show that 1 would need to be in the residual spectrum.]

18. Let T be a compact, positive operator on a complex Hilbert space H. Use the spectral theorem to show that there is a unique positive operator S on H such that $S^2 = T$. Show also that S is compact.

19. Give an example of a self-adjoint operator on a Hilbert space that has no eigenvalues. [Hint: consider $L^2(0,1)$ and define T by $Tf(x) = xf(x)$.]

20. Let H be a complex Hilbert space and $T : H \to H$ a linear operator that is not necessarily bounded. Show that if T is also formally self-adjoint, meaning that $(x, Ty) = (Tx, y)$ for all $x, y \in H$, then T must be bounded.

21. Let H be a complex Hilbert space and $T : H \to H$ a bounded linear operator. Note that both T^*T and TT^* are positive, self-adjoint operators. Determine their spectral properties, and in particular, show that $I + T^*T$ and $I + TT^*$ must be boundedly invertible.

22. Let H be a complex Hilbert space and A a bounded linear operator on H. A bounded linear operator U on H is a *partial isometry* if $\|Ux\| = \|x\|$ for all $x \in N(U)^\perp$ (i.e., U is an isometry except on its null space, where it is zero).

(a) Show that $|A| = (A^*A)^{1/2}$ is a well-defined, bounded linear, self-adjoint operator.

(b) Show that $\| \, |A|x \, \| = \|Ax\|$ for all $x \in H$.

(c) Show that $H = \overline{R(|A|)} \oplus N(|A|)$ and that $N(|A|) = N(A)$.

(d) Show that there exists a partial isometry U such that $A = U \, |A|$. [Hint: define $U : R(|A|) \to R(A)$ by $U(|A|x) = Ax$ (is this well-defined?) and extend U first to $\overline{R(|A|)}$ and then to all of H.]

23. Let H be a nontrivial complex Hilbert space and $T : H \to H$ a bounded, self-adjoint linear operator. Show that the spectrum of T is not empty. Moreover, if T is compact, show that T has at least one eigenvalue.

24. Let H be a separable Hilbert space and let $\{e_n\}_{n=1}^\infty$ be a maximal orthonormal set (i.e., a Hilbert basis). Let $\{\lambda_n\}_{n=1}^\infty$ be a bounded sequence of real numbers. For any $x \in H$, let the operator $A : H \to H$ be given by

$$Ax = \sum_{n=1}^\infty \lambda_n \langle x, e_n \rangle e_n.$$

(a) Show that A is a well-defined, bounded linear operator that is also self-adjoint.

(b) Find the point spectrum of A, and for each distinct eigenvalue, find its eigenspace.

(c) Find a reasonable condition on $\{\lambda_n\}_{n=1}^\infty$ so that A is surjective.

(d) Show that if $\lambda_n \to 0$, then A is compact.

25. Let T and S be positive operators on a complex Hilbert space. Prove that if T and S commute, then TS is also positive.

26. Let A be a self-adjoint, compact operator on a complex Hilbert space. Prove that there are positive operators P and N such that $A = P -$

N and $PN = 0$. A much more difficult exercise: prove the conclusion if A is merely self-adjoint. [Hint: use the fact that when T and S are positive commuting operators, then TS is also positive, and then consider the orthogonal projection onto the null space of $|A| + A$.]

27. Let P_1 and P_2 be two orthogonal projections on a Hilbert space H. Show that $P = P_2 - P_1$ is also a projection if and only if $P_1 \le P_2$.

28. Let H be a separable Hilbert space and T a positive operator on H. Let $\{e_n\}_{n=1}^\infty$ be an orthonormal basis for H and suppose that $\mathrm{tr}(T)$ is finite, where the *trace* of T is

$$\mathrm{tr}(T) = \sum_{n=1}^\infty (Te_n, e_n).$$

Show the same is true for any other orthonormal basis, and that the sum is independent of which basis is chosen. Show that this is not necessarily true if we omit the assumption that T is positive. [Hint: care must be taken when interchanging infinite sums, unless, say, all the terms are positive (so use the operator $T^{1/2}$ to resolve this issue).]

29. Let H be a Hilbert space and $S \in B(H, H)$. Recall that $|S| = \sqrt{S^*S}$. Extend the definition of trace to nonpositive operators by $\mathrm{tr}(S) = \mathrm{tr}(|S|)$. Show that the set of operators with finite trace form an ideal in $B(H, H)$.

30. Let H be a separable Hilbert space and $\{\phi_n\}_{n=1}^\infty$ an ON basis. We say that $A : H \to H$ is a *Hilbert-Schmidt operator* if A is bounded, linear, and

$$\|A\|_{\mathrm{tr}}^2 = \sum_{n=1}^\infty \|A\phi_n\|^2 < \infty$$

(which defines a norm). Let $HS_2 \subset B(H, H)$ be the linear subspace of Hilbert-Schmidt operators.

(a) Show that the definition of $\| \cdot \|_{\mathrm{tr}}$ is independent of the choice of ON basis.

(b) Show that $\|A\| \le \|A\|_{\mathrm{tr}}$ for all $A \in HS_2$.

31. Let H be a separable Hilbert space, $S \in H^*$, and $T : H \to H$ a self-adjoint, compact linear and strictly positive $((Tx, x) > 0$ unless $x = 0)$ operator.

(a) Why is S compact and T an injective map?

(b) Suppose the base field $\mathbb{F} = \mathbb{R}$ and define $F : H \to \mathbb{R}$ by

$$F(x) = (Tx, x)_H - STx.$$

Find an x so that $F(x)$ is minimal. [Hint: work with an ON basis of eigenfunctions for T.]

32. Derive a spectral theorem for compact normal operators on a Hilbert space. We say that $T \in B(H,H)$ is a *normal operator* if $T^*T = TT^*$.

33. If X and Y are Banach spaces, show that $E \subset B(X,Y)$ is equicontinuous if and only if there is an $M < \infty$ such that $\|T\| \le M$ for all $T \in E$.

34. Prove that an integral operator defined on L^2 is compact; that is, prove Corollary 4.33.

35. Define the operator $T : L^2(0,1) \to L^2(0,1)$ by

$$Tu(x) = \int_0^x u(y)\, dy.$$

Show that T is compact, and find the eigenvalues of the self-adjoint compact operator T^*T. [Hint: T^* involves integration, so differentiate twice to get a second-order ODE with two boundary conditions.]

36. Define the linear operator $T : L^2([0,1]) \to L^2([0,1])$ by

$$Tf(x) = \int_0^x \int_y^1 f(z)\, dz\, dy.$$

(a) Show that T is self-adjoint.

(b) Show that T is compact.

(c) Find an orthogonal basis for $L^2([0,1])$ based on the eigenvalues of this operator. [Hint: differentiate twice and consider carefully the boundary conditions that must be satisfied.]

37. Let the underlying field be real and let $V \in L^2((0,1) \times \Omega)$, where Ω is some bounded domain (connected open set) in $\mathbb{R}^d$. Suppose that we define $T : L^2(0,1) \to L^2(0,1)$ by

$$Tf(x) = \int_0^1 \int_\Omega V(x,\omega)\, V(y,\omega)\, f(y)\, d\omega\, dy.$$

(a) Justify that T is well-defined, compact, self-adjoint, and positive semi-definite.

(b) Show that we can express

$$V(x,\omega) = \sum_{\alpha \in \mathcal{I}} a_\alpha(\omega)\, v_\alpha(x)$$

for some orthonormal basis $\{v_\alpha(x)\}_{\alpha \in \mathcal{I}}$ of $L^2(0,1)$. Give an expression for the coefficients $a_\alpha(\omega)$.

38. Let $\Omega \subset \mathbb{R}^d$ be a domain, $K \in L^2(\Omega \times \Omega)$, and T the compact operator defined on $L^2(\Omega)$ by

$$Tf(x) = \int_\Omega K(x, y) f(y) \, dy.$$

Show that the null space of $T - I$ satisfies

$$\dim N(T - I) \le \|K\|^2_{L^2(\Omega \times \Omega)}.$$

39. For the differential operator

$$L = D^2 + xD,$$

meaning $L\phi(x) = \phi''(x) + x\phi'(x)$, find a multiplying factor w so that wL is formally self-adjoint. Find boundary conditions on $I = [0, 1]$ which make this operator into a regular Sturm-Liouville problem for which 0 is not an eigenvalue.

40. Give conditions under which the Sturm-Liouville operator

$$L = DpD + q,$$

defined over an interval $I = [a, b]$, is a positive operator.

41. Write the *Euler operator*

$$L = x^2 D^2 + xD$$

with the boundary conditions $u(1) = u(e) = 0$ on the interval $[1, e]$ as a regular Sturm-Liouville problem with an appropriate weight function w. Find the eigenvalues and eigenfunctions for this problem.

42. Consider the heat equation

$$\frac{\partial u}{\partial t} - \frac{\partial^2 u}{\partial x^2} - \frac{\partial^2 u}{\partial y^2} = 0, \qquad t > 0, \ 0 < x < 1, \ 0 < y < 1,$$

$$u(t, 0, y) = u(t, 1, y) = 0, \qquad t > 0, \ 0 < y < 1,$$

$$u(t, x, 0) = u(t, x, 1) = 0, \qquad t > 0, \ 0 < x < 1,$$

$$u(0, x, y) = g(x, y), \qquad 0 < x < 1, \ 0 < y < 1,$$

where $g \in L^2((0, 1)^2)$.

(a) Solve the problem for $u(t, x, y)$ using separation of variables techniques and Sturm-Liouville theory. [Hint: view the variable t as being fixed in the expansions and obtain Sturm-Liouville problems by viewing the equation as $\frac{\partial^2 u}{\partial x^2} + \frac{\partial^2 u}{\partial y^2} = f$.]

(b) Justify your final expression for the solution.

(c) Use your solution to show that $u(t, x, y) \to 0$ in $L^2((0,1)^2)$ as $t \to \infty$.

43. Let X be an infinite dimensional Banach space over $\mathbb{C}$, and assume that $f \in X^*$ and $f_1, \ldots, f_n \in X^*$. Suppose that there exists a constant $C > 0$ such that whenever $|f_j(x)| < C$ holds for all $j = 1, \ldots, n$, with $x \in X$, then $|f(x)| < 1$.

(a) Prove that $\bigcap_{j=1}^n N(f_j) \subset N(f)$.

(b) Prove that f is a linear combination of the f_j. [Hint: using (a), study the range $R(F) \subset \mathbb{C}^{n+1}$ of the map $F : X \to \mathbb{C}^{n+1}$ defined by $F(x) = (f_1(x), \ldots, f_n(x), f(x))$.]

44. Suppose that $H = L^2(\Omega)$ for a bounded domain $\Omega \subset \mathbb{R}^d$ and $\{\phi_n(x)\}_{n=1}^\infty$ is an ON basis.

(a) Show that for $K \in L^2(\Omega \times \Omega)$, we can write

$$K(x, y) = \sum_{n=1}^\infty \sum_{m=1}^\infty k_{m,n} \, \phi_m(x) \, \overline{\phi_n(y)}$$

for some $k_{m,n} \in \mathbb{F}$.

(b) Show that if $A \in B(L^2(\Omega), L^2(\Omega))$ is an integral operator, meaning that there is $K \in L^2(\Omega \times \Omega)$ such that

$$Af = \int_\Omega K(x, y) \, f(y) \, dy,$$

then A is a Hilbert-Schmidt operator (as defined above in Exercise 30).

CHAPTER 5

Distributions

The theory of *distributions*, also known as *generalized functions*, provides a wide-ranging setting within which differentiation is defined and can be exploited. It underlies the modern study of differential equations, optimization, the calculus of variations, and many other subjects where differentiation plays a significant role.

The gist of the idea behind generalized functions goes back to Sergei Sobolev in the 1930s, though there are earlier roots. After World War II, Laurent Schwartz provided a more systematic rendition of the theory of generalized function, or distributions as he named them. Since the mid-1950s, the theory itself has not changed all that much, but the range of its uses is still growing.

5.1 THE NOTION OF GENERALIZED FUNCTIONS

The classic definition of the derivative is rather restrictive. For example, consider the function defined by

$$f(x) = \begin{cases} x, & x \geq 0, \\ 0, & x < 0. \end{cases}$$

Then $f \in C^0(-\infty, \infty)$ and f is differentiable at every point except 0. The derivative of f is the *Heaviside function*

$$H(x) = \begin{cases} 1, & x > 0, \\ 0, & x < 0. \end{cases} \tag{5.1}$$

The nondifferentiability of f at 0 creates no particular problem, so should we consider f differentiable on $(-\infty, \infty)$? The derivative of H is also well-defined, except at 0. However, it would appear that

$$H'(x) = \begin{cases} 0, & x \neq 0, \\ +\infty, & x = 0, \end{cases}$$

DOI: 10.1201/9781003492139-5

at least in some sense. Is it possible to make a precise statement? That is, can we generalize the notion of function so that H' is well-defined?

A precise statement emerges if integration by parts is employed. Recall that if $u, \phi \in C^1([a, b])$, then

$$\int_a^b u' \phi \, dx = u\phi \Big|_a^b - \int_a^b u\phi' \, dx.$$

If $\phi \in C^1$ but $u \in C^0 \setminus C^1$, one could tentatively define "$\int_a^b u'v \, dx$" by the expression

$$u\phi \Big|_a^b - \int_a^b u\phi' \, dx.$$

If there are enough "test functions" $\phi \in C^1$, then perhaps the properties of the putative u' can be determined, rather like physicists ascertain the properties of a particle by observing its interaction with other particles. In practice, test functions ϕ are drawn from

$$C_0^\infty = C_0^\infty(-\infty, \infty)$$
$$= \{\phi \in C^\infty(-\infty, \infty) : \exists R > 0 \text{ for which } \phi(x) = 0 \ \forall \, |x| > R\}$$

so that the boundary terms vanish as $a \to -\infty$, $b \to +\infty$.

Returning to the example of the function f with which the discussion began, for any $\phi \in C_0^\infty$,

$$\int_{-\infty}^\infty f' \phi \, dx \equiv - \int_{-\infty}^\infty f\phi' \, dx$$
$$= - \int_0^\infty x\phi' \, dx$$
$$= -x\phi \Big|_0^\infty + \int_0^\infty \phi \, dx$$
$$= \int_{-\infty}^\infty H\phi \, dx.$$

As this holds for all $\phi \in C_0^\infty$ it seems reasonable to identify $f' = H$. Continuing in the vein, define

$$\text{``} \int_{-\infty}^\infty H'\phi \, dx \text{''} \equiv - \int_{-\infty}^\infty H\phi' \, dx = - \int_0^\infty \phi' \, dx = \phi(0),$$

and thus H' is identified with evaluation at the origin! The object $H'(x)$ is usually written $\delta_0(x)$ and is called the Dirac delta function. It is essentially zero everywhere except at the origin, where it must be infinite in some way that is not clear at first. It is *not* a function; it is a *generalized* function (or *distribution*).

This line of thought does not stop. For example,

$$\text{``} \int H'' \phi \, dx \text{''} = \text{``} \int \delta_0' \phi \, dx \text{''} = \text{``} -\int \delta_0 \phi' \, dx \text{''} = -\phi'(0).$$

Obviously, $H'' = \delta_0'$ has no well-defined value at the origin; nevertheless, we have a precise statement of the "integral" of δ_0' against any test function $\phi \in C_0^\infty$.

The process just described can be viewed advantageously as a *duality pairing* between function spaces. Introducing the standard notation

$$\mathcal{D} = \mathcal{D}(\mathbb{R}) = C_0^\infty(-\infty, \infty)$$

for this space of test functions, the objects

$$f, \quad f' = H, \quad H' = \delta_0, \quad \text{and} \quad H'' = \delta_0'$$

can be viewed as linear functionals on $\mathcal{D}$, since integrals are linear and map to $\mathbb{F}$. For any linear functional u, we imagine

$$u(\phi) = \text{``} \int u\phi \, dx, \text{''}$$

even when the integral is not in the Lebesgue sense, and define the derivative of u by

$$u'(\phi) = -u(\phi').$$

Then also

$$u''(\phi) = -u'(\phi') = u(\phi''),$$

and so on for higher derivatives. In our little example, precise statements are

$$f(\phi) = \int f\phi \, dx,$$

$$f'(\phi) = -f(\phi') = -\int f\phi' \, dx = \int H\phi \, dx = H(\phi),$$

$$H'(\phi) = -H(\phi') = -\int H\phi' \, dx = \phi(0) = \delta_0(\phi),$$

$$H''(\phi) = H(\phi'') = \int H\phi'' \, dx = -\phi'(0) = -\delta_0(\phi') = \delta_0'(\phi),$$

for any $\phi \in \mathcal{D}$.

Investigations involving derivatives often encounter limit processes. There will thus be needed a notion of continuity, which mandates a topology on $\mathcal{D}$. Unfortunately, a suitable topology on $\mathcal{D}$ is not so simple. For example, there is no Banach-space topology that works. A topology that does meet requirements is the next topic.

5.2 TEST FUNCTIONS

Throughout this chapter, $\Omega \subset \mathbb{R}^d$ will be a domain, i.e., a connected open subset of $\mathbb{R}^d$. Let $C^0(\Omega)$ denote the set of continuous functions mapping Ω into the ground field $\mathbb{F}$.

Definition. If $f \in C^0(\Omega)$, the *support* of f is

$$\mathrm{supp}(f) = \overline{\{x \in \Omega : |f(x)| > 0\}} \subset \Omega,$$

the *closure* (in Ω) of the set where f is nonzero. A *multi-index* $\alpha = (\alpha_1, \alpha_2, \ldots, \alpha_d) \in \mathbb{N}^d$ is an ordered d-tuple of nonnegative integers, and

$$|\alpha| = \alpha_1 + \alpha_2 + \cdots + \alpha_d.$$

The notation

$$\partial^\alpha = D^\alpha = \left(\frac{\partial}{\partial x_1}\right)^{\alpha_1} \cdots \left(\frac{\partial}{\partial x_d}\right)^{\alpha_d}$$

will be used for the partial differential operator. Notice that it has order $|\alpha|$. With this notation, the following function classes are written conveniently as follows:

$$C^n(\Omega) = \left\{f \in C^0(\Omega) : D^\alpha f \in C^0(\Omega) \ \forall \, |\alpha| \le n\right\};$$

$$C^\infty(\Omega) = \left\{f \in C^0(\Omega) : D^\alpha f \in C^0(\Omega) \ \forall \, \alpha\right\} = \bigcap_{n=1}^\infty C^n(\Omega);$$

$$\mathcal{D}(\Omega) = C_0^\infty(\Omega) = \left\{f \in C^\infty(\Omega) : \mathrm{supp}(f) \text{ is compact in } \Omega\right\}.$$

This latter collection is referred to as the set of *test functions* on Ω. If $K \subset\subset \Omega$ (K is *compactly contained in* Ω, i.e., K compact and $K \subset \Omega$), define

$$\mathcal{D}_K = \{f \in C_0^\infty(\Omega) : \mathrm{supp}(f) \subset K\}.$$

Proposition 5.1. *The sets $C^n(\Omega)$, $C^\infty(\Omega)$, $\mathcal{D}(\Omega)$, and $\mathcal{D}_K$ (for any $K \subset\subset \Omega$ with nonempty interior) are nontrivial vector spaces.*

Proof. It is trivial to verify that addition of functions and scalar multiplication are algebraically closed operations. Thus, each set is a vector subspace of the vector space of all functions from Ω to $\mathbb{F}$.

To see that these spaces are nontrivial, we construct an interesting element of $\mathcal{D}_K \subset \mathcal{D}(\Omega) \subset C^\infty(\Omega) \subset C^n(\Omega)$. Consider first *Cauchy's infinitely differentiable function* $\psi : \mathbb{R} \to \mathbb{R}$ given by

$$\psi(x) = \begin{cases} e^{-1/x^2}, & x > 0, \\ 0, & x \le 0. \end{cases} \tag{5.2}$$

This function is clearly infinitely differentiable for $x \neq 0$, and its mth derivative takes the form

$$\psi^{(m)}(x) = \begin{cases} R_m(x)e^{-1/x^2}, & x > 0, \\ 0, & x < 0, \end{cases}$$

where $R_m(x)$ is some polynomial divided by x to a power. But L'Hôpital's rule implies that

$$\lim_{x \to 0} R_m(x)e^{-1/x^2} = 0,$$

so in fact $\psi^{(m)}$ is continuous at 0 for all m, and thus ψ is infinitely differentiable.

Now let $\phi(x) = \psi(1-x)\,\psi(1+x)$. Then $\phi \in C_0^\infty(\mathbb{R})$ and $\mathrm{supp}(\phi) = [-1,1]$. Finally, for $x \in \mathbb{R}^d$,

$$\Phi(x) = \phi(x_1)\phi(x_2)\dots\phi(x_d) \in C^\infty(\mathbb{R}^d)$$

has support $[-1,1]^d$. By translation and dilation, an element of $\mathcal{D}_K$ may then be realized. $\quad\square$

Corollary 5.2. *There exist infinitely differentiable, nonanalytic functions.*

Another way to say this is that there are infinitely differentiable real-valued functions not given by their Taylor series. This is evident since the Taylor series of $\psi(x)$ about 0 is 0, but $\psi(x) \neq 0$ for $x > 0$.

The formula

$$\|\phi\|_{m,\infty,\Omega} = \sum_{|\alpha| \leq m} \|D^\alpha \phi\|_{L^\infty(\Omega)}$$

is the obvious first attempt to define a norm on $C^m(\Omega)$. Unfortunately, this norm is not finite for all $\phi \in C^m(\Omega)$. If we restrict to

$$C_B^m(\Omega) = \{\phi \in C^m(\Omega) : \|\phi\|_{m,\infty,\Omega} < \infty\},$$

then $C_B^m(\Omega)$ is a Banach space, since completeness follows from the fact that, on compact subsets, the uniform limit of continuous functions is continuous. Note that

$$\|\phi\|_{n,\infty,\Omega} \geq \|\phi\|_{m,\infty,\Omega} \quad \text{when } n \geq m,$$

so these comprise an increasing sequence of norms. They will be used to define convergence in $\mathcal{D}(\Omega)$. Care must be taken as the following example shows.

Example. Take any $\phi \in C_0^\infty(\mathbb{R})$ such that $\mathrm{supp}(\phi) = [0,1]$ and $\phi(x) > 0$ for $x \in (0,1)$ (for example, such a function can be constructed using Cauchy's infinitely differentiable function (5.2) as $\psi(x)\psi(1-x)$). Define, for any integer $n \geq 1$,

$$\psi_n(x) = \sum_{j=1}^n \frac{1}{j}\phi(x - j) \in C_0^\infty(\mathbb{R});$$

clearly $\operatorname{supp}(\psi_n) = [1, n+1]$. Define also

$$\psi(x) = \sum_{j=1}^{\infty} \frac{1}{j} \phi(x - j) \in C^{\infty}(\mathbb{R}) \smallsetminus C_0^{\infty}(\mathbb{R}).$$

It is straightforward to verify that for any $m \geq 0$,

$$D^m \psi_n \xrightarrow{L^{\infty}} D^m \psi;$$

that is,

$$\|\psi_n - \psi\|_{m,\infty,\mathbb{R}} \to 0$$

for each m, but $\psi \notin C_0^{\infty}(\mathbb{R})$.

To insure that $\mathcal{D}(\Omega)$ is complete, we will need both uniform convergence and a condition to force the limit to be compactly supported. The following definition suffices, and gives the usual topology on $C_0^{\infty}(\Omega)$. Strictly speaking, when the vector space $C_0^{\infty}(\Omega)$ is equipped with the topology indicated below, then it is denoted by $\mathcal{D}(\Omega)$.

Definition. Let $\Omega \subset \mathbb{R}^d$ be a domain in $\mathbb{R}^d$ and let $\mathcal{D}(\Omega)$ be the vector space $C_0^{\infty}(\Omega)$ endowed with the following notion of convergence: a sequence $\{\phi_j\}_{j=1}^{\infty} \subset \mathcal{D}(\Omega)$ *converges* to $\phi \in \mathcal{D}(\Omega)$ if and only if

(a) there is some fixed $K \subset\subset \Omega$ such that $\operatorname{supp}(\phi_j) \subset K$ for all j, and

(b) $\lim\limits_{j \to \infty} \|\phi_j - \phi\|_{n,\infty,\Omega} = 0$ for all n.

The sequence is *Cauchy* if $\operatorname{supp}(\phi_j) \subset K$ for all j and some fixed $K \subset\subset \Omega$ and, given $\epsilon > 0$ and $n \geq 0$, there exists $N = N(\epsilon, n) > 0$ such that for all $j, k \geq N$,

$$\|\phi_j - \phi_k\|_{n,\infty,\Omega} \leq \epsilon.$$

Thus, convergence in $\mathcal{D}(\Omega)$ means all the elements ϕ_j of the sequence are localized to a fixed compact set K, and each of their derivatives converges uniformly. Our definition does not identify open and closed sets; nevertheless, it does define a topology on $\mathcal{D}$. This will be explored in the problem set. Unfortunately, $\mathcal{D}(\Omega)$ is *not* metrizable! However it is not difficult to show that $\mathcal{D}(\Omega)$ is complete.

Theorem 5.3. *The linear space $\mathcal{D}(\Omega)$ is complete.*

For any $K \subset\subset \Omega$, the space $\mathcal{D}_K$ is complete. It is also metrizable. In fact, a metric can be given by

$$d(\phi, \psi) = \sum_{n=0}^{\infty} 2^{-n} \frac{\|\phi - \psi\|_{n,\infty,\Omega}}{1 + \|\phi - \psi\|_{n,\infty,\Omega}}.$$

The space $\mathcal{D}_K$ is a *Fréchet space*, which is a topological vector space with a metric that is complete in this metric. The metric defined above through an infinite sequence of seminorms $\|\cdot\|_{n,\infty,\Omega}$, $n = 0, 1, \ldots, \infty$, that *separate points* (which means that $\|\phi\|_{n,\infty,\Omega} = 0$ for all n if and only if $\phi = 0$) is typical.

5.3 DISTRIBUTIONS

Since $\mathcal{D}(\Omega)$ is not a metric space, continuity and sequential continuity need not be equivalent; however, it turns out that they are equivalent for linear functionals. We do not use or prove the following fact, but it does explain why the word "sequential" is often omitted before the word "continuous" in the present context.

Theorem 5.4. *If $T : \mathcal{D}(\Omega) \to Y$ is linear, where Y is an NLS, then T is continuous if and only if T is sequentially continuous.*

Definition. A *distribution* or *generalized function* on a domain Ω is a (sequentially) continuous linear functional on $\mathcal{D}(\Omega)$. The vector space of all distributions is denoted $\mathcal{D}'(\Omega)$ (or sometimes $\mathcal{D}(\Omega)^*$). When $\Omega = \mathbb{R}^d$, we will often write $\mathcal{D}$ for $\mathcal{D}(\mathbb{R}^d)$ and $\mathcal{D}'$ for $\mathcal{D}'(\mathbb{R}^d)$.

As for linear mappings of any linear space, the following result is valid (consider Proposition 2.6 and its proof).

Theorem 5.5. *If $T : \mathcal{D}(\Omega) \to \mathbb{F}$ is linear, then T is sequentially continuous if and only if T is sequentially continuous at $0 \in \mathcal{D}(\Omega)$.*

Here is a concrete condition that is equivalent to continuity of a linear map $T : \mathcal{D}(\Omega) \to \mathbb{F}$.

Theorem 5.6. *Suppose that $T : \mathcal{D}(\Omega) \to \mathbb{F}$ is linear. Then $T \in \mathcal{D}'(\Omega)$ (i.e., T is continuous) if and only if for every $K \subset\subset \Omega$, there are $n \geq 0$ and $C > 0$, depending on K, such that*

$$|T(\phi)| \leq C\|\phi\|_{n,\infty,\Omega}$$

for every $\phi \in \mathcal{D}_K$.

The minimal value of n needed above is called the *order* of the distribution.

Proof. Suppose that $T \in \mathcal{D}'(\Omega)$, but suppose also that the conclusion is false. Then there is some $K \subset\subset \Omega$ such that for every $n \geq 0$ and $m > 0$, we have some $\phi_{n,m} \in \mathcal{D}_K$ such that

$$|T(\phi_{n,m})| > m\|\phi_{n,m}\|_{n,\infty,\Omega}.$$

Rescale by setting $\hat{\phi}_j = \phi_{j,j}/(j\|\phi_{j,j}\|_{j,\infty,\Omega}) \in \mathcal{D}_K$. Then $|T(\hat{\phi}_j)| > 1$, but $\hat{\phi}_j \to 0$ in $\mathcal{D}(\Omega)$ (since $\|\hat{\phi}_j\|_{n,\infty,\Omega} \leq \|\hat{\phi}_j\|_{j,\infty,\Omega} = 1/j$ for $j \geq n$), contradicting the hypothesis.

For the converse, suppose that $\phi_j \to 0$ in $\mathcal{D}(\Omega)$. Then there is some $K \subset\subset \Omega$ such that $\mathrm{supp}(\phi_j) \subset K$ for all j, and, by hypothesis, some n and C such that

$$|T(\phi_j)| \leq C\|\phi_j\|_{n,\infty,\Omega} \longrightarrow 0.$$

That is, T is (sequentially) continuous at 0. ∎

Here are some important examples.

Definition.

$$L^1_{\text{loc}}(\Omega) = \left\{ f : \Omega \to \mathbb{F} \text{ measurable} \mid \text{for every } K \subset\subset \Omega, \int_K |f(x)| \, dx < \infty \right\}.$$

Note that $L^1(\Omega) \subset L^1_{\text{loc}}(\Omega)$. Any polynomial is in $L^1_{\text{loc}}(\Omega)$ but it may *not* lie in $L^1(\Omega)$, if Ω is unbounded. Elements of $L^1_{\text{loc}}(\Omega)$ are not too singular at a point, but they may grow very rapidly near the boundary of Ω.

Example. If $f \in L^1_{\text{loc}}(\Omega)$, define $\Lambda_f \in \mathcal{D}'(\Omega)$ by

$$\Lambda_f(\phi) = \int_\Omega f(x)\phi(x) \, dx$$

for every $\phi \in \mathcal{D}(\Omega)$. The mapping Λ_f is obviously well-defined and a linear functional; it is also continuous, since for $\phi \in \mathcal{D}_K$,

$$|\Lambda_f(\phi)| \le \int_K |f(x)| \, |\phi(x)| \, dx \le \left(\int_K |f(x)| \, dx \right) \|\phi\|_{0,\infty,\Omega},$$

showing that Λ_f satisfies the requirement of Theorem 5.6 with order 0.

The mapping $f \mapsto \Lambda_f$ is one-to-one in the following sense.

Lemma 5.7 (Lebesgue lemma). *Let $f, g \in L^1_{\text{loc}}(\Omega)$. Then $\Lambda_f = \Lambda_g$ if and only if $f = g$ almost everywhere.*

Proof. If $f = g$ a.e., then obviously $\Lambda_f = \Lambda_g$. Conversely, suppose $\Lambda_f = \Lambda_g$. Then $\Lambda_{f-g} = 0$ by linearity. Let

$$R = \left\{ x \in \mathbb{R}^d : a_i \le x_i \le b_i, \ i = 1, \ldots, d \right\} \subset \Omega$$

be an arbitrary closed rectangle, and let $\psi(x)$ be Cauchy's infinitely differentiable function on $\mathbb{R}$ given by (5.2). For $\epsilon > 0$, let

$$\phi_\epsilon(x) = \psi(\epsilon - x)\psi(x) \ge 0$$

and

$$\Phi_\epsilon(x) = \frac{\displaystyle\int_{-\infty}^{x} \phi_\epsilon(\xi) \, d\xi}{\displaystyle\int_{-\infty}^{\infty} \phi_\epsilon(\xi) \, d\xi}.$$

Then $\text{supp}(\phi_\epsilon) = [0, \epsilon]$, $0 \le \Phi_\epsilon(x) \le 1$, $\Phi_\epsilon(x) = 0$ for $x \le 0$, and $\Phi_\epsilon(x) = 1$ for $x \ge \epsilon$.

Now let

$$\Psi_\epsilon(x) = \prod_{i=1}^{d} \Phi_\epsilon(x_i - a_i)\Phi_\epsilon(b_i - x_i) \in \mathcal{D}_R.$$

The characteristic function of R is

$$\chi_R(x) = \begin{cases} 1, & x \in R, \\ 0, & x \notin R. \end{cases}$$

Pointwise, $\Psi_\epsilon(x) \to \chi_R(x)$ as $\epsilon \to 0$, and so Lebesgue's dominated convergence theorem implies that

$$(f - g)\Psi_\epsilon \longrightarrow (f - g)\chi_R$$

in $L^1(R)$. Thus

$$0 = \Lambda_{f-g}(\Psi_\epsilon) = \int_R (f - g)(x)\Psi_\epsilon(x)\,dx \longrightarrow \int_R (f - g)(x)\,dx$$

as $\epsilon \to 0$. So the integral of $f - g$ vanishes over any closed rectangle. The elementary theory of Lebesgue integration then implies that $f - g = 0$ almost everywhere. $\qquad\square$

A function $f \in L^1_{\mathrm{loc}}(\Omega)$ is identified with $\Lambda_f \in \mathcal{D}'(\Omega)$, calling the *function* f a *distribution* in this sense. There are distributions that do not arise this way. This observation leads to calling distributions *generalized functions*: functions are distributions but also more general objects are distributions.

Definition. For $T \in \mathcal{D}'(\Omega)$, if there is $f \in L^1_{\mathrm{loc}}(\Omega)$ such that $T = \Lambda_f$, then T is called a *regular* distribution. Otherwise T is a *singular* distribution.

Because the action of regular distributions is given by integration, people sometimes write, improperly but conveniently,

$$T(\phi) = \int_\Omega T\phi\,dx$$

for $T \in \mathcal{D}'(\Omega)$, $\phi \in \mathcal{D}(\Omega)$. We will favor the more precise notation

$$T(\phi) = \langle T, \phi \rangle = \langle T, \phi \rangle_{\mathcal{D}',\mathcal{D}},$$

where the notation $\langle \cdot, \cdot \rangle_{\mathcal{D}',\mathcal{D}}$ emphasizes the duality pairing, i.e., the dual nature of the pairing of elements of $\mathcal{D}'(\Omega)$ and $\mathcal{D}(\Omega)$. This is sometimes, but not always, ordinary integration on Ω (the standard $L^2(\Omega)$ inner-product if $\mathbb{F} = \mathbb{R}$). When the pairing is clear from context, one writes merely $\langle \cdot, \cdot \rangle$, omitting the cumbersome subscripts.

Example. Let $\delta_0 \in \mathcal{D}'(\Omega)$ be defined by

$$\langle \delta_0, \phi \rangle = \phi(0)$$

for every $\phi \in \mathcal{D}(\Omega)$. (Of course, if $0 \notin \Omega$, this map is trivial.) Linearity is clear, and

$$|\langle \delta_0, \phi \rangle| = |\phi(0)| \le \|\phi\|_{0,\infty,\Omega}$$

implies, by Theorem 5.6, that δ_0 is continuous. The mapping δ_0 is called the *Dirac distribution* at 0 (it is also known as the Dirac measure or mass, or the Dirac delta function). There is clearly no $f \in L^1_{loc}(\Omega)$ such that $\delta_0 = \Lambda_f$, so δ_0 is a singular distribution. If $x \in \Omega$, $\delta_x \in \mathcal{D}'(\Omega)$ is defined by

$$\langle \delta_x, \phi \rangle = \phi(x).$$

This is the Dirac mass at x. This generalized function is often written, improperly, as

$$\delta_x(y) = \delta_0(y - x) = \delta_0(x - y).$$

Remark. Here is a sketch of why $\mathcal{D}(\Omega)$ is *not* metrizable. The details are left to the reader. For $K \subset\subset \Omega$,

$$\mathcal{D}_K = \bigcap_{x \in \Omega \setminus K} \ker(\delta_x).$$

Since $\ker(\delta_x)$ is closed, so is $\mathcal{D}_K$ (in $\mathcal{D}(\Omega)$). It is straightforward to show that $\mathcal{D}_K$ has empty interior in $\mathcal{D}$. For an appropriate sequence $K_1 \subset K_2 \subset \cdots \subset \Omega$ of compact sets such that

$$\bigcup_{n=1}^{\infty} K_n = \Omega,$$

it will be the case that

$$\mathcal{D}(\Omega) = \bigcup_{n=1}^{\infty} \mathcal{D}_{K_n}.$$

Thus, $\mathcal{D}(\Omega)$ is seen to be a countable union of nowhere dense sets. But, it transpires that $\mathcal{D}(\Omega)$ is sequentially complete, so if it were metrizable, Baire's theorem would lead to a contradiction.

Example. If μ is either a complex Borel measure on Ω or a positive measure on Ω such that $\mu(K) < \infty$ for every $K \subset\subset \Omega$, then

$$\Lambda_\mu(\phi) = \int_\Omega \phi(x)\, d\mu(x)$$

defines a distribution, since for $\phi \in \mathcal{D}_K$,

$$|\Lambda_\mu(\phi)| \le |\mu(\text{supp}(\phi))|\, \|\phi\|_{0,\infty,\Omega} \le |\mu(K)|\, \|\phi\|_{0,\infty,\Omega}.$$

Example. Define the distribution $PV\frac{1}{x} \in \mathcal{D}'(\mathbb{R})$ by

$$\left\langle PV\frac{1}{x}, \phi \right\rangle = PV \int \frac{1}{x}\phi(x)\, dx \equiv \lim_{\epsilon \to 0^+} \int_{|x|>\epsilon} \frac{1}{x}\phi(x)\, dx.$$

This is usually called *Cauchy's principal value* of $1/x$. Since $1/x \notin L^1_{\text{loc}}(\mathbb{R})$, it must be verified that the limit is well-defined. Fix $\phi \in \mathcal{D}$ and $\epsilon > 0$. Then, integration by parts gives

$$\int_{|x|>\epsilon} \frac{1}{x}\phi(x)\,dx = [\phi(-\epsilon) - \phi(\epsilon)]\ln \epsilon - \int_{|x|>\epsilon} \ln|x|\phi'(x)\,dx.$$

The boundary terms tend to 0 as $\epsilon \to 0^+$ since

$$\lim_{\epsilon \to 0^+} [\phi(-\epsilon) - \phi(\epsilon)]\ln \epsilon = \lim_{\epsilon \to 0^+} 2\frac{\phi(-\epsilon) - \phi(\epsilon)}{2\epsilon}\epsilon \ln \epsilon = -2\phi'(0)\lim_{\epsilon \to 0^+}\epsilon\ln\epsilon = 0.$$

Thus, if $\text{supp}(\phi) \subset [-R, R] = K$, then

$$PV \int \frac{1}{x}\phi(x) = -\lim_{\epsilon \to 0^+}\int_{|x|>\epsilon}\ln|x|\phi'(x)\,dx = -\int_{-R}^{R}\ln|x|\phi'(x)\,dx$$

exists, and the inequality

$$\left|PV\int \frac{1}{x}\phi(x)\,dx\right| \leq \left(\int_{-R}^{R}|\ln|x||\,dx\right)\|\phi\|_{1,\infty,\mathbb{R}}$$

shows that $PV(1/x)$ is a distribution of order 1, since $\ln|x|$ is locally integrable.

5.4 OPERATIONS WITH DISTRIBUTIONS

A simple way to define a new distribution from an existing one is to use duality. If $T : \mathcal{D}(\Omega) \to \mathcal{D}(\Omega)$ is sequentially continuous and linear, then $T^* : \mathcal{D}'(\Omega) \to \mathcal{D}'(\Omega)$ satisfies

$$\langle u, T\phi \rangle = \langle T^*u, \phi \rangle$$

for all $u \in \mathcal{D}'(\Omega)$, $\phi \in \mathcal{D}(\Omega)$. Obviously $T^*u = u \circ T$ is sequentially continuous and linear.

Proposition 5.8. *If $u \in \mathcal{D}'(\Omega)$ and $T : \mathcal{D}(\Omega) \to \mathcal{D}(\Omega)$ is sequentially continuous and linear, then $T^*u = u \circ T \in \mathcal{D}'(\Omega)$.*

This proposition is used to conclude that various linear functionals are distributions; alternatively, we could have shown that the condition of Theorem 5.6 holds, as the reader may verify.

5.4.1 Multiplication by a smooth function

If $f \in C^\infty(\Omega)$, we can define $T_f : \mathcal{D}(\Omega) \to \mathcal{D}(\Omega)$ by $T_f(\phi) = f\phi$. Obviously T_f is linear and sequentially continuous on account of the product rule for differentiation. Thus, for any $u \in \mathcal{D}'(\Omega)$, $T_f^* u = u \circ T_f \in \mathcal{D}'(\Omega)$. But if $u = \Lambda_u$ is a regular distribution (so $u \in L_{\mathrm{loc}}^1(\Omega)$),

$$\begin{aligned} \langle T_f^* u, \phi \rangle = \langle u, T_f \phi \rangle &= \langle u, f\phi \rangle \\ &= \int_\Omega u(x) f(x) \phi(x)\, dx \\ &= \langle fu, \phi \rangle, \end{aligned}$$

for any $\phi \in \mathcal{D}(\Omega)$. For any $u \in \mathcal{D}'$ and $f \in C^\infty(\Omega)$, define a new distribution, denoted fu, as $fu = T_f^* u$. Its action on a test function ϕ is

$$\langle fu, \phi \rangle = \langle u, f\phi \rangle \quad \forall \phi \in \mathcal{D}(\Omega).$$

Thus, any distribution can be multiplied by a smooth function, and

$$f\Lambda_u = \Lambda_{fu}$$

for a regular distribution. This definition of multiplication of a distribution by a (smooth) function is consistent with the existing one of multiplying a function by another function.

5.4.2 Differentiation

Our most important example is differentiation. Note that $D^\alpha : \mathcal{D}(\Omega) \to \mathcal{D}(\Omega)$ is (sequentially) continuous for any multi-index α, so $(D^\alpha)^* u = u \circ D^\alpha \in \mathcal{D}'(\Omega)$. Moreover, for $\phi, \psi \in C_0^\infty(\Omega)$, integration by parts reveals that

$$\int D^\alpha \phi(x)\, \psi(x)\, dx = (-1)^{|\alpha|} \int \phi(x) D^\alpha \psi(x)\, dx.$$

Definition. If α is a multi-index and $u \in \mathcal{D}'(\Omega)$, define $D^\alpha u \in \mathcal{D}'(\Omega)$ by

$$\langle D^\alpha u, \phi \rangle = (-1)^{|\alpha|} \langle u, D^\alpha \phi \rangle \quad \forall \phi \in \mathcal{D}(\Omega);$$

that is, $\mathcal{D}^\alpha u = ((-1)^{|\alpha|} \mathcal{D}^\alpha)^* u$.

An issue of consistency arises: are distributional derivatives the same as classical derivatives when the latter exist?

Proposition 5.9. *Suppose $u \in C^n(\Omega)$ for some $n \geq 0$. Let α be a multi-index such that $|\alpha| \leq n$, and denote the classical αth partial derivative of u by $\partial^\alpha u = \partial^\alpha u / \partial x^\alpha$. Then*

$$D^\alpha u \equiv D^\alpha \Lambda_u = \Lambda_{\partial^\alpha u} \equiv \partial^\alpha u.$$

That is, the two distributions $D^\alpha \Lambda_u$ and $\Lambda_{\partial^\alpha u}$ agree.

Proof. For any $\phi \in \mathcal{D}(\Omega)$,

$$\langle D^\alpha \Lambda_u, \phi \rangle = (-1)^{|\alpha|} \langle \Lambda_u, D^\alpha \phi \rangle$$

$$= (-1)^{|\alpha|} \int u(x) D^\alpha \phi(x) \, dx$$

$$= \int \partial^\alpha u(x) \phi(x) \, dx$$

$$= \langle \partial^\alpha u, \phi \rangle,$$

where the third equality comes by the ordinary integration by parts formula. Since ϕ is arbitrary, $D^\alpha \Lambda_u = \partial^\alpha u$. □

Examples.

(1) If $H(x)$ is the Heaviside function (5.1), then $H \in L^1_{\text{loc}}(\mathbb{R})$ is also a distribution, and, for any $\phi \in \mathcal{D}(\mathbb{R})$,

$$\langle H', \phi \rangle = -\langle H, \phi' \rangle$$

$$= -\int_{-\infty}^{\infty} H(x) \phi'(x) \, dx$$

$$= -\int_0^{\infty} \phi'(x) \, dx$$

$$= \phi(0) = \langle \delta_0, \phi \rangle.$$

Thus $H' = \delta_0$, as *distributions*.

(2) Since $\ln|x| \in L^1_{\text{loc}}(\mathbb{R})$ is a distribution, its distributional derivative applied to $\phi \in \mathcal{D}$ is

$$\langle D \ln|x|, \phi \rangle = -\langle \ln|x|, D\phi \rangle$$

$$= -\int \ln|x| \phi'(x) \, dx$$

$$= -\lim_{\epsilon \to 0^+} \int_{|x|>\epsilon} \ln|x| \phi'(x) \, dx$$

$$= \lim_{\epsilon \to 0^+} \left[\int_{|x|>\epsilon} \frac{1}{x} \phi(x) \, dx + \big(\phi(\epsilon) - \phi(-\epsilon)\big) \ln|\epsilon| \right]$$

$$= \lim_{\epsilon \to 0^+} \int_{|x|>\epsilon} \frac{1}{x} \phi(x) \, dx.$$

Thus $D \ln|x| = PV(1/x)$.

Proposition 5.10. *If $u \in \mathcal{D}'(\Omega)$ and α and β are multi-indices, then*

$$D^\alpha D^\beta u = D^\beta D^\alpha u = D^{\alpha+\beta} u.$$

Proof. For $\phi \in \mathcal{D}(\Omega)$,

$$\langle D^\alpha D^\beta u, \phi \rangle = (-1)^{|\alpha|} \langle D^\beta u, D^\alpha \phi \rangle$$
$$= (-1)^{|\alpha|+|\beta|} \langle u, D^\beta D^\alpha \phi \rangle$$
$$= (-1)^{|\beta|+|\alpha|} \langle u, D^\alpha D^\beta \phi \rangle.$$

Thus α and β may be interchanged. Moreover,

$$\langle D^\alpha D^\beta u, \phi \rangle = (-1)^{|\alpha|+|\beta|} \langle u, D^\alpha D^\beta \phi \rangle$$
$$= (-1)^{|\alpha+\beta|} \langle u, D^{\alpha+\beta} \phi \rangle$$
$$= \langle D^{\alpha+\beta} u, \phi \rangle. \qquad \square$$

Lemma 5.11 (Leibniz rule). *Let $f \in C^\infty(\Omega)$, $u \in \mathcal{D}'(\Omega)$, and $\alpha \in \mathbb{N}^d$ a multi-index. Then*

$$D^\alpha(fu) = \sum_{\beta \le \alpha} \binom{\alpha}{\beta} D^{\alpha-\beta} f D^\beta u \in \mathcal{D}'(\Omega),$$

where

$$\binom{\alpha}{\beta} = \frac{\alpha!}{(\alpha-\beta)!\beta!},$$

$\alpha! = \alpha_1! \alpha_2! \cdots \alpha_d!$, and $\beta \le \alpha$ means that β is a multi-index with $\beta_i \le \alpha_i$ for $i = 1, \ldots, d$.

If $u \in C^\infty(\Omega)$, this is just the product rule for differentiation.

Proof. By the previous proposition, the result holds if it is true for multi-indices that have a single nonzero component, say the first component. We proceed by induction on $n = |\alpha|$. The result holds trivially for $n = 0$, but we will also need the result for $n = 1$. Denote D^α by D_1^n. When $n = 1$, write D_1 for D_1^1 and then for any $\phi \in \mathcal{D}(\Omega)$,

$$\langle D_1(fu), \phi \rangle = -\langle fu, D_1\phi \rangle$$
$$= -\langle u, f D_1\phi \rangle$$
$$= -\langle u, D_1(f\phi) - D_1 f \phi \rangle$$
$$= \langle D_1 u, f\phi \rangle + \langle u, D_1 f \phi \rangle$$
$$= \langle f D_1 u + D_1 f u, \phi \rangle,$$

and the result holds.

Now assume the result for derivatives up to order $n - 1$. Then

$$D_1^n(fu) = D_1 D_1^{n-1}(fu)$$

$$= D_1 \sum_{j=0}^{n-1} \binom{n-1}{j} D_1^{n-1-j} f D_1^j u$$

$$= \sum_{j=0}^{n-1} \binom{n-1}{j} (D_1^{n-j} f D_1^j u + D_1^{n-1-j} f D_1^{j+1} u)$$

$$= \sum_{j=0}^{n-1} \binom{n-1}{j} D_1^{n-j} f D_1^j u + \sum_{j=1}^{n} \binom{n-1}{j-1} D_1^{n-j} f D_1^j u$$

$$= \sum_{j=0}^{n} \binom{n}{j} D_1^{n-j} f D_1^j u,$$

where the last equality follows from the combinatorial identity

$$\binom{n}{j} = \binom{n-1}{j} + \binom{n-1}{j-1}.$$

So the induction proceeds. ∎

Example. Consider $f(x) = x \ln |x|$. Since $x \in C^\infty(\mathbb{R})$ and $\ln |x| \in \mathcal{D}'$, the simplest case of the last proposition implies that

$$D(x \ln |x|) = \ln |x| + x \, PV \frac{1}{x}.$$

But, for $\phi \in \mathcal{D}$, integration by parts yields

$$\langle D(x \ln |x|), \phi \rangle = -\langle x \ln |x|, D\phi \rangle$$

$$= -\int x \ln |x| \phi'(x) \, dx$$

$$= \int_0^\infty (\ln |x| + 1)\phi(x) \, dx + \int_{-\infty}^0 (\ln |x| + 1)\phi(x) \, dx$$

$$= \langle \ln |x| + 1, \phi \rangle.$$

Thus

$$x \, PV \frac{1}{x} = 1,$$

which the reader can prove directly quite easily.

5.4.3 Translations and dilations of $\mathbb{R}^d$

Assume $\Omega = \mathbb{R}^d$ and define for any fixed $x \in \mathbb{R}^d$ and $\lambda \in \mathbb{R}$, $\lambda \neq 0$, the maps $\tau_x : \mathcal{D} \to \mathcal{D}$ and $T_\lambda : \mathcal{D} \to \mathcal{D}$ by

$$\tau_x \phi(y) = \phi(y - x) \quad \text{and} \quad T_\lambda \phi(y) = \phi(\lambda y),$$

for $y \in \mathbb{R}^d$. These maps translate and dilate the domain. They are clearly linear and (sequentially) continuous maps of $\mathcal{D}$ to itself.

Given $u \in \mathcal{D}'$, define the distributions $\tau_x u$ and $T_\lambda u$ for $\phi \in \mathcal{D}$ by

$$\langle \tau_x u, \phi \rangle = \langle u, \tau_{-x}\phi \rangle \quad \text{and} \quad \langle T_\lambda u, \phi \rangle = \frac{1}{|\lambda|^d} \langle u, T_{1/\lambda}\phi \rangle.$$

These definitions are easily seen to be consistent with the usual change of variables formulas for integrals when u is a regular distribution.

5.4.4 Convolutions

If $f, g : \mathbb{R}^d \to \mathbb{F}$ are functions, the *convolution* of f and g, is the function denoted $f * g : \mathbb{R}^d \to \mathbb{F}$ defined by

$$(f * g)(x) = \int_{\mathbb{R}^d} f(y)\, g(x - y)\, dy = (g * f)(x),$$

provided the (Lebesgue) integral exists for almost every $x \in \mathbb{R}^d$. If τ_x denotes the spatial translation distribution defined above and R denotes reflection (i.e., $R = T_{-1}$ where the notation is from the previous subsection), then a complicated way to write the convolution is

$$f * g(x) = \int_{\mathbb{R}^d} f(y)\, (\tau_x R g)(y)\, dy.$$

This motivates the definition of the *convolution* of a distribution $u \in \mathcal{D}'(\mathbb{R}^d)$ with a test function $\phi \in \mathcal{D}(\mathbb{R}^d)$, namely

$$(u * \phi)(x) = \langle u, \tau_x R\phi \rangle = \langle R\tau_{-x}u, \phi \rangle \quad \text{for any } x \in \mathbb{R}^d.$$

Notice that for fixed $x \in \mathbb{R}^d$, $R\tau_{-x}u = u \circ \tau_x \circ R$ is a well-defined element of $\mathcal{D}'$.

Examples.

(1) If $\phi \in \mathcal{D}$ and $x \in \mathbb{R}^d$, then

$$\delta_0 * \phi(x) = \langle \delta_0, \tau_x R\phi \rangle = \phi(x).$$

(2) If $u \in \mathcal{D}'$, then
$$u * R\phi\,(0) = \langle u, \phi \rangle.$$

Proposition 5.12. *If $u \in \mathcal{D}'(\mathbb{R}^d)$ and $\phi \in \mathcal{D}(\mathbb{R}^d)$, then*

(a) *for any $x \in \mathbb{R}^d$,*

$$\tau_x(u * \phi) = (\tau_x u) * \phi = u * (\tau_x \phi),$$

(b) $u * \phi \in C^\infty(\mathbb{R}^d)$ and, for any multi-index α,

$$D^\alpha(u * \phi) = (D^\alpha u) * \phi = u * (D^\alpha \phi).$$

Remark. Since u could be an element of $L^1_{loc}(\mathbb{R}^d)$, these results hold for functions as well.

Proof. For (a), note that

$$\tau_x(u * \phi)(y) = (u * \phi)(y - x) = \langle u, \tau_{y-x} R\phi \rangle,$$
$$(\tau_x u) * \phi(y) = \langle \tau_x u, \tau_y R\phi \rangle = \langle u, \tau_{y-x} R\phi \rangle,$$
$$(u * \tau_x \phi)(y) = \langle u, \tau_y R\tau_x \phi \rangle = \langle u, \tau_{y-x} R\phi \rangle.$$

Part of (b) is straightforward, *viz.*

$$\begin{aligned}
D^\alpha u * \phi(x) &= \langle D^\alpha u, \tau_x R\phi \rangle \\
&= (-1)^{|\alpha|} \langle u, D^\alpha \tau_x R\phi \rangle \\
&= (-1)^{|\alpha|} \langle u, \tau_x D^\alpha R\phi \rangle \\
&= \langle u, \tau_x R D^\alpha \phi \rangle \\
&= u * D^\alpha \phi(x).
\end{aligned}$$

For $h \neq 0$ and $\mathbf{e} \in \mathbb{R}^d$ a unit vector, let

$$D_h = \frac{1}{h}(I - \tau_{h\mathbf{e}}).$$

Then, the agreeable result

$$\lim_{h \to 0} D_h \phi(x) = \frac{\partial \phi}{\partial e}(x)$$

pointwise is seen to be valid. In fact, we claim that the convergence is uniform since $\partial \phi / \partial e$ is uniformly continuous (it has a bounded gradient). Given $\epsilon > 0$, there is $\delta > 0$ such that

$$\left| \frac{\partial \phi}{\partial e}(x) - \frac{\partial \phi}{\partial e}(y) \right| \leq \epsilon$$

whenever $|x - y| < \delta$. Thus

$$\left| D_h \phi(x) - \frac{\partial \phi}{\partial e}(x) \right| = \left| \frac{1}{h} \int_{-h}^{0} \left(\frac{\partial \phi}{\partial e}(x + se) - \frac{\partial \phi}{\partial e}(x) \right) ds \right| \leq \epsilon$$

whenever $|h| < \delta$. Similarly

$$D^\alpha D_h \phi = D_h D^\alpha \phi \xrightarrow{L^\infty(\mathbb{R}^d)} D^\alpha \frac{\partial \phi}{\partial e},$$

so it is concluded that

$$D_h\phi \xrightarrow{\mathcal{D}} \frac{\partial \phi}{\partial e} \quad \text{as } h \to 0,$$

since $\text{supp}(D_h\phi) \subset \{x \in \mathbb{R}^d : \text{dist}(x, \text{supp}(\phi)) \le |h|\}$ is compact for finite values of h.

By part (a), for any $x \in \mathbb{R}^d$,

$$D_h(u * \phi)(x) = u * D_h\phi(x),$$

so

$$\lim_{h \to 0} D_h(u * \phi)(x) = \lim_{h \to 0} u * D_h\phi(x) = u * \frac{\partial \phi}{\partial e}(x),$$

since $u \circ \tau_x \circ R \in \mathcal{D}'$. Thus $\frac{\partial}{\partial e}(u * \phi)$ exists and equals $u * \frac{\partial \phi}{\partial e}$. By iterating this observation, (b) is established. ☐

If $\phi, \psi \in \mathcal{D}$, then $\phi * \psi \in \mathcal{D}$, since

$$\text{supp}(\phi * \psi) \subset \text{supp}(\phi) + \text{supp}(\psi).$$

Proposition 5.13. *If $\phi, \psi \in \mathcal{D}$, $u \in \mathcal{D}'$, then*

$$(u * \phi) * \psi = u * (\phi * \psi).$$

Proof. Since $\phi * \psi$ is uniformly continuous, we may approximate the convolution integral by a Riemann sum. For $h > 0$, if

$$r_h(x) = \sum_{k \in \mathbb{Z}^d} \phi(x - kh)\psi(kh)h^d,$$

then $r_h(x) \to \phi * \psi(x)$ uniformly in x as $h \to 0$. Moreover, for any multi-index α,

$$D^\alpha r_h \longrightarrow (D^\alpha \phi) * \psi = D^\alpha(\phi * \psi)$$

uniformly, and

$$\text{supp}(r_h) \subset \text{supp}(\phi) + \text{supp}(\psi).$$

We conclude that

$$r_h \xrightarrow{\mathcal{D}} \phi * \psi.$$

Thus

$$u * (\phi * \psi)(x) = \lim_{h \to 0^+} u * r_h(x)$$
$$= \lim_{h \to 0^+} \sum_{k \in \mathbb{Z}^d} u * \phi(x - kh)\psi(kh)h^d$$
$$= (u * \phi) * \psi(x). \qquad \square$$

5.5 CONVERGENCE OF DISTRIBUTIONS AND APPROXIMATIONS TO THE IDENTITY

The linear space $\mathcal{D}'(\Omega)$ is endowed with its weak topology. Although we will not prove or use the fact, $\mathcal{D}$ is reflexive, so the weak topology on $\mathcal{D}'(\Omega)$ is in fact the weak-* topology. The weak topology on $\mathcal{D}'(\Omega)$ is defined by the following notion of convergence: a sequence $\{u_j\}_{j=1}^{\infty} \subset \mathcal{D}'(\Omega)$ converges to $u \in \mathcal{D}'(\Omega)$ if and only if

$$\langle u_j, \phi \rangle \to \langle u, \phi \rangle \quad \forall \phi \in \mathcal{D}(\Omega).$$

As the following proposition assures, $\mathcal{D}'(\Omega)$ is (sequentially) complete.

Proposition 5.14. *If $\{u_n\}_{n=1}^{\infty} \subset \mathcal{D}'(\Omega)$ and $\{\langle u_n, \phi \rangle\}_{n=1}^{\infty} \subset \mathbb{F}$ is Cauchy for all $\phi \in \mathcal{D}(\Omega)$, then $u : \mathcal{D} \to \mathbb{F}$ defined by*

$$u(\phi) = \langle u, \phi \rangle = \lim_{n \to \infty} \langle u_n, \phi \rangle$$

is a distribution.

The proof requires a more general version of the uniform boundedness principle.

Theorem 5.15 (Banach-Steinhaus theorem). *Suppose X is a Fréchet space and Y is an NLS. If $\{T_\alpha\}_{\alpha \in \mathcal{I}}$ is a collection of continuous linear maps from X to Y such that $\sup_{\alpha \in \mathcal{I}} \|T_\alpha(x)\|_Y < \infty$ for each $x \in X$, then $\{T_\alpha\}_{\alpha \in \mathcal{I}}$ is* equicontinuous, *meaning that for every $r > 0$, there is a neighborhood $U \in X$ about 0 such that $\|T_\alpha(x)\|_Y < r$ for all $\alpha \in \mathcal{I}$ and $x \in U$.*

Note that the neighborhood U does not depend on $\alpha \in \mathcal{I}$. For a proof, please consult [21] or [28] for example.

Proof of Proposition 5.14. The existence and linearity of u is clear. We hypothesize pointwise convergence, so the continuity of u will follow from the Banach-Steinhaus theorem. It is sufficient to prove continuity at 0 (Theorem 5.5). Let $\phi_k \to 0$ in $\mathcal{D}(\Omega)$. Then there is some $K \subset\subset \Omega$ such that all $\phi_k \in \mathcal{D}_K$, which is a Fréchet space. Now $\sup_{n \geq 1} |u_n(\phi)| < \infty$ for all $\phi \in \mathcal{D}_K$ (since the sequence converges), so it follows that $\{u_n\}_{n=1}^{\infty}$ is equicontinuous. For $\epsilon > 0$, there is a neighborhood U about 0 in $\mathcal{D}_K$ such that $|u_n(\phi)| < \epsilon$ for all $n \geq 1$ and $\phi \in U$. Since eventually the tail of the sequence $\{\phi_k\}$ is contained in U, then $|u_n(\phi_k)| \leq \epsilon$ for k large enough and all n. Thus $|u(\phi_k)| = \lim_{n \to \infty} |u_n(\phi_k)| \leq \epsilon$ for arbitrary ϵ. We conclude that $|u(\phi_k)| \to 0$ as $k \to \infty$, and so u is a distribution. □

Lemma 5.16. *If $T : \mathcal{D}(\Omega) \to \mathcal{D}(\Omega)$ is continuous and linear, and if $u_n \to u$ in $\mathcal{D}'(\Omega)$, then $T^* u_n \to T^* u$.*

Proof. Simply compute, for $\phi \in \mathcal{D}(\Omega)$,

$$\langle T^* u_n, \phi \rangle = \langle u_n, T\phi \rangle \longrightarrow \langle u, T\phi \rangle = \langle T^* u, \phi \rangle. \qquad \square$$

In light of this lemma, all operations from the previous section are continuous, since each was defined as the dual of some continuous linear map $T : \mathcal{D}(\Omega) \to \mathcal{D}(\Omega)$.

Corollary 5.17. *If* $u_n \xrightarrow{\mathcal{D}'(\Omega)} u$ *and* α *is any multi-index, then* $D^\alpha u_n \xrightarrow{\mathcal{D}'(\Omega)} D^\alpha u.$

Of course, this corollary can be shown directly: for any $\phi \in \mathcal{D}$,

$$\langle D^\alpha u_n, \phi \rangle = (-1)^{|\alpha|} \langle u_n, D^\alpha \phi \rangle \longrightarrow (-1)^{|\alpha|} \langle u, D^\alpha \phi \rangle = \langle D^\alpha u, \phi \rangle.$$

We leave the following two propositions as exercises.

Proposition 5.18. *If* $u \in \mathcal{D}'(\Omega)$ *and* α *is a multi-index with* $|\alpha| = 1$, *then*

$$\lim_{h \to 0} \frac{1}{h}(u - \tau_{h\alpha} u) = D^\alpha u \quad \text{in } \mathcal{D}'(\Omega),$$

wherein the first α *is interpreted as a unit vector in* $\mathbb{R}^d$.

Proposition 5.19. *Let* $\chi_R(x)$ *denote the characteristic function of a set* $R \subset \mathbb{R}$. *For* $\epsilon > 0$,

$$\frac{1}{\epsilon} \chi_{[-\epsilon/2, \epsilon/2]} \xrightarrow{\mathcal{D}'(\mathbb{R})} \delta_0$$

as $\epsilon \to 0$.

Definition. Let $\varphi \in \mathcal{D}(\mathbb{R}^d)$ satisfy

(a) $\varphi \geq 0$,

(b) $\int \varphi(x) \, dx = 1$,

and, for $\epsilon > 0$, define

$$\varphi_\epsilon(x) = \frac{1}{\epsilon^d} \varphi\left(\frac{x}{\epsilon}\right).$$

Then $\{\varphi_\epsilon\}_{\epsilon > 0}$ is an *approximation to the identity*.

The following is easily verified.

Proposition 5.20. *If* $\{\varphi_\epsilon\}_{\epsilon > 0}$ *is an approximation to the identity, then*

$$\int \varphi_\epsilon(x) \, dx = 1 \quad \forall \epsilon > 0$$

and $\operatorname{supp}(\varphi_\epsilon) \to \{0\}$ *as* $\epsilon \to 0$ *in the sense that that*

$$\bigcap_{\epsilon > 0} \operatorname{supp}(\varphi_\epsilon) = \{0\}.$$

Theorem 5.21. *Let* $\{\varphi_\epsilon\}_{\epsilon > 0}$ *be an approximation to the identity.*

(a) *If $\psi \in \mathcal{D}$, then $\psi * \varphi_\epsilon \xrightarrow{\mathcal{D}} \psi$.*

(b) *If $u \in \mathcal{D}'$, then $u * \varphi_\epsilon \xrightarrow{\mathcal{D}'} u$. Moreover, $C^\infty(\mathbb{R}^d) \subset \mathcal{D}'$ is dense.*

Proof. (a) Let $\mathrm{supp}(\varphi) \subset \overline{B_R(0)}$ for some $R > 0$. First note that for $0 < \epsilon \leq 1$,

$$\mathrm{supp}(\psi * \varphi_\epsilon) \subset \mathrm{supp}(\psi) + \mathrm{supp}(\varphi_\epsilon) \subset \mathrm{supp}(\psi) + \overline{B_R(0)} = K$$

is contained in a fixed compact set. If $f \in C_0^\infty(\mathbb{R}^d)$, then

$$\begin{aligned} f * \varphi_\epsilon(x) &= \int f(x-y)\varphi_\epsilon(y)\,dy \\ &= \int f(x-y)\epsilon^{-d}\varphi(\epsilon^{-1}y)\,dy \\ &= \int f(x-\epsilon z)\varphi(z)\,dz \\ &= \int (f(x-\epsilon z) - f(x))\varphi(z)\,dz + f(x), \end{aligned}$$

and this converges uniformly to $f(x)$. Thus for any multi-index α,

$$D^\alpha(\psi * \varphi_\epsilon) = (D^\alpha \psi) * \varphi_\epsilon \xrightarrow{L^\infty} D^\alpha \psi;$$

that is, $\psi * \varphi_\epsilon \xrightarrow{\mathcal{D}_K} \psi$, and so also $\psi * \varphi_\epsilon \xrightarrow{\mathcal{D}} \psi$.

(b) Since convolution generates a (continuous) distribution for any fixed x, by (a) and Proposition 5.13, we have for $\psi \in \mathcal{D}$,

$$\begin{aligned} \langle u, \psi \rangle &= u * R\psi(0) \\ &= \lim_{\epsilon \to 0} u * (R\psi * \varphi_\epsilon)(0) \\ &= \lim_{\epsilon \to 0} (u * \varphi_\epsilon) * R\psi(0) \\ &= \lim_{\epsilon \to 0} \langle u * \varphi_\epsilon, \psi \rangle. \end{aligned}$$

The final remark regarding density follows since $u * \varphi_\epsilon \in C^\infty(\mathbb{R}^d)$. $\qquad\square$

Corollary 5.22. *If $\{\varphi_\epsilon\}_{\epsilon>0}$ be an approximation to the identity, then*

$$\varphi_\epsilon = \delta_0 * \varphi_\epsilon \xrightarrow{\mathcal{D}'} \delta_0.$$

5.6 SOME APPLICATIONS TO DIFFERENTIAL EQUATIONS

An operator $L : C^m(\mathbb{R}^d) \to C^0(\mathbb{R}^d)$ is called a *linear differential operator* if there are functions $a_\alpha \in C^0(\mathbb{R}^d)$ for all multi-indices α such that

$$L = \sum_{|\alpha| \leq m} a_\alpha D^\alpha.$$

The maximal $|\alpha|$ for which a_α is not identically zero is the *order* of L.

If $a_\alpha \in C^\infty(\mathbb{R}^d)$, then we can extend L to

$$L : \mathcal{D}' \to \mathcal{D}',$$

and this operator is linear and continuous. Given $f \in \mathcal{D}'$, consider the partial or ordinary differential equation

$$Lu = f \quad \text{in } \mathcal{D}'$$

for which we seek a *distributional solution* $u \in \mathcal{D}'$, which is to say an element $u \in \mathcal{D}'$ such that

$$\langle Lu, \phi \rangle = \langle f, \phi \rangle \quad \forall \phi \in \mathcal{D}.$$

We say that any such u is a *classical solution* if $u \in C^m(\mathbb{R}^d)$ satisfies the equation pointwise. If u is a regular distribution, then u is called a *weak solution* (so classical solutions are also weak solutions). Thus, if $u \in \mathcal{D}'$ solves the equation, it would fail to be a weak solution if u is a singular distribution.

5.6.1 Ordinary differential equations

Consider first the case when $d = 1$.

Lemma 5.23. *Let $\phi \in \mathcal{D}(\mathbb{R})$. Then $\int \phi(x)\,dx = 0$ if and only if there is some $\psi \in \mathcal{D}(\mathbb{R})$ such that $\phi = \psi'$.*

The proof is left to the reader. Clearly $\psi(x) = \int_{-\infty}^{x} \phi(\xi)\,d\xi$.

Definition. A distribution $v \in \mathcal{D}'(\mathbb{R})$ is a *primitive* of $u \in \mathcal{D}'(\mathbb{R})$ if $Dv \equiv v' = u$.

Theorem 5.24. *Every $u \in \mathcal{D}'(\mathbb{R})$ has infinitely many primitives, and any two differ by a constant.*

Proof. Let $\mathcal{D}_0$ be the vector subspace of $\mathcal{D}$ consisting of functions having integral zero, *viz.*

$$\mathcal{D}_0 = \left\{ \phi \in \mathcal{D}(\mathbb{R}) : \int \phi(x)\,dx = 0 \right\}$$
$$= \{ \phi \in \mathcal{D}(\mathbb{R}) : \text{there is } \psi \in \mathcal{D}(\mathbb{R}) \text{ such that } \psi' = \phi \}.$$

Now $v \in \mathcal{D}'$ is a primitive for u if and only if

$$\langle u, \psi \rangle = \langle v', \psi \rangle = -\langle v, \psi' \rangle \quad \forall \psi \in \mathcal{D}.$$

Thus, by the lemma, we require that

$$\langle v, \phi \rangle = -\left\langle u, \int_{-\infty}^{x} \phi(\xi)\, d\xi \right\rangle \quad \forall \phi \in \mathcal{D}_0, \tag{5.3}$$

and so $v : \mathcal{D}_0 \to \mathbb{F}$ is well-defined on the subspace. Extend v to all of $\mathcal{D}$ as follows. Fix $\tilde{\phi} \in \mathcal{D}$ such that $\int \tilde{\phi}(x)\, dx = 1$. Then any $\psi \in \mathcal{D}$ is uniquely decomposed as

$$\psi = \phi + \langle 1, \psi \rangle \tilde{\phi},$$

where $\phi \in \mathcal{D}_0$. Choose $c \in \mathbb{F}$ and define v_c for $\psi \in \mathcal{D}$ by

$$\langle v_c, \psi \rangle = \langle v_c, \phi \rangle + \langle 1, \psi \rangle \langle v_c, \tilde{\phi} \rangle \equiv \langle v, \phi \rangle + c \langle 1, \psi \rangle.$$

Clearly v_c is linear and $v_c|_{\mathcal{D}_0} = v$. We claim that v_c is continuous. If $\psi_n \xrightarrow{\mathcal{D}} 0$, then $\langle 1, \psi_n \rangle \xrightarrow{\mathbb{F}} 0$ and it is easy to see also that

$$\phi_n = \psi_n - \langle 1, \psi_n \rangle \tilde{\phi} \xrightarrow{\mathcal{D}} 0.$$

Since the ϕ_n all lie in $\mathcal{D}_0$ and have support in, say, $[a, b]$, it follows that $\int_{-\infty}^{x} \phi_n(\xi)\, d\xi \in \mathcal{D}$ has support also in $[a, b]$ and that

$$\left\| \int_{-\infty}^{x} \phi_n(\xi)\, d\xi \right\|_{0,\infty,\mathbb{R}} \leq (b - a) \| \phi_n \|_{0,\infty,\mathbb{R}} \longrightarrow 0.$$

Therefore $\int_{-\infty}^{x} \phi_n(\xi)\, d\xi \xrightarrow{\mathcal{D}} 0$, and consequently

$$\langle v, \phi_n \rangle = -\left\langle u, \int_{-\infty}^{x} \phi_n(\xi)\, d\xi \right\rangle \longrightarrow 0, \quad \text{whence} \quad \langle v_c, \psi_n \rangle \to 0.$$

Thus, for each $c \in \mathbb{F}$, v_c is a distribution and $v_c' = u$.

If $v, w \in \mathcal{D}'$ are primitives of u, then for any $\psi \in \mathcal{D}$ expanded as above relative to a fixed choice of $\tilde{\phi}$, using (5.3),

$$\langle v - w, \psi \rangle = \langle v - w, \phi \rangle + \langle v - w, \langle 1, \psi \rangle \tilde{\phi} \rangle$$

$$= \left\langle u - u, \int_{-\infty}^{x} \phi(\xi)\, d\xi \right\rangle + \langle 1, \psi \rangle \langle v - w, \tilde{\phi} \rangle$$

$$= \langle \langle v - w, \tilde{\phi} \rangle, \psi \rangle,$$

and so

$$v - w = \langle v - w, \tilde{\phi} \rangle \in \mathbb{F}. \qquad \square$$

Corollary 5.25. *If $u' = 0$ in $\mathcal{D}'(\mathbb{R})$, then u is constant.*

Corollary 5.26. *If $a \in \mathbb{F}$, then $u' = au$ in $\mathcal{D}'(\mathbb{R})$ has only the classical solutions given by*

$$u(x) = Ce^{ax}$$

for some $C \in \mathbb{F}$.

Proof. We have the existence of at least the solutions Ce^{ax}. Let u be any distributional solution. Note that $e^{-ax} \in C^{\infty}(\mathbb{R})$, so $v = e^{-ax}u \in \mathcal{D}'$ and the Leibniz rule implies

$$v' = -ae^{-ax}u + e^{-ax}u' = e^{-ax}(u' - au) = 0.$$

Thus $v = C$, a constant, and $u = Ce^{ax}$. □

Corollary 5.27. *Let $a(x), b(x) \in C^{\infty}(\mathbb{R})$. Then the differential equation*

$$u' + a(x)u = b(x) \quad in \ \mathcal{D}'(\mathbb{R}) \tag{5.4}$$

possesses only the classical solutions

$$u = e^{-A(x)}\left[\int_0^x e^{A(\xi)}b(\xi)\,d\xi + C\right]$$

for any $C \in \mathbb{F}$, where A is any primitive of a (i.e., $A' = a$).

Proof. If $u, v \in \mathcal{D}'$ solve the equation, then their difference solves the homogeneous equation

$$w' + a(x)w = 0 \quad in \ \mathcal{D}'(\mathbb{R}).$$

But, as in the proof of the last corollary, such solutions must have the form

$$w = Ce^{-A(x)}$$

since $(e^{A(x)}w)' = e^{A(x)}w' + a(x)e^{A(x)}w = 0$. Thus any solution of the nonhomogeneous equation (5.4) has the form

$$u = Ce^{-A(x)} + v$$

where v is any particular solution of the nonhomogeneous problem. Since

$$v = e^{-A(x)}\int_0^x e^{A(\xi)}b(\xi)\,d\xi$$

is such a solution, the result follows. □

Not all equations are quite so simple.

Example. Consider the equation

$$xu' = 1 \quad in \ \mathcal{D}'(\mathbb{R}).$$

We know $u = \ln|x| \in L^1_{\text{loc}}(\mathbb{R})$ is a weak solution, since $(\ln|x|)' = PV(1/x)$ and $xPV(1/x) = 1$. All other solutions are given by adding any solution to

$$xv' = 0 \quad \text{in } \mathcal{D}'(\mathbb{R}).$$

Since $v' \in \mathcal{D}'(\mathbb{R})$ may not be a regular distribution, we must *not* divide by x to conclude v is a constant (since $x = 0$ is possible). In fact,

$$v = c_1 + c_2 H(x),$$

for constants $c_1, c_2 \in \mathbb{F}$, where $H(x)$ is the Heaviside function. To see this, consider

$$xw = 0 \quad \text{in } \mathcal{D}'.$$

For $\psi \in \mathcal{D}$,

$$0 = \langle xw, \psi \rangle = \langle w, x\psi \rangle,$$

so we wish to write a general $\phi \in \mathcal{D}$ in terms of $x\psi$ for some $\psi \in \mathcal{D}$. To this end, let $r \in \mathcal{D}$ be any function that is 1 for $-\epsilon < x < \epsilon$ for some $\epsilon > 0$ (such functions were constructed earlier). Then

$$\phi(x) = \phi(0)r(x) + \big(\phi(x) - \phi(0)r(x)\big)$$

$$= \phi(0)r(x) + \int_0^x \big(\phi'(y) - \phi(0)r'(y)\big)\, dy$$

$$= \phi(0)r(x) + x \int_0^1 \big(\phi'(xz) - \phi(0)r'(xz)\big)\, dz$$

$$= \phi(0)r(x) + x\psi(x),$$

where

$$\psi(x) = \int_0^1 \big(\phi'(xz) - \phi(0)r'(xz)\big)\, dz$$

clearly has compact support and $\psi \in C^\infty$, since differentiation and integration commute in this case. Thus

$$\langle w, \phi \rangle = \langle w, \phi(0)r \rangle + \langle w, x\psi \rangle = \phi(0)\langle w, r \rangle;$$

that is, with $c = \langle w, r \rangle$,

$$w = c\delta_0.$$

Finally, then $v' = c_2 \delta_0$ and $v = c_1 + c_2 H$. Our general solution is the two-parameter family

$$u = \ln|x| + c_1 + c_2 H(x).$$

These are not classical solutions but merely weak solutions.

5.6.2 Partial differential equations and fundamental solutions

Attention is now turned to higher dimensional problems where $d > 1$, but restricted to the case of constant coefficients, *viz.*

$$L = \sum_{|\alpha| \leq m} c_\alpha D^\alpha,$$

where $c_\alpha \in \mathbb{F}$. Associated to L is the polynomial

$$p(x) = \sum_{|\alpha| \leq m} c_\alpha x^\alpha,$$

where $x^\alpha = x_1^{\alpha_1} x_2^{\alpha_2} \cdots x_d^{\alpha_d}$; thus,

$$L = p(D)$$

in an obvious notation. The operator L is the adjoint of

$$\mathcal{L} = \sum_{|\alpha| \leq m} (-1)^{|\alpha|} c_\alpha D^\alpha,$$

since $\langle u, \mathcal{L}\phi \rangle = \langle \mathcal{L}^* u, \phi \rangle = \langle Lu, \phi \rangle$ for any $u \in \mathcal{D}'$, $\phi \in \mathcal{D}$.

Example. Suppose L is the one-dimensional *wave operator*

$$L = \frac{\partial^2}{\partial t^2} - c^2 \frac{\partial^2}{\partial x^2}$$

for $(t, x) \in \mathbb{R}^2$ and $c > 0$ (the variables t and x are considered to represent time and space, respectively, and the equation is described as being one-dimensional in terms of the space variable). This choice of L is self-adjoint and for every $g \in C^2(\mathbb{R})$, $f(t, x) \equiv g(x - ct)$ is a classical solution of the wave equation $Lf = 0$. If instead, $g \in L^1_{\text{loc}}$, then $g(x - ct)$ is a weak solution. In fact, $f(t, x) = \delta_0(x - ct)$ is a distributional solution, although we need to be a little more precise. Let $u \in \mathcal{D}'(\mathbb{R}^2)$ be defined by

$$\langle u, \phi \rangle = \langle \delta_0(x - ct), \phi(t, x) \rangle \equiv \int_{-\infty}^{\infty} \phi(t, ct) \, dt \quad \forall \phi \in \mathcal{D}(\mathbb{R}^2).$$

(It is a simple exercise to verify that u is a well-defined element of $\mathcal{D}'$.) Calculate Lu thusly:

$$\langle Lu, \phi \rangle = \langle u, L\phi \rangle$$

$$= \left\langle u, \left(\frac{\partial^2}{\partial t^2} - c^2 \frac{\partial^2}{\partial x^2} \right) \phi \right\rangle$$

$$= \left\langle u, \left(\frac{\partial}{\partial t} + c \frac{\partial}{\partial x} \right) \left(\frac{\partial}{\partial t} - c \frac{\partial}{\partial x} \right) \phi \right\rangle$$

$$= \left\langle u, \left(\frac{\partial}{\partial t} + c \frac{\partial}{\partial x} \right) \psi \right\rangle,$$

where $\psi \in \mathcal{D}$. Continuing, we see that

$$\langle Lu, \phi \rangle = \int_{-\infty}^{\infty} \left(\frac{\partial}{\partial t} + c \frac{\partial}{\partial x} \right) \psi(t, ct)\, dt = \int_{-\infty}^{\infty} \frac{d}{dt} \psi(t, ct)\, dt = 0.$$

Definition. If $Lu = \delta_0$ for some $u \in \mathcal{D}'$, then u is called a *fundamental solution* of L.

If a fundamental solution u exists, it is *not* in general unique, since any solution to $Lv = 0$ gives another fundamental solution $u + v$. The reason for the name and its importance is explained in the following theorem.

Theorem 5.28. *If $f \in \mathcal{D}$ and $E \in \mathcal{D}'$ is a fundamental solution for L, then $E * f$ is a solution to*

$$Lu = f.$$

Proof. Since $LE = \delta_0$, then also

$$(LE) * f = \delta_0 * f = f.$$

But

$$(LE) * f = L(E * f). \qquad \square$$

Theorem 5.29 (Malgrange-Ehrenpreis theorem). *Every constant coefficient linear partial differential operator on $\mathbb{R}^d$ has a fundamental solution.*

A proof can be found in [28] and [21]. The solution has a simple interpretation, referred to as the *principle of superposition*. If one considers f as a source of disturbance to the system described by the linear partial differential operator L, then E is the response to a unit point disturbance at the origin (i.e., to δ_0). Translating this remark, $E(x - y)$ is the response to a unit point disturbance at y (i.e., to $\delta_y = \delta_0(x - y)$). If we multiply by the correct magnitude $f(y)$ of the disturbance corresponding to the unit disturbance at y, we see that the response to the actual point disturbance at y is $E(x - y) f(y)$, again by linearity of the operator L. These responses are now "summed" in the sense of integration theory, to obtain

$$u(x) = \int E(x - y) f(y)\, dy,$$

which is the entire response of the system. That is, the point solutions are *superimposed* to obtain the entire solution.

Example. A fundamental solution of

$$L = \frac{\partial^2}{\partial t^2} - c^2 \frac{\partial^2}{\partial x^2},$$

where $c > 0$, is

$$E(t, x) = \frac{1}{2c} H(ct - |x|) = \frac{1}{2c} H(ct - x) H(ct + x),$$

where H is the Heaviside function. To see this, it must be shown that

$$\langle LE, \phi \rangle = \phi(0,0) \quad \forall \phi \in \mathcal{D}(\mathbb{R}^2).$$

For convenience, let $D_\pm = \frac{\partial}{\partial t} \pm c\frac{\partial}{\partial x}$, so $L = D_+D_- = D_-D_+$, and

$$\langle LE, \phi \rangle = \langle E, D_+D_-\phi \rangle = \iint \frac{1}{2c} H(ct - |x|) D_+D_-\phi \, dt \, dx$$

$$= \frac{1}{2c} \left\{ \int_0^\infty \int_{x/c}^\infty D_+D_-\phi \, dt \, dx + \int_{-\infty}^0 \int_{-x/c}^\infty D_-D_+\phi \, dt \, dx \right\}.$$

Now

$$\int_0^\infty \int_{x/c}^\infty D_+D_-\phi \, dt \, dx = \int_0^\infty \int_0^\infty (D_+D_-\phi)(t + x/c, x) \, dt \, dx$$

$$= c \int_0^\infty \int_0^\infty \frac{d}{dx}(D_-\phi)(t + x/c, x) \, dx \, dt$$

$$= -c \int_0^\infty D_-\phi(t, 0) \, dt,$$

whereas

$$\int_{-\infty}^0 \int_{-x/c}^\infty D_-D_+\phi \, dt \, dx = \int_{-\infty}^0 \int_0^\infty (D_-D_+\phi)(t - x/c, x) \, dt \, dx$$

$$= -c \int_0^\infty \int_{-\infty}^0 \frac{d}{dx}(D_+\phi)(t - x/c, x) \, dx \, dt$$

$$= -c \int_0^\infty D_+\phi(t, 0) \, dt.$$

In consequence,

$$\langle LE, \phi \rangle = -\frac{1}{2} \int_0^\infty \left[D_-\phi(t, 0) + D_+\phi(t, 0) \right] dt$$

$$= -\int_0^\infty \frac{\partial}{\partial t}\phi(t, 0) \, dt = \phi(0,0) = \langle \delta_0, \phi \rangle.$$

Since $E(t, x) = \frac{1}{2c} H(ct - |x|)$ is a fundamental solution, it follows that a solution to $Lu = f \in \mathcal{D}$ is given by

$$u(t, x) = E * f(t, x) = \frac{1}{2c} \iint_{c(t-s)-|x-y|>0} f(s, y) \, dy \, ds$$

$$= \frac{1}{2c} \int_{-\infty}^t \int_{x-c(t-s)}^{x+c(t-s)} f(s, y) \, dy \, ds.$$

From this formula, it is apparent that the solution at a point (t, x) depends on the values of f only in the cone of points

$$\{(s, y) : -\infty < s \le t \text{ and } x - c(t - s) \le y \le x + c(t - s)\},$$

which is called the *domain of dependence* of the point (t, x). If f were to be changed outside this cone, the solution at (t, x) would be unchanged. Conversely, note that a point (s, y) will influence the solution in the cone

$$\{(t, x) : 0 \le s \le t < \infty \text{ and } y - c(t - s) \le x \le y + c(t - s)\},$$

called the *domain of influence* of the point (s, y). These notions can be generalized to singular distributions, but this point is left aside.

Example. The *Laplace operator* or *Laplacian* is

$$\Delta = \frac{\partial^2}{\partial x_1^2} + \cdots + \frac{\partial^2}{\partial x_d^2} = \nabla \cdot \nabla = \nabla^2.$$

It gives rise to *Laplace's equation* $\Delta u = 0$ and *Poisson's equation* $\Delta u = f$. A fundamental solution is

$$E(x) = \begin{cases} \dfrac{1}{2}|x|, & d = 1, \\[2mm] \dfrac{1}{2\pi} \ln|x|, & d = 2, \\[2mm] \dfrac{1}{d\omega_d} \dfrac{|x|^{2-d}}{2 - d}, & d > 2, \end{cases} \tag{5.5}$$

where

$$\omega_d = \frac{2\pi^{d/2}}{d\Gamma(d/2)}$$

is the hyper-volume of the unit ball in $\mathbb{R}^d$. (As a side remark, the hyper-area of the unit sphere in $\mathbb{R}^d$ is $d\omega_d$.) It is trivial to verify the claim if $d = 1$: $D^2 \frac{1}{2}|x| = D\frac{1}{2}(2H(x) - 1) = H' = \delta_0$. For $d \ge 2$, it is necessary to show that

$$\langle \Delta E, \phi \rangle = \langle E, \Delta \phi \rangle = \phi(0) \quad \forall \phi \in \mathcal{D}(\mathbb{R}^d).$$

It is important to recognize that E is a regular distribution, i.e., $E \in L^1_{\text{loc}}(\mathbb{R}^d)$. This is clear everywhere except possibly near $x = 0$. For $1 > r > 0$ and $d = 2$, a change of variables to polar coordinates yields

$$\int_{B_r(0)} \left| \frac{1}{2\pi} \ln|x| \right| dx = - \int_0^{2\pi} \int_0^r \frac{1}{2\pi} \ln r \, r dr \, d\theta = -\frac{1}{2} r^2 \ln r + \frac{1}{4} r^2 < \infty$$

and, for $d > 2$, the analogous change of variables produces the formula

$$\int_{B_r(0)} |E(x)| \, dx = - \int_{S_1(0)} \int_0^r \frac{r^{2-d}}{d\omega_d(2 - d)} r^{d-1} \, dr \, d\sigma = \frac{r^2}{2(d - 2)} < \infty,$$

where $S_1(0)$ is the unit sphere and σ is the surface measure on $S_1(0)$. To show that

$$\int E(x) \Delta \phi(x) \, dx = \phi(0) \quad \forall \phi \in \mathcal{D},$$

let $R > 0$ be such that $\text{supp}(\phi) \subset B_R(0)$ and let $0 < \epsilon < R$. Then

$$\int_{\epsilon < |x| < R} E\Delta\phi \, dx = -\int_{\epsilon < |x| < R} \nabla E \cdot \nabla\phi \, dx + \int_{|x|=\epsilon} E\nabla\phi \cdot \nu \, d\sigma,$$

by the divergence theorem, where $\nu \in \mathbb{R}^d$ is the unit vector normal to the surface $|x| = \epsilon$ pointing toward 0 (*out* of the set $\epsilon < |x| < R$). Another application of the divergence theorem produces

$$\int_{\epsilon < |x| < R} E\Delta\phi \, dx = \int_{\epsilon < |x| < R} \Delta E\phi \, dx - \int_{|x|=\epsilon} \nabla E \cdot \nu\phi \, d\sigma + \int_{|x|=\epsilon} E\nabla\phi \cdot \nu \, d\sigma.$$

It is an exercise to verify that $\Delta E = 0$ for $x \neq 0$. Moreover,

$$\int_{|x|=\epsilon} E\nabla\phi \cdot \nu \, d\sigma = \int_{S_1(0)} \frac{1}{d\omega_d} \frac{\epsilon^{2-d}}{2-d} \nabla\phi \cdot \nu \epsilon^{d-1} \, d\sigma \longrightarrow 0$$

as $\epsilon \to 0^+$ for $d > 2$. The same conclusion holds when $d = 2$. Another calculation reveals that

$$-\int_{|x|=\epsilon} \nabla E \cdot \nu\phi \, d\sigma = \int_{S_1(0)} \frac{\partial E}{\partial r}(\epsilon, \sigma)\phi(\epsilon, \sigma)\epsilon^{d-1} \, d\sigma$$

$$= \int_{S_1(0)} \frac{1}{d\omega_d} \epsilon^{1-d}\phi(\epsilon, \sigma)\epsilon^{d-1} \, d\sigma \longrightarrow \phi(0)$$

as $\epsilon \to 0^+$. Thus

$$\int E\Delta\phi \, dx = \lim_{\epsilon \to 0^+} \int_{\epsilon < |x| < R} E\Delta\phi \, dx = \phi(0),$$

which is the advertised conclusion.

If $f \in \mathcal{D}$, then
$$\Delta u = f$$
is solved by $u = E * f$. Note that in this case, a change in f on a set of nontrivial measure will change the solution everywhere. This result can be extended to a class of $f \in L^1(\mathbb{R}^d)$-functions as is seen next.

Theorem 5.30. *If $E(x)$ is the fundamental solution to the Laplacian displayed in (5.5) and $f \in L^1(\mathbb{R}^d)$ is such that for almost every $x \in \mathbb{R}^d$,*

$$E(x - y)f(y) \in L^1(\mathbb{R}^d)$$

(as a function of y), then
$$u = E * f$$
is well-defined in $L^1_{\text{loc}}(\mathbb{R}^d)$ and

$$\Delta u = f \quad \text{in } \mathcal{D}'.$$

Proof. For any $r > 0$, Fubini's theorem implies that

$$
\int_{B_r(0)} |u(x)| \, dx \leq \int_{B_r(0)} \int |E(x-y)\, f(y)| \, dy \, dx
$$

$$
= \int \int_{B_r(0)} |E(x-y)| \, dx \, |f(y)| \, dy
$$

$$
= \int \int_{B_r(-y)} |E(z)| \, dz \, |f(y)| \, dy
$$

$$
\leq \int \int_{B_r(0)} |E(z)| \, dz \, |f(y)| \, dy < \infty,
$$

since E decreases radially, $E \in L^1_{\text{loc}}$ and $f \in L^1$. Thus $u \in L^1_{\text{loc}}$.

For $\phi \in \mathcal{D}$, using again Fubini's theorem, it follows that

$$
\langle \Delta u, \phi \rangle = \langle u, \Delta \phi \rangle
$$

$$
= \int u \Delta \phi \, dx = \iint E(x-y) f(y) \Delta \phi(x) \, dy \, dx
$$

$$
= \iint E(x-y) \Delta \phi(x) \, dx \, f(y) \, dy
$$

$$
= \int E * \Delta \phi(y) f(y) \, dy
$$

$$
= \int \phi(y) f(y) \, dy = \langle f, \phi \rangle,
$$

since $E(x-y) = E(y-x)$ and

$$
E * \Delta \phi = \Delta E * \phi = \delta_0 * \phi = \phi.
$$

Thus $\Delta u = f$ in $\mathcal{D}'$ as claimed. □

5.7 LOCAL STRUCTURE OF $\mathcal{D}'$

The following structure theorem for elements of $\mathcal{D}'$ is stated for the reader's information. See [21, p. 154] for a proof.

Theorem 5.31. *Let Ω be a domain in $\mathbb{R}^d$. If $u \in \mathcal{D}'(\Omega)$, then there exist continuous functions g_α, one for each multi-index α, such that*

(a) *each $K \subset\subset \Omega$ intersects the supports of only finitely many of the g_α*

and

(b) $u = \sum_\alpha D^\alpha g_\alpha.$

The conclusion here is that $\mathcal{D}'(\Omega)$ consists of nothing more than sums of (distributional) derivatives of continuous functions, such that locally on any compact set, the sum is finite. We certainly wanted continuous functions to belong to our class of generalized functions and we demanded that derivatives of elements of the class remain in the class. Since addition of generalized functions must again be in the class, all elements as just described in Theorem 5.31 must be generalized functions. It turns out that the rather elaborate definition of $\mathcal{D}'(\Omega)$ has given us only these objects.

5.8   EXERCISES

1. Let $\psi \in \mathcal{D}$ be fixed and define $T : \mathcal{D} \to \mathcal{D}$ by $T(\phi) = \int \phi(\xi) \, d\xi \, \psi$. Show that T is a sequentially continuous linear map.

2. Show that if $\phi \in \mathcal{D}(\mathbb{R})$, then $\int \phi(x) \, dx = 0$ if and only if there is $\psi \in \mathcal{D}(\mathbb{R})$ such that $\phi = \psi'$.

3. Prove that $\mathcal{D}(\Omega)$ is *not* metrizable. [Hint: see the sketch of the proof given in Section 5.3.]

4. Prove that $\mathcal{D}(\Omega)$ is (sequentially) complete.

5. Prove directly that $x \, \mathrm{PV}(1/x) = 1$.

6. Let $T : \mathcal{D}(\mathbb{R}) \to \mathbb{R}$.

   (a) If $T(\phi) = |\phi(0)|$, show T is *not* a distribution.

   (b) If $T(\phi) = \sum\limits_{n=0}^{\infty} \phi(n)$, show T is a distribution.

   (c) If $T(\phi) = \sum\limits_{n=0}^{\infty} D^n \phi(n)$, show T is a distribution.

7. Is it true that $\delta_{1/n} \to \delta_0$ in $\mathcal{D}'$? Why or why not?

8. Show that the following are or are not distributions.

   (a) $\sum\limits_{n=1}^{\infty} n\delta_n = \lim\limits_{N \to \infty} \sum\limits_{n=1}^{N} n\delta_n$ is a distribution.

   (b) $\sum\limits_{n=1}^{\infty} \frac{1}{n}\delta_{1/n} = \lim\limits_{N \to \infty} \sum\limits_{n=1}^{N} \frac{1}{n}\delta_{1/n}$ is *not* a distribution.

   (c) $\sum\limits_{n=1}^{\infty} \frac{(-1)^n}{n}\delta_{1/n} = \lim\limits_{N \to \infty} \sum\limits_{n=1}^{N} \frac{(-1)^n}{n}\delta_{1/n}$ is a distribution (of order 1).

9. Let $\Omega \subset \mathbb{R}^d$ be open and let $\{a_n\}_{n=1}^{\infty}$ be a sequence from Ω with no accumulation point in Ω. For $\phi \in \mathcal{D}(\Omega)$, define

$$T(\phi) = \sum_{n=1}^{\infty} \lambda_n \phi(a_n),$$

where $\{\lambda_n\}_{n=1}^{\infty}$ is a sequence of complex numbers. Show that $T \in \mathcal{D}'(\Omega)$.

10. Let Ω be a domain in $\mathbb{R}^d$. A set $U \subset \mathcal{D}(\Omega)$ is called open if whenever $\{\phi_n\}_{n=1}^{\infty}$ is a sequence converging to an element $\phi \in U$, there is an N such that for all $n \geq N$, $\phi_n \in U$. Show that this defines a topology on $\mathcal{D}(\Omega)$. Characterize the closed sets in this topology. Show also that a linear mapping $T : \mathcal{D}(\Omega) \to \mathbb{F}$ is continuous for this topology if and only if it is sequentially continuous for the $\mathcal{D}(\Omega)$-convergence as defined in this chapter.

11. Formulate and prove a result analogous to that in the last problem for $\mathcal{D}'$.

12. Prove the *Plemelij-Sochozki formula* $\dfrac{1}{x + i0} = \mathrm{PV}(1/x) - i\pi\delta_0(x)$; that is, for $\phi \in \mathcal{D}$,

$$\lim_{r \to 0} \left[\lim_{\epsilon \to 0^+} \int_{|x| \geq \epsilon} \frac{1}{x + ir} \phi(x)\,dx \right] = \lim_{\epsilon \to 0^+} \int_{|x| \geq \epsilon} \frac{1}{x} \phi(x)\,dx - i\pi\phi(0).$$

13. Let τ_h be the translation operator on $\mathcal{D}(\mathbb{R})$: $\tau_h\phi(x) = \phi(x - h)$. Show that for any $\phi \in \mathcal{D}(\mathbb{R})$,

$$\lim_{h \to 0} \frac{1}{h}(\phi - \tau_h\phi) = \phi' \quad \text{in } \mathcal{D}(\mathbb{R}).$$

14. Show the following in $\mathcal{D}'(\mathbb{R})$:

   (a) $\lim\limits_{n \to \infty} \cos(nx)\,\mathrm{PV}(1/x) = 0$;

   (b) $\lim\limits_{n \to \infty} \sin(nx)\,\mathrm{PV}(1/x) = \pi\delta_0$;

   (c) $\lim\limits_{n \to \infty} e^{inx}\,\mathrm{PV}(1/x) = i\pi\delta_0$.

15. Suppose that $T_n : \mathcal{D} \to \mathcal{D}$ is continuous and linear. Suppose also that $T_n\phi \to T\phi$ in $\mathcal{D}$ for each $\phi \in \mathcal{D}$ and that T is also continuous (and linear).

   (a) For any $u \in \mathcal{D}'$, show that $T_n^* u \to T^* u$ in $\mathcal{D}'$. Why is $T^* u \in \mathcal{D}'$?

   (b) Let $d = 1$. For $h = 1/n$, let $T_h = \dfrac{1}{h^2}(\tau_h + \tau_{-h} - 2I)$. Show that $T_h : \mathcal{D} \to \mathcal{D}$ is continuous and linear and find the limit in $\mathcal{D}$ of $T_h\phi$ for $\phi \in \mathcal{D}$. Use this limit to show that $T_h u \to u''$ in $\mathcal{D}'(\mathbb{R})$.

16. Prove that the trigonometric series $\displaystyle\sum_{n=-\infty}^{\infty} a_n e^{inx}$ converges in $\mathcal{D}'(\mathbb{R})$ if there exists a constant $A > 0$ and an integer $N \geq 0$ such that $|a_n| \leq A|n|^N$.

17. Let $T : \mathcal{D}((-1, 1)^2) \to \mathcal{D}(-1, 1)$ be defined by $T\varphi(x, y) = \varphi(x, 0)$.

   (a) Show that T is a (sequentially) continuous linear operator.

   (b) Note that $T' : \mathcal{D}'(-1, 1) \to \mathcal{D}'((-1, 1)^2)$. Determine $T'(\delta_0)$ and $T'(\delta_0')$, where δ_0 is the usual Dirac distribution at 0 in one space dimension.

18. Let $\psi \in \mathcal{D}(\mathbb{R}^d)$ be such that $\int \psi(x)\, dx = 0$ and $\int x\psi(x)\, dx = \eta \in \mathbb{R}^d$. For $\epsilon > 0$, let

$$\psi_\epsilon(x) = \frac{1}{\epsilon^{d+1}} \psi\left(\frac{x}{\epsilon}\right).$$

Show that as $\epsilon \to 0$, ψ_ϵ converges as a distribution and find the action of the limit distribution on $\phi \in \mathcal{D}(\mathbb{R}^d)$.

19. Let u be a distribution on $\mathbb{R}$ such that $|\langle u, \phi \rangle| \le \|\phi\|_{L^2}$ for any test function ϕ.

(a) Show that u can be represented by a function in L^2 with norm less than or equal to 1. [Hint: use the Riesz representation theorem.]

(b) Assuming also that $|\langle u, \phi' \rangle| \le \|\phi\|_{L^1}$ for any test function ϕ, show that u can be represented by a Lipschitz function with Lip seminorm (L^∞ norm of the derivative) less than or equal to 1.

20. The norm

$$\|\phi\|_1 = \|\phi\|_{L^1(-1,1)} + \|\phi'\|_{L^1(-1,1)}$$

is defined for $\phi \in \mathcal{D}(-1,1)$, and in fact for any function f in $X = \{f \in L^1(-1,1) : f'$ is a regular distribution and $f' \in L^1(-1,1)\}$. Show that $\delta_0 \in \mathcal{D}'(-1,1)$ is well-defined on X. $\Big[$Hint: use density and note that

$$\phi(0) = \phi(x) - \int_0^x \phi'(y)\, dy.\Big]$$

21. Prove that the set of functions $\phi * \psi$, for ϕ and ψ in $\mathcal{D}$, is dense in $\mathcal{D}$.

22. Suppose that $u \in \mathcal{D}'$ and that there is some compact $K \subset\subset \mathbb{R}^d$ such that $u(\phi) = 0$ for all $\phi \in \mathcal{D}$ with $\mathrm{supp}(\phi) \subset K^c$. (We say that u has *compact support* in this case.) Show that for any $\phi \in \mathcal{D}$, $u * \phi \in C^\infty$ has compact support. For any $v \in \mathcal{D}'$, show that $v * (u * \phi)$ is well-defined. Further define the convolution $v * u$, show that it is in $\mathcal{D}'$, and that $(v * u) * \phi = v * (u * \phi)$.

23. Find the general solution to the differential equations:

(a) $D^2 u = 0$ in $\mathcal{D}'(\mathbb{R})$;

(b) $x D^2 u = 0$ in $\mathcal{D}'(\mathbb{R})$.

24. Verify that $\Delta E = 0$ for $x \ne 0$, where E is the fundamental solution to the Laplacian given in the text.

25. Find a fundamental solution for the operator $-D^2 + I$ on $\mathbb{R}$.

26. On $\mathbb{R}^3$, show that the operator

$$T(\phi) = \lim_{\epsilon \to 0^+} \int_{|x| \ge \epsilon} \frac{1}{4\pi|x|} e^{-|kx|} \phi(x)\, dx$$

is a fundamental solution to the *Helmholtz operator* $-\Delta + k^2 I$.

27. The biharmonic operator is $\Delta^2 = \Delta\Delta$.

  (a) One can solve the biharmonic equation $\Delta^2 u = f$ in two parts by solving $\Delta v = f$ and then $\Delta u = v$, for $f \in \mathcal{D}$. Describe informally the solution to this equation in terms of E, a fundamental solution to the Laplace operator Δ. Is it obvious that this solution is well-defined? Why or why not?

  (b) In dimension $d = 3$, show that in fact $U(x) = (8\pi)^{-1}|x|$ is a fundamental solution to the biharmonic operator. [Hint: it is easier if you can relate U to $E = (4\pi|x|)^{-1}$.]

28. Suppose that $d \geq 1$ is an integer and $u \in \mathcal{D}'(\mathbb{R}^d)$.

  (a) Suppose that $D^\alpha u = 0$ for all $|\alpha| = 1$. Show that u is a constant. [Hint: let $\{\varphi_\epsilon\}_{\epsilon > 0}$ be an approximation to the identity and consider $u_\epsilon = u * \varphi_\epsilon$.]

  (b) Suppose that $D^\alpha u = 0$ for all $|\alpha| = m$, where $m \geq 1$ is an integer. Show that u is a polynomial of degree at most $m - 1$ in d variables.

CHAPTER 6

The Fourier Transform

Though the Swiss mathematician Leonard Euler and other 17th-century mathematicians had made use of what we now call Fourier series, the modern theory of Fourier analysis began with the early 18th-century research of Jean-Baptiste-Joseph Fourier. Fourier was concerned, off and on throughout his life, with the propagation of heat and put forward what we now call Fourier series. Indeed, he used a Fourier series representation to express solutions of the linear heat equation. His work was greeted with suspicion or outright disbelief by some of his well-known contemporaries. However, it turned out to be a seminal contribution. Indeed, in his study of heat conduction, Fourier was the first to formulate the greenhouse effect, so well known now as a major cause of global warming. It is amusing to note that when the Eiffel tower was built in Paris (1887–1889), Gustave Eiffel himself chose to inscribe 72 names of French scientists, engineers, and mathematicians whose work in the preceding century, 1789–1889, was considered to be groundbreaking. Fourier's name appears there along with some of his famous critics.

6.1 MOTIVATION FOR FOURIER ANALYSIS

The paradigm that Fourier put forward has proved to be a central conception in analysis and in the theory of differential equations. The idea is this. Consider for example the linear, one-space-dimensional heat equation

$$\begin{cases} \dfrac{\partial u}{\partial t} = \dfrac{\partial^2 u}{\partial x^2}, & 0 < x < 1,\ t > 0, \\ u(0,t) = u(1,t) = 0, \\ u(x,0) = \varphi(x), \end{cases} \tag{6.1}$$

for the temperature distribution $u(x,t)$ in a perfectly conducting rod in which the two ends are held at the constant temperature 0, and the initial temperature distribution $\varphi(x)$ is given. This might look difficult to solve, so let us try the special cases

$$\varphi(x) = \sin(n\pi x), \quad n = 1, 2, \ldots.$$

These initial conditions are consistent with the problem in that they satisfy the given end conditions, called boundary conditions, at $x = 0$ and 1. Try for a solution of the form

$$u_n(x, t) = U_n(t) \sin(n\pi x).$$

Then U_n has to satisfy

$$U_n' \sin(n\pi x) = -n^2\pi^2 U_n \sin(n\pi x),$$

so

$$U_n' = -n^2\pi^2 U_n, \quad n = 1, 2, \ldots. \tag{6.2}$$

This simple equation has the solution

$$U_n(t) = U_n(0)e^{-n^2\pi^2 t},$$

so the complete solution for this initial temperature distribution is

$$u_n(x, t) = U_n(0)e^{-n^2\pi^2 t} \sin(n\pi x).$$

Now, and here is Fourier's great conception, suppose we can decompose a more general initial temperature distribution φ into a sum of these special solutions $\{\sin(n\pi x)\}_{n=1}^{\infty}$ with initial conditions

$$\varphi(x) = \sum_{n=1}^{\infty} \varphi_n \sin(n\pi x). \tag{6.3}$$

Then, at least formally, there would follow a representation of the solution of (6.1), namely

$$u(x, t) = \sum_{n=1}^{\infty} u_n(x, t) = \sum_{n=1}^{\infty} \varphi_n e^{-n^2\pi^2 t} \sin(n\pi x).$$

This representation of u in terms of simple harmonic functions $\{\sin(n\pi x)\}_{n=1}^{\infty}$ converts the partial differential equation (PDE) (6.1) into the system of ordinary differential equations (ODEs) (6.2), which are easily solved. Under suitable hypotheses on φ, the Sturm-Liouville theory of Section 4.7 can be used to justify the expansion (6.3), thereby giving rigorous credence to the formal solution.

Suppose now the rod is infinitely long, so we want to solve

$$\begin{cases} u_t = u_{xx}, & -\infty < x < \infty, \ t > 0, \\ u(x, 0) = \varphi(x), \\ u(x, t) \to 0 & \text{as } x \to \pm\infty. \end{cases} \tag{6.4}$$

Guided by the case of a rod of finite length, try to represent φ in terms of harmonic functions, e.g.,

$$\varphi(x) = \sum_{n=-\infty}^{\infty} \varphi_n e^{-inx}.$$

Any such function is periodic of period 2π, however. It turns out that to represent a general function defined on all of $\mathbb{R}$, one needs the uncountable class

$$\{e^{-i\lambda x}\}_{\lambda \in \mathbb{R}}.$$

These cannot be summed, but we might be able to integrate them, to wit,

$$\varphi(x) = \int_{-\infty}^{\infty} e^{-i\lambda x} \rho(\lambda) \, d\lambda,$$

say, for some density ρ. Suppose we could. We are led to search for a solution in the form

$$U(x,t) = \int_{-\infty}^{\infty} e^{-i\lambda x} \rho(\lambda, t) \, d\lambda.$$

If this is to satisfy (6.4), then

$$\int_{-\infty}^{\infty} e^{-i\lambda x} \frac{\partial \rho}{\partial t}(\lambda, t) \, d\lambda = -\int_{-\infty}^{\infty} e^{-i\lambda x} \lambda^2 \rho(\lambda, t) \, d\lambda,$$

or

$$\int_{-\infty}^{\infty} e^{-i\lambda x} \left[\frac{\partial \rho}{\partial t} + \lambda^2 \rho \right] d\lambda = 0,$$

for all x, t. As x is allowed to wander over all of $\mathbb{R}$, we conclude that this will hold only when

$$\frac{\partial \rho}{\partial t} + \lambda^2 \rho = 0 \quad \forall \lambda \in \mathbb{R}.$$

The solution of this collection of ODEs is

$$\rho(\lambda, t) = \rho(\lambda, 0) e^{-\lambda^2 t}.$$

Thus, formally, the full solution has the integral representation

$$u(x,t) = \int_{-\infty}^{\infty} e^{-i\lambda x} e^{-\lambda^2 t} \rho(\lambda, 0) \, d\lambda.$$

The observations that potentially,

(1) functions can be represented in terms of harmonic functions, and

(2) in this representation, PDEs may be reduced in complexity to ODEs,

is already enough to warrant further study. The crux of the formula above for u is ρ: what is ρ so that

$$\varphi(x) = \int_{-\infty}^{\infty} e^{-i\lambda x} \rho(\lambda) \, d\lambda?$$

Is there such a ρ, and if so, how do we find it? This leads us directly to the study of the Fourier transform.

The Fourier transform is a linear operator that can be defined naturally for any function in $L^1(\mathbb{R}^d)$. The definition can be extended to apply to functions in $L^2(\mathbb{R}^d)$. As extended, this transform maps $L^2(\mathbb{R}^d)$ onto itself and has some very helpful properties which are explored in what follows. A further extension allows the Fourier transform to be applied to some, but unfortunately not all, distributions. The class of objects where the Fourier transform makes sense are called tempered distributions.

Unless stated explicitly to the contrary, the underlying ground field $\mathbb{F}$ is always taken to be the set of complex numbers $\mathbb{C}$ in this chapter.

6.2   THE $L^1(\mathbb{R}^d)$ THEORY

If $\xi \in \mathbb{R}^d$, the function

$$\varphi_\xi(x) = e^{-ix\cdot\xi} = \cos(x \cdot \xi) - i\sin(x \cdot \xi), \quad x \in \mathbb{R}^d,$$

is a plane wave in the direction ξ. Its period in the jth direction is $2\pi/\xi_j$. These functions have useful algebraic and differential properties.

Proposition 6.1. *The following hold:*

(a) $|\varphi_\xi| = 1$ *and* $\bar{\varphi}_\xi = \varphi_{-\xi}$ *for any* $\xi \in \mathbb{R}^d$;

(b) $\varphi_\xi(x+y) = \varphi_\xi(x)\varphi_\xi(y)$ *for any* $x, y, \xi \in \mathbb{R}^d$;

(c) $-\Delta\varphi_\xi = |\xi|^2 \varphi_\xi$ *for any* $\xi \in \mathbb{R}^d$.

These are easily verified. Note that part (c) says that φ_ξ is an eigenfunction of the Laplace operator with eigenvalue $-|\xi|^2$. The reader might wish to confirm that, in fact, $\{\varphi_\xi\}_{\xi\in\mathbb{R}^d}$ is an orthogonal set in $L^2([0, 2\pi]^d)$.

If $f(x)$ is periodic, then f may be expanded in a Fourier series using commensurate waves $e^{-ix\cdot\xi}$ (i.e., waves of the same period) as discussed above and in Chapter 4. As already mentioned, if f is not periodic, then waves of all periods, not just the commensurate ones, are needed in their representation. Thus we are led to the Fourier transform, which has algebraic and differential properties related to those listed above for $e^{-ix\cdot\xi}$.

Definition. If $f \in L^1(\mathbb{R}^d)$, the *Fourier transform* of f is

$$\mathcal{F}f(\xi) = \hat{f}(\xi) = (2\pi)^{-d/2} \int_{\mathbb{R}^d} f(x)e^{-ix\cdot\xi} \, dx.$$

Both the notations $\mathcal{F}f$ and $\hat{f}$ will be convenient in our development. The Fourier transform is well-defined since

$$|f(x)e^{-ix\cdot\xi}| = |f(x)| \in L^1(\mathbb{R}^d). \tag{6.5}$$

The definition just given is not universal. You can also find the Fourier transform defined as

$$\int_{\mathbb{R}^d} f(x)e^{\pm 2\pi ix\cdot\xi}\,dx, \quad \int_{\mathbb{R}^d} f(x)e^{\pm ix\cdot\xi}\,dx, \quad \text{or} \quad (2\pi)^{-d/2}\int_{\mathbb{R}^d} f(x)e^{\pm ix\cdot\xi}\,dx.$$

The definition of choice affects the detailed form of the results that follow, but not their substance. Different authors make different choices, but it is straightforward to translate results obtained using one definition into corresponding results using another.

Proposition 6.2. *The Fourier transform*

$$\mathcal{F}: L^1(\mathbb{R}^d) \to L^\infty(\mathbb{R}^d)$$

is a bounded linear operator, and

$$\|\hat{f}\|_{L^\infty(\mathbb{R}^d)} \le (2\pi)^{-d/2}\|f\|_{L^1(\mathbb{R}^d)}.$$

The proof is an easy exercise using (6.5).

Example. Consider the characteristic function f of $[-1,1]^d$, which is

$$f(x) = \begin{cases} 1 & \text{if } -1 < x_j < 1, \ j = 1,\ldots,d, \\ 0 & \text{otherwise.} \end{cases}$$

Its Fourier transform is

$$\begin{aligned}
\hat{f}(\xi) &= (2\pi)^{-d/2}\int_{-1}^{1}\cdots\int_{-1}^{1} e^{-ix\cdot\xi}\,dx \\
&= \prod_{j=1}^{d}(2\pi)^{-1/2}\int_{-1}^{1} e^{-ix_j\xi_j}\,dx_j \\
&= \prod_{j=1}^{d}(2\pi)^{-1/2}\frac{-1}{i\xi_j}(e^{-i\xi_i} - e^{i\xi_j}) \\
&= \prod_{j=1}^{d}\sqrt{\frac{2}{\pi}}\frac{\sin\xi_j}{\xi_j}.
\end{aligned}$$

The following properties of the Fourier transform follow from those in Proposition 6.1 together with appropriate changes of variables.

Proposition 6.3. *If $f \in L^1(\mathbb{R}^d)$ and τ_y is translation by y, i.e., $\tau_y\varphi(x) = \varphi(x - y)$, then*

(a) $(\tau_y f)^\wedge(\xi) = e^{-iy\cdot\xi}\hat{f}(\xi)$ $\forall y \in \mathbb{R}^d$;

(b) $(e^{ix\cdot y}f)^\wedge(\xi) = \tau_y\hat{f}(\xi)$ $\forall y \in \mathbb{R}^d$;

(c) if $r_1, \ldots, r_d > 0$ are given,

$$\mathcal{F}\big(f(r_1x_1, \ldots, r_dx_d)\big)(\xi) = \frac{\hat{f}(\xi_1/r_1, \ldots, \xi_d/r_d)}{r_1\cdots r_d}$$

(so $\widehat{f(rx)}(\xi) = r^{-d}\hat{f}(r^{-1}\xi)$);

(d) $\hat{\bar{f}}(\xi) = \overline{\hat{f}(-\xi)}$.

While the Fourier transform maps $L^1(\mathbb{R}^d)$ into $L^\infty(\mathbb{R}^d)$, it does not map onto. Its range is not so well understood, but it is known to be contained in a set we will call $C_v(\mathbb{R}^d)$.

Definition. A continuous function f on $\mathbb{R}^d$ is said to *vanish at infinity* if for any $\epsilon > 0$ there is $K \subset\subset \mathbb{R}^d$ such that

$$|f(x)| < \epsilon \quad \forall x \notin K.$$

The subspace of all such continuous functions is denoted

$$C_v(\mathbb{R}^d) = \{f \in C^0(\mathbb{R}^d) : f \text{ vanishes at infinity}\}.$$

Proposition 6.4. *The space $C_v(\mathbb{R}^d)$ is a closed linear subspace of $L^\infty(\mathbb{R}^d)$.*

Proof. It is easy to show that $C_v(\mathbb{R}^d)$ is a linear subspace, so we show only that it is closed. Suppose that $\{f_n\}_{n=1}^\infty \subset C_v(\mathbb{R}^d)$ and that

$$f_n \xrightarrow{L^\infty} f.$$

Then f is continuous since the uniform limit of continuous functions is continuous. Now let $\epsilon > 0$ be given and fix N such that $\|f - f_N\|_{L^\infty} < \epsilon/2$. Let $K \subset\subset \mathbb{R}^d$ such that $|f_N(x)| < \epsilon/2$ for $x \notin K$. Then, for $x \notin K$,

$$|f(x)| \leq |f(x) - f_N(x)| + |f_N(x)| < \epsilon,$$

so $f \in C_v(\mathbb{R}^d)$.   □

Lemma 6.5 (Riemann-Lebesgue lemma). *The Fourier transform*

$$\mathcal{F} : L^1(\mathbb{R}^d) \to C_v(\mathbb{R}^d) \subsetneq L^\infty(\mathbb{R}^d).$$

Thus for $f \in L^1(\mathbb{R}^d)$,

$$\lim_{|\xi|\to\infty} |\hat{f}(\xi)| = 0 \quad and \quad \hat{f} \in C^0(\mathbb{R}^d).$$

Proof. Let $f \in L^1(\mathbb{R}^d)$. There is a sequence of simple functions $\{f_n\}_{n=1}^{\infty}$ such that $f_n \to f$ in $L^1(\mathbb{R}^d)$. Recall that a simple function is a finite linear combination of characteristic functions of rectangles. If $\hat{f}_n \in C_{\text{v}}(\mathbb{R}^d)$, we are done since

$$\hat{f}_n \xrightarrow{L^{\infty}} \hat{f}$$

and $C_{\text{v}}(\mathbb{R}^d)$ is a closed subspace. We saw above that the Fourier transform of the characteristic function of $[-1, 1]^d$ is

$$\prod_{j=1}^{d} \sqrt{\frac{2}{\pi}} \frac{\sin \xi_j}{\xi_j} \in C_{\text{v}}(\mathbb{R}^d).$$

By Proposition 6.3, translation and dilation of this cube shows that the characteristic function of any rectangle is in $C_{\text{v}}(\mathbb{R}^d)$, and hence also any finite linear combination of them. □

Some helpful properties of the Fourier transform are given in the following result.

Proposition 6.6. *If* $f, g \in L^1(\mathbb{R}^d)$, *then*

(a) $\displaystyle \int \hat{f}(x) \, g(x) \, dx = \int f(x) \, \hat{g}(x) \, dx,$

(b) $f * g \in L^1(\mathbb{R}^d)$ *and* $\widehat{f * g} = (2\pi)^{d/2} \hat{f}\hat{g},$

where the convolution of f *and* g *is*

$$f * g(x) = \int f(x - y) \, g(y) \, dy,$$

which is defined for almost every $x \in \mathbb{R}^d$.

Proof. For (a), note that $\hat{f} \in L^{\infty}$ and $g \in L^1$ implies $\hat{f}g \in L^1$, so the integrals are well-defined. Fubini's theorem gives the result, *viz.*

$$\int \hat{f}(x) \, g(x) \, dx = (2\pi)^{-d/2} \iint f(y) e^{-ix \cdot y} g(x) \, dy \, dx$$

$$= (2\pi)^{-d/2} \iint f(y) e^{-ix \cdot y} g(x) \, dx \, dy$$

$$= \int f(y) \, \hat{g}(y) \, dy.$$

Part (b) follows similarly, using Fubini's theorem and a change of variables, once we know that $f * g \in L^1(\mathbb{R}^d)$. This fact is established next, in more generally than needed at the moment. □

Theorem 6.7 (generalized Young's inequality). *Suppose $K(x, y)$ is measurable on $\mathbb{R}^d \times \mathbb{R}^d$ and there is some $M > 0$ such that*

$$\int |K(x, y)| \, dx \leq M \quad \text{for almost every } y \in \mathbb{R}^d$$

and

$$\int |K(x, y)| \, dy \leq M \quad \text{for almost every } x \in \mathbb{R}^d.$$

Define the operator T by

$$Tf(x) = \int K(x, y) f(y) \, dy.$$

If $1 \leq p \leq \infty$, then $T : L^p(\mathbb{R}^d) \to L^p(\mathbb{R}^d)$ is a bounded linear map with operator norm $\|T\| \leq M$.

Corollary 6.8 (Young's inequality). *If $1 \leq p \leq \infty$, $f \in L^p(\mathbb{R}^d)$, and $g \in L^1(\mathbb{R}^d)$, then $f * g \in L^p(\mathbb{R}^d)$ and*

$$\|f * g\|_{L^p(\mathbb{R}^d)} \leq \|f\|_{L^p(\mathbb{R}^d)} \|g\|_{L^1(\mathbb{R}^d)}.$$

Proof. Just take $K(x, y) = g(x - y)$. □

Corollary 6.9. *The space $L^1(\mathbb{R}^d)$ is an algebra with multiplication defined by the convolution operation.*

Proof of the generalized Young's inequality. If $p = \infty$, the result is trivial (and, in fact, we need not assume that $\int |K(x, y)| \, dx \leq M$). If $p < \infty$, let q be the conjugate exponent defined by $\frac{1}{q} + \frac{1}{p} = 1$. Then, calculate as follows:

$$|Tf(x)| \leq \int |K(x, y)|^{1/q} |K(x, y)|^{1/p} |f(y)| \, dy$$

$$\leq \left(\int |K(x, y)| \, dy \right)^{1/q} \left(\int |K(x, y)| \, |f(y)|^p \, dy \right)^{1/p}$$

by Hölder's inequality. Taking the pth power of this inequality and integrating leads to

$$\|Tf\|_{L^p}^p \leq M^{p/q} \iint |K(x, y)| \, |f(y)|^p \, dy \, dx$$

$$= M^{p/q} \iint |K(x, y)| \, dx \, |f(y)|^p \, dy$$

$$\leq M^{(p/q)+1} \int |f(y)|^p \, dy$$

$$= M^p \|f\|_{L^p}^p,$$

and the theorem follows since T is clearly linear. □

An unresolved question is: given f, what does $\hat{f}$ look like? Partial answers are provided by the Riemann-Lebesgue lemma, and the following theorem.

Theorem 6.10 (Paley-Wiener theorem). *If $f \in C_0^\infty(\mathbb{R}^d)$, then $\hat{f}$ extends to an entire holomorphic function on $\mathbb{C}^d$.*

Proof. The function

$$\xi \longmapsto e^{-ix \cdot \xi}$$

is an entire function for $x \in \mathbb{R}^d$ fixed. The Riemann sums approximating

$$\hat{f}(\xi) = (2\pi)^{-d/2} \int f(x) e^{-ix \cdot \xi} \, dx$$

are entire, and they converge uniformly on compact sets since $f \in C_0^\infty(\mathbb{R}^d)$. Thus it is concluded that $\hat{f}$ is entire. $\qquad\square$

See [21] for the converse. Since holomorphic functions do *not* have compact support, we see that functions which are localized in space are *not* localized in Fourier space (the converse will follow after we develop the inverse Fourier transform). This implies that the Fourier transform cannot be realized as a mapping of $\mathcal{D}$ into $\mathcal{D}$. This state of affairs led Laurent Schwartz to introduce a larger class of test functions better suited to the Fourier transform.

6.3   THE SCHWARTZ SPACE THEORY

Since $L^2(\mathbb{R}^d)$ is not contained in $L^1(\mathbb{R}^d)$, we restrict to a suitable subspace $\mathcal{S} \subset L^2(\mathbb{R}^d) \cap L^1(\mathbb{R}^d)$ on which to define the Fourier transform before attempting its definition on $L^2(\mathbb{R}^d)$.

Definition. The *Schwartz space* or space of *functions of rapid decrease* is

$$\mathcal{S} = \mathcal{S}(\mathbb{R}^d) = \left\{ \phi \in C^\infty(\mathbb{R}^d) : \sup_{x \in \mathbb{R}^d} |x^\alpha D^\beta \phi(x)| < \infty \right.$$

$$\left. \text{for all multi-indices } \alpha \text{ and } \beta \right\}.$$

In other words, ϕ and all its derivatives tend to 0 at infinity faster than any polynomial. As an example, consider $\phi(x) = p(x) e^{-a|x|^2}$ for any $a > 0$ and any polynomial $p(x)$.

Proposition 6.11. *The following inclusions hold:*

$$C_0^\infty(\mathbb{R}^d) \subsetneq \mathcal{S} \subsetneq L^1(\mathbb{R}^d) \cap L^\infty(\mathbb{R}^d).$$

Hence it is also the case that $\mathcal{S}(\mathbb{R}^d) \subset L^p(\mathbb{R}^d)$ for all p with $1 < p < \infty$.

Proof. The only nontrivial statement is that $\mathcal{S} \subset L^1$. For $\phi \in \mathcal{S}$,

$$\int |\phi(x)| \, dx = \int_{B_1(0)} |\phi(x)| \, dx + \int_{|x| \geq 1} |\phi(x)| \, dx.$$

The first integral on the right-hand side is finite, so consider the second one. Since $\phi \in \mathcal{S}$, there is a $C > 0$ such that $|x|^{d+1}|\phi(x)| < C$ for all $|x| > 1$. Then

$$\int_{|x| \geq 1} |\phi(x)|\, dx = \int_{|x| \geq 1} |x|^{-d-1}(|x|^{d+1}|\phi(x)|)\, dx$$

$$\leq C \int_{|x| \geq 1} |x|^{-d-1}\, dx$$

$$= C\, d\omega_d \int_1^\infty r^{-d-1} r^{d-1}\, dr$$

$$= C\, d\omega_d \int_1^\infty r^{-2}\, dr < \infty,$$

where $d\omega_d$ is the measure of the unit sphere in $\mathbb{R}^d$. □

For $n = 0, 1, 2, \ldots$ and $\phi \in \mathcal{S}$, define

$$\rho_n(\phi) = \sup_{|\beta| \leq n} \sup_{x \in \mathbb{R}^d} (1 + |x|^2)^{n/2} |D^\beta \phi(x)|$$

$$= \sup_{|\beta| \leq n} \|(1 + |\cdot|^2)^{n/2} D^\beta \phi(\cdot)\|_{L^\infty(\mathbb{R}^d)}.$$

Each ρ_n is a norm on $\mathcal{S}$ and $\rho_n(\phi) \leq \rho_m(\phi)$ whenever $n \leq m$.

Proposition 6.12. *With*

$$\omega_{\alpha\beta}(\phi) = \sup_x |x^\alpha D^\beta \phi(x)| = \|(\cdot)^\alpha D^\beta \phi(\cdot)\|_{L^\infty(\mathbb{R}^d)},$$

the Schwartz class is identified as

$$\mathcal{S} = \{\phi \in C^\infty(\mathbb{R}^d) : \omega_{\alpha\beta}(\phi) < \infty \text{ for all multi-indices } \alpha \text{ and } \beta\}$$
$$= \{\phi \in C^\infty(\mathbb{R}^d) : \rho_n(\phi) < \infty \text{ for all } n\}.$$

Proof. The expression $\rho_n(\phi)$ is bounded by sums of terms of the form $\omega_{\alpha\beta}(\phi)$, and $\rho_n(\phi)$ bounds $\omega_{\alpha\beta}(\phi)$ for $n = \max(|\alpha|, |\beta|)$. □

Proposition 6.13. *The Schwartz class $\mathcal{S}$ is a linear vector space and a complete metric space for the topology generated by the metric*

$$d(\phi_1, \phi_2) = \sum_{n=0}^\infty 2^{-n} \frac{\rho_n(\phi_1 - \phi_2)}{1 + \rho_n(\phi_1 - \phi_2)}.$$

Notice that a sequence $\{\phi_j\}_{j=1}^\infty$ in $\mathcal{S}$ converges to ϕ for the metric d if and only if $\rho_n(\phi - \phi_j) \to 0$ for all n if and only if $\omega_{\alpha\beta}(\phi - \phi_j) \to 0$ for all α and β. The Schwartz class is an example of a *Fréchet space*.

Proof. Clearly $\mathcal{S}$ is a vector space and d is a metric provided the triangle inequality holds. The latter follows since the function $g(x) = x/(1 + x)$ has $g''(x) \leq 0$ for $x \geq 0$.

To show completeness, let $\{\phi_j\}_{j=1}^{\infty}$ be a Cauchy sequence in $\mathcal{S}$. As already remarked, this means that

$$\rho_n(\phi_j - \phi_k) \longrightarrow 0 \quad \text{as } j, k \to \infty \ \forall n,$$

which is to say, for any α and $n \geq |\alpha|$,

$$\{(1 + |x|^2)^{n/2} D^{\alpha} \phi_j\}_{j=1}^{\infty} \quad \text{is Cauchy in } C^0(\mathbb{R}^d). \tag{6.6}$$

For any fixed $\xi \in \mathcal{D} = \mathcal{D}(\mathbb{R}^d)$, $\xi\phi_j$ is Cauchy in $\mathcal{D}$, so the completeness of $\mathcal{D}$ implies that $\xi\phi_j \xrightarrow{\mathcal{D}} \psi_\xi$ for some $\psi_\xi \in \mathcal{D}$. For any compact $K \subset \mathbb{R}^d$, there are elements $\xi \in \mathcal{D}$ such that $\xi|_K = 1$. It follows that there is a C^{∞} function ψ to which $\{\phi_j\}_{j=1}^{\infty}$ converges pointwise.

Because of (6.6) and the completeness of C^0, there is a $\psi_{n,\alpha} \in C^0(\mathbb{R}^d)$ such that

$$(1 + |x|^2)^{n/2} D^{\alpha} \phi_j \xrightarrow{L^{\infty}} \psi_{n,\alpha},$$

and so

$$D^{\alpha} \phi_j \xrightarrow{L^{\infty}} \frac{\psi_{n,\alpha}}{(1 + |x|^2)^{n/2}} \in C^0(\mathbb{R}^d).$$

But, $\phi_j \xrightarrow{L^{\infty}} \psi_{0,0} = \psi \in C^{\infty}$, so as distributions $D^{\alpha}\phi_j \xrightarrow{\mathcal{D}'} D^{\alpha}\psi$. In consequence, it is determined that $\psi_{n,\alpha} = (1 + |x|^2)^{n/2} D^{\alpha}\psi \in L^{\infty}(\mathbb{R}^d)$, and further that $\rho_n(\psi) < \infty$ and $\rho_n(\phi_j - \psi) \to 0$ for all n. That is, $\psi \in \mathcal{S}$ and $\phi_j \xrightarrow{\mathcal{S}} \psi$. $\square$

Proposition 6.14. *If $p(x)$ is a polynomial, $g \in \mathcal{S}$, and α a multi-index, then each of the three mappings*

$$f \mapsto pf, \quad f \mapsto gf, \quad \text{and} \quad f \mapsto D^{\alpha}f$$

is a continuous linear map from $\mathcal{S}$ to $\mathcal{S}$.

Proof. By the Leibnitz formula, the range of the first two maps is $\mathcal{S}$ and this fact is obvious for the third map. These maps are easily seen to be sequentially continuous, and thus continuous. $\square$

The proof of the following result is left to the reader.

Proposition 6.15. *If $\{f_j\}_{j=1}^{\infty} \subset \mathcal{S}$, $f_j \xrightarrow{\mathcal{S}} f$, and $1 \leq p \leq \infty$, then $f_j \xrightarrow{L^p} f$.*

Since $\mathcal{S} \subset L^1(\mathbb{R}^d)$, we can take the Fourier transform of functions in $\mathcal{S}$.

Theorem 6.16. *If $f \in \mathcal{S}$ and α is a multi-index, then $\hat{f} \in C^{\infty}(\mathbb{R}^d)$ and*

(a) $(D^{\alpha}f)^{\wedge}(\xi) = (i\xi)^{\alpha}\hat{f}(\xi)$,

(b) $D^{\alpha}\hat{f}(\xi) = ((-ix)^{\alpha}f(x))^{\wedge}(\xi)$.

Proof. For (a), remark that

$$(2\pi)^{d/2}(D^\alpha f)^\wedge(\xi) = \int D^\alpha f(x)e^{-ix\cdot\xi}\,dx$$

$$= \lim_{r\to\infty} \int_{B_r(0)} D^\alpha f(x)e^{-ix\cdot\xi}\,dx$$

$$= \lim_{r\to\infty} \left\{ \int_{B_r(0)} f(x)(i\xi)^\alpha e^{-ix\cdot\xi}\,dx + \text{(boundary terms)} \right\},$$

via integration by parts. There are finitely many boundary terms, each evaluated at vectors x with $|x| = r$ and the absolute value of any such boundary term is bounded by a constant times $|D^\beta f(x)|$ for some multi-index $\beta \le \alpha$. Since $f \in \mathcal{S}$, each of these tends to zero faster than the measure of $\partial B_r(0) = S_r(0)$ (i.e., faster than r^{d-1}), so each boundary term vanishes in the limit as $r \to \infty$. Taking this limit yields

$$(D^\alpha f)^\wedge(\xi) = (2\pi)^{-d/2} \int f(x)(i\xi)^\alpha e^{-ix\cdot\xi}\,dx = (i\xi)^\alpha \hat{f}(\xi).$$

For (b), we need to interchange integration and differentiation, since, provided this can be done and the derivatives and integrals exist, then

$$(2\pi)^{d/2}D^\alpha \hat{f}(\xi) = D^\alpha \int f(x)e^{-ix\cdot\xi}\,dx$$

$$= \int f(x)D^\alpha e^{-ix\cdot\xi}\,dx = \int f(x)(-ix)^\alpha e^{-ix\cdot\xi}\,dx.$$

Consider a single derivative in the jth direction, *viz.*

$$(2\pi)^{d/2}D_j\hat{f}(\xi) = \lim_{h\to 0} \int f(x)e^{-ix\cdot\xi}\frac{e^{-ix_j h} - 1}{h}\,dx.$$

Since

$$\left|\frac{e^{-i\theta} - 1}{\theta}\right|^2 = 2\left|\frac{1 - \cos\theta}{\theta^2}\right| \le 1,$$

it follows that

$$\left| ix_j f(x)e^{-ix\cdot\xi}\frac{e^{-ix_j h} - 1}{ix_j h} \right| \le |x_j f(x)| \in L^1,$$

independently of h. The dominated convergence theorem applies and its application yields that

$$(2\pi)^{d/2}D_j\hat{f}(\xi) = \int \lim_{h\to 0} ix_j f(x)\, e^{-ix\cdot\xi}\frac{e^{-ix_j h} - 1}{ix_j h}\,dx$$

$$= \int -ix_j f(x)e^{-ix\cdot\xi}\,dx$$

$$= (2\pi)^{d/2}\big(-ix_j f(x)\big)^\wedge(\xi).$$

Thus the limit exists, and, by iteration, the result for general multi-indices α obtains. Combining this with the Riemann-Lebesgue lemma (Lemma 6.5) reveals that $\hat{f} \in C^\infty(\mathbb{R}^d)$. □

Lemma 6.17. *The Fourier transform $\mathcal{F} : \mathcal{S} \to \mathcal{S}$ is continuous and linear.*

Proof. We first show that the range of $\mathcal{F}$ lies in $\mathcal{S}$. For $f \in \mathcal{S}$, $x^\alpha D^\beta f \in L^\infty$ for any multi-indices α and β. But then,

$$\xi^\alpha D^\beta \hat{f} = \xi^\alpha \big((-ix)^\beta f\big)^\wedge = (-1)^{|\beta|} i^{|\beta|-|\alpha|} \big(D^\alpha(x^\beta f)\big)^\wedge,$$

and so

$$\|\xi^\alpha D^\beta \hat{f}\|_{L^\infty} \le (2\pi)^{-d/2}\|D^\alpha(x^\beta f)\|_{L^1} < \infty,$$

since $D^\alpha(x^\beta f)$ rapidly decreases. It is concluded that $\hat{f} \in \mathcal{S}$.

The linearity of $\mathcal{F}$ is clear. If $\{f_j\}_{j=1}^\infty \subset \mathcal{S}$ and $f_j \xrightarrow{\mathcal{S}} f$, then also $f_j \xrightarrow{L^1} f$. Since $\mathcal{F}$ is continuous on L^1, $\hat{f}_j \xrightarrow{L^\infty} \hat{f}$. Using Proposition 6.14, it is similarly concluded that for any multi-indices α and β, $D^\alpha x^\beta f_j \xrightarrow{L^1} D^\alpha x^\beta f$, whence

$$(D^\alpha x^\beta f_j)^\wedge \xrightarrow{L^\infty} (D^\alpha x^\beta f)^\wedge.$$

Applying Theorem 6.16, it is deduced that

$$\xi^\alpha D^\beta \hat{f}_j \xrightarrow{L^\infty} \xi^\alpha D^\beta \hat{f},$$

which is to say, $\hat{f}_j \xrightarrow{\mathcal{S}} \hat{f}$, and the Fourier transform is continuous. □

In fact, the following lemma will lead to the conclusion that $\mathcal{F} : \mathcal{S} \to \mathcal{S}$ is one-to-one and maps onto $\mathcal{S}$.

Lemma 6.18. *If $\phi(x) = e^{-|x|^2/2}$, then $\phi \in \mathcal{S}$ and $\hat{\phi}(\xi) = \phi(\xi)$.*

Proof. It is clear that $\phi \in \mathcal{S}$. Since

$$\hat{\phi}(\xi) = (2\pi)^{-d/2} \int_{\mathbb{R}^d} e^{-|x|^2/2} e^{-ix\cdot\xi}\, dx$$

$$= \prod_{j=1}^d (2\pi)^{-1/2} \int_{\mathbb{R}} e^{-x_j^2/2} e^{-ix_j\xi_j}\, dx_j,$$

we need only show the result for $d = 1$. This can be accomplished directly using complex contour integration and Cauchy's theorem. An alternate proof is to note that for $d = 1$, $\phi(x)$ solves

$$y' + xy = 0$$

whereas, upon taking the Fourier transform we see that $\hat{\phi}(\xi)$ solves

$$0 = \widehat{y'} + \widehat{xy} = i\xi\hat{y} + i\hat{y}',$$

which is the same equation. Thus $\hat{\phi}/\phi$ is constant. But $\phi(0) = 1$ and $\hat{\phi}(0) = (2\pi)^{-1/2}\int e^{-x^2/2}\, dx = 1$, so $\hat{\phi} = \phi$. □

The following result is the centrally important Plancherel theorem restricted to $\mathcal{S}(\mathbb{R}^d)$.

Theorem 6.19. *The Fourier transform $\mathcal{F} : \mathcal{S} \to \mathcal{S}$ is a continuous, linear, one-to-one map of $\mathcal{S}$ onto $\mathcal{S}$ with a continuous inverse. The map $\mathcal{F}$ has period 4, and in fact $\mathcal{F}^2$ is reflection about the origin. If $f \in \mathcal{S}$, then*

$$f(x) = (2\pi)^{-d/2} \int \hat{f}(\xi) e^{ix\cdot\xi}\, d\xi. \tag{6.7}$$

Moreover, if $f \in L^1(\mathbb{R}^d)$ and $\hat{f} \in L^1(\mathbb{R}^d)$, then (6.7) holds for almost every $x \in \mathbb{R}^d$.

Sometimes we write $\check{g} = \mathcal{F}^{-1}(g)$ for the inverse Fourier transform, *viz.*

$$\check{g}(x) = \mathcal{F}^{-1}(g)(x) = (2\pi)^{-d/2} \int g(\xi) e^{ix\cdot\xi}\, d\xi = \hat{g}(-x).$$

Proof. The first step is to show that (6.7) holds for $f \in \mathcal{S}$. Let $\phi \in \mathcal{S}$ and $\epsilon > 0$. Then

$$\int f(x)\epsilon^{-d}\hat{\phi}(\epsilon^{-1}x)\, dx = \int f(\epsilon y)\hat{\phi}(y)\, dy \longrightarrow f(0)\int \hat{\phi}(y)\, dy$$

as $\epsilon \to 0^+$ by the dominated convergence theorem since $f(\epsilon y) \to f(0)$ uniformly (so pointwise) and $f \in L^\infty(\mathbb{R}^d)$. (We have just shown that $\epsilon^{-d}\hat{\phi}(\epsilon^{-1}x)$ converges to a multiple of δ_0 in $\mathcal{S}'$, the dual of $\mathcal{S}$, which will be defined later.) But also, by Proposition 6.3(c),

$$\int f(x)\epsilon^{-d}\hat{\phi}(\epsilon^{-1}x)\, dx = \int \hat{f}(x)\phi(\epsilon x)\, dx \longrightarrow \phi(0)\int \hat{f}(x)\, dx,$$

so

$$f(0)\int \hat{\phi}(y)\, dy = \phi(0)\int \hat{f}(x)\, dx.$$

Take

$$\phi(x) = e^{-|x|^2/2} \in \mathcal{S}$$

to see by the last lemma that

$$f(0) = (2\pi)^{-d/2} \int \hat{f}(\xi)\, d\xi,$$

which is (6.7) for $x = 0$. The general result follows by translation:

$$f(x) = (\tau_{-x}f)(0)$$
$$= (2\pi)^{-d/2} \int (\tau_{-x}f)^{\wedge}(\xi)\, d\xi$$
$$= (2\pi)^{-d/2} \int e^{ix\cdot\xi}\hat{f}(\xi)\, d\xi.$$

We saw earlier that $\mathcal{F} : \mathcal{S} \to \mathcal{S}$ is continuous and linear; it is one-to-one by (6.7). Moreover,

$$\mathcal{F}^2 f(x) = f(-x)$$

follows as a simple computation since $\mathcal{F}$ and $\mathcal{F}^{-1}$ are so similar. Thus $\mathcal{F}$ maps onto $\mathcal{S}$, $\mathcal{F}^4 = I$, and $\mathcal{F}^{-1} = \mathcal{F}^3$ is continuous.

It remains to extend (6.7) to $L^1(\mathbb{R}^d)$. If $f, \hat{f} \in L^1(\mathbb{R}^d)$, then define

$$f_0(x) = \check{\hat{f}}(x) = (2\pi)^{-d/2} \int \hat{f}(\xi) e^{ix \cdot \xi} \, d\xi.$$

For $\phi \in \mathcal{S}$,

$$\int f(x) \, \hat{\phi}(x) \, dx = \int \hat{f}(x) \, \phi(x) \, dx$$

$$= (2\pi)^{-d/2} \int \hat{f}(x) \int \hat{\phi}(\xi) e^{ix \cdot \xi} \, d\xi \, dx$$

$$= (2\pi)^{-d/2} \iint \hat{f}(x) e^{ix \cdot \xi} \hat{\phi}(\xi) \, dx \, d\xi$$

$$= \int f_0(\xi) \, \hat{\phi}(\xi) \, d\xi,$$

and it is concluded that

$$\int \psi(x) \big(f(x) - f_0(x) \big) = 0$$

for an arbitrary member $\psi(x)$ of $\mathcal{S}$, because $\mathcal{F}$ is a surjection. This in turn implies that $f(x) = f_0(x)$ for almost every $x \in \mathbb{R}^d$. $\quad\square$

We conclude the theory in $\mathcal{S}$ with a result about convolutions.

Theorem 6.20. *If $f, g \in \mathcal{S}$, then $f * g \in \mathcal{S}$ and*

$$(2\pi)^{d/2} (fg)^{\wedge} = \hat{f} * \hat{g}.$$

Proof. The L^1 theory has revealed that

$$(f * g)^{\wedge} = (2\pi)^{d/2} \hat{f} \hat{g},$$

so

$$(\hat{f} * \hat{g})^{\wedge} = (2\pi)^{d/2} \hat{\hat{f}} \hat{\hat{g}} = (2\pi)^{d/2} (fg)^{\check{\wedge}},$$

since $\mathcal{F}^2$ is reflection. Taking the inverse Fourier transform of this relation then gives

$$\hat{f} * \hat{g} = (2\pi)^{d/2} (fg)^{\wedge}.$$

We saw in Proposition 6.14 that $\check{f} \check{g} \in \mathcal{S}$, so also

$$f * g = \hat{\check{f}} * \hat{\check{g}} = (2\pi)^{d/2} (\check{f} \check{g})^{\wedge} \in \mathcal{S}. \quad\square$$

6.4 THE $L^2(\mathbb{R}^d)$ THEORY

Recall from Proposition 6.6 that for $f, g \in \mathcal{S}$,

$$\int f\hat{g} = \int \hat{f}g.$$

Corollary 6.21 (Parseval identity). *If $f, g \in \mathcal{S}$, then*

$$\int f(x)\,\overline{g(x)}\,dx = \int \hat{f}(\xi)\,\overline{\hat{g}(\xi)}\,d\xi.$$

Proof. The computation

$$\int f\bar{g} = \int f\hat{\check{\bar{g}}} = \int \hat{f}\check{\bar{g}} = \int \hat{f}\bar{\hat{g}},$$

holds true since $\check{\bar{g}} = \bar{\hat{g}}$ is readily verified. □

Thus $\mathcal{F}$ preserves the L^2 inner-product on $\mathcal{S}$. Since $C_0^\infty \subset \mathcal{S} \subset L^2(\mathbb{R}^d)$ is dense, the mapping $\mathcal{F} : \mathcal{S}$ (with L^2 topology) $\to L^2$ may be extended to a mapping $\mathcal{F} : L^2 \to L^2$ by way of the following general result.

Theorem 6.22. *Suppose X and Y are metric spaces, Y is complete, and $A \subset X$ is dense. If $T : A \to Y$ is uniformly continuous, then there is a unique extension $\tilde{T} : X \to Y$ which is uniformly continuous.*

Proof. Given $x \in X$, take $\{x_j\}_{j=1}^\infty \subset A$ such that $x_j \xrightarrow{X} x$. Let $y_j = T(x_j)$. Since T is uniformly continuous, $\{y_j\}_{j=1}^\infty$ is Cauchy in Y. Since Y is complete, there is $y \in Y$ such that $y_j \xrightarrow{Y} y$. *Define*

$$\tilde{T}(x) = y = \lim_{j \to \infty} T(x_j).$$

The mapping $\tilde{T}$ is well-defined. For if we have two sequences drawn from A, say $\{x_j\}_{j=1}^\infty$ and $\{z_j\}_{j=1}^\infty$ such that $x_j \to x$ and $z_j \to x$ as $j \to \infty$, consider the sequence obtained from these by alternately choosing elements from each. That is, consider the sequence $\{w_j\}_{j=1}^\infty$ with $w_{2j} = x_j$ and $w_{2j+1} = z_j, j = 1, 2, \dots$. This also converges to x, so $\{Tw_j\}_{j=1}^\infty$ converges and therefore the even and odd subsequences converge to the same thing. In other words, $\{Tx_j\}_{j=1}^\infty$ and $\{Tz_j\}_{j=1}^\infty$ have the same limit.

That $\tilde{T}$ extends T is clear, since if $x \in A$, then take the approximating sequence $\{x_j\}_{j=1}^\infty$ with $x_j = x$ for all j to see that $\tilde{T}x = Tx$. If $\tilde{T}$ is continuous on X, then any other continuous extension would necessarily agree with $\tilde{T}$ since they would agree on a dense subset, so $\tilde{T}$ would be unique.

To see that indeed $\tilde{T}$ is continuous, let $\epsilon > 0$ be given. Since T is uniformly continuous, there is $\delta > 0$ such that for all $x, z \in A$,

$$d_Y(T(x), T(z)) < \epsilon \quad \text{whenever } d_X(x, z) < \delta.$$

Now let $x, z \in X$ be such that $d_X(x, z) < \delta/3$. Choose $\{x_j\}_{j=1}^\infty$ and $\{z_j\}_{j=1}^\infty$ in A such that $x_j \xrightarrow{X} x$ and $z_j \xrightarrow{X} z$, and choose N large enough that for $j \geq N$,

$$d_X(x, x_j) \leq \frac{\delta}{3} \quad \text{and} \quad d_X(z, z_j) \leq \frac{\delta}{3}$$

as well as

$$d_Y(\tilde{T}x, Tx_j) \leq \epsilon \quad \text{and} \quad d_Y(\tilde{T}z, Tz_j) \leq \epsilon.$$

Then, $d_X(x_N, z_N) \leq \delta$ so $d_Y(Tx_N, Tz_N) \leq \epsilon$, whence

$$d_Y(\tilde{T}(x), \tilde{T}(z))$$
$$\leq d_Y(\tilde{T}(x), T(x_N)) + d_Y(T(x_N), T(z_N)) + d_Y(T(z_N), \tilde{T}(z)) < 3\epsilon.$$

Thus, $\tilde{T}$ is continuous, and in fact uniformly so. □

Corollary 6.23. *If X is an NLS and Y is a Banach space, $A \subset X$ is a dense subspace, and $T : A \to Y$ is continuous and linear, then there is a unique continuous linear extension $\tilde{T} : X \to Y$.*

Proof. A continuous linear map is uniformly continuous, and the extension, defined by continuity, is necessarily linear. □

Theorem 6.24 (Plancherel theorem). *The Fourier transform extends to a unitary isomorphism of $L^2(\mathbb{R}^d)$ to itself. That is,*

$$\mathcal{F} : L^2(\mathbb{R}^d) \to L^2(\mathbb{R}^d)$$

is a bounded linear, one-to-one, and onto map with a bounded linear inverse such that the $L^2(\mathbb{R}^d)$ inner-product is preserved:

$$\int f(x)\overline{g(x)}\,dx = \int \hat{f}(\xi)\overline{\hat{g}(\xi)}\,d\xi. \tag{6.8}$$

Moreover, $\mathcal{F}^* = \mathcal{F}^{-1}$, $\|\mathcal{F}\| = 1$,

$$\|f\|_{L^2} = \|\hat{f}\|_{L^2} \quad \forall f \in L^2(\mathbb{R}^d),$$

and $\mathcal{F}^2$ is reflection.

Proof. Note that $\mathcal{S}$ (in fact C_0^∞) is dense in $L^2(\mathbb{R}^d)$, and that Corollary 6.21 (i.e., (6.8) on $\mathcal{S}$) implies uniform continuity of $\mathcal{F}$ on $\mathcal{S}$, because

$$\|\hat{f}\|_{L^2(\mathbb{R}^d)} = \left(\int \hat{f}\bar{\hat{f}}\,dx \right)^{1/2} = \left(\int f\bar{f}\,dx \right)^{1/2} = \|f\|_{L^2(\mathbb{R}^d)} \quad \forall f \in \mathcal{S}.$$

Thus, $\mathcal{F}$ is a bounded linear operator on $\mathcal{S}$. Extend $\mathcal{F}$ uniquely as a continuous linear operator to $L^2(\mathbb{R}^d)$ by Corollary 6.23 . By continuity, (6.8) on $\mathcal{S}$ continues to hold on all of $L^2(\mathbb{R}^d)$, and so $\|f\|_{L^2} = \|\hat{f}\|_{L^2}$ and $\|\mathcal{F}\| = 1$.

Similarly, extend $\mathcal{F}^{-1} : \mathcal{S} \to L^2$ to L^2. For $f \in L^2$ and $\{f_j\}_{j=1}^\infty \subset \mathcal{S}$ with $f_j \to f$ in L^2, we have

$$\mathcal{F}\mathcal{F}^{-1} f = \lim_{j \to \infty} \mathcal{F}\mathcal{F}^{-1} f_j = \lim_{j \to \infty} f_j = f$$

and similarly $\mathcal{F}^{-1}\mathcal{F}f = f$. Thus $\mathcal{F}$ is one-to-one, onto, continuous, and linear. Since $\mathcal{F}^2$ is reflection on $\mathcal{S}$, it is so on $L^2(\mathbb{R}^d)$ by continuity (or by the uniqueness of the extension, since reflection on $\mathcal{S}$ extends to reflection on $L^2(\mathbb{R}^d)$). Finally, (6.8) implies that $\mathcal{F}^* = \mathcal{F}^{-1}$. □

The definition of $\mathcal{F}$ as the continuous extension from $\mathcal{S}$ to L^2 implies that many useful properties of $\mathcal{F}$ on $\mathcal{S}$ extend immediately to $L^2(\mathbb{R}^d)$.

Corollary 6.25. *For all $f, g \in L^2(\mathbb{R}^d)$,*

$$\int f\hat{g}\,dx = \int \hat{f}g\,dx.$$

Proof. Extend Proposition 6.6. □

The following lemma allows us to compute Fourier transforms of L^2 functions.

Lemma 6.26. *Let $f \in L^2(\mathbb{R}^d)$.*

(a) *If $f \in L^1(\mathbb{R}^d)$ as well, then the L^2 Fourier transform of f is*

$$\hat{f}(\xi) = (2\pi)^{-d/2} \int_{\mathbb{R}^d} f(x)e^{-ix\cdot\xi}\,dx$$

(i.e., the L^1 and L^2 Fourier transforms agree).

(b) *If $R > 0$, then*

$$\hat{f}(\xi) = \lim_{R \to \infty} (2\pi)^{-d/2} \int_{|x| \leq R} f(x)e^{-ix\cdot\xi}\,dx.$$

That is, $(\chi_{B_R(0)} f)^\wedge \xrightarrow{L^2} \hat{f}$ as $R \to \infty$, where $\chi_{B_R(0)}$ is the characteristic function of $B_R(0)$.

Similar statements hold for $\mathcal{F}^{-1}$.

Proof. Let $R > 0$ be given and denote the L^1 and L^2 Fourier transforms by $\mathcal{F}_1$ and $\mathcal{F}_2$, respectively. For $f \in L^2$, the observation that

$$\|\mathcal{F}_2(\chi_{B_R(0)}f) - \mathcal{F}_2 f\|_{L^2} = \|\chi_{B_R(0)}f - f\|_{L^2} \longrightarrow 0 \quad \text{as } R \to \infty$$

shows that $\mathcal{F}_2(\chi_{B_R(0)}f) \to \mathcal{F}_2 f$. Therefore, part (b) will follow from (a), since $\chi_{B_R(0)}f \in L^1$.

Now suppose that $f \in L^1 \cap L^2$ and take $f_j^R \in C_0^\infty(B_{R+1}(0)) \subset \mathcal{S}$ such that $f_j^R \to \chi_{B_R(0)}f$ in L^2 as $j \to \infty$; this implies also convergence in L^1. Notice that $\mathcal{F}_2 f_j^R \to \mathcal{F}_2(\chi_{B_R(0)}f)$ in L^2 and $\mathcal{F}_1 f_j^R \to \mathcal{F}_1(\chi_{B_R(0)}f)$ in L^∞. But by definition of $\mathcal{F}_2$ on $\mathcal{S}$, $\mathcal{F}_2 f_j^R = \mathcal{F}_1 f_j^R$, so also $\mathcal{F}_2(\chi_{B_R(0)}f) = \mathcal{F}_1(\chi_{B_R(0)}f)$. Letting $R \to \infty$, yields the advertised results. □

6.5 THE $\mathcal{S}'$ THEORY

The Fourier transform cannot be defined on all distributions, but it can be defined on a subset $\mathcal{S}'$ of $\mathcal{D}'$. Here, $\mathcal{S}'$ is the dual of $\mathcal{S}$. Before attempting the definition of the Fourier transform on $\mathcal{S}'$, a preliminary study of $\mathcal{S}$ and $\mathcal{S}'$ is helpful. The first result says that the "identity" map i that "includes" $\mathcal{D}$ within $\mathcal{S}$ is continuous. We say that i induces a *continuous embedding* of $\mathcal{D}$ into $\mathcal{S}$ and connote this fact notationally by $\mathcal{D} \hookrightarrow \mathcal{S}$.

Proposition 6.27. *The inclusion map* $i : \mathcal{D} \to \mathcal{S}$ *is (sequentially) continuous (i.e.,* $\mathcal{D} \hookrightarrow \mathcal{S}$). *Moreover,* $\mathcal{D}$ *is dense in* $\mathcal{S}$.

Proof. Suppose that $\{\phi_j\}_{j=1}^{\infty} \subset \mathcal{D}$ and $\phi_j \to \phi$ in $\mathcal{D}$. Then there is a compact set K such that the supports of all ϕ_j and ϕ all lie in K, and $\|D^\alpha(\phi_j - \phi)\|_{L^\infty} \to 0$ for every multi-index α. But this immediately implies that in $\mathcal{S}$,

$$\rho_n(i(\phi_j) - i(\phi)) = \sup_{|\alpha| \le n} \sup_{x \in K} (1 + |x|^2)^{n/2} |D^\alpha(\phi_j(x) - \phi(x))|$$
$$\le \Big(\sup_{x \in K} (1 + |x|^2)^{n/2} \Big) \sup_{|\alpha| \le n} \|D^\alpha(\phi_j - \phi)\|_{L^\infty} \longrightarrow 0,$$

since K is bounded, which shows that $i(\phi_j) \to i(\phi)$ in $\mathcal{S}$.

Let $f \in \mathcal{S}$ and $\phi \in \mathcal{D}$ be such that $\phi \equiv 1$ on $B_1(0)$. For $\epsilon > 0$, set

$$f_\epsilon(x) = \phi(\epsilon x) f(x) \in \mathcal{D}.$$

We claim that $f_\epsilon \xrightarrow{\mathcal{S}} f$, so showing that $\mathcal{D}$ is dense in $\mathcal{S}$. It is required to establish that for any multi-indices α and β,

$$\|x^\alpha D^\beta(f - f_\epsilon)\|_{L^\infty} \longrightarrow 0 \quad \text{as } \epsilon \to 0^+.$$

Now $f_\epsilon(x) = f(x)$ for $|x| < 1/\epsilon$, so consider $|x| \ge 1/\epsilon$. By the Leibniz rule,

$$|x^\alpha D^\beta(f - f_\epsilon)| = \Big| x^\alpha \sum_{\gamma \le \beta} \binom{\beta}{\gamma} D^{\beta-\gamma} f \, D^\gamma (1 - \phi(\epsilon x)) \Big|$$
$$\le \sum_{\gamma \le \beta} \binom{\beta}{\gamma} \|x^{\alpha+\delta} D^{\beta-\gamma} f\|_{L^\infty} \|D^\gamma(1 - \phi(\epsilon x))\|_{L^\infty} \epsilon^{|\delta|}$$

for any multi-index δ. For ϵ small enough, the quantity on the right-hand side can be made as small as we please; the result follows. $\quad\square$

Corollary 6.28. *If* $\phi_j \xrightarrow{\mathcal{D}} \phi$, *then* $\phi_j \xrightarrow{\mathcal{S}} \phi$.

Proof. This follows since i is continuous. $\quad\square$

Definition. The dual of $\mathcal{S} = \mathcal{S}(\mathbb{R}^d)$ is the space of continuous linear functionals on $\mathcal{S}$. Continuity here means weak-$*$ continuity. That is, if $\{u_j\}_{j=1}^{\infty} \subset \mathcal{S}'$, then $u_j \to u$ in $\mathcal{S}'$ if $u_j(\phi) \to u(\phi)$ for all $\phi \in \mathcal{S}$. This dual space is denoted $\mathcal{S}' = \mathcal{S}'(\mathbb{R}^d)$ and called the space of *tempered distributions*.

Proposition 6.29. *Every tempered distribution $u \in \mathcal{S}'$ can be identified naturally with a unique distribution $v \in \mathcal{D}'$ by the relation*

$$v = u \circ i = u|_{\mathcal{D}} = i' \circ u.$$

(This is simply the assertion that the dual operator $i' : \mathcal{S}' \hookrightarrow \mathcal{D}'$ is the operator that restricts the domain from $\mathcal{S}$ to $\mathcal{D}$.) The mapping i' is one-to-one.

Proof. If we define $v = u \circ i$, then $v \in \mathcal{D}'$, since i is continuous and linear. If $u, w \in \mathcal{S}'$ and $u \circ i = w \circ i$, then in fact $u = w$ since $\mathcal{D}$ is dense in $\mathcal{S}$. $\qquad\square$

Corollary 6.30. *The dual space $\mathcal{S}'$ is precisely the vector subspace of $\mathcal{D}'$ consisting of those functionals that have continuous extensions from $\mathcal{D}$ to $\mathcal{S}$. Moreover, these extensions are unique.*

Example. If α is any multi-index, then we claim that

$$D^\alpha \delta_0 \in \mathcal{S}'.$$

The fact that $D^\alpha \delta_0$ is continuous is easily discerned. Let $\psi \in \mathcal{D}$ be identically one in a neighborhood of 0. Then for $\phi \in \mathcal{S}$,

$$D^\alpha \delta_0(\psi\phi) = (-1)^{|\alpha|} D^\alpha \phi(0)$$

is well-defined (i.e., independent of the choice of ψ), so $D^\alpha \delta_0 : \mathcal{S} \to \mathbb{F}$ is the composition of multiplication by ψ (taking $\mathcal{S}$ to $\mathcal{D}$) and $D^\alpha \delta_0 : \mathcal{D} \to \mathbb{F}$. The latter is continuous. For the former, if $\phi_j \xrightarrow{\mathcal{S}} \phi$, then each $\psi\phi_j$ is supported in supp(ψ) and $D^\beta(\psi\phi_j) \xrightarrow{L^\infty} D^\beta(\psi\phi)$ for all β. Thus $\psi\phi_j \xrightarrow{\mathcal{D}} \psi\phi$, so multiplication by ψ is a continuous operation.

The elements of $\mathcal{S}'$ are characterized in the next result, which is similar to Theorem 5.6 for $\mathcal{D}'$.

Theorem 6.31. *Let u be a linear functional on $\mathcal{S}$. Then $u \in \mathcal{S}'$ if and only if there is a constant $C > 0$ and an integer $N \geq 0$ such that*

$$|u(\phi)| \leq C\rho_N(\phi) \quad \forall \phi \in \mathcal{S}.$$

Proof. By linearity, u is continuous if and only if it is continuous at 0. If $\{\phi_j\}_{j=1}^\infty \subset \mathcal{S}$ converges to 0 and we assume the existence of $C > 0$ and $N \geq 0$ as in the statement of the theorem, then

$$|u(\phi_j)| \leq C\rho_N(\phi_j) \longrightarrow 0,$$

and so u is continuous.

Conversely, suppose that no such $C > 0$ and $N \geq 0$ exist. Then for each $j > 0$, there is a $\psi_j \in \mathcal{S}$ such that $\rho_j(\psi_j) = 1$ and

$$|u(\psi_j)| \geq j.$$

Let $\phi_j = \psi_j/j$, so that $\phi_j \to 0$ in $\mathcal{S}$ (since the ρ_n are increasing with n, for any fixed n and all $j \geq n$, $\rho_n(\phi_j) \leq \rho_j(\phi_j) \leq 1/j$). But then, u continuous implies that $|u(\phi_j)| \to 0$, which contradicts the previous fact that $|u(\phi_j)| = |u(\psi_j)|/j \geq 1$. □

Example (Tempered L^p). If for some $N > 0$ and $1 \leq p \leq \infty$,

$$\frac{f(x)}{(1 + |x|^2)^{N/2}} \in L^p(\mathbb{R}^d),$$

then we say that $f(x)$ is a *tempered L^p function* (if $p = \infty$, we also say that f is *slowly increasing*). Define $\Lambda_f \in \mathcal{S}'$ by

$$\Lambda_f(\phi) = \int f(x)\phi(x)\,dx, \quad \forall \phi \in \mathcal{S}. \tag{6.9}$$

This is well-defined since by Hölder's inequality,

$$
\begin{aligned}
|\Lambda_f(\phi)| &= \left| \int \frac{f(x)}{(1+|x|^2)^{N/2}}(1+|x|^2)^{N/2}\phi(x)\,dx \right| \\
&\leq \left\| \frac{f(x)}{(1+|x|^2)^{N/2}} \right\|_{L^p} \|(1+|x|^2)^{N/2}\phi(x)\|_{L^q},
\end{aligned}
$$

where q is the conjugate exponent with $1/p + 1/q = 1$. The right-hand side is clearly finite if $q = \infty$ (i.e., $p = 1$), while for $q < \infty$,

$$
\begin{aligned}
\|(1+|x|^2)^{N/2}\phi\|_{L^q}^q &= \int (1+|x|^2)^{Nq/2}|\phi(x)|^q\,dx \\
&= \int (1+|x|^2)^{Nq/2-M}(1+|x|^2)^M|\phi(x)|^q\,dx \\
&\leq \left(\int (1+|x|^2)^{Nq/2-M}\,dx \right) \|(1+|x|^2)^{M/q}\phi\|_{L^\infty}^q \\
&\leq \left(C\rho_{2M/q}(\phi) \right)^q
\end{aligned}
$$

is finite provided M is large enough. By the previous theorem, Λ_f is also continuous, so indeed $\Lambda_f \in \mathcal{S}'$. As with distributions, we simply say that $f \in \mathcal{S}'$, with the tacit understanding that the action of f is given by Λ_f. Since each of the following spaces is in tempered L^p for some p, we have shown:

(a) $L^p(\mathbb{R}^d) \subset \mathcal{S}'$ for all $1 \leq p \leq \infty$;

(b) $\mathcal{S} \subset \mathcal{S}'$;

(c) a polynomial, and more generally any measurable function majorized by a polynomial, is a tempered distribution.

Example. Not every function in $L^1_{\text{loc}}(\mathbb{R}^d)$ is in $\mathcal{S}'$. The reader can readily verify that $e^{|x|^2} \notin \mathcal{S}'$ by considering $\phi \in \mathcal{S}$ with asymptotic behavior like $e^{-|x|}$ for large values of its argument.

Proposition 6.32. *For any $1 \leq p \leq \infty$, $L^p \hookrightarrow S'$ (L^p is continuously embedded in S').*

Proof. We need to show that if $f_j \xrightarrow{L^p} f$, then

$$\int (f_j - f)\phi \, dx \longrightarrow 0 \quad \forall \phi \in S,$$

which is true by Hölder's inequality. □

As with distributions, operations on tempered distributions may be defined by duality. If $T : S \to S$ is continuous, and linear, then so is $T' : S' \to S'$. Since $\mathcal{F} : S \to S$ is continuous and linear, the Fourier transform on S' may be defined in this way.

Proposition 6.33. *If α is a multi-index, $x \in \mathbb{R}^d$, and $f \in C^\infty(\mathbb{R}^d)$ is such that $D^\beta f$ grows at most polynomially for all β, then for $u \in S'$ and all $\phi \in S$, the following hold:*

(a) $\langle D^\alpha u, \phi \rangle = (-1)^{|\alpha|} \langle u, D^\alpha \phi \rangle$ *defines* $D^\alpha u \in S'$;

(b) $\langle fu, \phi \rangle = \langle u, f\phi \rangle$ *defines* $fu \in S'$;

(c) $\langle \tau_x u, \phi \rangle = \langle u, \tau_{-x}\phi \rangle$ *defines* $\tau_x u \in S'$;

(d) $\langle Ru, \phi \rangle = \langle u, R\phi \rangle$, *where R is reflection about $x = 0$, defines $Ru \in S'$;*

(e) $\langle \hat{u}, \phi \rangle = \langle u, \hat{\phi} \rangle$ *defines* $\hat{u} \in S'$;

(f) $\langle \check{u}, \phi \rangle = \langle u, \check{\phi} \rangle$ *defines* $\check{u} \in S'$.

Moreover, these operations are continuous on S'.

Proof. Result (a) follows from Proposition 6.14. For (b), just note that the map $\phi \mapsto f\phi$ is a continuous linear map from S to itself. The reader can readily verify that $\tau_x : S \to S$ and $R : S \to S$ are continuous and linear, so (c) and (d) follow. The continuity of $\mathcal{F}$ and $\mathcal{F}^{-1}$ gives (e) and (f). □

Convolution between $u \in S'$ and $\phi \in S$ may be defined by the formula

$$(u * \phi)(x) = \langle u, \tau_x R\phi \rangle.$$

Proposition 6.34. *For $u \in S'$ and $\phi \in S$,*

(a) $u * \phi \in C^\infty$ *and*

$$D^\alpha(u * \phi) = (D^\alpha u) * \phi = u * D^\alpha \phi \quad \forall \alpha,$$

(b) $u * \phi \in S'$ *(in fact, $u * \phi$ grows at most polynomially).*

Proof. The proof of (a) is similar to the case of distributions (see Proposition 5.12) and left to the reader. For (b), note that

$$1 + |x + y|^2 \le 2(1 + |x|^2)(1 + |y|^2),$$

so

$$\rho_N(\tau_x \phi) \le 2^{N/2}(1 + |x|^2)^{N/2}\rho_N(\phi).$$

Since $u \in S'$, so there are values $C > 0$ and $N \in \mathbb{N}$ such that

$$|u(\phi)| \le C\rho_N(\phi).$$

In consequence, we see that

$$|u * \phi(x)| = |u(\tau_x R\phi)| \le C2^{N/2}(1 + |x|^2)^{N/2}\rho_N(\phi),$$

so $u * \phi$ grows at most polynomially, and also $u * \phi \in S'$. $\square$

The Fourier transform of tempered distributions is now examined in a little more detail. Recall that a tempered L^p function f is a member of S' when realized as the map Λ_f (see (6.9)). The next result shows that the L^1 and L^2 definitions of the Fourier transform of such a function are consistent with the S' definition.

Proposition 6.35. *If $f \in L^1 \cup L^2$, then $\hat{\Lambda}_f = \Lambda_{\hat{f}}$ and $\check{\Lambda}_f = \Lambda_{\check{f}}$.*

Proof. For $\phi \in S$,

$$\langle \hat{\Lambda}_f, \phi \rangle = \langle \Lambda_f, \hat{\phi} \rangle = \int f\hat{\phi} = \int \hat{f}\phi = \langle \Lambda_{\hat{f}}, \phi \rangle,$$

so $\hat{\Lambda}_f = \Lambda_{\hat{f}}$. A similar computation gives the result for the inverse Fourier transform. $\square$

Proposition 6.36. *If $u \in S'$ and R is the reflection operator, then*

(a) $\check{\hat{u}} = u,$

(b) $\hat{\check{u}} = u,$

(c) $\hat{\check{u}} = Ru,$

(d) $\hat{u} = (Ru)^{\vee} = R\check{u}.$

Proof. These hold by definition, since they hold on S. $\square$

Theorem 6.37 (Plancherel theorem). *The Fourier transform is a continuous, linear, one-to-one mapping of S' onto S', of period 4 (in fact, $\mathcal{F}^2 = R$), with a continuous inverse.*

Proof. Linearity of the Fourier transform is clear, and we already know that $\mathcal{F} : \mathcal{S}' \to \mathcal{S}'$. If $u_j \xrightarrow{\mathcal{S}'} u$ (i.e., $\langle u_j, \phi \rangle \to \langle u, \phi \rangle$ for all $\phi \in \mathcal{S}$), then

$$\langle \hat{u}_j, \phi \rangle = \langle u_j, \hat{\phi} \rangle \longrightarrow \langle u, \hat{\phi} \rangle = \langle \hat{u}, \phi \rangle,$$

so $\hat{u}_j \to \hat{u}$, which is to say, the Fourier transform is continuous on $\mathcal{S}'$. Now

$$\mathcal{F}^2 u = \hat{\hat{u}} = Ru,$$

so

$$\mathcal{F}^4 u = R^2 u = u = \mathcal{F}(\mathcal{F}^3)u = (\mathcal{F}^3)\mathcal{F}u$$

shows that $\mathcal{F}$ has period 4 and has a continuous inverse $\mathcal{F}^{-1} = \mathcal{F}^3$. □

Example. Consider $\delta_0 \in \mathcal{S}'$. For $\phi \in \mathcal{S}$,

$$\langle \hat{\delta}_0, \phi \rangle = \langle \delta_0, \hat{\phi} \rangle = \hat{\phi}(0) = (2\pi)^{-d/2} \int \phi(x) \, dx = \langle (2\pi)^{-d/2}, \phi \rangle,$$

so

$$\hat{\delta}_0 = (2\pi)^{-d/2}$$

is a constant function. Conversely, by Proposition 6.36(d),

$$\delta_0 = \mathcal{F}^{-1}(2\pi)^{-d/2} = \mathcal{F}(2\pi)^{-d/2},$$

so

$$\hat{1} = (2\pi)^{d/2} \delta_0.$$

Proposition 6.38. *If $u \in \mathcal{S}'$, $y \in \mathbb{R}^d$, and α is a multi-index, then*

(a) $(\tau_y u)^{\wedge}(\xi) = e^{-iy \cdot \xi} \hat{u}(\xi)$,

(b) $\tau_y \hat{u}(\xi) = (e^{iy \cdot x} u)^{\wedge}(\xi)$,

(c) $(D^\alpha u)^{\wedge}(\xi) = (i\xi)^\alpha \hat{u}(\xi)$,

(d) $D^\alpha \hat{u}(\xi) = ((-ix)^\alpha u)^{\wedge}(\xi)$,

wherein x is a dummy placeholder variable.

Proposition 6.33(b) implies that the products in (a)–(d) involving tempered distributions are well-defined in $\mathcal{S}'$.

Proof. For (a), consider $\phi \in \mathcal{S}$ and calculate thusly:

$$\langle (\tau_y u)^{\wedge}, \phi \rangle = \langle \tau_y u, \hat{\phi} \rangle = \langle u, \tau_{-y} \hat{\phi} \rangle = \langle u, \widehat{e^{-iy \cdot \xi} \phi} \rangle = \langle \hat{u}, e^{-iy \cdot \xi} \phi \rangle = \langle e^{-iy \cdot \xi} \hat{u}, \phi \rangle.$$

Results (b)–(d) are shown similarly. □

Proposition 6.39. *If $u \in \mathcal{S}'$ and $\phi, \psi \in \mathcal{S}$, then*

(a) $(u * \phi)^\wedge = (2\pi)^{d/2}\hat{\phi}\hat{u}$,

(b) $\hat{u} * \hat{\phi} = (2\pi)^{d/2}(\phi u)^\wedge$,

(c) $(u * \phi) * \psi = u * (\phi * \psi)$.

Proof. Note first that Proposition 6.33(b), Proposition 6.34(b), and Theorem 6.20 imply that $\hat{\phi}\hat{u} \in \mathcal{S}$, $u * \phi \in \mathcal{S}'$, and $\phi * \psi \in \mathcal{S}$, so the expressions in (a) and (c) are well-defined; similarly for (b). Let $\psi \in \mathcal{S}$ and choose $\psi_j \in \mathcal{D}$ with support in K_j such that $\psi_j \xrightarrow{\mathcal{S}} \psi$ (so that $\check{\psi}_j \xrightarrow{\mathcal{S}} \check{\psi}$). Notice that

$$\langle (u * \phi)^\wedge, \check{\psi}_j \rangle = \langle u * \phi, \psi_j \rangle = \int u * \phi(x)\psi_j(x)\, dx,$$

since $u * \phi \in C^\infty$ and has polynomial growth. Continuing, the right-hand side of the last display is

$$\int_{K_j} \langle u, \tau_x R\phi \rangle \psi_j(x)\, dx = \left\langle u, \int_{K_j} \tau_x R\phi\psi_j(x)\, dx \right\rangle,$$

which we see by approximating the integral by Riemann sums and using the linearity and continuity of u. Calculating further reveals that

$$\left\langle u, \int \phi(x - y)\psi_j(x)\, dx \right\rangle = \langle u, R\phi * \psi_j \rangle$$
$$= \langle \hat{u}, (R\phi * \psi_j)^\vee \rangle$$
$$= (2\pi)^{d/2}\langle \hat{u}, (R\phi)^\vee \check{\psi}_j \rangle$$
$$= (2\pi)^{d/2}\langle \hat{\phi}\hat{u}, \check{\psi}_j \rangle$$
$$\longrightarrow (2\pi)^{d/2}\langle \hat{\phi}\hat{u}, \check{\psi} \rangle.$$

That is, for all $\check{\psi} \in \mathcal{S}$,

$$\langle (u * \phi)^\wedge, \check{\psi} \rangle = \langle (2\pi)^{d/2}\hat{\phi}\hat{u}, \check{\psi} \rangle,$$

and (a) follows. Using the fact that $\mathcal{F}^2 = R$ and Proposition 6.36(d), (b) follows from (a).

Finally, (c) follows from (a) through the calculations

$$((u * \phi) * \psi)^\wedge = (2\pi)^{d/2}\hat{\psi}(u * \phi)^\wedge = (2\pi)^d \hat{\psi}\hat{\phi}\hat{u}$$

and

$$(u * (\phi * \psi))^\wedge = (2\pi)^{d/2}(\phi * \psi)^\wedge \hat{u} = (2\pi)^d \hat{\phi}\hat{\psi}\hat{u}.$$

Thus it transpires that

$$((u * \phi) * \psi)^\wedge = (u * (\phi * \psi))^\wedge,$$

and taking the inverse Fourier transform establishes (c). $\qquad\square$

6.6   SOME APPLICATIONS

The Fourier transform is an important tool in many applications. As examples, applications to two important differential equations and to signal processing are developed next.

6.6.1   The heat equation

The *heat operator* is

$$\frac{\partial}{\partial t} - \Delta.$$

The heat equation with an initial condition is an initial-value problem (IVP),

$$\begin{cases} \dfrac{\partial u}{\partial t} - \Delta u = 0, & (x,t) \in \mathbb{R}^d \times (0,\infty), \\ u(x,0) = f(x), & x \in \mathbb{R}^d, \end{cases}$$

associated with the heat operator, and it will be considered here. The initial value f is taken to be given. In Fourier's work on heat conduction, the function $u = u(x,t)$ is interpreted as the temperature of the medium at the point x at time t. Thus, f is the initial temperature distribution, which is assumed known.

To find a solution, we proceed *formally* (i.e., without rigor). Assume that the solution is at least a tempered distribution. Take the Fourier transform in x only, for each fixed t, to obtain that

$$\begin{cases} \widehat{\dfrac{\partial u}{\partial t}} - \widehat{\Delta u} = \dfrac{\partial}{\partial t}\hat{u} + |\xi|^2\hat{u} = 0, \\ \hat{u}(\xi,0) = \hat{f}(\xi). \end{cases}$$

(At this stage in the computations, it is unclear if transposing the Fourier transform and partial differentiation with respect to t holds true, but we assume so in formal terms.) For each fixed $\xi \in \mathbb{R}^d$, this is a first-order, ordinary differential equation with an initial condition. Its solution is

$$\hat{u}(\xi,t) = \hat{f}(\xi)e^{-|\xi|^2 t}.$$

Thus, $u(x,t) = \left(\hat{f}e^{-|\xi|^2 t}\right)^{\vee}$. Now

$$e^{-|\xi|^2 t} = e^{-|\xi\,(2t)^{1/2}|^2/2} = \left((2t)^{-d/2}e^{-|x|^2/4t}\right)^{\wedge},$$

by Proposition 6.3(c), so Lemma 6.18 and Proposition 6.39 imply that

$$u(x,t) = \left[\hat{f}\left(\frac{1}{(2t)^{d/2}}e^{-|x|^2/4t}\right)^{\wedge}\right]^{\vee} = (2\pi)^{-d/2}f * \left(\frac{1}{(2t)^{d/2}}e^{-|x|^2/4t}\right).$$

The function

$$K(x,t) = \frac{1}{(4\pi t)^{d/2}}e^{-|x|^2/4t}$$

is called the *Gaussian*, or *heat kernel*. The formal calculation above indicates that

$$u(x,t) = (f * K(\cdot, t))(x)$$

should be a solution to our IVP.

This model of heat flow has many interesting properties. First, the solution is *self-similar*, meaning that it is given by scaling a function of a single variable. To see this, just remark that

$$K(x,t) = t^{-d/2} K(t^{-1/2}x, 1) = t^{-d/2} \frac{1}{(4\pi)^{d/2}} e^{-|t^{-1/2}x|^2/4}.$$

A moment's contemplation of this formula reveals that K approximates δ_0 as $t \to 0$, and so at least in $\mathcal{S}'$, $u(x,t) \to f * \delta_0(x) = f(x)$ as $t \to 0^+$. This indicates that K itself should solve the IVP with $f = \delta_0$. Hence, it is a type of fundamental solution. It describes the diffusion in time of an initial unit amount of heat energy concentrated at the origin. For positive times, K controls how the initial unit of energy (initial heat distribution) dissipates, or *diffuses,* with time. At time $t > 0$, the maximal temperature is $(4\pi t)^{-d/2}$, and it decreases with time. The self-similar structure predicts that the distance through the origin between points of half this value, say, will increase on the order of $\sqrt{t}$; that is, the heat front will spread at a rate proportional to $\sqrt{t}$. Second, even though $K(x,t) \to 0$ as $t \to \infty$ for each fixed $x \neq 0$, our model *conserves* heat energy, since for all time $t \geq 0$,

$$\int K(x,t)\,dx = \hat{K}(0,t) = 1.$$

Finally, the model predicts an infinite speed of propagation of information, since $K(x,t) > 0$ for all x whenever $t > 0$. This property is perhaps unsatisfactory from the physical point of view.

To remove the formality of the above calculation, start with $K(x,t)$ as defined, and note that for $f \in \mathcal{D}$, $t > 0$, and $u = f * K$,

$$u_t - \Delta u = f * (K_t - \Delta K) = f * 0 = 0,$$

which follows from previous theory. To extend to $f \in L^p$, simply use the fact that $\mathcal{D}$ is dense in L^p (see [6, p. 190] for details).

6.6.2   The Schrödinger equation

In quantum mechanics theory, certain systems are governed by the time-dependent *Schrödinger equation,*

$$i\hbar \frac{\partial u}{\partial t} - Hu = 0,$$

where $\hbar$ is the fundamental Planck's constant divided by 2π, H is a given self-adjoint operator (the Hamiltonian operator), and u is the unknown wave function of the system. One interprets $|u|^2$ as the probability density of the system.

Rescale time to dispense with $\hbar$ and consider the simple Schrödinger IVP

$$\begin{cases} i\dfrac{\partial u}{\partial t} + \Delta u = 0, & (x,t) \in \mathbb{R}^d \times (0, \infty), \\ u(x,0) = f(x), & x \in \mathbb{R}^d, \end{cases}$$

where $f(x)$ is again assumed to be known. It resembles the heat equation except for the imaginary coefficient and the fact that solutions are complex valued. The Fourier transform implies that

$$i\widehat{u}_t - |\xi|^2 \widehat{u} = 0, \tag{6.10}$$

so that, after multiplying by $\bar{\widehat{u}}$, there obtains the equation

$$i\widehat{u}_t \bar{\widehat{u}} - |\xi|^2 |\widehat{u}|^2 = 0.$$

Since we expect that $\widehat{u}_t = \widehat{u}_t$ and by considering real and imaginary parts of $\widehat{u}$, one tentatively concludes that the real part of $\widehat{u}_t \bar{\widehat{u}}$ is $\frac{1}{2}|\widehat{u}|^2_t$. Separating real and imaginary parts of the equation shows that

$$\tfrac{1}{2}|\widehat{u}|^2_t = 0.$$

This implies that for all t,

$$|\widehat{u}(\xi,t)|^2 = |\widehat{f}(\xi)|^2,$$

which follows by invoking the initial condition. Integrating in space and applying the Plancherel theorem leads to the conservation principle

$$\|u(\cdot,t)\|_{L^2(\mathbb{R}^d)} = \|f\|_{L^2(\mathbb{R}^d)}.$$

This is interpreted as saying that the total probability remains constant (equal to the value one, if f is properly normalized).

Moreover, solving (6.10) yields

$$\widehat{u}_t = -i|\xi|^2 \widehat{u}.$$

As for the heat equation, it is deduced that

$$u(x,t) = f * K(x,t),$$

where the *Schrödinger kernel* is

$$K(x,t) = (2\pi)^{-d/2}\big((e^{-i|\xi|^2 t})^\vee\big)(x) = (4i\pi t)^{-d/2} e^{i|x|^2/4t}.$$

This inverse Fourier transform can be computed as follows. For $\epsilon \geq 0$, let

$$g_\epsilon(\xi) = e^{-(it+\epsilon)|\xi|^2}.$$

For $\epsilon > 0$, this is a well-behaved function. The dominated convergence theorem implies that $g_\epsilon \to g_0$ in $\mathcal{S}'$ as $\epsilon \to 0^+$, and so $\check{g}_\epsilon \to \check{g}_0$ in $\mathcal{S}'$ as well. Now, completing the square and using a change of variable, one finds that

$$
(2\pi)^{d/2}\check{g}_\epsilon(x) = \int e^{-(it+\epsilon)|\xi|^2 + ix\cdot\xi} \, d\xi
$$

$$
= \int e^{-(it+\epsilon)[|\xi - ix/2(it+\epsilon)|^2 + |x|^2/4(it+\epsilon)^2]} \, d\xi
$$

$$
= e^{-|x|^2/4(it+\epsilon)} \int e^{-(it+\epsilon)|\eta|^2} \, d\eta
$$

$$
= e^{-|x|^2/4(it+\epsilon)} \prod_{j=1}^{d} \int_{-\infty}^{\infty} e^{-(it+\epsilon)\eta_j^2} \, d\eta_j.
$$

A change of the contour of integration leads to

$$
(2\pi)^{d/2}\check{g}_\epsilon(x) = e^{-|x|^2/4(it+\epsilon)} \prod_{j=1}^{d} \int_{-\infty}^{\infty} \frac{e^{-z_j^2}}{(it+\epsilon)^{1/2}} \, dz_j
$$

$$
= \left(\frac{\pi}{it+\epsilon}\right)^{d/2} e^{-|x|^2/4(it+\epsilon)},
$$

wherein we take the branch of the square root giving positive real part. The formula for K now follows upon taking the limit $\epsilon \to 0^+$.

6.6.3 Signal processing and translation invariance

For the present purposes, a *signal* will be defined for all time $t \in \mathbb{R}$ as a function $f : \mathbb{R} \to \mathbb{C}$. A *filter* is an operator T that maps signals to signals. Since the set of signals is a vector space, it makes sense to discuss *linear filters*, which are simply linear operators on the class of signals. In certain commonly occurring circumstances, electronic or other devices that implement a filter on an electrical signal have the property that they behave the same tomorrow as they do today. We therefore define a *translation invariant filter* as a linear filter that commutes with translation, i.e.,

$$
\tau_a T = T \tau_a \quad \forall a \in \mathbb{R}.
$$

Some of the simplest signals are the plane waves $e^{i\xi t}$, ξ fixed, $t \in \mathbb{R}$, that we have been studying in this chapter.

Proposition 6.40. *If T is a linear, translation invariant filter and $\xi \in \mathbb{R}$, then there exists a function $\psi : \mathbb{R} \to \mathbb{C}$ such that*

$$
T(e^{i\xi t}) = \psi(\xi)e^{i\xi t}.
$$

In other words, the plane wave is an eigenfunction of T with eigenvalue $\psi(\xi)$. The filter T transforms the wave into itself with a possibly different amplitude that depends on ξ.

Proof. Let $\psi_\xi(t) = T(e^{i\xi t})$ and compute, for any $a \in \mathbb{R}$,

$$\psi_\xi(t - a) = \tau_a \psi_\xi(t) = \tau_a T(e^{i\xi t}) = T(\tau_a e^{i\xi t})$$
$$= T(e^{i\xi(t-a)}) = T(e^{-i\xi a} e^{i\xi t}) = e^{-i\xi a} T(e^{i\xi t}).$$

Thus $T(e^{i\xi t}) = \psi_\xi(t - a)e^{i\xi a}$ for any t and a. Take $a = t$ to obtain the desired result with $\psi(\xi) = \psi_\xi(0)$. $\qquad\square$

It is sometimes convenient to restrict the signals of interest to lie within a Lebesgue space or the Schwartz space. Of course, the plane wave is not in most of these spaces, so some care will be needed. For generality, we also extend the domain to $\mathbb{R}^d$.

Lemma 6.41. *Suppose that $T : L^p(\mathbb{R}^d) \to L^q(\mathbb{R}^d)$, $1 \le p, q \le \infty$, is a continuous, linear, translation invariant operator. Then for any $\phi \in \mathcal{S}(\mathbb{R}^d)$, $T D^\alpha \phi = D^\alpha T \phi \in L^q(\mathbb{R}^d)$ for all multi-indices α.*

Proof. Let $\phi \in \mathcal{S}$ and let $\mathbf{e}_i$ be the standard unit vector in the ith direction. Then

$$\lim_{h \to 0} \frac{1}{h}(\phi - \tau_{h\mathbf{e}_i}\phi) = \frac{\partial \phi}{\partial x_i} \quad \text{in } \mathcal{S},$$

and thus also in L^p and L^q by Proposition 6.15. Since T is continuous and commutes with translation, it is also the case that

$$T\left(\frac{\partial \phi}{\partial x_i}\right) = T\left(\lim_{h \to 0} \frac{1}{h}(\phi - \tau_{h\mathbf{e}_i}\phi)\right) = \lim_{h \to 0} \frac{1}{h}(T\phi - \tau_{h\mathbf{e}_i}T\phi) = \frac{\partial T\phi}{\partial x_i} \in L^q.$$

Iterating the one derivative case yields the general result. $\qquad\square$

Theorem 6.42. *Suppose that $T : L^p(\mathbb{R}^d) \to L^q(\mathbb{R}^d)$, $1 \le p, q \le \infty$, is a continuous linear translation invariant operator. Then there is a unique $u \in \mathcal{S}'(\mathbb{R}^d)$ such that T restricted to $\mathcal{S}(\mathbb{R}^d)$ satisfies*

$$T\phi = u * \phi \quad \forall \phi \in \mathcal{S}(\mathbb{R}^d).$$

Proof. By the lemma, if $\phi \in \mathcal{S}$, then $D^\alpha T\phi \in L^q$ for all multi-indices α. We invoke a special case of the Sobolev embedding theorem (Theorem 7.22 in Chapter 7) which enables us to conclude that $T\phi$ is almost everywhere equal to a continuous function g_ϕ, and that there is a constant C such that for any $y \in \mathbb{R}^d$,

$$|g_\phi(y)| \le C \sum_{|\alpha| \le d+1} \|D^\alpha T\phi\|_{L^q} = C \sum_{|\alpha| \le d+1} \|T D^\alpha \phi\|_{L^q}$$
$$\le C\|T\| \sum_{|\alpha| \le d+1} \|D^\alpha \phi\|_{L^q} < \infty.$$

This implies by Theorem 6.31 that the linear functional $\phi \mapsto g_\phi(y)$ is continuous on $\mathcal{S}$ for any y. Fix a y where $g_\phi(y) = T\phi(y)$ and call this functional $u_1 \in \mathcal{S}'$; that is,

$$u_1(\phi) = g_\phi(y) = T\phi(y).$$

Let $u = \tau_y R u_1 \in \mathcal{S}'$, where R is reflection. Then, simply calculate as follows:

$$u * \phi(x) = \langle u, \tau_x R\phi \rangle = \langle \tau_y R u_1, \tau_x R\phi \rangle = \langle u_1, R\tau_{x-y} R\phi \rangle$$
$$= \langle u_1, \tau_{y-x}\phi \rangle = T(\tau_{y-x}\phi)(y) = \tau_{y-x}T\phi(y) = T\phi(x).$$

Uniqueness is clear. □

Definition. If $T : L^2(\mathbb{R}^d) \to L^2(\mathbb{R}^d)$ is defined for $m \in L^\infty(\mathbb{R}^d)$ by

$$T\phi = (m\hat{\phi})^\vee,$$

then T is called a *multiplier operator* with *multiplier m*.

It is easy to verify that multiplier operators are linear and translation invariant. It is a consequence of the next theorem that the set of continuous, linear, translation invariant operators on $L^2(\mathbb{R}^d)$ is exactly the set of multiplier operators. By the previous theorem, these are convolution operators when restricted to $\mathcal{S}(\mathbb{R}^d)$.

Theorem 6.43. *Let $u \in \mathcal{S}'$ and define $T : \mathcal{S}(\mathbb{R}^d) \to L^2(\mathbb{R}^d)$ by*

$$T\phi = u * \phi \quad \forall \phi \in \mathcal{S}(\mathbb{R}^d).$$

Then, there is a constant C such that

$$\|T\phi\|_{L^2(\mathbb{R}^d)} \leq C\|\phi\|_{L^2(\mathbb{R}^d)} \quad \forall \phi \in \mathcal{S}(\mathbb{R}^d)$$

if and only if

$$\hat{u} \in L^\infty(\mathbb{R}^d).$$

Moreover, in that case, T can be extended continuously to $L^2(\mathbb{R}^d)$ by the formula

$$Tf = (2\pi)^{d/2}(\hat{u}\hat{f})^\vee \quad \forall f \in L^2(\mathbb{R}^d),$$

and $\|T\| = (2\pi)^{d/2}\|\hat{u}\|_{L^\infty(\mathbb{R}^d)}$. If also $u \in L^1(\mathbb{R}^d)$, then $Tf = u * f$ for all $f \in L^2(\mathbb{R})$.

Proof. Recall that for $\phi \in \mathcal{S}$, $\widehat{T\phi} = (u * \phi)^\wedge = (2\pi)^{d/2}\hat{u}\hat{\phi}$ and $u * \phi \in C^\infty$ can grow at most polynomially. If $\hat{u} \in L^\infty(\mathbb{R}^d)$, the Plancherel theorem gives us that

$$\|T\phi\|_{L^2} = \|\widehat{T\phi}\|_{L^2} = (2\pi)^{d/2}\|\hat{u}\hat{\phi}\|_{L^2} \leq (2\pi)^{d/2}\|\hat{u}\|_{L^\infty}\|\phi\|_{L^2},$$

so providing the converse implication.

Now suppose T is bounded on $\mathcal{S}$ in the L^2-norm. Let $\phi_0 = \hat{\phi}_0 = e^{-|x|^2/2} \in \mathcal{S}$. Then

$$\hat{u}\hat{\phi}_0 = (2\pi)^{-d/2}(u * \phi_0)^\wedge = (2\pi)^{-d/2}\widehat{T\phi_0} \in L^2,$$

consequently,

$$\hat{u} = e^{|x|^2/2}(2\pi)^{-d/2}(u * \phi_0)^\wedge \in L^1_{\text{loc}}.$$

But, we know that

$$(2\pi)^{-d/2}\|T\check{\phi}\|_{L^2} = \|\hat{u}\phi\|_{L^2} \leq C(2\pi)^{-d/2}\|\phi\|_{L^2} \quad \forall \phi \in \mathcal{S},$$

so it is concluded that $\hat{u} \in L^\infty$ since C is independent of ϕ. Moreover, T extends by density to L^2, and $C = (2\pi)^{d/2}\|\hat{u}\|_{L^\infty} = \|T\|$. Finally, if $u \in L^1(\mathbb{R}^d)$ and $f \in L^2(\mathbb{R}^d)$, then $\hat{u} \in L^\infty(\mathbb{R}^d)$ and so $Tf = (2\pi)^{d/2}(\hat{u}\hat{f})^\vee = u * f$. □

Corollary 6.44. *If $T : L^2(\mathbb{R}^d) \to L^2(\mathbb{R}^d)$, then T is a continuous, linear, translation invariant operator if and only if T is a multiplier operator.*

Proof. Suppose that T is a continuous, linear, translation invariant operator. By Theorem 6.42, T restricted to $\mathcal{S}$ is a convolution operator. The previous theorem tells us that it can be extended continuously to a multiplier operator on L^2. Since $\mathcal{S} \subset L^2$ is dense, the continuous extension of $T|_\mathcal{S}$ to L^2 must be the original continuous operator T as defined on L^2. The proof of the converse is straightforward. □

It has now been determined that a translation invariant filter on the set of $L^2(\mathbb{R})$ signals is a multiplier operator. A natural question to ask is whether we can design a filter that will remove high frequencies. For the cut-off frequency $\xi_c > 0$, a natural choice for such a filter T would be associated with the multiplier

$$m(\xi) = \begin{cases} 1 & \text{for } |\xi| \leq \xi_c, \\ 0 & \text{for } |\xi| > \xi_c, \end{cases}$$

and so be defined by $Tf = (m\hat{f})^\vee$. But, direct calculation shows that

$$\frac{1}{\sqrt{2\pi}}\check{m}(t) = \frac{\sin(\xi_c t)}{\pi t},$$

and so, for f nice enough,

$$Tf(t) = \int_{-\infty}^\infty \frac{\sin(\xi_c(t-s))}{\pi(t-s)} f(s)\,ds.$$

If we are considering realistic signals with time going forward, this is problematic, as it implies that the output signal Tf at, say $t = 0$, depends on the input signal for $t > 0$! Even worse, consider the signal $f(t) = \chi_{[0,1]}(t)$, the characteristic function of the unit interval. In this case,

$$Tf(t) = T\chi_{[0,1]}(t) = \frac{1}{\pi}\int_{\xi_c(t-1)}^{\xi_c t} \frac{\sin s}{s}\,ds$$

is nonzero except at isolated points. Therefore, $Tf \neq 0$ for $t < 0$, which means that the output signal occurs before the input signal arrives at $t = 0$.

We conclude that our desired linear filter cannot be constructed for realistic applications; however, it may be implementable for other applications or by nonlinear components.

A *causal filter* is a filter for which the output signal begins only after an input signal is given. It is clear that $Tf = u * f$ is causal if and only if $u = 0$ for $t < 0$.

In practice, a signal may naturally have no large frequencies. We say that a signal is *band limited* if there is $\xi_c > 0$ such that $\hat{f}(\xi) = 0$ for all $|\xi| > \xi_c$.

Theorem 6.45 (Shannon-Whittaker sampling theorem). *Suppose that* $f \in L^2(\mathbb{R})$ *is band limited by* ξ_c *and* $\hat{f}(\xi)$ *is piecewise smooth and continuous. Then* f *is completely determined by its values at* $j\pi/\xi_c$ *for integers* j. *Moreover,*

$$f(t) = \sum_{j=-\infty}^{\infty} f\left(\frac{j\pi}{\xi_c}\right) \frac{\sin(\xi_c t - j\pi)}{\xi_c t - j\pi}, \tag{6.11}$$

and the series converges uniformly.

Proof. Since $\hat{f}$ is supported on $|\xi| \le \xi_c$, its Fourier series there is

$$\hat{f}(\xi) = \sum_{k=-\infty}^{\infty} c_k e^{i\pi k\xi/\xi_c} \chi_{|\xi| \le \xi_c}(\xi),$$

where

$$c_k = \frac{1}{2\xi_c} \int_{-\xi_c}^{\xi_c} \hat{f}(\xi) e^{-i\pi k\xi/\xi_c} \, d\xi.$$

Thus, the coefficient c_k can be interpreted as an inverse Fourier transform $c_k = (\sqrt{2\pi}/2\xi_c)f(-k\pi/\xi_c)$. Taking $j = -k$, this amounts to

$$\hat{f}(\xi) = \sum_{j=-\infty}^{\infty} \frac{\sqrt{2\pi}}{2\xi_c} f\left(\frac{j\pi}{\xi_c}\right) e^{-i\pi j\xi/\xi_c} \chi_{|\xi| \le \xi_c}(\xi),$$

and this is uniformly convergent since $\hat{f}$ is piecewise smooth. Finally, inverting gives

$$f(t) = \sum_{j=-\infty}^{\infty} \frac{\sqrt{2\pi}}{2\xi_c} f\left(\frac{j\pi}{\xi_c}\right) \mathcal{F}^{-1}\left(e^{-i\pi j\xi/\xi_c} \chi_{|\xi| \le \xi_c}(\xi)\right),$$

which is the stated result. $\qquad\qquad\qquad\qquad\qquad\qquad\qquad\qquad\square$

Perceived sound is generally band limited, since the human ear can detect only frequencies between about 10 and 20,000 Hertz. Thus, multiple audio signals can be sent over a single transmission line, as long as each receiver samples the signal only at its appropriate times $j\pi/\xi_c$. The signal can be reconstructed via (6.11). Note that the series dies away for large j, so it can be truncated without appreciably degrading the reconstructed signal.

6.7 EXERCISES

1. Compute the Fourier transform of $e^{-|x|}$ for $x \in \mathbb{R}$.

2. Compute the Fourier transform of $e^{-a|x|^2}$, $a > 0$, directly, where $x \in \mathbb{R}$. You will need to use the Cauchy theorem.

3. If $f \in L^1(\mathbb{R}^d)$ and $f > 0$, show that for every $\xi \neq 0$, $|\hat{f}(\xi)| < \hat{f}(0)$.

4. If $f \in L^1(\mathbb{R}^d)$ and $f(x) = g(|x|)$ for some g, show that $\hat{f}(\xi) = h(|\xi|)$ for some h. Can you relate g and h?

5. Let $1 \leq p < \infty$ and suppose $f \in L^p(\mathbb{R})$. Let $g(x) = \displaystyle\int_x^{x+1} f(y)\, dy$. Prove that $g \in C_v(\mathbb{R})$.

6. Show that the Fourier transform $\mathcal{F} : L^1(\mathbb{R}^d) \to C_v(\mathbb{R}^d)$ is not onto. Show, however, that $\mathcal{F}(L^1(\mathbb{R}^d))$ is dense in $C_v(\mathbb{R}^d)$. [Hint: in one dimension, let f_n be the characteristic function of $[-n, n]$ and consider $f_n * f_1$ (which is in $C_v(\mathbb{R})$ by Exercise 5). Take $n \to \infty$.]

7. Consider the function $f(x) = |x|^{-1/2} e^{-|x|}$. Show that f is in $L^1(\mathbb{R})$ but not in $L^2(\mathbb{R})$. Also, find the Fourier transform of f.

8. Give an example of a function $f \in L^2(\mathbb{R}^d)$ which is not in $L^1(\mathbb{R}^d)$, but such that $\hat{f} \in L^1(\mathbb{R}^d)$. Under what circumstances can this happen?

9. Suppose that $f \in L^p(\mathbb{R}^d)$ for some p between 1 and 2.

   (a) Show that there are $f_1 \in L^1(\mathbb{R}^d)$ and $f_2 \in L^2(\mathbb{R}^d)$ such that $f = f_1 + f_2$.

   (b) Define $\hat{f} = \hat{f}_1 + \hat{f}_2$. Show that this definition is independent of the choice of f_1 and f_2.

10. The Young and Hausdorff-Young Inequalities.

   (a) Prove the following generalization of Young's inequality. For $r \geq 1$, suppose $K(x, y)$ is continuous (measurable would be enough) on $\mathbb{R}^d \times \mathbb{R}^d$ and there is some constant $C > 0$ such that

   $$\int_{\mathbb{R}^d} |K(x, y)|^r \, dx \leq C^r \quad \text{and} \quad \int_{\mathbb{R}^d} |K(x, y)|^r \, dy \leq C^r.$$

   Let the operator T be defined by

   $$Tf(x) = \int_{\mathbb{R}^d} K(x, y)\, f(y)\, dy.$$

   If $p \geq 1$ and $q \geq 1$ satisfy $1/p + 1/r = 1/q + 1$, then $T : L^p(\mathbb{R}^d) \to L^q(\mathbb{R}^d)$ is a bounded linear map with norm $\|T\| \leq C$.

   (b) Use this result to show the following special case of the Hausdorff-Young inequality: the Fourier transform $\mathcal{F} : L^{4/3}(\mathbb{R}^d) \to L^4(\mathbb{R}^d)$. [Hint:

note that $\|\hat{f}\|_{L^4}^2 = \|\hat{f}^2\|_{L^2}$, and $\hat{f}^2$ can be written as the Fourier transform of a convolution.]

11. Let the ground field be $\mathbb{C}$ and define $T : L^2(\mathbb{R}^d) \to L^2(\mathbb{R}^d)$ by

$$Tf(x) = \int e^{-|x-y|^2/2} f(y) \, dy.$$

Use the Fourier transform to show that T is a positive, injective operator, but that T is *not* surjective.

12. Suppose that f and g are in $L^2(\mathbb{R}^d)$. The convolution $f * g$ is in $L^\infty(\mathbb{R}^d)$ but it need not be in L^2. So it may not have a classically defined Fourier transform. Nevertheless, prove that $f * g = (2\pi)^{d/2}(\hat{f}\hat{g})^\vee$ is well-defined, and that the inverse Fourier transform is given by the usual integration formula.

13. Find the four possible eigenvalues of the Fourier transform: $\hat{f} = \lambda f$. For each possible eigenvalue, show that there is at least one eigenfunction. [Hint: when $d = 1$, consider $p(x)e^{-x^2/2}$, where p is a polynomial.]

14. Prove that if $\{f_j\}_{j=1}^\infty \subset \mathcal{S}$ and $f_j \xrightarrow{\mathcal{S}} f$, then for any $1 \leq p \leq \infty$, $f_j \xrightarrow{L^p} f$.

15. Let $\varphi \in \mathcal{S}(\mathbb{R}^d)$, $\hat{\varphi}(0) = (2\pi)^{-d/2}$, and $\varphi_\epsilon(x) = \epsilon^{-d}\varphi(x/\epsilon)$. Prove that $\varphi_\epsilon \to \delta_0$ and $\hat{\varphi}_\epsilon \to (2\pi)^{-d/2}$ as $\epsilon \to 0^+$. In what sense do these convergences take place?

16. Is it possible for there to be a continuous function f defined on $\mathbb{R}^d$ with the following two properties?

    (a) There is no polynomial P in d variables such that $|f(x)| \leq P(x)$ for all $x \in \mathbb{R}^d$.

    (b) The distribution $\phi \mapsto \int f\phi \, dx$ is tempered.

17. When is $\displaystyle\sum_{k=1}^\infty a_k \delta_k \in \mathcal{S}'(\mathbb{R})$? (Here, δ_k is the point mass centered at $x = k$.)

18. For $f \in \mathcal{S}(\mathbb{R})$, define the Hilbert transform of f by $Hf = \mathrm{PV}\left(\dfrac{1}{\pi x}\right) * f$, where the convolution uses ordinary Lebesgue measure.

    (a) Show that $\mathrm{PV}(1/x) \in \mathcal{S}'$.

    (b) Show that $\mathcal{F}(\mathrm{PV}(1/x)) = -i\sqrt{\pi/2}\,\mathrm{sgn}(\xi)$, where $\mathrm{sgn}(\xi)$ is the sign of ξ. [Hint: recall that $x\mathrm{PV}(1/x) = 1$.]

    (c) Show that $\|Hf\|_{L^2} = \|f\|_{L^2}$ and $HHf = -f$, for $f \in \mathcal{S}(\mathbb{R})$.

    (d) Extend H to $L^2(\mathbb{R})$.

19. The gamma function is $\Gamma(s) = \displaystyle\int_0^\infty t^{s-1}e^{-t} \, dt$. Let $\phi \in \mathcal{S}(\mathbb{R}^d)$ and $0 < \alpha < d$.

(a) Show that $|\xi|^{-\alpha} \in L^1_{\text{loc}}(\mathbb{R}^d)$ and $|\xi|^{-\alpha}\hat{\phi} \in L^1(\mathbb{R}^d)$.

(b) Let $c_\alpha = 2^{\alpha/2}\Gamma(\alpha/2)$. Show that

$$\left(|\xi|^{-\alpha}\hat{\phi}\right)^{\vee}(x) = \frac{c_{d-\alpha}}{(2\pi)^{d/2}c_\alpha}\int_{\mathbb{R}^d}|x-y|^{\alpha-d}\phi(y)\,dy.$$

[Hint: first show that

$$c_\alpha|\xi|^{-\alpha} = \int_0^\infty t^{\alpha/2-1}e^{-|\xi|^2 t/2}\,dt,$$

and then recall that $e^{-|\xi|^2 t/2} = t^{-d/2}\left(e^{-|x|^2/2t}\right)^{\wedge}.]$

20. Compute the Fourier transforms of the following functions, considered as tempered distributions:

(a) $f(x) = x^n$ for $x \in \mathbb{R}$ and for integer $n \geq 0$;

(b) $g(x) = e^{-|x|}$ for $x \in \mathbb{R}$;

(c) $h(x) = e^{i|x|^2}$ for $x \in \mathbb{R}^d$;

(d) $\sin x$ and $\cos x$ for $x \in \mathbb{R}$.

21. Give a careful argument that $\mathcal{D}(\mathbb{R}^d)$ is dense in $\mathcal{S}$. Show also that $\mathcal{S}'$ is dense in $\mathcal{D}'$ and that distributions with compact support are dense in $\mathcal{S}'$ (see Exercise 22 in Chapter 5 for the definition of a distribution with compact support).

22. Make an argument that there is no simple way to define the Fourier transform on $\mathcal{D}'$ in the way we have for $\mathcal{S}'$.

23. Let

$$H = \left\{\text{measurable functions } f : \mathbb{R} \to \mathbb{C} \mid f \in \mathcal{S}', \hat{f} \text{ is a measurable}\right.$$

$$\left.\text{function, and } \|f\|_H^2 \equiv \int |\hat{f}(\xi)|^2\, m(\xi)\, d\xi < \infty\right\},$$

where $m(\xi) \geq m_* > 0$ is also measurable (but possibly unbounded). The space H is endowed with the inner-product

$$(f,g)_H \equiv \int \hat{f}(\xi)\,\overline{\hat{g}(\xi)}\, m(\xi)\, d\xi.$$

(a) Show carefully that H is complete (and thus a Hilbert space).

(b) Prove that H is continuously imbedded in $L^2(\mathbb{R})$.

(c) If $m(\xi) \geq \alpha|\xi|^2$, for some $\alpha > 0$, prove that for $f \in H$, the (tempered) distributional derivative $f' \in L^2(\mathbb{R})$.

24. Let $s \in \mathbb{R}$ and

$$H^s(\mathbb{R}^d) = \{f \in L^2(\mathbb{R}^d) : (1 + |\xi|^2)^{s/2} |\hat{f}(\xi)| \in L^2(\mathbb{R}^d)\}.$$

(a) Show that there is some $s_0 \in \mathbb{R}$ such that for all $f \in H^s(\mathbb{R}^d)$, $\hat{f}(\xi) \in L^1(\mathbb{R}^d)$ for $s > s_0$.

(b) Apply the Riemann-Lebesgue lemma to $\hat{f}(\xi)$ to show that for $s > s_0$, there is some continuous function g such that $f = g$ almost everywhere.

25. Use the Fourier transform to find a solution to

$$u - \frac{\partial^2 u}{\partial x_1^2} - \frac{\partial^2 u}{\partial x_2^2} = e^{-x_1^2 - x_2^2}.$$

Can you find a fundamental solution to the differential operator? [Hint: write your answer in terms of a suitable inverse Fourier transform and a convolution.]

26. Consider the partial differential equation

$$\begin{cases} \dfrac{\partial^2 u}{\partial x^2} + \dfrac{\partial^2 u}{\partial y^2} = 0, \quad -\infty < x < \infty, \ 0 < y < \infty, \\ u(x,0) = f(x) \quad \text{and} \quad u(x,y) \to 0 \quad \text{as } x^2 + y^2 \to \infty, \end{cases}$$

for the unknown function $u(x,y)$, where f is a nice function.

(a) Find the Fourier transform of $e^{-|x|}$.

(b) Using your answer from (a), find the Fourier transform of $\dfrac{1}{1 + x^2}$.

(c) Find a function $g(x,y)$ such that

$$u(x,y) = f * g(x,y) = \int f(z)\, g(x - z, y)\, dz.$$

27. Consider the *telegrapher's equation*

$$u_{tt} + 2u_t + u = c^2 u_{xx} \quad \text{for } x \in \mathbb{R} \text{ and } t > 0,$$

where also

$$u(x,0) = f(x) \quad \text{and} \quad u_t(x,0) = g(x)$$

are given in $L^2(\mathbb{R})$.

(a) Use the Fourier transform (in x only) and its inverse to find an explicit representation of the solution.

(b) Justify that your representation is indeed a solution.

(c) Show that the solution can be viewed as a sum of two wave packets, one moving to the right with a given constant speed, and the other moving to the left with the same speed.

28. Use the Fourier transform and its inverse to find a representation of solutions to the *Klein-Gordon equation*

$$u_{tt} - \Delta u + u = 0, \quad x \in \mathbb{R}^d \text{ and } t > 0,$$

where $u(x,0) = f(x)$ and $u_t(x,0) = g(x)$ for given $f, g \in L^2(\mathbb{R}^d)$. Leave your answer in terms of an inverse Fourier transform.

29. Consider the problem $u'''' + u = f$ (4 derivatives). Up to calculating an integral, find a fundamental solution that is real. Using the fundamental solution, find a solution to the original problem.

30. Consider the linear problem

$$\Delta^2 u + u = f(x) \quad \text{for } x \in \mathbb{R}^d,$$

where $\Delta^2 u = \Delta \Delta u$ (Δ^2 is the *biharmonic* operator).

(a) Suppose that $f \in \mathcal{S}'(\mathbb{R}^d)$. Use the Fourier transform to find a solution $u \in \mathcal{S}'$ to the problem. Leave your answer as a multiplier operator, i.e., $u = \left(m(\xi)\hat{f}\right)^{\vee}$ for some $m(\xi)$, but justify each of your steps.

(b) Is your solution unique? Why or why not?

(c) Suppose now that $f \in \mathcal{S}(\mathbb{R}^d)$. Write the solution as a convolution operator (you may leave your answer in terms of an inverse Fourier transform).

(d) Show that $\check{m} \in L^2(\mathbb{R}^d)$ for some range of d, and use this to extend your convolution solution to $f \in L^1(\mathbb{R}^d)$. In that case, in what $L^p(\mathbb{R}^d)$ space is u?

31. Consider $L^2(0, \infty)$ as a real Hilbert space. Let $A : L^2(0, \infty) \to L^2(-\infty, \infty)$ be defined by

$$Au(x) = u(x) + \int_0^\infty w(x - y)\, u(y)\, dy,$$

where $w \in L^1(0, \infty) \cap L^2(0, \infty) \cap C^2(0, \infty)$ is nonnegative, decreasing, convex, and extended to $\mathbb{R}$ as an even function.

(a) Show that A maps into $L^2(-\infty, \infty)$. [Hint: extend u by zero outside $(0, \infty)$ in the integral.]

(b) Show that A is symmetric on $L^2(0, \infty)$.

(c) Show that $\hat{w} \geq 0$.

(d) Use the Fourier transform to show that A is strictly positive definite on $L^2(0, \infty)$.

(e) Use the Fourier transform to solve $Au = f$ for $f \in L^2(-\infty, \infty)$.

(f) Why is the solution unique?

32. Let T be a multiplier operator mapping $L^2(\mathbb{R}^d)$ into itself, so there is a bounded measurable function $m(\xi)$ such that $\widehat{Tf}(\xi) = m(\xi)\hat{f}(\xi)$ for all $f \in L^2(\mathbb{R}^d)$. Show that T commutes with translation and $\|T\| = \|m\|_{L^\infty}$.

33. Let $u \in L^2(\mathbb{R}^d)$ and define the operator $T : L^1(\mathbb{R}^d) \to L^2(\mathbb{R}^d)$ by convolution with u. Show that T is a bounded linear operator, a multiplier operator, and translation invariant.

CHAPTER 7

Sobolev Spaces

In this chapter, we define and study some important families of Banach spaces of measurable functions with distributional derivatives that lie in some Lebesgue space L^p ($1 \leq p \leq \infty$). We include spaces of "fractional order" consisting of functions having smoothness between integral numbers of derivatives, as well as their dual spaces.

Work on what we now call Sobolev spaces started in the 19th century with attempts by the Italian school to understand obtaining solutions of Laplace's equation by way of a minimization problem, the so-called Dirichlet principle. Solving this problem will be treated in the next chapter in fact, but by more modern methods. It was Hilbert in 1900 who first showed that Dirichlet's principle can be used to solve Laplace's equation. The Göttingen school around Hilbert, Richard Courant, Kurt Friedrichs, Franz Rellich, Hermann Weyl, and others, introduced and studied various function classes in their pursuit of what we now call the calculus of variations (Chapter 10), as did the Polish-American mathematician Otton Nikodym and the French mathematician Jean Leray. It would turn out many of these classes were in fact the Sobolev space $W^{1,2}$ to be introduced presently here. (The early theories were all based on L^2-spaces. It was Charles Murray and Nikodym who brought in L^p-spaces.)

From the middle 1930s onward, but before World War II broke out, Sergei Sobolev studied what we now call Sobolev spaces as mathematical objects in their own right. It is hard to overestimate how influential this line of development has been, especially on the theory of partial differential equations.

While such spaces arise in a number of contexts, one fundamental motivation for their study is to understand the trace of a function. Consider a domain $\Omega \subset \mathbb{R}^d$ and its boundary $\partial\Omega$. If $f \in C^0(\bar{\Omega})$, then its trace $f|_{\partial\Omega}$ is well-defined and $f|_{\partial\Omega} \in C^0(\partial\Omega)$. However, if merely $f \in L^2(\Omega)$, then $f|_{\partial\Omega}$ is *not* defined, since $\partial\Omega$ has measure zero in $\mathbb{R}^d$. The "function" f is actually the equivalence class of all functions on Ω that differ on a set of measure zero from any other function in the class; thus, $f|_{\partial\Omega}$ can be chosen arbitrarily from the equivalence class. As part of what we will see, if $f \in L^2(\Omega)$ and $\partial f/\partial x_i \in L^2(\Omega)$ for $i = 1,\ldots,d$, then $f|_{\partial\Omega}$ can be defined uniquely and—a

DOI: 10.1201/9781003492139-7

fact that is not immediately obvious—it has half a derivative when restricted
to the boundary.

7.1 DEFINITIONS AND BASIC PROPERTIES

We begin by defining Sobolev spaces of functions with an integral number of
derivatives.

Definition. Let $\Omega \subset \mathbb{R}^d$ be a domain, $1 \leq p \leq \infty$, and $m \geq 0$ be an integer.
The *Sobolev space* of m derivatives in $L^p(\Omega)$ is

$$W^{m,p}(\Omega) = \{f \in L^p(\Omega) : D^\alpha f \in L^p(\Omega) \text{ for all}$$
$$\text{multi-indices } \alpha \text{ with } |\alpha| \leq m\}.$$

Of course, the elements are equivalence classes of functions that differ only
on a set of measure zero. The derivatives are taken in the sense of distributions.
These spaces are clearly linear vector spaces.

Example. The reader can verify that when Ω is bounded and $0 \in \Omega$, $f(x) = |x|^r \in W^{m,p}(\Omega)$ if and only if $(r - m)p + d > 0$.

Definition. For $f \in W^{m,p}(\Omega)$, the $W^{m,p}(\Omega)$-*norm* is

$$\|f\|_{W^{m,p}(\Omega)} = \left\{ \sum_{|\alpha| \leq m} \|D^\alpha f\|_{L^p(\Omega)}^p \right\}^{1/p} \quad \text{if } p < \infty$$

and

$$\|f\|_{W^{m,\infty}(\Omega)} = \max_{|\alpha| \leq m} \|D^\alpha f\|_{L^\infty(\Omega)} \quad \text{if } p = \infty.$$

Proposition 7.1. *The following hold:*

(a) $\| \cdot \|_{W^{m,p}(\Omega)}$ *is indeed a norm;*

(b) $W^{0,p}(\Omega) = L^p(\Omega)$;

(c) $W^{m,p}(\Omega) \hookrightarrow W^{k,p}(\Omega)$ *for all* $m \geq k \geq 0$ *(i.e.,* $W^{m,p}$ *is continuously embedded in* $W^{k,p}$*);*

(d) $D^\alpha : W^{m,p}(\Omega) \to W^{m-|\alpha|,p}(\Omega)$ *is a continuous linear operator for all multi-indices* α *such that* $|\alpha| \leq m$.

The proof is straightforward and left to the reader as an exercise at the
end of the chapter.

Proposition 7.2. *The space* $W^{m,p}(\Omega)$ *is a Banach space.*

Proof. It only remains to show that $W^{m,p}(\Omega)$ is complete. Let $\{u_j\}_{j=1}^{\infty} \subset W^{m,p}(\Omega)$ be Cauchy. Then $\{D^{\alpha}u_j\}_{j=1}^{\infty}$ is Cauchy in $L^p(\Omega)$ for all $|\alpha| \leq m$, and, $L^p(\Omega)$ being complete, there are functions $u_{\alpha} \in L^p(\Omega)$ such that

$$D^{\alpha}u_j \xrightarrow{L^p} u_{\alpha} \quad \text{as } j \to \infty.$$

If $u = u_0$, the claim is that $D^{\alpha}u = u_{\alpha}$. To see this, let $\phi \in \mathcal{D}(\Omega)$ and note that

$$\langle D^{\alpha}u_j, \phi \rangle \longrightarrow \langle u_{\alpha}, \phi \rangle$$

and

$$\langle D^{\alpha}u_j, \phi \rangle = (-1)^{|\alpha|} \langle u_j, D^{\alpha}\phi \rangle \longrightarrow (-1)^{|\alpha|} \langle u, D^{\alpha}\phi \rangle = \langle D^{\alpha}u, \phi \rangle.$$

Thus $u_{\alpha} = D^{\alpha}u$ as distributions, and so also as $L^p(\Omega)$ functions. We conclude that

$$D^{\alpha}u_j \xrightarrow{L^p} D^{\alpha}u \quad \forall |\alpha| \leq m;$$

which is to say,

$$u_j \xrightarrow{W^{m,p}} u. \qquad \square$$

Certain basic properties of L^p spaces hold for $W^{m,p}$ spaces.

Proposition 7.3. *The space $W^{m,p}(\Omega)$ is separable if $1 \leq p < \infty$ and reflexive if $1 < p < \infty$.*

Proof. We use strongly the same result known for $L^p(\Omega)$, i.e., $m = 0$. Let N denote the number of multi-indices of order less than or equal to m. Let

$$L_N^p = \underbrace{L^p(\Omega) \times \cdots \times L^p(\Omega)}_{N \text{ times}} = \overset{N}{\underset{j=1}{\times}} L^p(\Omega)$$

and define the norm for $u \in L_N^p$ by

$$\|u\|_{L_N^p} = \left\{ \sum_{j=1}^{N} \|u_j\|_{L^p(\Omega)}^p \right\}^{1/p}.$$

It is trivial to verify that L_N^p is a Banach space with properties similar to those of L^p: L_N^p is separable, and reflexive if $p > 1$, since $(L_N^p)^* = L_N^q$ where $1/p + 1/q = 1$.

Define $T : W^{m,p}(\Omega) \to L_N^p$ by

$$(Tu)_j = D^{\alpha}u,$$

where α is the jth multi-index. Then T is linear and

$$\|Tu\|_{L_N^p} = \|u\|_{W^{m,p}(\Omega)}.$$

That is, T is an isometric isomorphism of $W^{m,p}(\Omega)$ onto a subspace W of L_N^p. Since $W^{m,p}(\Omega)$ is complete, W is closed. Finally, L_N^p is separable and reflexive for $1 < p < \infty$, so the same is true of W. $\qquad \square$

When $p = 2$, we have a Hilbert space.

Definition. The mth order Sobolev space based on $L^2(\Omega)$ is denoted by

$$H^m(\Omega) = W^{m,2}(\Omega).$$

Proposition 7.4. *The space $H^m(\Omega) = W^{m,2}(\Omega)$ is a separable Hilbert space with the inner-product*

$$(u,v)_{H^m(\Omega)} = \sum_{|\alpha|\leq m} (D^\alpha u, D^\alpha v)_{L^2(\Omega)},$$

where $(f,g)_{L^2(\Omega)} = \int_\Omega f(x)\,\overline{g(x)}\,dx$ is the usual $L^2(\Omega)$ inner-product.

When $p < \infty$, a very useful fact about Sobolev spaces is that C^∞ functions form a dense subset. In fact, one can *define* $W^{m,p}(\Omega)$ to be the *completion* (i.e., the set of "limits" of Cauchy sequences, see, e.g., [1], [19], [20]) of $C^\infty(\Omega)$ (or even $C^m(\Omega)$) with respect to the $W^{m,p}(\Omega)$-norm.

Theorem 7.5. *If $1 \leq p < \infty$, then*

$$\{f \in C^\infty(\Omega) : \|f\|_{W^{m,p}(\Omega)} < \infty\} = C^\infty(\Omega) \cap W^{m,p}(\Omega)$$

is dense in $W^{m,p}(\Omega)$.

A few preliminary results are needed before we prove this theorem.

Lemma 7.6. *Suppose that $1 \leq p < \infty$ and $\varphi \in C_0^\infty(\mathbb{R}^d)$ is an approxima-tion to the identity supported in the unit ball about the origin (i.e., $\varphi \geq 0$, $\int \varphi(x)\,dx = 1$, $\mathrm{supp}(\varphi) \subset B_1(0)$, and $\varphi_\epsilon(x) = \epsilon^{-d}\varphi(\epsilon^{-1}x)$ for $\epsilon > 0$). If $f \in L^p(\Omega)$ is extended by 0 to $\mathbb{R}^d$ (if necessary), then*

(a) $\varphi_\epsilon * f \in L^p(\mathbb{R}^d) \cap C^\infty(\mathbb{R}^d)$,

(b) $\|\varphi_\epsilon * f\|_{L^p(\mathbb{R}^d)} \leq \|f\|_{L^p(\Omega)}$,

(c) $\varphi_\epsilon * f \xrightarrow{L^p(\mathbb{R}^d)} f$ *as $\epsilon \to 0^+$.*

Proof. Conclusions (a) and (b) follow from Young's inequality, Corollary 6.8, and Proposition 5.12. For (c), use the fact that continuous functions with compact support are dense in $L^p(\mathbb{R}^d)$. Let $\eta > 0$ and choose $g \in C_0(\mathbb{R}^d)$ such that

$$\|f - g\|_{L^p} \leq \eta/3.$$

Then, because of (b),

$$\|\varphi_\epsilon * f - f\|_{L^p} \leq \|\varphi_\epsilon * (f-g)\|_{L^p} + \|\varphi_\epsilon * g - g\|_{L^p} + \|g - f\|_{L^p}$$
$$\leq 2\eta/3 + \|\varphi_\epsilon * g - g\|_{L^p}.$$

Since g has compact support, it is uniformly continuous. Now $\text{supp}(g) \subset B_R(0)$ for some $R > 0$, so $\text{supp}(\varphi_\epsilon * g - g) \subset B_{R+2}(0)$ for all $\epsilon \le 1$. Choose $0 < \epsilon \le 1$ such that

$$|g(x) - g(y)| \le \frac{\eta}{3|B_{R+2}(0)|^{1/p}}$$

whenever $|x - y| < 2\epsilon$, where $|B_{R+2}(0)|$ is the measure of the ball. Then for $x \in B_{R+2}(0)$,

$$(\varphi_\epsilon * g - g)(x) = \int \varphi_\epsilon(x - y)(g(y) - g(x)) \, dy$$

$$\le \sup_{|x-y|<2\epsilon} |g(y) - g(x)| \le \frac{\eta}{3|B_{R+2}(0)|^{1/p}},$$

so $\|\varphi_\epsilon * g - g\|_{L^p} \le \eta/3$ and $\|\varphi_\epsilon * f - f\|_{L^p} \le \eta$ is as small as we like. □

Corollary 7.7. *If $\Omega' \subset\subset \Omega$ or $\Omega' = \Omega = \mathbb{R}^d$, then*

$$\varphi_\epsilon * f \xrightarrow{W^{m,p}(\Omega')} f \quad \forall f \in W^{m,p}(\Omega).$$

Proof. Extend f by 0 to $\mathbb{R}^d$ if necessary. For any multi-index α with $|\alpha| \le m$,

$$D^\alpha(\varphi_\epsilon * f) = \varphi_\epsilon * D^\alpha f,$$

since $\varphi_\epsilon \in \mathcal{D}(\mathbb{R}^d)$ and $f \in \mathcal{D}'(\mathbb{R}^d)$. The subtlety above is whether $D^\alpha f$, on $\mathbb{R}^d$ after extension of f, has a jump on $\partial\Omega$, i.e., has some kind of δ-like function there. Restriction to Ω' removes any such difficulty:

$$\varphi_\epsilon * D^\alpha f \xrightarrow{L^p(\Omega')} D^\alpha f,$$

since eventually as $\epsilon \to 0$, $\varphi_\epsilon * D^\alpha f = \int_{B_\epsilon(x)} \varphi(x - y) \, D^\alpha f(y) \, dy$ involves only values of $D^\alpha f$ strictly supported in Ω. □

Proof of Theorem 7.5. Define $\Omega_0 = \Omega_{-1} = \emptyset$ and for integers $k \ge 1$

$$\Omega_k = \{x \in \Omega : |x| < k \text{ and } \text{dist}(x, \partial\Omega) > 1/k\}.$$

Let $\phi_k \in C_0^\infty(\Omega)$ be such that $0 \le \phi_k \le 1$, $\phi_k \equiv 1$ on Ω_k, and $\phi_k \equiv 0$ on Ω_{k+1}^c. Let $\psi_1 = \phi_1$ and $\psi_k = \phi_k - \phi_{k-1}$ for $k \ge 2$, so $\psi_k \ge 0$, $\psi_k \in C_0^\infty(\Omega)$, $\text{supp}(\psi_k) \subset \overline{\Omega}_{k+1} \setminus \Omega_{k-1}$, and

$$\sum_{k=1}^\infty \psi_k(x) = 1 \quad \forall x \in \Omega.$$

At each $x \in \Omega$, this sum has at most two nonzero terms. (This is an example of what is called a *locally finite partition of unity*, which is a partition of unity such that at each x, only a finite number of terms in the representation of 1 are nonzero.)

Now let $\epsilon > 0$ be given and φ be an approximation to the identity as in Lemma 7.6. For $f \in W^{m,p}(\Omega)$, choose in Corollary 7.7 values of $\epsilon_k > 0$ small enough that $\epsilon_k \leq \frac{1}{2} \operatorname{dist}(\Omega_{k+1}, \partial \Omega_{k+2})$ and

$$\|\varphi_{\epsilon_k} * (\psi_k f) - \psi_k f\|_{W^{m,p}} \leq \epsilon 2^{-k}.$$

Then $\operatorname{supp}(\varphi_{\epsilon_k} * (\psi_k f)) \subset \overline{\Omega}_{k+2} \setminus \Omega_{k-2}$, so set

$$g = \sum_{k=1}^{\infty} \varphi_{\epsilon_k} * (\psi_k f) \in C^\infty,$$

which is a finite sum at any point $x \in \Omega$. Note that

$$\|f - g\|_{W^{m,p}(\Omega)} \leq \sum_{k=1}^{\infty} \|\psi_k f - \varphi_{\epsilon_k} * (\psi_k f)\|_{W^{m,p}} \leq \epsilon \sum_{k=1}^{\infty} 2^{-k} = \epsilon. \qquad \square$$

The space $C_0^\infty(\Omega) = \mathcal{D}(\Omega)$ is dense in a generally smaller Sobolev space.

Definition. We let $W_0^{m,p}(\Omega)$ be the closure in $W^{m,p}(\Omega)$ of $C_0^\infty(\Omega)$.

The proof of the following result should be clear.

Proposition 7.8. *If $1 \leq p < \infty$ and $m \geq 0$, then*

(a) $W_0^{m,p}(\mathbb{R}^d) = W^{m,p}(\mathbb{R}^d)$,

(b) $W_0^{m,p}(\Omega) \hookrightarrow W^{m,p}(\Omega)$ *(continuously embedded)*,

(c) $W_0^{0,p}(\Omega) = L^p(\Omega)$;

(d) $D^\alpha : W_0^{m,p}(\Omega) \to W_0^{m-|\alpha|,p}(\Omega)$ *is a continuous linear operator for all multi-indices α such that $|\alpha| \leq m$.*

The dual of $L^p(\Omega)$ is $L^q(\Omega)$ when $1 \leq p < \infty$ and $1/p + 1/q = 1$. Since $W^{m,p}(\Omega) \hookrightarrow L^p(\Omega)$, $L^q(\Omega) \subset (W^{m,p}(\Omega))^*$. In general, the dual of $W^{m,p}(\Omega)$ is much larger than $L^q(\Omega)$, and consists of objects that are more general than distributions. To stay within the class of distributions, we restrict attention to the dual of $W_0^{m,p}(\Omega)$.

Definition. For $1 \leq p < \infty$, $1/p + 1/q = 1$, and $m \geq 0$ an integer, let

$$(W_0^{m,p}(\Omega))^* = W^{-m,q}(\Omega).$$

Of course, the norm of $v \in W^{-m,q}(\Omega)$, when $m \geq 0$, is

$$\|v\|_{W^{-m,q}(\Omega)} = \sup_{u \in W_0^{m,p}(\Omega), u \neq 0} \frac{\langle u, v \rangle}{\|u\|_{W^{m,p}(\Omega)}}.$$

Proposition 7.9. *For $1 \leq p < \infty$, $1/p + 1/q = 1$, and $m \geq 0$ an integer, $W^{-m,q}(\Omega)$ consists of distributions that have unique, continuous extensions from $\mathcal{D}(\Omega)$ to $W_0^{m,p}(\Omega)$.*

Proof. Note that $\mathcal{D}(\Omega) \hookrightarrow W_0^{m,p}(\Omega)$, since inclusion $i : \mathcal{D}(\Omega) \to W^{m,p}(\Omega)$ is clearly (sequentially) continuous. Thus, given $T \in W^{-m,q}(\Omega) = (W_0^{m,p}(\Omega))^*$, $T \circ i \in \mathcal{D}'(\Omega)$, so $T \circ i$ has an extension to $W_0^{m,p}(\Omega)$. That this extension is unique is due to Corollary 6.23, since $\mathcal{D}(\Omega)$ is dense in $W_0^{m,p}(\Omega)$. □

Extensions of distributions from $\mathcal{D}(\Omega)$ to $W^{m,p}(\Omega)$ are *not* necessarily unique, since $\mathcal{D}(\Omega)$ is not necessarily dense. Thus $(W^{m,p}(\Omega))^*$ may indeed contain objects that are *not* distributions.

Proposition 7.10. *Let $1 < p < \infty$, q the conjugate exponent, m an integer, and α a multi-index. Let $W_{(0)}^{m,p}(\Omega)$ denote $W_0^{m,p}(\Omega)$ when $m \geq 0$ and $W^{m,p}(\Omega)$ when $m < 0$. Then*

$$D^\alpha : W_{(0)}^{m,p}(\Omega) \to W_{(0)}^{m-|\alpha|,p}(\Omega)$$

is a continuous linear operator. Moreover, whenever $u \in W_0^{m,p}(\Omega)$ and $v \in W_0^{|\alpha|-m,q}(\Omega)$,

$$\langle D^\alpha u, v \rangle = (-1)^{|\alpha|}\langle u, D^\alpha v \rangle. \tag{7.1}$$

Proof. It is enough to show the results for $|\alpha| = 1$ and all $m \in \mathbb{Z}$. For simplicity, we write D in place of D^α.

We have observed the continuity of differentiation when $m - 1 \geq 0$ in Proposition 7.8, so consider now $m \leq 0$. If $u \in W^{m,p}(\Omega)$, then $u \in \mathcal{D}'(\Omega)$ and $Du \in \mathcal{D}'(\Omega)$. Let $v \in W_0^{1-m,q}(\Omega)$ and $\psi_j \in \mathcal{D}(\Omega)$ such that $\psi_j \to v$ in $W_0^{1-m,q}(\Omega)$ by density, since $q < \infty$. Now

$$\langle Du, \psi_j \rangle = -\langle u, D\psi_j \rangle \longrightarrow -\langle u, Dv \rangle,$$

by Proposition 7.9, so we can define $\Lambda \in W^{m-1,p}(\Omega) = (W_0^{1-m,q}(\Omega))^*$ by

$$\langle \Lambda, v \rangle = -\langle u, Dv \rangle,$$

which agrees with Du on $\mathcal{D}(\Omega)$. We can therefore extend the definition of Du to $W^{m-1,p}(\Omega)$ as a continuous linear operator by taking $Du = \Lambda$, which satisfies the desired action.

The above construction shows the validity of the integration by parts formula (7.1) for $m \leq 0$. For $m \geq 1$ and $u \in W_0^{m,p}(\Omega)$, there are $\phi_j \in \mathcal{D}(\Omega)$ such that $\phi_j \to u$ in $W_0^{m,p}(\Omega)$. But then also $D\phi_j \to Du$ in $W_0^{m-1,p}(\Omega)$. For $v \in W^{1-m,q}(\Omega)$, $v \in \mathcal{D}'(\Omega)$ and

$$\langle D\phi_j, v \rangle = -\langle \phi_j, Dv \rangle.$$

The limit on j is well-defined and gives the integration by parts formula for $m \geq 1$. □

Corollary 7.11. *Let $1 < p < \infty$, q the conjugate exponent, m an integer, and α a multi-index. Then*

$$D^\alpha : W^{m,p}(\Omega) \to W^{m-|\alpha|,p}(\Omega).$$

is a continuous linear operator.

Proof. The one-derivative operator D acting on $W^{m,p}(\Omega)$ maps as claimed for $m \geq 1$, and also for $m \leq 0$, since $L^p(\Omega) = W^{0,p}(\Omega) = W_0^{0,p}(\Omega)$. □

7.2 EXTENSIONS FROM Ω TO $\mathbb{R}^d$

If $\Omega \subsetneq \mathbb{R}^d$, how are $W^{m,p}(\Omega)$ and $W^{m,p}(\mathbb{R}^d)$ related? It would seem plausible that $W^{m,p}(\Omega)$ is exactly the set of restrictions to Ω of functions in $W^{m,p}(\mathbb{R}^d)$. However, the boundary of Ω, $\partial\Omega$, plays a subtle role, and our conjecture is true only for reasonable Ω, as we will see in this section.

The converse to our question is: given $f \in W^{m,p}(\Omega)$, can we find $\tilde{f} \in W^{m,p}(\mathbb{R}^d)$ such that $f = \tilde{f}$ on Ω? The existence of such an *extension* $\tilde{f}$ of f can be very useful, so this issue is investigated further now. The development begins with a very simple case.

Lemma 7.12. *If Ω is a half space in $\mathbb{R}^d$, $1 \leq p < \infty$, and $m \geq 0$ is fixed, then there is a bounded linear extension operator*

$$E : W^{m,p}(\Omega) \to W^{m,p}(\mathbb{R}^d);$$

that is, for $f \in W^{m,p}(\Omega)$, $Ef|_\Omega = f$ and there is some $C > 0$ such that

$$\|Ef\|_{W^{m,p}(\mathbb{R}^d)} \leq C\|f\|_{W^{m,p}(\Omega)}.$$

Note that in fact

$$\|f\|_{W^{m,p}(\Omega)} \leq \|Ef\|_{W^{m,p}(\mathbb{R}^d)} \leq C\|f\|_{W^{m,p}(\Omega)},$$

so $\|f\|_{W^{m,p}(\Omega)}$ and $\|Ef\|_{W^{m,p}(\mathbb{R}^d)}$ are comparable.

Proof. Choose a coordinate system so that

$$\Omega = \{x \in \mathbb{R}^d : x_d > 0\} \equiv \mathbb{R}_+^d.$$

If f is defined (almost everywhere) on $\mathbb{R}_+^d$, we extend f to the rest of $\mathbb{R}^d$ by reflection about $x_d = 0$. A simple reflection would not have derivatives that are continuous across the boundary, but a more complicated construction saves the day. For almost every $x \in \mathbb{R}^d$, let

$$Ef(x) = \begin{cases} f(x) & \text{if } x_d > 0, \\ \sum_{j=1}^{m+1} \lambda_j f(x_1, \ldots, x_{d-1}, -jx_d) & \text{if } x_d < 0, \end{cases}$$

where the numbers λ_j are defined below.

If $f \in C^m(\overline{\mathbb{R}_+^d}) \cap W^{m,p}(\mathbb{R}_+^d)$, then for any integer k between 0 and m,

$$D_d^k Ef(x) = \begin{cases} D_d^k f(x_1, \ldots, x_{d-1}, x_d) & \text{if } x_d > 0 \\ \sum_{j=1}^{m+1} (-j)^k \lambda_j D_d^k f(x_1, \ldots, x_{d-1}, -jx_d) & \text{if } x_d < 0. \end{cases}$$

We claim that the λ_j can be chosen so that

$$\sum_{j=1}^{m+1} (-j)^k \lambda_j = 1, \quad k = 0, 1, \ldots, m. \tag{7.2}$$

If so, then $D_d^k E f(x)$ is continuous as $x_d \to 0$, i.e., $Ef \in C^m(\mathbb{R}^d)$. Moreover, for $|\alpha| \leq m$,

$$\|D^\alpha E f\|_{L^p(\mathbb{R}^d)}^p$$

$$= \|D^\alpha f\|_{L^p(\mathbb{R}_+^d)}^p + \int_{\mathbb{R}_+^d} \left| \sum_{j=1}^{m+1} (-j)^{\alpha_d} \lambda_j D^\alpha f(x_1, \ldots, x_{d-1}, jx_d) \right|^p dx \tag{7.3}$$

$$\leq C_{m,p} \|D^\alpha f\|_{L^p(\mathbb{R}_+^d)}^p,$$

for some constant $C_{m,p} \geq 0$.

Let now $f \in W^{m,p}(\mathbb{R}_+^d) \cap C^\infty(\mathbb{R}_+^d)$, extended by zero. For $t > 0$, let τ_t be translation by t in the $(-e_d)$-direction, viz.

$$\tau_t f(x) = f(x + te_d).$$

Translation is continuous in $L^p(\mathbb{R}^d)$, so

$$D^\alpha \tau_t f = \tau_t D^\alpha f \xrightarrow{L^p} D^\alpha f \quad \text{as } t \to 0^+,$$

or what is the same,

$$\tau_t f \xrightarrow{W^{m,p}(\mathbb{R}_+^d)} f.$$

But $\tau_t f \in C^\infty(\overline{\mathbb{R}_+^d})$, so in fact $C^\infty(\overline{\mathbb{R}_+^d}) \cap W^{m,p}(\mathbb{R}_+^d)$ is dense in $W^{m,p}(\mathbb{R}_+^d)$. Thus (7.3) extends to all of $W^{m,p}(\mathbb{R}_+^d)$ by Corollary 6.23.

It is left to prove that $\{\lambda_j\}_{j=1}^{m+1}$ satisfying (7.2) can be found. Let the $(m+1) \times (m+1)$ matrix M be defined by

$$M_{ij} = (-j)^{i-1}, \quad 1 \leq i, j \leq m+1.$$

Then (7.2) is the linear system $M\lambda = e$, where λ is the $(m+1)$-vector of the λ_j's and e is the $(m+1)$-vector of 1s. Suppose M^T is singular. Then there is a vector $c \neq 0$ such that $M^T c = 0$. Consider the polynomial $p(x) = \sum_{i=1}^{m+1} c_i x^{i-1}$. Now M^T is a Vandermonde matrix, and the jth entry of $M^T c$ is

$$p(-j) = \sum_{i=1}^{m+1} c_i(-j)^{i-1} = 0.$$

Thus p has degree m, but has at least $m+1$ distinct roots, and so it must be identically zero. This means that $c = 0$, and we conclude that M^T, and so also M, is nonsingular. Therefore the λ_j's exist (uniquely, in fact), and independently of f, so E is linear. $\quad\square$

This lemma may be generalized via a smooth distortion of the boundary. We first define what is meant by a smooth boundary.

Definition. For integer $m \geq 0$, the domain $\Omega \subset \mathbb{R}^d$ has a *locally $C^{m,1}$-boundary* (or a *locally Lipschitz boundary* when $m = 0$) if, for every $x \in \partial\Omega$, there exists an open set $\Omega_x \subset \mathbb{R}^d$ and a map $\psi_x : \Omega_x \to B_1(0)$ such that

(a) ψ_x is one-to-one and onto, and both ψ_x and ψ_x^{-1} are of class $C^{m,1}$, i.e., $\psi_x \in C^{m,1}(\Omega_x)$ and $\psi_x^{-1} \in C^{m,1}(B_1(0))$;

(b) $\psi_x(\Omega_x \cap \Omega) = B^+ \equiv B_1(0) \cap \mathbb{R}_+^d$ and $\psi_x(\Omega_x \cap \partial\Omega) = B^+ \cap \partial\mathbb{R}_+^d$.

Note that $\psi \in C^{m,1}(\Omega)$ means that $\psi \in C^m(\Omega)$ and, for all $|\alpha| = m$, there is some *Lipschitz constant* $M > 0$ such that

$$|D^\alpha \psi(x) - D^\alpha \psi(y)| \leq M |x - y| \quad \forall x, y \in \Omega;$$

that is, $D^\alpha \psi$ is *Lipschitz continuous*. More simply, the definition states that any point $x \in \partial\Omega$ of a locally Lipschitz domain is covered by an open set Ω_x that can be smoothly distorted by ψ into a ball with $\partial\Omega$ distorted to the plane $x_d = 0$, and such that the domain is exactly on only one side of its boundary.

If Ω is bounded, then $\partial\Omega$ is compact, and the construction requires only a finite set of points. That is, we only require that there exist open sets $\Omega_j \subset \mathbb{R}^d$ and $\psi_j : \Omega_j \to B_1(0)$ for $j = 1, \ldots, N$ for some finite N, such that $\partial\Omega \subset \bigcup_j \Omega_j$ and the two properties of the definition are satisfied. If Ω is unbounded, we need to add a uniformity to the above construction. The following definition suffices, most of which holds trivially for a bounded domain.

Definition ($C^{m,1}$-boundary). For integer $m \geq 0$, the domain $\Omega \subset \mathbb{R}^d$ has a $C^{m,1}$-*boundary* (or a *Lipschitz boundary* if $m = 0$) if there exist open sets $\Omega_j \subset \mathbb{R}^d$ and functions $\psi_j : \Omega_j \to B_1(0)$ for $j = 1, \ldots, N$ with N possibly $+\infty$, with the following properties:

(a) $\overline{\Omega}_j \subset \mathbb{R}^d$ is compact and $\partial\Omega \subset \bigcup_{j=1}^{N} \Omega_j$;

(b) ψ_j is one-to-one and onto such that ψ_j and ψ_j^{-1} are of class $C^{m,1}$, i.e., $\psi_j \in C^{m,1}(\Omega_j)$ and $\psi_j^{-1} \in C^{m,1}(B_1(0))$;

(c) $\|\psi_j\|_{W^{m,\infty}}$, $\|\psi_j^{-1}\|_{W^{m,\infty}}$, and the Lipschitz constants for $D^\alpha \psi_j$ and $D^\alpha \psi_j^{-1}$, $|\alpha| = m$, are all bounded by some constant $M > 0$ independent of j;

(d) $\psi_j(\Omega_j \cap \Omega) = B^+ \equiv B_1(0) \cap \mathbb{R}_+^d$ and $\psi_j(\Omega_j \cap \partial\Omega) = B_1(0) \cap \partial\mathbb{R}_+^d$;

(e) for some $r \in \mathbb{N}$, every intersection of $r + 1$ of the sets Ω_j is empty;

(f) for some fixed $\delta > 0$, $\bigcup_{j=1}^{N} \psi_j^{-1}(B_{1/2}(0)) \supset \Omega^\delta$, where

$$\Omega^\delta = \{x \in \Omega : \text{dist}(x, \partial\Omega) < \delta\}.$$

The definition is satisfied by commonly encountered simple domains with nondegenerate corners and uniform geometry, such as a finite or infinite rectangle. On domains with a $C^{m,1}$-boundary, one can construct an important decomposition tool, which is the next topic.

First, the notion of a partition of unity is expanded here. For the reader's convenience, the whole idea is reviewed.

Definition. Let $\Omega \subset \mathbb{R}^d$ be a domain. A *partition of unity* is a collection of nonnegative functions $\{\phi_j\}_{j \in \mathcal{I}}$ defined on $\mathbb{R}^d$ such that

$$\sum_{j \in \mathcal{I}} \phi_j(x) = 1 \quad \forall x \in \Omega.$$

The partition of unity is *locally finite* if, at each point $x \in \Omega$, there is a neighborhood containing x on which all but a finite number of the functions are nonzero. If Ω has an open cover $\{\Omega_k\}_{k \in \mathcal{J}}$ (i.e., $\Omega \subset \bigcup_{k \in \mathcal{J}} \Omega_k$), then the partition of unity is *subordinate to the cover* provided that each ϕ_j has support completely within some Ω_k.

Normally, the functions will also possess some smoothness property, such as being continuous or perhaps lying in $C^\infty(\mathbb{R}^d)$.

Lemma 7.13. *Let $m \geq 1$ and $\Omega \subset \mathbb{R}^d$ be a domain with a $C^{m-1,1}$-boundary. Let $\Omega_j \subset \mathbb{R}^d$ and $\psi_j : \Omega_j \to B_1(0)$, for $j = 1,\ldots,N$, be as given in the definition of such a boundary, and define $\Omega_0 = \Omega \setminus \overline{\Omega^\delta}$. Then there exists a locally finite partition of unity $\{\phi_j\}_{j=0}^N \subset C^{m-1,1}(\mathbb{R}^d)$ such that $\operatorname{supp}(\phi_j) \subset \Omega_j$ for $j \geq 1$. Moreover, there is a constant C such that $\|\phi_j\|_{W^{m,\infty}} \leq C$ for all j.*

Proof. Note that $\Omega \subset \bigcup_{j=0}^N \Omega_j$ and $\operatorname{dist}(\Omega_0, \partial\Omega) = \delta$. Let $\psi(x)$ be Cauchy's infinitely differentiable function (5.2), and define

$$\phi(x) = \frac{\psi(1 - |x|)\,\psi(1 + |x|)}{\displaystyle\int_{B_1(0)} \psi(1 - |y|)\,\psi(1 + |y|)\,dy},$$

which is infinitely differentiable, nonnegative, has support $\overline{B_1(0)}$, decreases monotonically in the radial direction, and integrates to one. Let the set $\Omega_0' = \Omega \setminus \overline{\Omega^{3\delta/4}} \supset \Omega_0$ and $\chi_{\Omega_0'}$ denote its characteristic function. Define

$$\tilde{\phi}_0(x) = \chi_{\Omega_0'} * \phi_{\delta/8}(x) = \left(\frac{8}{\delta}\right)^d \int_{\Omega \setminus \overline{\Omega^{3\delta/4}}} \phi(8(x - y)/\delta)\,dy,$$

which is in C^∞, strictly positive on its support $\Omega \setminus \overline{\Omega^{\delta/2}}$, equal to one on Ω_0, and has uniformly bounded derivatives (in x, not δ). Also define

$$\tilde{\phi}_j(x) = \phi(\psi_j(x)),$$

extended by zero outside Ω_j, which also has uniformly bounded derivatives. Finally, the normalized functions

$$\phi_j(x) = \frac{\tilde{\phi}_j(x)}{\tilde{\phi}_0(x) + \sum_{k=1}^{N} \tilde{\phi}_k(x)}$$

clearly form a $C^{m-1,1}(\mathbb{R}^d)$ locally finite partition of unity with the desired supports.

The bound on the derivatives follows from the uniform bound on the derivatives of the $\tilde{\phi}_j$ and the fact that the denominator is uniformly bounded above and below. To see the upper bound, note that

$$\tilde{\phi}_0(x) + \sum_{k=1}^{N} \tilde{\phi}_k(x) \leq 1 + r\,\phi(0).$$

For the lower bound, since $\bigcup_{j=1}^{N} \psi_j^{-1}(B_{1/2}(0)) \supset \Omega^\delta$, we see that

$$\tilde{\phi}_0(x) + \sum_{k=1}^{N} \tilde{\phi}_k(x) \geq \min\left(1, \phi(1/2)\right). \qquad \square$$

Theorem 7.14. *Let $m \geq 0$, $1 \leq p < \infty$, and $\Omega \subset \mathbb{R}^d$ be a domain such that when $m \geq 1$, Ω has a $C^{m-1,1}$-boundary. Then there exists a bounded linear extension operator*

$$E : W^{m,p}(\Omega) \to W^{m,p}(\mathbb{R}^d).$$

Proof. If $m = 0$, Ω may be any domain and we can extend by zero. If $m \geq 1$, let $\{\Omega_j\}_{j=1}^{N}$ and $\{\psi_j\}_{j=1}^{N}$ be as in the definition of a $C^{m-1,1}$-boundary, where $N = +\infty$ is possible. Take from Lemma 7.13 the open set $\Omega_0 \subset \Omega$ and the locally finite $C^{m-1,1}$ partition of unity $\{\phi_j\}_{j=0}^{N}$. For $f \in W^{m,p}(\Omega)$, note that $\phi_j f \in W^{m,p}(\Omega_j \cap \Omega)$ has support inside Ω_j and can be extended to Ω by zero, maintaining smoothness. Let E_0 be the extension operator given in Lemma 7.12. If $j \neq 0$,

$$E_0\big((\phi_j f) \circ \psi_j^{-1}\big) \in W_0^{m,p}(B_1(0)),$$

so

$$E_0\big((\phi_j f) \circ \psi_j^{-1}\big) \circ \psi_j \in W_0^{m,p}(\Omega_j).$$

Extend this by zero to all of $\mathbb{R}^d$. Define E by

$$Ef = \phi_0 f + \sum_{j=1}^{N} E_0\big((\phi_j f) \circ \psi_j^{-1}\big) \circ \psi_j,$$

which is a linear operator.

On Ω, $E_0\big((\phi_j f)\circ\psi_j^{-1}\big)\circ\psi_j = \big((\phi_j f)\circ\psi_j^{-1}\big)\circ\psi_j = \phi_j f$, so $Ef = f$. Observe that derivatives of Ef are in $L^p(\mathbb{R}^d)$ because the ϕ_j, ψ_j, and $\psi_j^{-1} \in C^{m-1,1}$ uniformly with respect to j (i.e., derivatives up to order m of these functions are uniformly bounded), and so $Ef \in W^{m,p}(\mathbb{R}^d)$, $Ef|_\Omega = f$, and

$$\|Ef\|_{W^{m,p}(\mathbb{R}^d)} \leq C\|f\|_{W^{m,p}(\Omega)},$$

where $C \geq 0$ depends on m, p, and Ω through the bounds on derivatives of ϕ_j, ψ_j, and ψ_j^{-1}. $\qquad\square$

Remark. If $\bar{\Omega} \subset\subset \tilde{\Omega} \subset \mathbb{R}^d$, then we can assume that $Ef \in W_0^{m,p}(\tilde{\Omega})$. To see this, take any $\phi \in C_0^\infty(\tilde{\Omega})$ with $\phi \equiv 1$ on $\bar{\Omega}$, and define a new bounded extension operator by ϕE.

Many generalizations of this result are possible. In 1961, Alberto P. Calderón (see [1]) gave a proof assuming only that Ω is Lipschitz. In 1970, Elias M. Stein [24] gave a proof where a single operator E can be used for *any* values of m and p (and Ω is merely Lipschitz). Accepting the extension to Lipschitz domains, we have the following characterization of $W^{m,p}(\Omega)$.

Corollary 7.15. *If Ω has a Lipschitz boundary, $1 \leq p < \infty$, and $m \geq 0$, then*

$$W^{m,p}(\Omega) = \{f|_\Omega : f \in W^{m,p}(\mathbb{R}^d)\}.$$

If we restrict to the $W_0^{m,p}(\Omega)$ spaces, extension by zero gives a bounded extension operator, even if $\partial\Omega$ is ill-behaved.

Theorem 7.16. *Suppose $\Omega \subset \mathbb{R}^d$, $1 \leq p < \infty$, and $m \geq 0$. Let E be defined on $W_0^{m,p}(\Omega)$ as the operator that extends the domain of the function to $\mathbb{R}^d$ by zero; that is, for $f \in W_0^{m,p}(\Omega)$,*

$$Ef(x) = \begin{cases} f(x) & \text{if } x \in \Omega, \\ 0 & \text{if } x \notin \Omega. \end{cases}$$

Then $E : W_0^{m,p}(\Omega) \to W^{m,p}(\mathbb{R}^d)$ is a bounded linear extension operator.

Of course, then

$$\|f\|_{W_0^{m,p}(\Omega)} = \|Ef\|_{W^{m,p}(\mathbb{R}^d)}.$$

Proof. If $f \in W_0^{m,p}(\Omega)$, then there is a sequence $\{f_j\}_{j=1}^\infty \subset C_0^\infty(\Omega)$ such that

$$f_j \xrightarrow{\ W^{m,p}(\Omega)\ } f.$$

Let $\phi \in \mathcal{D}(\mathbb{R}^d)$. Then as distributions and for $|\alpha| \leq m$,

$$\int_\Omega D^\alpha f \, \phi \, dx \longleftarrow \int_\Omega D^\alpha f_j \, \phi \, dx = (-1)^{|\alpha|} \int_\Omega f_j \, D^\alpha \phi \, dx$$

$$\longrightarrow (-1)^{|\alpha|} \int_\Omega f \, D^\alpha \phi \, dx$$

$$= (-1)^{|\alpha|} \int_{\mathbb{R}^d} E f \, D^\alpha \phi \, dx$$

$$= \int_{\mathbb{R}^d} D^\alpha E f \, \phi \, dx,$$

so $E D^\alpha f = D^\alpha E f$ in $\mathcal{D}'$. The former is an L^1_{loc} function on $\mathbb{R}^d$, so the Lebesgue lemma (Lemma 5.7) implies that the two agree as functions. Thus

$$\|f\|_{W^{m,p}(\Omega)} = \left\{ \sum_{|\alpha| \leq m} \int_{\mathbb{R}^d} |E D^\alpha f|^p \, dx \right\}^{1/p}$$

$$= \left\{ \sum_{|\alpha| \leq m} \int_{\mathbb{R}^d} |D^\alpha E f|^p \, dx \right\}^{1/p} = \|E f\|_{W^{m,p}(\mathbb{R}^d)}. \qquad \square$$

7.3 THE SOBOLEV EMBEDDING THEOREM

A measurable function f fails to lie in some L^p space either because it blows up at some point or its tail fails to converge to zero fast enough (consider $|x|^{-r}$ for either $|x|$ near 0 or $|x| > R > 0$). However, if Ω is bounded and $f \in W^{m,p}(\Omega)$, $m \geq 1$, the derivative is well behaved, so the function cannot blow up as fast as an arbitrary function. Indeed, a moment's consideration leads to the expectation that such an f will lie in $L^q(\Omega)$ for some $q > p$.

Example. Consider $\Omega = (0, 1/2)$ and

$$f(x) = \frac{1}{\log x} \quad \text{for which} \quad f'(x) = \frac{-1}{x(\log x)^2}.$$

The change of variable $y = -\log x$ ($x = e^{-y}$) shows $f \in W^{1,1}(\Omega)$. In fact, $f' \in L^p(\Omega)$ only for $p = 1$. But $f \in L^p(\Omega)$ for any $p \geq 1$.

In this section, a precise statement is provided for this idea of trading derivatives for bounds in higher index L^p spaces. Surprisingly, if we have enough derivatives, the function will not only lie in L^∞, but will in fact be continuous. We begin with an important estimate.

Theorem 7.17 (Sobolev inequality). *If $1 \leq p < d$ and*

$$q = \frac{dp}{d - p},$$

then there is a constant $C = C(d, p)$ such that

$$\|u\|_{L^q(\mathbb{R}^d)} \leq C \|\nabla u\|_{(L^p(\mathbb{R}^d))^d} \quad \forall u \in C_0^1(\mathbb{R}^d). \tag{7.4}$$

The proof will make extensive use of the following worthwhile result.

Lemma 7.18 (generalized Hölder's inequality). *If $\Omega \subset \mathbb{R}^d$, $1 \le p_i \le \infty$ for $i = 1, \ldots, m$, and*

$$\sum_{i=1}^{m} \frac{1}{p_i} = 1,$$

then for $f_i \in L^{p_i}(\Omega)$, $i = 1, \ldots, m$,

$$\int_\Omega f_1(x) \cdots f_m(x) \, dx \le \|f_1\|_{L^{p_1}(\Omega)} \cdots \|f_m\|_{L^{p_m}(\Omega)}.$$

Proof. The case $m = 1$ is clear. We proceed by induction on m, using the already established Hölder inequality for pairs of functions lying in conjugate L^p classes. Let p'_m be conjugate to p_m (i.e., $1/p_m + 1/p'_m = 1$), where we reorder if necessary so $p_m \ge p_i$ for all $i < m$. Then

$$\int_\Omega f_1 \cdots f_m \, dx \le \|f_1 \cdots f_{m-1}\|_{L^{p'_m}} \|f_m\|_{L^{p_m}}.$$

Now $p_1/p'_m, \ldots, p_{m-1}/p'_m$ lie in the range from 1 to ∞, and

$$\frac{p'_m}{p_1} + \cdots + \frac{p'_m}{p_{m-1}} = 1,$$

so the induction hypothesis can be applied to see that

$$\|f_1 \cdots f_{m-1}\|_{L^{p'_m}} = \left\{ \int |f_1|^{p'_m} \cdots |f_{m-1}|^{p'_m} \, dx \right\}^{1/p'_m}$$

$$\le \left\{ \left(\int |f_1|^{p_1} \, dx \right)^{p'_m/p_1} \cdots \left(\int |f_{m-1}|^{p_{m-1}} \, dx \right)^{p'_m/p_{m-1}} \right\}^{1/p'_m}$$

$$= \|f_1\|_{L^{p_1}} \cdots \|f_{m-1}\|_{L^{p_{m-1}}}. \qquad \square$$

Proof of the Sobolev inequality, Theorem 7.17. Let $D_i = \partial/\partial x_i$, $i = 1, \ldots, d$. We begin with the case $p = 1 < d$. For $u \in C_0^1(\mathbb{R}^d)$,

$$|u(x)| = \left| \int_{-\infty}^{x_i} D_i u(x) \, dx_i \right| \le \int_{-\infty}^{\infty} |D_i u| \, dx_i \quad \forall i,$$

and so

$$|u(x)|^{d/(d-1)} \le \prod_{i=1}^{d} \left(\int_{-\infty}^{\infty} |D_i u| \, dx_i \right)^{1/(d-1)}.$$

We proceed in each variable separately. Let

$$f_i = \int_{-\infty}^{\infty} |D_i u| \, dx_i,$$

which does not depend on x_i. Successively for $i = 1, 2, \ldots, d$, integrate over $\mathbb{R}$ in x_i and use the generalized Hölder inequality for the $d - 1$ functions that actually depend on that variable, each with Lebesgue exponent $d - 1$. For x_1,

$$\int_{\mathbb{R}} |u(x)|^{d/(d-1)} \, dx_1 \leq \int_{\mathbb{R}} \prod_{i=1}^{d} f_i^{1/(d-1)} \, dx_1$$

$$= f_1^{1/(d-1)} \int_{\mathbb{R}} \prod_{i=2}^{d} f_i^{1/(d-1)} \, dx_1$$

$$\leq f_1^{1/(d-1)} \prod_{i=2}^{d} \left(\int_{\mathbb{R}} f_i \, dx_1 \right)^{1/(d-1)}.$$

Continuing for x_2, we have

$$\int_{\mathbb{R}^2} |u(x)|^{d/(d-1)} \, dx_1 \, dx_2$$

$$\leq \left(\int_{\mathbb{R}} f_2 \, dx_1 \right)^{1/(d-1)} \int_{\mathbb{R}} f_1^{1/(d-1)} \prod_{i=3}^{d} \left(\int_{\mathbb{R}} f_i \, dx_1 \right)^{1/(d-1)} dx_2$$

$$\leq \left(\int_{\mathbb{R}} f_2 \, dx_1 \right)^{1/(d-1)} \left(\int_{\mathbb{R}} f_1 \, dx_2 \right)^{1/(d-1)} \prod_{i=3}^{d} \left(\int_{\mathbb{R}^2} f_i \, dx_1 \, dx_2 \right)^{1/(d-1)}.$$

If $d > 2$, we continue for the other variables and obtain

$$\int_{\mathbb{R}^d} |u(x)|^{d/(d-1)} \, dx \leq \left(\prod_{i=1}^{d} \int_{\mathbb{R}^{d-1}} f_i \, dx_1 \cdots dx_{i-1} \, dx_{i+1} \cdots dx_d \right)^{1/(d-1)}$$

$$= \left(\prod_{i=1}^{d} \int_{\mathbb{R}^d} |D_i u| \, dx \right)^{1/(d-1)}.$$

For nonnegative numbers $a_1, \ldots, a_n$, the geometric mean is bounded by the arithmetic mean, *viz.*

$$\left(\prod_{i=1}^{n} a_i \right)^{1/n} \leq \frac{1}{n} \sum_{i=1}^{n} a_i;$$

moreover,

$$\left(\sum_{i=1}^{n} a_i \right)^2 \leq n \sum_{i=1}^{n} a_i^2$$

(i.e., in $\mathbb{R}^n$, $|\mathbf{a}|_{\ell^1} \leq \sqrt{n}|\mathbf{a}|_{\ell^2}$). Thus, it transpires that

$$
\int_{\mathbb{R}^d} |u(x)|^{d/(d-1)} \, dx \leq \left(\frac{1}{d} \sum_{i=1}^{d} \int_{\mathbb{R}^d} |D_i u| \, dx \right)^{d/(d-1)}
$$

$$
= \left(\frac{1}{d} \int_{\mathbb{R}^d} |\nabla u|_{\ell^1} \, dx \right)^{d/(d-1)}
$$

$$
\leq \left(\frac{1}{\sqrt{d}} \int_{\mathbb{R}^d} |\nabla u|_{\ell^2} \, dx \right)^{d/(d-1)},
$$

and so for C_d a constant depending on d,

$$
\|u\|_{L^{d/(d-1)}} \leq C_d \|\nabla u\|_{(L^1)^d}. \tag{7.5}
$$

For $p \neq 1$, apply (7.5) to $|u|^\gamma$ for an appropriate choice of $\gamma > 0$ to determine that

$$
\left\| |u|^\gamma \right\|_{L^{d/(d-1)}} \leq \gamma C_d \left\| |u|^{\gamma-1} |\nabla u| \right\|_{(L^1)^d} \leq \gamma C_d \left\| |u|^{\gamma-1} \right\|_{L^{p'}} \|\nabla u\|_{(L^p)^d},
$$

where $1/p + 1/p' = 1$. Choose γ so that

$$
\frac{\gamma d}{d-1} = (\gamma - 1)p';
$$

that is,

$$
\gamma = \frac{(d-1)p}{d-p} > 0,
$$

whence

$$
\frac{\gamma d}{d-1} = (\gamma - 1)p' = \frac{dp}{d-p} = q.
$$

Consequently,

$$
\|u\|_{L^q}^\gamma \leq \gamma C_d \|u\|_{L^q}^{\gamma-1} \|\nabla u\|_{(L^p)^d},
$$

and the result follows. □

A better result obtains if $p > d$.

Lemma 7.19. *If $p > d$, then there is a constant $C = C(d, p)$ such that*

$$
\|u\|_{L^\infty(\mathbb{R}^d)} \leq C(\text{diam}(\Omega)^d)^{\frac{1}{d}-\frac{1}{p}} \|\nabla u\|_{(L^p(\mathbb{R}^d))^d} \quad \forall u \in C_0^1(\mathbb{R}^d), \tag{7.6}
$$

where $\Omega = \text{supp}(u)$ and $\text{diam}(\Omega)$ is the diameter of Ω.

Proof. Suppose $u \in C_0^1(\mathbb{R}^d)$. For any unit vector $\mathbf{e}$,

$$
u(x) = \int_0^\infty \frac{\partial u}{\partial e}(x - r\mathbf{e}) \, dr = \int_0^\infty \nabla u(x - r\mathbf{e}) \cdot \mathbf{e} \, dr.
$$

Integrate over $\mathbf{e} \in S_1(0)$, the unit sphere, to come to the formula

$$d\omega_d\, u(x) = \int_{S_1(0)} \int_0^\infty \nabla u(x - r\mathbf{e}(\Theta)) \cdot \mathbf{e}(\Theta)\, dr\, d\Theta$$

$$= \int_{\mathbb{R}^d} \nabla u(x - y) \cdot \frac{y}{|y|} \frac{1}{|y|^{d-1}}\, dy,$$

where ω_d is the d-dimensional measure of the unit ball.

Now suppose $\operatorname{supp}(u) \subset B_1(0)$. Then for $1/p + 1/p' = 1$,

$$|u(x)| \leq \frac{1}{d\omega_d} \|\nabla u\|_{(L^p)^d} \big\| |y|^{1-d} \big\|_{L^{p'}(B_2(0))}$$

and

$$\big\| |y|^{1-d} \big\|_{L^{p'}}^{p'} = \int_{B_2(0)} |y|^{(1-d)p'}\, dy$$

$$= d\omega_d \int_0^2 r^{(1-d)p'+d-1}\, dr$$

$$= \frac{d\omega_d}{(1-d)p'+d}\, r^{(1-d)p'+d}\, \Big|_0^2 < \infty$$

provided $(1-d)p' + d > 0$, i.e., $p > d$. So there is $C_{d,p} > 0$ for which

$$|u(x)| \leq C_{d,p} \|\nabla u\|_{(L^p)^d}.$$

If $\Omega = \operatorname{supp}(u) \not\subset B_1(0)$, for $x \in \Omega$, consider the change of variable

$$y = \frac{x - \bar{x}}{\operatorname{diam}(\Omega)} \in B_1(0),$$

where $\bar{x}$ is the average of x on Ω, and apply the above result to

$$\tilde{u}(y) = u\big(\operatorname{diam}(\Omega)y + \bar{x}\big). \qquad \square$$

We summarize and extend the two previous results in the next lemma.

Lemma 7.20. *Let $\Omega \subset \mathbb{R}^d$ and $1 \leq p < \infty$.*

(a) *If $1 \leq p < d$ and $q = dp/(d-p)$, then there is a constant $C > 0$ independent of Ω such that for all $u \in W_0^{1,p}(\Omega)$,*

$$\|u\|_{L^q(\Omega)} \leq C \|\nabla u\|_{(L^p(\Omega))^d}. \tag{7.7}$$

(b) *If $p = d$ and Ω is bounded, then there is a constant $C_{\Omega,q} > 0$ such that for all $u \in W_0^{1,d}(\Omega)$,*

$$\|u\|_{L^q(\Omega)} \leq C_{\Omega,q} \|\nabla u\|_{(L^d(\Omega))^d} \quad \forall\, 1 \leq q < \infty, \tag{7.8}$$

where $C_{\Omega,q}$ depends on the measure of Ω and on q. In addition, if $p = d = 1$, $q = \infty$ is allowed.

(c) *If $d < p < \infty$ and Ω is bounded, then there is a constant $C > 0$ independent of Ω such that for all $u \in W_0^{1,p}(\Omega)$,*

$$\|u\|_{L^\infty(\Omega)} \le C\big(\operatorname{diam}(\Omega)^d\big)^{\frac{1}{d}-\frac{1}{p}} \|\nabla u\|_{(L^p(\Omega))^d}. \qquad (7.9)$$

Moreover, $W_0^{1,p}(\Omega) \subset C(\bar{\Omega})$, meaning that each equivalence class of functions in $W_0^{1,p}(\Omega)$ contains a function that is continuous on $\bar{\Omega}$.

Proof. For (7.7) and (7.9), just extend (7.4) and (7.6) by density. Note that a sequence in $C_0^\infty(\Omega)$, Cauchy in $W_0^{1,p}(\Omega)$, is also Cauchy in $L^q(\Omega)$ if $1 \le p < d$ and in $C^0(\bar{\Omega})$ if $p > d$. Just apply (7.4) or (7.6) to the difference of elements of the sequence. Moreover, when $p > d$ and Ω is bounded, the uniform limit of continuous functions in $C_0^\infty(\Omega) \subset C(\bar{\Omega})$ is continuous on $\bar{\Omega}$, so $W_0^{1,p}(\Omega) \subset C(\bar{\Omega})$.

Consider (7.8). The case $d = 1$ is a consequence of the fundamental theorem of calculus and left to the reader. Since Ω is bounded, the Hölder inequality implies $L^{p_1}(\Omega) \subset L^{p_2}(\Omega)$ whenever $p_1 \ge p_2$. Thus if $p = d > 1$ and $u \in W_0^{1,d}(\Omega)$, also $u \in W_0^{1,p^-}(\Omega)$ for any $1 \le p^- < p = d$. Apply (7.7) to obtain that

$$\|u\|_{L^q(\Omega)} \le C\|\nabla u\|_{(L^{p^-}(\Omega))^d} \le C|\Omega|^{(d-p^-)/d}\|\nabla u\|_{(L^d(\Omega))^d}$$

for $q \le dp^-/(d - p^-)$, which can be made as large as we like by taking p^- close to d. $\qquad \square$

Corollary 7.21 (Poincaré-Friedrichs inequality)**.** *If $\Omega \subset \mathbb{R}^d$ is bounded, $m \ge 0$ and $1 \le p < \infty$, then the norm on $W_0^{m,p}(\Omega)$ is equivalent to*

$$|u|_{W^{m,p}(\Omega)} = \left\{ \sum_{|\alpha|=m} \|D^\alpha u\|_{L^p(\Omega)}^p \right\}^{1/p},$$

which is the $W^{m,p}(\Omega)$-seminorm of $u \in W^{m,p}(\Omega)$; that is, there is some constant $C > 0$ such that

$$|u|_{W^{m,p}(\Omega)} \le \|u\|_{W^{m,p}(\Omega)} \le C|u|_{W^{m,p}(\Omega)} \quad \forall u \in W_0^{m,p}(\Omega).$$

Proof. Repeatedly use the Sobolev inequality (7.7), (7.8), or (7.9), depending on p, and the fact that $L^q(\Omega) \subset L^p(\Omega)$ for $q \ge p$. $\qquad \square$

The corollary shows that only the highest order derivatives are needed in the $W_0^{m,p}(\Omega)$-norm. This is an important result that will find repeated use later when studying boundary value problems in Chapter 8.

Recall that

$$C_B^n(\Omega) = \{u \in C^n(\Omega) : D^\alpha u \in L^\infty(\Omega) \,\forall\, |\alpha| \le n\}.$$

This is a Banach space containing $C^n(\bar{\Omega})$. We come now to the central result of this chapter.

Theorem 7.22 (Sobolev embedding theorem). *Let $\Omega \subset \mathbb{R}^d$ be a domain, $j \geq 0$ and $m \geq 1$ integers, and $1 \leq p < \infty$. The following continuous embeddings hold.*

(a) *If $mp \leq d$, then*

$$W_0^{j+m,p}(\Omega) \hookrightarrow W_0^{j,q}(\Omega) \quad \forall \ finite \ q \leq \frac{dp}{d-mp}$$

with $q \geq p$ if Ω is unbounded and $q \geq 1$ otherwise.

(b) *If $mp > d$ and Ω bounded, then*

$$W_0^{j+m,p}(\Omega) \hookrightarrow C_B^j(\Omega).$$

Moreover, if Ω has a bounded extension operator on $W^{j+m,p}(\Omega)$, or if $\Omega = \mathbb{R}^d$, then the following hold.

(c) *If $mp \leq d$, then*

$$W^{j+m,p}(\Omega) \hookrightarrow W^{j,q}(\Omega) \quad \forall \ finite \ q \leq \frac{dp}{d-mp}$$

with $q \geq p$ if Ω unbounded and $q \geq 1$ otherwise.

(d) *If $mp > d$, then*

$$W^{j+m,p}(\Omega) \hookrightarrow C_B^j(\Omega).$$

Proof. We begin with some remarks that simplify our task.

Note that the results for $j = 0$ extend immediately to the case for $j > 0$. We claim the results for $m = 1$ also extend by iteration to the case $m > 1$. Pair the number of derivatives m with the Lebesgue exponent q_m in a chain, i.e.,

$$(m,p) = (m, q_m) \mapsto (m-1, q_{m-1}) \mapsto \cdots \mapsto (0, q_0),$$

where the critical exponent q_k separates case (a) from (b) or case (c) from (d) when there is only one derivative. It is given by

$$q_k = \frac{d q_{k+1}}{d - q_{k+1}}, \quad k = 0, 1, \ldots, m-1.$$

It is easy to check that

$$q_k = \frac{dp}{d-(m-k)p} \quad and \quad d - kq_k = \frac{d}{d-(m-k)p}(d-mp).$$

Thus if $mp \leq d$, so also $kq_k \leq d$ and we stay in case (a) or (c) as k is reduced. If $mp > d$, we have case (b) or (d) throughout the chain. We can then apply the $m = 1$ result through the chain to derive the general result.

We also claim that the results for $\Omega = \mathbb{R}^d$ imply the results for $\Omega \neq \mathbb{R}^d$ through the bounded extension operator E. If $u \in W^{m,p}(\Omega)$, then $Eu \in W^{m,p}(\mathbb{R}^d)$ and we apply the result to Eu. The boundedness of E allows us to restrict back to Ω. For the $W_0^{m,p}(\Omega)$ spaces, we have E defined by extension by 0 for any domain, so the argument can be applied to this case as well.

Our task is thus simplified to the case where $\Omega = \mathbb{R}^d$, $m = 1$, and $j = 0$.

Consider the case of $p \leq d$, and take any $v \in W^{1,p}(\mathbb{R}^d)$ such that $\|v\|_{W^{1,p}(\mathbb{R}^d)} \leq 1$. We wish to apply (7.7) or (7.8) to v. To apply the latter, we must restrict to a bounded domain and lie in $W_0^{1,p}$.

Let $R = (-1,1)^d$ be a cube centered at 0, and $\tilde{R} = (-2,2)^d \supset\supset \bar{R}$. Let $\beta \in \mathbb{Z}^d$ be any vector with integer components. Clearly we can decompose

$$\mathbb{R}^d = \bigcup_{\beta \in \mathbb{Z}^d} (R + \beta) = \bigcup_{\beta \in \mathbb{Z}^d} (\tilde{R} + \beta)$$

into bounded domains; however, $v|_{R+\beta}$ does not lie in $W_0^{1,p}(R + \beta)$. Let

$$E : W^{1,p}(R) \to W_0^{1,p}(\tilde{R})$$

be a bounded extension operator with bounding constant C_E. By translation we define the extension operator

$$E_\beta : W^{1,p}(R + \beta) \to W_0^{1,p}(\tilde{R} + \beta),$$

i.e., by

$$E_\beta(\psi) = E(\tau_\beta \psi) = E(\psi(\cdot - \beta)).$$

Obviously the bounding constant for E_β can also be taken to be C_E.

Now apply (7.7) or (7.8) to $E_\beta(v|_{R+\beta})$ to obtain, for appropriate q,

$$\|E_\beta(v|_{R+\beta})\|_{L^q(\tilde{R}+\beta)} \leq C_S \|\nabla E_\beta(v|_{R+\beta})\|_{(L^p(\tilde{R}+\beta))^d},$$

where C_S is independent of β. Thus

$$\begin{aligned}
\|v\|_{L^q(R+\beta)}^q &\leq \|E_\beta(v|_{R+\beta})\|_{L^q(\tilde{R}+\beta)}^q \\
&\leq C_S^q \|\nabla E_\beta(v|_{R+\beta})\|_{(L^p(\tilde{R}+\beta))^d}^q \\
&\leq C_S^q C_E^q \|v\|_{W^{1,p}(R+\beta)}^q \\
&\leq C_S^q C_E^q \|v\|_{W^{1,p}(R+\beta)}^p,
\end{aligned}$$

since $p \leq q$ and $\|v\|_{W^{1,p}(\mathbb{R}^d)} \leq 1$. Summing over β gives

$$\|v\|_{L^q(\mathbb{R}^d)} \leq C$$

for some $C > 0$, since the union of the $\tilde{R} + \beta$ cover $\mathbb{R}^d$ a finite number of times.

If now $u \in W^{1,p}(\mathbb{R}^d)$, $u \neq 0$, let

$$v = \frac{u}{\|u\|_{W^{1,p}(\mathbb{R}^d)}}$$

to obtain

$$\|u\|_{L^q(\mathbb{R}^d)} \leq C\|u\|_{W^{1,p}(\mathbb{R}^d)};$$

thus, (a) and (c) follow.

Finally the argument for $p > d$, i.e., (b) and (d), is similar, since again our bounding constant in (7.9) is independent of β. This completes the proof. $\square$

Remark. The extension operator need only work for $W^{1,p}(\Omega)$, since we iterated the one derivative case. Thus Lipschitz domains satisfy the requirements. Most domains of interest (e.g., any polygon or polytope) have Lipschitz boundaries.

7.4 COMPACTNESS

Here is an important compactness result for Sobolev spaces. Denote that a space A is compactly embedded in a space B by writing $A \overset{c}{\hookrightarrow} B$.

Theorem 7.23 (Rellich-Kondrachov theorem). *Let $\Omega \subset \mathbb{R}^d$ be a bounded domain, $1 \leq p < \infty$, and $j \geq 0$ and $m \geq 1$ be integers.*

(a) *If $mp \leq d$, then $W_0^{j+m,p}(\Omega)$ is compactly embedded in $W_0^{j,q}(\Omega)$ for all $1 \leq q < dp/(d - mp)$. This could be denoted*

$$W_0^{j+m,p}(\Omega) \overset{c}{\hookrightarrow} W_0^{j,q}(\Omega) \quad \forall 1 \leq q < \frac{dp}{d - mp}.$$

(b) *If $mp > d$, then $W_0^{j+m,p}(\Omega)$ is compactly embedded in $C^j(\bar{\Omega})$, i.e.,*

$$W_0^{j+m,p}(\Omega) \overset{c}{\hookrightarrow} C^j(\bar{\Omega}).$$

Moreover, if $\Omega \subset \mathbb{R}^d$ has a bounded extension operator, then similar statements hold for $W^{j+m,p}(\Omega)$.

Proof. We only sketch the proof and leave it to reader to fill out the argument (more details may be found in, e.g., [8, p. 167–8] or [1, p. 144–8]). We need only show the result for $j = 0$, $m = 1$, and $W_0^{1,p}(\Omega)$. The result for general j and m follows from an iteration of this case. The result for $W^{j+m,p}(\Omega)$ can be obtained from bounded extension to $\tilde{\Omega} \supset \Omega$ and the result for $W_0^{j+m,p}(\tilde{\Omega})$.

We wish to apply the Ascoli-Arzelà theorem (Theorem 4.31) to a bounded set A in $W_0^{1,p}(\Omega)$. By density we may assume $A \subset C_0^1(\Omega) \subset C^1(\bar{\Omega})$.

First consider the case $p > d$. Lemma 7.19 gives us that A is bounded in $C^0(\bar{\Omega})$. For equicontinuity of A, consider $u \in A$, extended by zero to $\mathbb{R}^d$. Let $\epsilon > 0$ be given, and take any x and y such that $|x - y| < \epsilon$. Fix any ball

$B = B_{\epsilon/2}$ of radius $\epsilon/2$ containing both x and y. In a manner similar to the proof of Lemma 7.19, we have that for $z \in B$,

$$u(x) - u(z) = \int_0^{|x-z|} \nabla u(x - r\mathbf{e}_z) \cdot \mathbf{e}_z \, dr,$$

where $\mathbf{e}_z = \dfrac{x - z}{|x - z|}$. With $u_B = \dfrac{1}{|B|} \displaystyle\int_B u(x) \, dx$, integration in z over B gives

$$|u(x) - u_B| = \frac{1}{|B|} \left| \int_B \int_0^{|x-z|} \nabla u(x - r\mathbf{e}_z) \cdot \mathbf{e}_z \, dr \, dz \right|,$$

$$\leq \frac{1}{|B|} \int_{B_\epsilon(x)} \int_0^\epsilon |\nabla u(x - r\mathbf{e}_z)| \, dr \, dz$$

$$= \frac{1}{|B|} \int_{B_\epsilon(0)} \int_0^\epsilon |\nabla u(x - rz/|z|)| \, dr \, dz$$

$$= \frac{1}{|B|} \int_0^\epsilon \int_{S_1(0)} \int_0^\epsilon |\nabla u(x - r\mathbf{e}(\Theta))| \, dr \, d\Theta \, \rho^{d-1} d\rho.$$

Integrating first in ρ leads to

$$|u(x) - u_B| = \frac{\epsilon^d}{d|B|} \int_{B_\epsilon(0)} |\nabla u(x - z)| \, |z|^{1-d} \, dz$$

$$\leq \frac{\epsilon^d}{d|B|} \|\nabla u\|_{L^p(\Omega)} \, \| \, |z|^{1-d} \|_{L^{p'}(B_\epsilon(0))}$$

$$\leq C\epsilon^{1-d/p},$$

where p' is the conjugate exponent to p and C is independent of x, y, u, and ϵ, since $|B| = \omega_d(\epsilon/2)^d$. Now

$$|u(x) - u(y)| \leq |u(x) - u_B| + |u(y) - u_B| \leq 2C\epsilon^{1-d/p},$$

and equicontinuity of A follows for $p > d$. The Ascoli-Arzelà theorem implies compactness of A in $C^0(\bar{\Omega})$.

Now consider the case $p \leq d$, and assume initially that $q = 1$. For $\varphi \in C_0^\infty(B_1(0))$ an approximation to the identity and $\epsilon > 0$, let

$$A_\epsilon = \{u * \varphi_\epsilon : u \in A\} \subset C^0(\bar{\Omega}).$$

We estimate $u * \varphi_\epsilon$ and $\nabla(u * \varphi_\epsilon) = u * \nabla\varphi_\epsilon$ using the Hölder inequality to see that A_ϵ is bounded and equicontinuous in $C^0(\bar{\Omega})$ (although not uniformly so in ϵ). Thus A_ϵ is precompact in $C^0(\bar{\Omega})$ by the Ascoli-Arzelà theorem, and so also precompact in $L^1(\Omega)$, since Ω is bounded. Next, we estimate

$$\int_\Omega |u(x) - u * \varphi_\epsilon(x)| \, dx \leq \epsilon \int_\Omega |\nabla u| \, dx \leq C\epsilon,$$

so $u * \varphi_\epsilon$ is uniformly close to u in $L^1(\Omega)$. It follows that A is precompact in $L^1(\Omega)$ as well.

For $1 < q < dp/(d-p)$, use Hölder's inequality with exponents $1/(\lambda q)$ and $1/(1 - \lambda q)$ for some λ, and assume for the moment that $0 < \lambda q < 1$. Then apply (7.7) or (7.8) to show that

$$
\|u\|_{L^q(\Omega)} = \left(\int_\Omega |u|^{\lambda q} |u|^{q(1-\lambda)} \, dx \right)^{1/q}
$$
$$
\leq \|u\|_{L^1(\Omega)}^\lambda \|u\|_{L^{q(1-\lambda)/(1-\lambda q)}(\Omega)}^{1-\lambda} \leq C \|u\|_{L^1(\Omega)}^\lambda \|\nabla u\|_{(L^p(\Omega))^d}^{1-\lambda},
$$

where $q(1 - \lambda)/(1 - \lambda q) = dp/(d - p)$, which is the same as $\lambda = \big(dp - (d - p)q\big)/\big(q(dp + p - d)\big)$. Note that the required range $0 < \lambda q < 1$ holds (only) for $q < dp/(d - p)$ (when $d - p \geq 0$). Finallly, boundedness in $W_0^{1,p}(\Omega)$ and convergence in $L^1(\Omega)$ implies convergence in $L^q(\Omega)$. $\qquad \square$

Corollary 7.24. *If $\Omega \subset \mathbb{R}^d$ is bounded and has a Lipschitz boundary, $1 \leq p < \infty$, and $\{u_j\}_{j=1}^\infty \subset W^{j+m,p}(\Omega)$ is a bounded sequence, then there exists a subsequence $\{u_{j_k}\}_{k=1}^\infty \subset \{u_j\}_{j=1}^\infty$ which converges in $W^{j,q}(\Omega)$ for $q < dp/(d - mp)$ if $mp \leq d$, and in $C^j(\bar\Omega)$ if $mp > d$.*

This result is often used in the following way. Suppose

$$
u_j \xrightarrow{\;\;W^{m,p}(\Omega)\;\;} u \quad \text{as } j \to \infty, \text{ weakly.}
$$

Then $\{u_j\}$ is bounded, so there is a subsequence for which

$$
u_{j_k} \xrightarrow{\;\;W^{m-1,p}(\Omega)\;\;} u \quad \text{as } k \to \infty, \text{ strongly.}
$$

7.5 THE H^s SOBOLEV SPACES

In this section we give an alternate definition of $W^{m,2}(\mathbb{R}^d) = H^m(\mathbb{R}^d)$ which has a natural extension to noninteger values of m. These fractional order spaces will be useful in the next section on traces.

If $f \in \mathcal{S}(\mathbb{R})$, then

$$
\widehat{Df} = i\xi \hat{f}.
$$

This is an example of a *multiplier operator* $T : \mathcal{S} \to \mathcal{S}$ defined by

$$
T(f) = (m(\xi)\hat{f}(\xi))^\vee,
$$

where $m(\xi)$, called the *symbol* of the operator, is in $C^\infty(\mathbb{R})$ and has polynomial growth. For $T = D$, $m(\xi) = i\xi$. While $i\xi$ is smooth, it is *not* invertible, so D is a troublesome operator. However $T = 1 - D^2$ has

$$
((1 - D^2)f)^\wedge = (1 + \xi^2)\hat{f}(\xi),
$$

and $(1 + \xi^2)$ is well behaved, even though it involves two derivatives of f. What is the square root of this operator? Let $f, g \in \mathcal{S}$ and compute using the L^2 inner-product

$$(Tf, g) = (\widehat{Tf}, \hat{g}) = ((1 + \xi^2)\hat{f}, \hat{g}) = ((1 + \xi^2)^{1/2}\hat{f}, (1 + \xi^2)^{1/2}\hat{g}).$$

Thus $T = S^2$ where
$$(Sf)^\wedge = (1 + \xi^2)^{1/2}\hat{f}(\xi),$$

and $S = (1 - D^2)^{1/2}$ is like D.

We are thus led to consider in $\mathbb{R}^d$ the symbol for $(I - \Delta)^{1/2}$, which is

$$b_1(\xi) = (1 + |\xi|^2)^{1/2} \in \mathcal{S}'(\mathbb{R}^d).$$

Then $b_1(\xi)$ is like D in $\mathbb{R}^d$. For other order derivatives, we generalize for $s \in \mathbb{R}$ to

$$b_s(\xi) = (1 + |\xi|^2)^{s/2} \in \mathcal{S}'(\mathbb{R}^d).$$

In fact $b_s(\xi) \in C^\infty(\mathbb{R}^d)$ and all derivatives grow at most polynomially. Thus we can multiply tempered distributions by $b_s(\xi)$ by Proposition 6.33.

Definition. For $s \in \mathbb{R}$, let $\Lambda^s : \mathcal{S}' \to \mathcal{S}'$ be given by

$$(\Lambda^s u)^\wedge(\xi) = (1 + |\xi|^2)^{s/2}\hat{u}(\xi)$$

for all $u \in \mathcal{S}'$. We call Λ^s the *Bessel potential* of order s.

Remark. If $u \in \mathcal{S}$, then

$$\Lambda^s u(x) = (2\pi)^{-d/2}\check{b}_s * u(x).$$

Proposition 7.25. *For any* $s \in \mathbb{R}$, $\Lambda^s : \mathcal{S}' \to \mathcal{S}'$ *is a continuous, linear, one-to-one, and onto map. Moreover*

$$\Lambda^{s+t} = \Lambda^s \Lambda^t \quad \forall s, t \in \mathbb{R} \quad and \quad (\Lambda^s)^{-1} = \Lambda^{-s}.$$

The reader can provide the proof.

Definition. For $s \in \mathbb{R}$, let

$$H^s(\mathbb{R}^d) = \{u \in \mathcal{S}' : \Lambda^s u \in L^2(\mathbb{R}^d)\},$$

and for $u \in H^s(\mathbb{R}^d)$, let

$$\|u\|_{H^s} = \|\Lambda^s u\|_{L^2(\mathbb{R}^d)}.$$

There is a question of consistency when $s = m$ is an integer. We defined $H^m(\mathbb{R}^d)$ previously as $W^{m,2}(\mathbb{R}^d)$ when $m \geq 0$ and as $(W_0^{-m,2}(\mathbb{R}^d))^*$ when $m < 0$. Our definitions do coincide, although the norms are only equivalent, not identical. We will need a bit more development before we can validate this assertion, so in the meantime, the reader should consider $H^m(\mathbb{R}^d)$ as defined above in this section.

Proposition 7.26. *For all $s \in \mathbb{R}$, $\| \cdot \|_{H^s}$ is a norm, and for $u \in H^s$,*

$$\|u\|_{H^s} = \|\Lambda^s u\|_{L^2} = \left\{ \int_{\mathbb{R}^d} (1 + |\xi|^2)^s |\hat{u}(\xi)|^2 \, d\xi \right\}^{1/2}.$$

Moreover, $H^0 = L^2$.

Proof. Apply the Plancherel theorem. □

Proposition 7.27. *A compatible inner-product on $H^s(\mathbb{R}^d)$ for any $s \in \mathbb{R}$ is given by*

$$(u, v)_{H^s} = (\Lambda^s u, \Lambda^s v)_{L^2} = \int \Lambda^s u \, \overline{\Lambda^s v} \, dx = \int (1 + |\xi|^2)^s \hat{u}(\xi) \, \overline{\hat{v}(\xi)} \, d\xi$$

for all $u, v \in H^s(\mathbb{R}^d)$. Moreover, $\mathcal{S} \subset H^s$ is dense and H^s is a Hilbert space.

Proof. It is easy to verify that $(u, v)_{H^s}$ is an inner-product, and that the last expression for it holds by the Plancherel theorem. Moreover,

$$\|u\|_{H^s}^2 = (u, u)_{H^s} \quad \forall u \in H^s.$$

Given $\epsilon > 0$ and $u \in H^s$, there is $f \in \mathcal{S}$ such that

$$\|(1 + |\xi|^2)^{s/2} \hat{u} - f\|_{L^2} < \epsilon,$$

since $\mathcal{S}$ is dense in L^2. But

$$g = (1 + |\xi|^2)^{-s/2} f \in \mathcal{S},$$

so

$$\|u - \check{g}\|_{H^s} = \|(1 + |\xi|^2)^{s/2} (\hat{u} - g)\|_{L^2} < \epsilon,$$

showing that $\mathcal{S}$ is dense in H^s.

Finally, if $\{u_j\}_{j=1}^\infty \subset H^s$ is Cauchy, then

$$f_j = (1 + |\xi|^2)^{s/2} \hat{u}_j$$

gives a Cauchy sequence in L^2. Let $f_j \xrightarrow{L^2} f$ and let

$$u = \left((1 + |\xi|^2)^{-s/2} f\right)^{\vee} \in H^s.$$

Then

$$\|u_j - u\|_{H^s} = \|f_j - f\|_{L^2} \longrightarrow 0$$

as $j \to \infty$. Thus H^s is complete. □

These Hilbert spaces form a one-parameter family $\{H^s\}_{s \in \mathbb{R}}$. They are also nested, and the positive index spaces consist of ordinary L^2-functions.

Proposition 7.28. *If $s \geq t$, then $H^s \subset H^t$ is continuously embedded (i.e., $H^s \hookrightarrow H^t$). Moreover, $H^s \subset H^0 = L^2$ for all $s \geq 0$.*

Proof. If $u \in H^s$, then

$$\|u\|_{H^t}^2 = \int (1 + |\xi|^2)^t |\hat{u}(\xi)|^2 \, d\xi \leq \int (1 + |\xi|^2)^s |\hat{u}(\xi)|^2 \, dx = \|u\|_{H^s}^2. \qquad \square$$

Let us now settle the question of consistency between $H^m(\mathbb{R}^d)$ and $W^{m,2}(\mathbb{R}^d)$ when $s = m$ is an integer. We need the following technical result.

Lemma 7.29. *For integer $m \geq 0$, there are constants $C_1, C_2 > 0$ such that*

$$C_1 (1 + x^2)^{m/2} \leq \sum_{k=0}^{m} x^k \leq C_2 (1 + x^2)^{m/2} \quad \forall x \geq 0.$$

Proof. We need constants $c_1, c_2 > 0$ such that

$$c_1 (1 + x^2)^m \leq \left(\sum_{k=0}^{m} x^k \right)^2 \leq c_2 (1 + x^2)^m \quad \forall x \geq 0.$$

Consider

$$f(x) = \frac{\left(\sum_{k=0}^{m} x^k \right)^2}{(1 + x^2)^m} \in C^0([0, \infty)).$$

Since $f(0) = 1$ and $\lim_{x \to \infty} f(x) = 1$, $f(x)$ has a maximum on $[0, \infty)$, which gives c_2. Similarly $g(x) = 1/f(x)$ has a maximum, giving c_1. $\qquad \square$

Theorem 7.30. *If $m \geq 0$ is an integer, then*

$$H^m(\mathbb{R}^d) = W^{m,2}(\mathbb{R}^d).$$

Proof. If $u \in W^{m,2}(\mathbb{R}^d)$, then $D^\alpha u \in L^2$ for all $|\alpha| \leq m$. But then

$$|\xi|^k |\hat{u}(\xi)| \in L^2 \quad \forall k \leq m,$$

which is equivalent by the lemma to saying that

$$(1 + |\xi|^2)^{m/2} |\hat{u}(\xi)| \in L^2.$$

That is, $u \in H^m(\mathbb{R}^d)$. For $u \in H^m$, reverse the steps above to conclude that $u \in W^{m,2}$. Moreover, we have shown that the norms are equivalent. $\qquad \square$

We turn next to the negative index spaces, showing first that they are dual to the positive ones.

Proposition 7.31. *If $s \in \mathbb{R}$, then we may identify $(H^s)^*$ with H^{-s} by the pairing*

$$\langle u, v \rangle_{H^s, H^{-s}} \equiv (\Lambda^s u, \Lambda^{-s} v)_{L^2}$$

for all $u \in H^s$ and $v \in H^{-s}$. Moreover, $\|v\|_{H^{-s}} = \|v\|_{(H^s)^*}$.

Proof. By the Riesz theorem, $(H^s)^*$ is isomorphic to H^s by the pairing

$$\langle u, w \rangle_{H^s, H^s} = (u, w)_{H^s} = (\Lambda^s u, \Lambda^s w)_{L^2}$$

for all $u \in H^s$ and $w \in H^s \cong (H^s)^*$. But then $\Lambda^s w = \Lambda^{-s} \Lambda^{2s} w \in L^2$, so there is a one-to-one correspondence between H^s and H^{-s} given by

$$v = \Lambda^{2s} w = \left((1 + |\xi|^2)^s \hat{w} \right)^{\vee} \in H^{-s} \quad \forall w \in H^s.$$

We also have the desired action, since

$$\langle u, v \rangle_{H^s, H^{-s}} = (\Lambda^s u, \Lambda^{-s} v)_{L^2} = (\Lambda^s u, \Lambda^s w)_{L^2} = \langle u, w \rangle_{H^s, H^s}.$$

Moreover,

$$\|v\|_{H^{-s}} = \|w\|_{H^s},$$

so we have H^{-s} isomorphic to $H^s \cong (H^s)^*$. □

Corollary 7.32. *For all integers* m, $H^m(\mathbb{R}^d) = W^{m,2}(\mathbb{R}^d)$.

Proof. Theorem 7.30 gives the result for $m \geq 0$, so suppose $m < 0$. Then $W^{m,2} = (W_0^{-m,2})^* = (W^{-m,2})^*$, since our domain is all of $\mathbb{R}^d$, and thus $W^{m,2} = (H^{-m})^* = H^m$. □

We close this vein of development by considering restriction to a domain $\Omega \subset \mathbb{R}^d$.

Definition. If $\Omega \subset \mathbb{R}^d$ is a domain and $s \geq 0$, let

$$H^s(\Omega) = \{u|_\Omega : u \in H^s(\mathbb{R}^d)\}.$$

Let $H_0^s(\Omega)$ be constructed as follows. Map functions in $C_0^\infty(\Omega)$ to $C_0^\infty(\mathbb{R}^d)$ by extending by zero. Take the closure of this space in $H^s(\mathbb{R}^d)$. Finally, restrict back to Ω. We say more concisely but imprecisely that $H_0^s(\Omega)$ is the completion in $H^s(\mathbb{R}^d)$ of $C_0^\infty(\Omega)$.

Let us elaborate on our definition of $H^s(\Omega)$, $s \geq 0$. We have the following general construction for a Banach space H and $Z \subset H$ a closed linear subspace. Define the *quotient space*

$$H/Z = \{x + Z \subset H : x \in H\};$$

that is, for $x \in H$, let

$$\hat{x} = x + Z \subset H$$

be the *coset* of x, and then H/Z is the set of cosets (or equivalence classes where $x, y \in H$ are equivalent if $x - y \in Z$, so $\hat{x} = \hat{y}$). Then H/Z is a vector space with a norm given by

$$\|\hat{x}\|_{H/Z} = \inf_{y \in \hat{x}} \|y\|_H = \inf_{z \in Z} \|x + z\|_H,$$

and H/Z is complete in this norm. If H is a Hilbert space, the construction is simpler. Let $P_Z^\perp$ be H-orthogonal projection onto $Z^\perp$. Then $P_Z^\perp x \in \hat{x} = x + Z$ and

$$\|\hat{x}\|_{H/Z} = \|P_Z^\perp x\|_H.$$

In the Hilbert space setting, there is also a natural inner-product defined by

$$(\hat{x}, \hat{y})_{H/Z} = (P_Z^\perp x, P_Z^\perp y)_H.$$

If the field $\mathbb{F} = \mathbb{R}$, this is more easily computed as

$$(\hat{x}, \hat{y})_{H/Z} = \frac{1}{4}\left(\|\hat{x} + \hat{y}\|^2_{H/Z} - \|\hat{x} - \hat{y}\|^2_{H/Z}\right),$$

which is a modified application of Proposition 3.4. Moreover, H/Z is a Hilbert space, isomorphic to $Z^\perp$. We leave these facts for the reader to verify.

For $H = H^s(\mathbb{R}^d)$, let

$$Z = \{u \in H^s(\mathbb{R}^d) : u|_\Omega = 0\},$$

which is a closed subspace, so we have the quotient space

$$H^s(\mathbb{R}^d)/Z = \{u + Z : u \in H^s(\mathbb{R}^d)\}$$

as outlined above. Now define

$$\pi : H^s(\mathbb{R}^d)/Z \to H^s(\Omega)$$

by

$$\pi\hat{u} = \pi(u + Z) = u|_\Omega.$$

This map is well-defined, since if $\hat{u} = \hat{v}$, then $u|_\Omega = v|_\Omega$. Moreover, π is linear, one-to-one, and onto. So for $u, v \in H^s(\Omega)$, define

$$\|u\|_{H^s(\Omega)} = \|\pi^{-1}u\|_{H^s(\mathbb{R}^d)/Z} = \inf_{\substack{v \in H^s(\mathbb{R}^d) \\ v|_\Omega = u}} \|v\|_{H^s(\mathbb{R}^d)} = \|P_Z^\perp u\|_{H^s(\mathbb{R}^d)},$$

and $H^s(\Omega)$ is isometrically isomorphic to $H^s(\mathbb{R}^d)/Z$; that is, $H^s(\Omega)$ becomes a Hilbert space with inner-product

$$(u, v)_{H^s(\Omega)} = (\pi^{-1}u, \pi^{-1}v)_{H^s(\mathbb{R}^d)/Z} = (P_Z^\perp u, P_Z^\perp v)_{H^s(\mathbb{R}^d)},$$

which, if $\mathbb{F} = \mathbb{R}$, can be computed as

$$(u, v)_{H^s(\Omega)} = \frac{1}{4}\left(\|u + v\|^2_{H^s(\Omega)} - \|u - v\|^2_{H^s(\Omega)}\right).$$

Proposition 7.33. *If $\Omega \subset \mathbb{R}^d$ is a domain and $s \geq 0$, then $H^s(\Omega)$ is a Hilbert space. Moreover, for any constant $C > 1$, given $u \in H^s(\Omega)$, there is $\tilde{u} \in H^s(\mathbb{R}^d)$ such that $\tilde{u}|_\Omega = u$ and*

$$\|\tilde{u}\|_{H^s(\mathbb{R}^d)} \leq C\|u\|_{H^s(\Omega)};$$

that is, there is a bounded extension operator $E : H^s(\Omega) \to H^s(\mathbb{R}^d)$ with $\|E\| \leq C$.

If $s = m \geq 0$ is an integer, then we had previously defined $H^m(\Omega)$ as $W^{m,2}(\Omega)$. If Ω has a Lipschitz boundary, the sets of functions coincide and the norms are equivalent, but not equal. This can be seen by considering the bounded extension operator

$$E : W^{m,2}(\Omega) \to W^{m,2}(\mathbb{R}^d),$$

for which $u \in W^{m,2}(\Omega)$ implies

$$\|Eu\|_{W^{m,2}(\mathbb{R}^d)} \leq C\|u\|_{W^{m,2}(\Omega)} \leq C\|Eu\|_{W^{m,2}(\mathbb{R}^d)}.$$

Since $W^{m,2}(\mathbb{R}^d)$ is the same as $H^m(\mathbb{R}^d)$, with equivalent norms,

$$\|u\|_{H^m(\Omega)} = \inf_{\substack{v \in H^m(\mathbb{R}^d) \\ v|_\Omega = u}} \|v\|_{H^m(\mathbb{R}^d)} \leq \|Eu\|_{H^m(\mathbb{R}^d)} \leq C_1\|Eu\|_{W^{m,2}(\mathbb{R}^d)}$$

$$\leq C_2\|u\|_{W^{m,2}(\Omega)} \leq C_2 \inf_{\substack{v \in W^{m,2}(\mathbb{R}^d) \\ v|_\Omega = u}} \|v\|_{W^{m,2}(\mathbb{R}^d)}$$

$$\leq C_3 \inf_{\substack{v \in H^m(\mathbb{R}^d) \\ v|_\Omega = u}} \|v\|_{H^m(\mathbb{R}^d)} = C_3\|u\|_{H^m(\Omega)}.$$

Thus our two definitions of $H^m(\Omega)$ are consistent, and, depending on the norm used, the constant in the previous proposition may be different than described (i.e., not necessarily any $C > 1$). Summarizing, there emerges the following result.

Proposition 7.34. *If $\Omega \subset \mathbb{R}^d$ has a Lipschitz boundary and $m \geq 0$ is an integer, then*

$$H^m(\Omega) = W^{m,2}(\Omega)$$

and the $H^m(\Omega)$ and $W^{m,2}(\Omega)$ norms are equivalent.

7.6   TRACE THEOREMS

Given a domain $\Omega \subset \mathbb{R}^d$ and a function $f : \Omega \to \mathbb{R}$, the *trace* of f is its value on the boundary of Ω; i.e., the trace is $f|_{\partial\Omega}$, provided this makes sense. We give a precise meaning and construction when f belongs to a Sobolev space.

We begin by restricting functions to lower dimensional hypersurfaces. Let $0 < k < d$ be an integer, and decompose

$$\mathbb{R}^d = \mathbb{R}^{d-k} \times \mathbb{R}^k.$$

If $\phi \in C^0(\mathbb{R}^d)$, then the restriction map

$$R : C^0(\mathbb{R}^d) \to C^0(\mathbb{R}^{d-k})$$

is defined by

$$R\phi(x') = \phi(x', 0) \quad \forall x' \in \mathbb{R}^{d-k},$$

wherein $0 \in \mathbb{R}^k$.

Theorem 7.35. *Let k and d be integers with $0 < k < d$. The restriction map R extends to a bounded linear map from $H^s(\mathbb{R}^d)$ onto $H^{s-k/2}(\mathbb{R}^{d-k})$, provided that $s > k/2$.*

Proof. Since $\mathcal{S}$ is dense in our two Sobolev spaces, it is enough to consider $u \in \mathcal{S}(\mathbb{R}^d)$ where R is well-defined. Let $v = Ru \in \mathcal{S}(\mathbb{R}^{d-k})$.

The Sobolev norm involves the Fourier transform, so we compute for $y \in \mathbb{R}^{d-k}$

$$v(y) = (2\pi)^{-(d-k)/2} \int_{\mathbb{R}^{d-k}} e^{i\eta \cdot y} \hat{v}(\eta) \, d\eta.$$

But, with $\xi = (\eta, \zeta) \in \mathbb{R}^{d-k} \times \mathbb{R}^k$, this is

$$v(y) = u(y, 0) = (2\pi)^{-d/2} \int_{\mathbb{R}^d} e^{i\xi \cdot (y,0)} \hat{u}(\xi) \, d\xi$$

$$= (2\pi)^{-(d-k)/2} \int_{\mathbb{R}^{d-k}} e^{i\eta \cdot y} \left[(2\pi)^{-k/2} \int_{\mathbb{R}^k} \hat{u}(\eta, \zeta) \, d\zeta \right] d\eta.$$

Comparing these two expressions, we see that

$$\hat{v}(\eta) = (2\pi)^{-k/2} \int_{\mathbb{R}^k} \hat{u}(\eta, \zeta) \, d\zeta.$$

Introduce $(1 + |\eta|^2 + |\zeta|^2)^{s/2}(1 + |\eta|^2 + |\zeta|^2)^{-s/2}$ into the integral above and apply Hölder's inequality to obtain

$$|\hat{v}(\eta)|^2 \le (2\pi)^{-k} \int_{\mathbb{R}^k} |\hat{u}(\eta, \zeta)|^2 (1 + |\eta|^2 + |\zeta|^2)^s \, d\zeta \int_{\mathbb{R}^k} (1 + |\eta|^2 + |\zeta|^2)^{-s} \, d\zeta.$$

The second integral on the right-hand side is

$$\int_{\mathbb{R}^k} (1 + |\eta|^2 + |\zeta|^2)^{-s} \, d\zeta = k\omega_k \int_0^\infty (1 + |\eta|^2 + r^2)^{-s} r^{k-1} \, dr,$$

where ω_k is the measure of the unit ball in k-dimensions. With the change of variable

$$(1 + |\eta|^2)^{1/2} \rho = r,$$

this is

$$k\omega_k (1 + |\eta|^2)^{k/2-s} \int_0^\infty (1 + \rho^2)^{-s} \rho^{k-1} \, d\rho,$$

which is finite provided $-2s + k - 1 < -1$, i.e., $s > k/2$. Combining, we have shown that there is a constant $C > 0$ such that

$$|\hat{v}(\eta)|^2 (1 + |\eta|^2)^{s-k/2} \le C^2 \int_{\mathbb{R}^k} |\hat{u}(\eta, \zeta)|^2 (1 + |\eta|^2 + |\zeta|^2)^s \, d\zeta.$$

Integrating in η gives the bound

$$\|v\|_{H^{s-k/2}(\mathbb{R}^{d-k})} \le C \|u\|_{H^s(\mathbb{R}^d)}.$$

Thus R is a bounded linear operator mapping into $H^{s-k/2}(\mathbb{R}^{d-k})$.

To see that R maps onto $H^{s-k/2}(\mathbb{R}^{d-k})$, proceed one spatial dimension at a time. To this end, fix $\varphi \in C_0^\infty(\mathbb{R})$ such that $\int_\mathbb{R} \varphi(t)\,dt = 1$. Let $v_k \in H^{s-k/2}(\mathbb{R}^{d-k})$ and, for $j = k, k-1, \ldots, 1$, extend v_j to $v_{j-1} \in H^{s-(j-1)/2}(\mathbb{R}^{d-(j-1)})$ as follows. For $(\eta, \zeta) \in \mathbb{R}^{d-(j-1)} \times \mathbb{R}$, define

$$\hat{v}_{j-1}(\eta, \zeta) = \hat{v}_j(\eta)\, \varphi\left(\frac{\zeta}{\sqrt{1+|\eta|^2}}\right) \frac{1}{\sqrt{1+|\eta|^2}}.$$

The change of variables $\zeta = t\sqrt{1+|\eta|^2}$ shows that

$$\int_\mathbb{R} \hat{v}_{j-1}(\eta, \zeta)\,d\zeta = \int_\mathbb{R} \hat{v}_j(\eta)\, \varphi\left(\frac{\zeta}{\sqrt{1+|\eta|^2}}\right) \frac{1}{\sqrt{1+|\eta|^2}}\,d\zeta$$

$$= \int_\mathbb{R} \hat{v}_j(\eta)\, \varphi(t)\,dt = \hat{v}_j(\eta).$$

Apply the inverse Fourier transform to the above result to see that $v_{j-1}(y, 0) = v_j(y)$. Now the same change of variables shows that

$$\int_\mathbb{R} |\hat{v}_{j-1}(\eta, \zeta)|^2 (1+|\eta|^2+|\zeta|^2)^{s-(j-1)/2}\,d\zeta$$

$$= \int_\mathbb{R} |\hat{v}_j(\eta)|^2 (1+|\eta|^2+|\zeta|^2)^{s-(j-1)/2} \left|\varphi\left(\frac{\zeta}{\sqrt{1+|\eta|^2}}\right)\right|^2 \frac{1}{1+|\eta|^2}\,d\zeta$$

$$= |\hat{v}_j(\eta)|^2 (1+|\eta|^2)^{s-j/2} \int_\mathbb{R} (1+t^2)^{s-(j-1)/2} |\varphi(t)|^2\,dt,$$

and integration in η shows that $v_{j-1} \in H^{s-(j-1)/2}(\mathbb{R}^{d-(j-1)})$. For a given $v = v_k \in H^{s-k/2}(\mathbb{R}^{d-k})$, $u = v_0 \in H^s(\mathbb{R}^d)$ satisfies $Ru = v$. □

Remark. We saw in the Sobolev embedding theorem that

$$H^s(\mathbb{R}^d) \hookrightarrow C_B^0(\mathbb{R}^d)$$

for $s > d/2$. Thus we can even restrict to a point ($k = d$ above).

Now consider $\Omega \subset \mathbb{R}^d$ such that $\partial\Omega$ is $C^{0,1}$ smooth (i.e., Lipschitz). Our goal is to define the trace of $u \in H^1(\Omega)$ on $\partial\Omega$. Let $\{\Omega_j\}_{j=1}^N$ and $\{\psi_j\}_{j=1}^N$ be as in the definition of a $C^{0,1}$-boundary given earlier in the chapter. Take Ω_0 open such that $\bar{\Omega}_0 \subset \Omega \subset \bigcup_{j=0}^N \Omega_j$. Let $\{\phi_k\}_{k=1}^M$ be a locally finite C^∞ partition of unity subordinate to the cover such that $\|\phi_k\|_{W^{1,\infty}(\Omega)}$ is uniformly bounded for all k (such as was constructed in Lemma 7.13). Then $\operatorname{supp}(\phi_k) \subset \Omega_{j_k}$ for some j_k. Let E be the extension operator from the proof of Lemma 7.12. This operator was constructed so that if the support of v is contained in $B^+ = B_1(0) \cap \mathbb{R}_+^d$, then the support of Ev is contained in $B_1(0)$; that is, $E : H_0^1(B^+) \to H_0^1(B_1(0))$.

For $u \in H^1(\Omega)$, define

$$u_k = E\big((\phi_k u) \circ \psi_{j_k}^{-1}\big) : B_1(0) \to \mathbb{F},$$

and note that $u_k \in H_0^1(B_1(0))$. Restrict u to $\partial\Omega$ by restricting u_k to the interface $S \equiv B_1(0) \cap \{x_d = 0\}$. Since $\operatorname{supp}(u_k) \subset\subset B_1(0)$, we can extend by zero and apply Theorem 7.35 to obtain

$$\|u_k\|_{H^{1/2}(S)} \leq C_1\|u_k\|_{H^1(B_1(0))}.$$

We need to combine the u_k and change variables back to Ω and $\partial\Omega$.

Summing on k, there obtains

$$\sum_{k=1}^{M}\|u_k\|_{H^{1/2}(S)}^2 \leq C_1^2 \sum_{k=1}^{M}\|u_k\|_{H^1(B_1(0))}^2 \leq C_2 \sum_{k=1}^{M}\|(\phi_k u)\circ\psi_{j_k}^{-1}\|_{H^1(B+)}^2,$$

using the bound on E. The final norm merely involves L^2 norms of (weak) derivatives of $(\phi_k u)\circ\psi^{-1}$. The Leibniz rule, chain rule, and change of variables imply that each such norm is bounded by the $H^1(\Omega)$ norm of u, i.e.,

$$\|(\phi_k u)\circ\psi_{j_k}^{-1}\|_{H^1(B+)} \leq C_3\|\phi_k u\|_{H^1(\Omega_{j_k})} \leq C_4\|u\|_{H^1(\Omega_{j_k})},$$

so

$$\sum_{k=1}^{M}\|u_k\|_{H^{1/2}(S)}^2 \leq C_2 \sum_{k=1}^{M}\|(\phi_k u)\circ\psi_{j_k}^{-1}\|_{H^1(B+)}^2$$

$$\leq C_2 C_4 \sum_{k=1}^{M}\|u\|_{H^1(\Omega_{j_k})}^2 \leq C_5\|u\|_{H^1(\Omega)}^2.$$

Let the *trace* of u, $\gamma_0 u$, be defined for a.e. $x \in \partial\Omega$ by

$$\gamma_0 u(x) = \sum_{k=1}^{M}\left(E(\phi_k u)\circ\psi_{j_k}^{-1}\right)\left(\psi_{j_k}(x)\right) = \sum_{k=1}^{M} u_k(\psi_{j_k}(x)),$$

which exists by the previous theorem. After change of variables,

$$\|\gamma_0 u\|_{L^2(\partial\Omega)}^2 \leq C_6 \sum_{k=1}^{M}\|u_k\|_{L^2(S)}^2 \leq C_7 \sum_{k=1}^{M}\|u_k\|_{H^{1/2}(S)}^2 \leq C_7 C_5\|u\|_{H^1(\Omega)}^2.$$

$$(7.10)$$

In summary, for $u \in H^1(\Omega)$, we can define its trace $\gamma_0 u$ on $\partial\Omega$ as a function in $L^2(\partial\Omega)$, and

$$\gamma_0 : H^1(\Omega) \to L^2(\partial\Omega)$$

is a well-defined, bounded linear operator.

The above computations carry over to $u \in H^s(\Omega)$ for nonintegral $s > 1/2$, as can be seen by using the equivalent norms of the next section. Since we do not prove that those norms are indeed equivalent, we have restricted to integral $s = 1$ here (and used the ordinary chain rule rather than requiring some generalization to fractional derivatives). What we have proven is sufficient for the next chapter, where s will take only integer values.

While $L^2(\partial\Omega)$ is well-defined (given the Lebesgue measure on the manifold $\partial\Omega$), we do not yet have a definition of the Sobolev spaces on $\partial\Omega$. For $s > 1/2$, let

$$Z = \{u \in H^s(\Omega) : \gamma_0 u = 0 \text{ on } \partial\Omega\};$$

this set is well-defined by (7.10) (at least we have proven this for $s \geq 1$), and is in fact closed in $H^s(\Omega)$. We therefore define

$$H^{s-1/2}(\partial\Omega) = \{\gamma_0 u : u \in H^s(\Omega)\} \subset L^2(\partial\Omega),$$

which is isomorphic to $H^s(\Omega)/Z$, a Hilbert space. While $H^{s-1/2}(\partial\Omega) \subset L^2(\partial\Omega)$, we expect that such functions are in fact smoother. A norm is given by

$$\|u\|_{H^{s-1/2}(\partial\Omega)} = \inf_{\substack{v \in H^s(\Omega) \\ \gamma_0 v = u}} \|v\|_{H^s(\Omega)}.$$

Note that this construction gives immediately the trace theorem

$$\|\gamma_0 u\|_{H^{s-1/2}(\partial\Omega)} \leq C\|u\|_{H^s(\Omega)},$$

where one can show that $C = 1$. If an equivalent norm is used for $H^{s-1/2}(\partial\Omega)$, $C \neq 1$ is likely. While we do not have a constructive definition of $H^{s-1/2}(\partial\Omega)$ and its norm that allow us to see explicitly the smoothness of such functions, by analogy to Theorem 7.35 for $\Omega = \mathbb{R}^d_+$, we recognize that $H^{s-1/2}(\partial\Omega)$ functions have intermediate smoothness. The equivalent norm of the next section gives a constructive sense to this statement. In summary, the following overarching result has been established.

Theorem 7.36 (trace theorem 1). *Let a domain $\Omega \subset \mathbb{R}^d$ have a Lipschitz boundary. The trace operator $\gamma_0 : C^0(\bar{\Omega}) \to C^0(\partial\Omega)$ defined by restriction, i.e., $(\gamma_0 u)(x) = u(x)$ for all $x \in \partial\Omega$, extends to a surjective, bounded linear map*

$$\gamma_0 : H^s(\Omega) \xrightarrow{onto} H^{s-1/2}(\partial\Omega)$$

for any $s \geq 1$ (actually, $s > 1/2$).

We can extend this result to higher order derivatives. Tangential derivatives of $\gamma_0 u$ are well-defined, since if D_τ is any derivative in a direction tangential to $\partial\Omega$, then for $u \in C^0(\bar{\Omega})$,

$$D_\tau \gamma_0 u = D_\tau E u = E D_\tau u = \gamma_0 D_\tau u,$$

where E is a bounded extension operator. However, derivatives normal to $\partial\Omega$ are more delicate.

Definition. Let $\nu \in \mathbb{R}^d$ be the unit outward normal vector to $\partial\Omega$. Then for $u \in C^1(\bar{\Omega})$,

$$D_\nu u = \frac{\partial u}{\partial \nu} = \nabla u \cdot \nu \quad \text{on } \partial\Omega$$

is the *normal derivative* of u on $\partial\Omega$. If $j \geq 0$ is an integer and $u \in C^j(\bar{\Omega})$, let

$$\gamma_j u = D_\nu^j u = \frac{\partial^j u}{\partial \nu^j}.$$

We state and prove the following theorem for integers $s = m$, though it actually holds for appropriate nonintegral values as well.

Theorem 7.37 (trace theorem 2). *Let $m \geq 0$ be an integer and $\Omega \subset \mathbb{R}^d$ be a domain with a $C^{m-1,1}$ boundary if $m \geq 1$ and a Lipschitz ($C^{0,1}$) boundary if $m = 0$. The map $\gamma : C^m(\bar{\Omega}) \to (C^0(\partial\Omega))^{m+1}$ defined by*

$$\gamma u = (\gamma_0 u, \gamma_1 u, \ldots, \gamma_m u)$$

extends to a surjective, bounded linear map

$$\gamma : H^{m+1}(\Omega) \xrightarrow{onto} \overset{m}{\underset{j=0}{\times}} H^{m-j+1/2}(\partial\Omega).$$

Proof. Let $u \in H^{m+1}(\Omega) \cap C^\infty(\bar{\Omega})$, which is dense because of the existence of an extension operator. Then iterate the single derivative result for γ_0:

$$\gamma_0 u \in H^{m+1/2}(\partial\Omega),$$
$$\gamma_1 u = \gamma_0(\nabla u \cdot \nu) \in H^{m-1/2}(\partial\Omega),$$
$$\gamma_2 u = \gamma_0(\nabla(\nabla u \cdot \nu) \cdot \nu) \in H^{m-3/2}(\partial\Omega), \quad \text{etc.},$$

wherein we require $\partial\Omega$ to be smooth eventually so that derivatives of ν can be taken, and wherein we have assumed that the vector field ν on $\partial\Omega$ has been extended locally into Ω (that this can be done follows from the tubular neighborhood theorem, see, e.g., [9]).

It remains to show that γ maps onto its range. Suppose first that $\Omega = \mathbb{R}_+^d = \{(x, y) \in \mathbb{R}^{d-1} \times \mathbb{R} : y > 0\} \subset \mathbb{R}^d$ and $\partial\Omega = \mathbb{R}^{d-1} \times \{0\} = \mathbb{R}^{d-1}$. For each $j = 0, 1, \ldots, m$, define the linear maps $Z_j : H^{m-j+1/2}(\mathbb{R}^{d-1}) \to L^2(\mathbb{R}_+^d)$ for w in the domain of Z_j by

$$Z_j w = u \quad \text{where} \quad \hat{u}(\xi, y) = y^j e^{-(1+|\xi|)y} \hat{w}(\xi)$$

and the Fourier transform is taken only in the first $d - 1$ variables. We claim that Z_j maps into $H^{m+1}(\mathbb{R}_+^d)$. For $(\alpha, \beta) \in \mathbb{N}^{d-1} \times \mathbb{N}$ a multi-index of length d such that $|\alpha| + \beta \leq m + 1$, $x \in \mathbb{R}^{d-1}$, and $y \in \mathbb{R}$, consider a derivative of $Z_j w$ extended to $\mathbb{R}^d$, i.e.,

$$U(x, y) = \begin{cases} D_x^\alpha D_y^\beta u(x, y) & \text{for } y > 0, \\ 0 & \text{for } y < 0. \end{cases}$$

Represent the full Fourier transform on $\mathbb{R}^d$ by $\mathcal{F}$, so

$$\mathcal{F}U(\xi, \eta) = \frac{(i\xi)^\alpha}{\sqrt{2\pi}} \int_0^\infty e^{-i\eta y} D_y^\beta \hat{u}(\xi, y) \, dy.$$

The Leibniz rule shows that this is a sum of terms of the form

$$W_k(\xi, \eta) = C_0 \xi^\alpha \, \hat{w}(\xi) \big(-(1+|\xi|) \big)^{\beta-k} \int_0^\infty e^{-i\eta y} \, y^{j-k} \, e^{-(1+|\xi|)y} \, dy$$

$$= C_1 \xi^\alpha \, \hat{w}(\xi)(1+|\xi|)^{\beta-k}(1+|\xi|+i\eta)^{k-j-1},$$

where C_0 and C_1 are complex constants depending on the indices and $k = 0, 1, \ldots, \min(\beta, j)$. Now

$$\int_{\mathbb{R}^d} |W_k(\xi, \eta)|^2 \, d\xi \, d\eta$$

$$= |C_1|^2 \int_{\mathbb{R}^{d-1}} |\xi|^{2|\alpha|} \, |\hat{w}(\xi)|^2 \, (1+|\xi|)^{2\beta-2k} \int_{\mathbb{R}} \frac{d\eta}{\big((1+|\xi|)^2 + \eta^2\big)^{j+1-k}} \, d\xi,$$

and

$$\int_{\mathbb{R}} \frac{d\eta}{\big((1+|\xi|)^2 + \eta^2\big)^{j+1-k}} = (1+|\xi|)^{2k-2j-1} \int_{\mathbb{R}} \frac{dr}{(1+r^2)^{j+1-k}} < \infty,$$

so for some constants C_2 and C_3,

$$\int_{\mathbb{R}^d} |W_k(\xi, \eta)|^2 \, d\xi \, d\eta \leq C_2 \int_{\mathbb{R}^{d-1}} |\xi|^{2|\alpha|} \, |\hat{w}(\xi)|^2 \, (1+|\xi|)^{2\beta-2j-1} \, d\xi$$

$$\leq C_2 \int_{\mathbb{R}^{d-1}} |\hat{w}(\xi)|^2 \, (1+|\xi|)^{2|\alpha|+2\beta-2j-1} \, d\xi$$

$$\leq C_2 \int_{\mathbb{R}^{d-1}} |\hat{w}(\xi)|^2 \, (1+|\xi|)^{2(m+1)-2j-1} \, d\xi$$

$$\leq C_3 \int_{\mathbb{R}^{d-1}} |\hat{w}(\xi)|^2 \, (1+|\xi|^2)^{m-j+1/2} \, d\xi < \infty,$$

since $w \in H^{m-j+1/2}(\mathbb{R}^{d-1})$, and the claim that Z_j maps into $H^{m+1}(\mathbb{R}^d_+)$ is established.

Let

$$v \in \underset{j=0}{\overset{m}{\times}} H^{m-j+1/2}(\mathbb{R}^{d-1}),$$

We construct $u \in H^{m+1}(\mathbb{R}^d)$ such that $\gamma u = v$ as follows. Let $u_0 = Z_0 v_0$ and define recursively

$$u_j = u_{j-1} + \frac{1}{j!} Z_j(v_j - D_y^j u_{j-1}),$$

wherein the derivatives are given technically by the trace operator γ_j. Noting that for any w

$$D_y^k Z_j w(x, 0) = \begin{cases} 0 & \text{if } k < j, \\ j! \, w(x) & \text{if } k = j, \end{cases}$$

one can verify that

$$D_y^j u_j = v_j \quad \text{and} \quad D_y^k u_j = D_y^k u_{j-1} \quad \forall k < j.$$

The function $u = u_m$ satisfies the desired properties.

If $\partial\Omega$ is curved, we decompose $\partial\Omega$ and map it according to the definition of a $C^{m-1,1}$-boundary and then apply the above construction. □

Recall that $H_0^m(\Omega) = W_0^{m,2}(\Omega)$ is the closure of $C_0^\infty(\Omega)$ in $W^{m,2}(\Omega)$. Since $\gamma u = 0$ for $u \in C_0^\infty(\Omega)$, the same is true for any $u \in H_0^m(\Omega)$. That is, u and its $m-1$ derivatives (normal and/or tangential) vanish on $\partial\Omega$.

Theorem 7.38. *If $m \geq 1$ is an integer and a domain $\Omega \subset \mathbb{R}^d$ has a $C^{m-1,1}$ boundary, then*

$$H_0^m(\Omega) = \{u \in H^m(\Omega) : \gamma u = 0\}$$
$$= \{u \in H^m(\Omega) : \gamma_j u = 0 \ \forall j \leq m-1\} = \ker(\gamma).$$

Proof. As mentioned above, $H_0^m(\Omega) \subset \ker(\gamma)$. We would like to show the opposite inclusion. Again, by a mapping argument of the $C^{m-1,1}$ boundary, it is only necessary to consider the case $\Omega = \mathbb{R}_+^d$.

We saw earlier in the proof of Lemma 7.12 that $C_0^\infty(\overline{\mathbb{R}_+^d}) \cap H^m(\mathbb{R}_+^d)$ is dense in $H^m(\mathbb{R}_+^d)$, which we recast as density of $C_0^\infty(\mathbb{R}^d)|_{\mathbb{R}_+^d}$ in $H_0^m(\mathbb{R}_+^d)$. So it is enough to consider

$$u \in \ker(\gamma) \cap C_0^\infty(\mathbb{R}^d)|_{\mathbb{R}_+^d}.$$

Let $\psi \in C^\infty(\mathbb{R})$ be such that $\psi(t) = 1$ for $t > 2$ and $\psi(t) = 0$ for $t < 1$. For $n \geq 1$, let

$$\psi_n(t) = \psi(nt),$$

which converges to 1 on $\{t > 0\}$ as $n \to \infty$. Then $\psi_n(x_d)u(x) \in C_0^\infty(\mathbb{R}_+^d)$. We claim that

$$\psi_n(x_d)u(x) \xrightarrow{H^m(\mathbb{R}_+^d)} u(x) \quad \text{as } n \to \infty.$$

If so, then $u \in H_0^m(\mathbb{R}_+^d)$ as desired.

Let $\alpha \in \mathbb{N}^d$ be a multi-index such that $|\alpha| \leq m$ and let $\alpha = (\beta, \ell)$ where $\beta \in \mathbb{N}^{d-1}$ and $\ell \geq 0$. Then

$$D^\alpha(\psi_n u - u) = D^\beta D_d^\ell(\psi_n u - u) = \sum_{k=0}^\ell \binom{\ell}{k} D_d^{\ell-k}(\psi_n - 1)D^\beta D_d^k u,$$

and we need to show that this tends to 0 in $L^2(\mathbb{R}_+^d)$ as $n \to \infty$. It is enough to show this for each term

$$D_d^{\ell-k}(\psi_n - 1)D^\beta D_d^k u = n^{\ell-k}D_d^{\ell-k}(\psi - 1)|_{nx_d}D^\beta D_d^k u,$$

which is clear if $k = \ell$, since the measure of $\{x : \psi_n(x) - 1 > 0\}$ tends to 0. If $k < \ell$, our expression is supported in $\{x \in \mathbb{R}_+^d : 1/n < x_d < 2/n\}$, whence

$$\|D_d^{\ell-k}(\psi_n - 1)D^\beta D_d^k u\|_{L^2(\mathbb{R}_+^d)}^2$$
$$\leq C_1 n^{2(\ell-k)} \int_{\mathbb{R}^{d-1}} \int_{1/n}^{2/n} |D^\beta D_d^k u(x', x_d)|^2 \, dx_d \, dx'.$$

Taylor's theorem implies that for $x = (x', x_d) \in \mathbb{R}^d$,

$$D_d^k u(x', x_d) = D_d^k u(x', 0) + \cdots + \frac{1}{(\ell - k - 1)!} D_d^{\ell - 1} u(x', 0)$$
$$+ \frac{1}{(\ell - k - 1)!} \int_0^{x_d} (x_d - t)^{\ell - k - 1} D_d^\ell u(x', t) \, dt,$$

which reduces to the last term since $\ell \leq m$ and $\gamma u = 0$. Thus

$$\|D_d^{\ell - k}(\psi_n - 1) D^\beta D_d^k u\|_{L^2(\mathbb{R}_+^d)}^2$$

$$\leq C_2 n^{2(\ell - k)} \int_{\mathbb{R}^{d-1}} \int_{1/n}^{2/n} \left| \int_0^{x_d} (x_d - t)^{\ell - k - 1} D^\beta D_d^\ell u(x', t) \, dt \right|^2 dx_d \, dx'$$

$$\leq C_3 n^{2(\ell - k)} n^{-2(\ell - k - 1)} \int_{\mathbb{R}^{d-1}} \int_{1/n}^{2/n} \left(\int_0^{2/n} |D^\beta D_d^\ell u(x', t)| \, dt \right)^2 dx_d \, dx'$$

$$\leq C_4 n^2 \int_{\mathbb{R}^{d-1}} \frac{1}{n^2} \int_0^{2/n} |D^\beta D_d^\ell u(x', t)|^2 \, dt \, dx'$$

$$\longrightarrow 0 \quad \text{as } n \to \infty$$

since the measure of the inner integral tends to 0. Thus the claim is established and the proof is complete. $\qquad \square$

7.7 THE $W^{s,p}(\Omega)$ SOBOLEV SPACES

The $L^2(\Omega)$ results of the last two sections can be generalized to $L^p(\Omega)$, and to noninteger numbers of derivatives. We do not go into detail about this matter, but summarize a few of the important results. See, e.g., [1] for details and precise statements.

Definition. Suppose $\Omega \subset \mathbb{R}^d$, $1 \leq p \leq \infty$, and $s > 0$ such that $s = m + \sigma$ where $0 < \sigma < 1$ and m is an integer. Then we define for a smooth function u,

$$\|u\|_{W^{s,p}(\Omega)} = \left\{ \|u\|_{W^{m,p}(\Omega)}^p + \sum_{|\alpha| = m} \int_\Omega \int_\Omega \frac{|D^\alpha u(x) - D^\alpha u(y)|^p}{|x - y|^{d + \sigma p}} \, dx \, dy \right\}^{1/p}$$

if $p < \infty$, and otherwise

$$\|u\|_{W^{s,\infty}(\Omega)} = \max \left\{ \|u\|_{W^{m,\infty}(\Omega)}, \max_{|\alpha| = m} \operatorname{ess\,sup}_{\substack{x, y \in \Omega \\ x \neq y}} \frac{|D^\alpha u(x) - D^\alpha u(y)|}{|x - y|^\sigma} \right\}.$$

Proposition 7.39. *For any* $1 \leq p \leq \infty$, $\| \cdot \|_{W^{s,p}(\Omega)}$ *is a norm.*

Definition. We let $W^{s,p}(\Omega)$ be the completion of $C^\infty(\Omega)$ under the norm $\| \cdot \|_{W^{s,p}(\Omega)}$, and $W_0^{s,p}(\Omega)$ is the completion of $C_0^\infty(\Omega)$.

Clearly $W^{s,p}(\Omega)$ and $W_0^{s,p}(\Omega)$ are Banach spaces.

Proposition 7.40. *If* $\Omega = \mathbb{R}^d$ *or* Ω *has a Lipschitz boundary, then*

$$W^{s,2}(\Omega) = H^s(\Omega) \quad and \quad W_0^{s,2}(\Omega) = H_0^s(\Omega).$$

Thus we have an equivalent norm on $H^s(\Omega)$ *given above.*

If $1 \le p < \infty$ and $m = s$ is not an integer, then we have analogues of the Sobolev embedding theorem, the Rellich-Kondrachov theorem, and the trace theorem. For the trace theorem, every time a trace is taken on a hypersurface of one less dimension (as from Ω to $\partial\Omega$), $1/p$ derivative is lost, rather than $1/2$.

7.8 EXERCISES

1. Prove Proposition 7.1.

2. Prove directly that for $f \in H^1(\mathbb{R}^d)$, $\|f\|_{H^1(\mathbb{R}^d)}$ is equivalent to

$$\left\{ \int_{\mathbb{R}^d} (1 + |\xi|^2) |\hat{f}(\xi)|^2 \, d\xi \right\}^{1/2}.$$

Can you generalize this result and your proof to $H^k(\mathbb{R}^d)$?

3. For $f \in H_0^1(0,1)$, use the fundamental theorem of calculus to show that there is some constant $C > 0$ such that

$$\|f\|_{L^2(0,1)} \le C \|f'\|_{L^2(0,1)}.$$

If instead $f \in \{g \in H^1(0,1) : \int_0^1 g(x)\,dx = 0\}$, prove a similar estimate.

4. Prove that $\delta_0 \notin (H^1(\mathbb{R}^d))^*$ for $d \ge 2$, but that $\delta_0 \in (H^1(\mathbb{R}))^*$. You will need to define what δ_0 applied to $f \in H^1(\mathbb{R})$ means.

5. Prove directly that $H^1(0,1)$ is continuously embedded in $C_B^0(0,1)$. Recall that $C_B^0(0,1)$ is the set of bounded and continuous functions on $(0,1)$.

6. Suppose that $\Omega \subset \mathbb{R}^d$ is a bounded set and $\{U_j\}_{j=1}^N$ is a finite collection of open sets in $\mathbb{R}^d$ that cover the closure of Ω (i.e., $\bar{\Omega} \subset \bigcup_{j=1}^N U_j$). Prove that there exists a finite C^∞ partition of unity in Ω subordinate to the cover. That is, construct $\{\phi_k\}_{k=1}^M$ such that $\phi_k \in C_0^\infty(\mathbb{R}^d)$, $\text{supp}(\phi_k) \subset U_{j_k}$ for some j_k, and

$$\sum_{k=1}^M \phi_k(x) = 1.$$

7. Suppose that $\Omega \subset \mathbb{R}^d$ is a (possibly unbounded) domain and $\{U_\alpha\}_{\alpha \in \mathcal{I}}$ is a collection of open sets in $\mathbb{R}^d$ that cover Ω (i.e., $\Omega \subset \bigcup_{\alpha \in \mathcal{I}} U_\alpha$). Prove that

there exists a locally finite C^∞ partition of unity in Ω subordinate to the cover. That is, there exists a sequence $\{\psi_j\}_{j=1}^\infty \subset C_0^\infty(\mathbb{R}^d)$ such that

(i) for every K compactly contained in Ω, all but finitely many of the ψ_j vanish on K;

(ii) each $\psi_j \geq 0$ and $\displaystyle\sum_{j=1}^\infty \psi_j(x) = 1$ for every $x \in \Omega$;

(iii) for each j, the support of ψ_j is contained in some U_{α_j}, $\alpha_j \in \mathcal{I}$.

[Hint: let S be a countable dense subset of Ω (e.g., points with rational coordinates). Consider the countable collection of balls $\mathcal{B} = \{B_r(x) \subset \mathbb{R}^d :$ r is rational, $x \in S$, and $B_r(x) \subset U_\alpha$ for some $\alpha \in \mathcal{I}\}$. Order the balls and construct on $B_j = B_{r_j}(x_j)$ a function $\phi_j \in C_0^\infty(B_j)$ such that $0 \leq \phi_j \leq 1$ and $\phi_j = 1$ on $B_{r_j/2}(x_j)$. Then $\psi_1 = \phi_1$ and $\psi_j = (1 - \phi_1) \cdots (1 - \phi_{j-1})\phi_j$ should work.]

8. Let $u \in \mathcal{D}'(\mathbb{R}^d)$ and $\phi \in \mathcal{D}(\mathbb{R}^d)$. For $y \in \mathbb{R}^d$, the translation operator τ_y is defined by $\tau_y\phi(x) = \phi(x - y)$.

(a) Show that

$$u(\tau_y\phi) - u(\phi) = \int_0^1 \sum_{j=1}^d y_j \frac{\partial u}{\partial x_j}(\tau_{ty}\phi)\, dt.$$

(b) Apply this to

$$f \in W_{\text{loc}}^{1,1}(\mathbb{R}^d) = \left\{ f \in L_{\text{loc}}^1(\mathbb{R}^d) : \frac{\partial f}{\partial x_j} \in L_{\text{loc}}^1(\mathbb{R}^d) \text{ for all } j \right\}$$

to show that, for a.e. x,

$$f(x + y) - f(x) = \int_0^1 y \cdot \nabla f(x + ty)\, dt.$$

(c) Let the locally Lipschitz functions be defined as

$$C_{\text{loc}}^{0,1}(\mathbb{R}^d) = \{f \in C^0(\mathbb{R}^d) : \forall R > 0, \text{there is some } L_{R,f} \text{ depending on } R$$
$$\text{and } f \text{ such that } |f(x) - f(y)| \leq L_{R,f}|x - y| \,\forall x, y \in B_R(0)\}.$$

Conclude that $W_{\text{loc}}^{1,\infty}(\mathbb{R}^d) \subset C_{\text{loc}}^{0,1}(\mathbb{R}^d)$.

9. Show the following counterexamples.

(a) There is no embedding of $W^{1,p}(\Omega) \hookrightarrow L^q(\Omega)$ for $1 \leq p < d$ and $q > dp/(d - p)$. Let $\Omega \subset \mathbb{R}^d$ be bounded and contain 0, and let $f(x) = |x|^r$. Find r so that $f \in W^{1,p}(\Omega)$ but $f \notin L^q(\Omega)$.

(b) There is no embedding of $W^{1,p}(\Omega) \hookrightarrow C_B^0(\Omega)$ for $1 \leq p < d$. Note that in the previous case, f is not bounded. What can you say about which (negative) Sobolev spaces the Dirac mass lies in?

(c) There is no embedding of $W^{1,p}(\Omega) \hookrightarrow L^\infty(\Omega)$ for $1 < p = d$. Let $\Omega \subset \mathbb{R}^d = B_R(0)$ and let $f(x) = \log(\log(4R/|x|))$. Show $f \in W^{1,p}(B_R(0))$.

(d) The set $C^\infty \cap W^{1,\infty}$ is not dense in $W^{1,\infty}$. Show that if $\Omega = (-1,1)$ and $u(x) = |x|$, then $u \in W^{1,\infty}$ but $u(x)$ is not the limit of C^∞ functions in the $W^{1,\infty}$-norm.

10. If $R = (-1,1)^d$ is the cube centered at 0, verify that

$$\mathbb{R}^d = \bigcup_{\beta \in \mathbb{Z}^d} (R + \beta).$$

Can you determine the maximal overlap of these translated cubes as a function of d?

11. Suppose that $f_j \in H^2(\Omega)$ for $j = 1, 2, \ldots$, $f_j \rightharpoonup f$ weakly in $H^1(\Omega)$, and $D^\alpha f_j \rightharpoonup g_\alpha$ weakly in $L^2(\Omega)$ for all multi-indices α such that $|\alpha| = 2$. Show that $f \in H^2(\Omega)$, $D^\alpha f = g_\alpha$, and, for a subsequence, $f_j \to f$ strongly in $H^1(\Omega)$.

12. Suppose that $\Omega \subset \mathbb{R}^d$ is bounded with a Lipschitz boundary and $f_j \rightharpoonup f$ and $g_j \rightharpoonup g$ weakly in $H^1(\Omega)$. Show that, for a subsequence, $\nabla(f_j g_j) \to \nabla(fg)$ as a distribution. Find all p in $[1, \infty]$ such that the convergence can be taken weakly in $L^p(\Omega)$.

13. Interpolation inequalities.

(a) Show that for $f \in H^1(\mathbb{R}^d)$ and $0 \leq s \leq 1$,

$$\|f\|_{H^s(\mathbb{R}^d)} \leq \|f\|_{H^1(\mathbb{R}^d)}^s \|f\|_{L^2(\mathbb{R}^d)}^{1-s}.$$

Can you generalize this result to $f \in H^r(\mathbb{R}^d)$ for $r > 0$?

(b) If Ω is bounded and $\partial\Omega$ is smooth, show that there is a constant C such that for all $f \in H^1(\Omega)$,

$$\|f\|_{L^2(\partial\Omega)} \leq C\|f\|_{H^1(\Omega)}^{1/2} \|f\|_{L^2(\Omega)}^{1/2}.$$

[Hint: show this for $d = 1$ on $(0,1)$ by considering

$$f(0)^2 = f(x)^2 - \int_0^x \frac{d}{dt} f(t)^2 \, dt.$$

For $d > 1$, flatten out $\partial\Omega$ and use a one-dimensional type of argument in the normal direction.]

14. Suppose that $\Omega \subset \mathbb{R}^d$ is a bounded domain with Lipschitz boundary and $\{u_k\} \subset H^{2+\epsilon}(\Omega)$ is a bounded sequence, where $\epsilon > 0$. For this problem, you will need to use the fact that the Rellich-Kondrachov theorem extends to fractional order spaces.

(a) Show that there is $u \in H^2(\Omega)$ such that, for a subsequence, $u_j \to u$ in $H^2(\Omega)$.

(b) Find all q and $s \geq 0$ such that, for a subsequence, $u_j \to u$ in $W^{s,q}(\Omega)$.

(c) For a subsequence, $|u_j|^r \nabla u_j \to |u|^r \nabla u$ in $(L^2(\Omega))^d$ for certain $r \geq 1$. For fixed d, how big can r be? Justify your answer.

15. Prove that $H^s(\mathbb{R}^d)$ is continuously embedded in $C_B^0(\mathbb{R}^d)$ if $s > d/2$ by completing the following outline.

(a) Show that $\int_{\mathbb{R}^d} (1 + |\xi|^2)^{-s} d\xi < \infty$.

(b) If $\phi \in \mathcal{S}$ and $x \in \mathbb{R}^d$, write $\phi(x)$ as the Fourier inversion integral of $\hat{\phi}$. Introduce $1 = (1 + |\xi|^2)^{s/2}(1 + |\xi|^2)^{-s/2}$ into the integral and apply Hölder's inequality to obtain the result for Schwartz class functions.

(c) Use density to extend the above result to $H^s(\mathbb{R}^d)$.

16. Suppose $f \in L^2(\mathbb{R})$ and $\hat{\omega}(\xi) = \sqrt{|\xi|}$. Make sense of the definition $g = \omega * f$, and determine s such that $g \in H^s(\mathbb{R})$.

17. Suppose that $\omega \in L^1(\mathbb{R})$, $\omega(x) > 0$, and ω is even. Moreover, for $x > 0$, $\omega \in C^2[0, \infty)$, $\omega'(x) < 0$, and $\omega''(x) > 0$. Consider the equation for u,

$$\omega * u - u'' = f \in L^2(\mathbb{R}).$$

(a) Show that $\hat{\omega}(\xi) > 0$. [Hint: use that ω is even, integrate by parts, and consider subintervals of size $2\pi/|\xi|$.]

(b) Find a fundamental solution to the differential equation (i.e., replace f by δ_0). You may leave your answer in terms of an inverse Fourier transform.

(c) For the original problem, find the solution operator as a convolution operator.

(d) Show that the solution $u \in H^2(\mathbb{R})$.

18. Let H be a Hilbert space and Z a closed linear subspace. Prove that H/Z has the inner-product

$$(\hat{x}, \hat{y})_{H/Z} = (P_Z^\perp x, P_Z^\perp y)_H,$$

and that H/Z is isomorphic to $Z^\perp$.

19. An *equivalence relation* on a set X is a binary relation, often denoted $\sim$, satisfying for all $x, y, z \in X$:

(i) $x \sim x$ (reflexivity);

(ii) $x \sim y$ if and only if $y \sim x$ (symmetry);

(iii) $x \sim y$ and $y \sim z$ implies $x \sim z$ (transitivity).

When $x \sim y$, we say that x is equivalent to y.

Let X be a vector space and $Z \subset X$ be a subspace. The *coset* of $x \in X$ is $\hat{x} = x + Z \subset X$. Show that the relation $x \sim y$ if and only if $\hat{x} = \hat{y}$ is an equivalence relation.

20. Elliptic regularity theory shows that if the domain $\Omega \subset \mathbb{R}^d$ has a smooth boundary and $f \in H^s(\Omega)$, then $-\Delta u = f$ in Ω, $u = 0$ on $\partial\Omega$, has a unique solution $u \in H^{s+2}$. For what values of s will u be continuous? Can you be sure that a fundamental solution is continuous? The answers depend on d.

21. Consider $x \in \mathbb{R}^{n \times n}$ with associated ℓ^2-norm $\|x\| = \left(\sum_{j=1}^{n} \sum_{k=1}^{n} |x_{j,k}|^2 \right)^{1/2}$

and total variation seminorm

$$|x|_{TV} = \sum_{j=1}^{n} \sum_{k=1}^{n-1} |x_{j,k+1} - x_{j,k}| + \sum_{j=1}^{n-1} \sum_{k=1}^{n} |x_{j+1,k} - x_{j,k}|.$$

(a) Why is $|\cdot|_{TV}$ not a norm?

(b) Prove directly the following Sobolev inequality: if $x_{1,j} = x_{j,1} = 0$ for all $1 \le j \le n$, then

$$\|x\| \le \frac{1}{2} |x|_{TV}.$$

[Hint: find two ways to write $x_{j,k}$ as a sum of discrete differences.]

(c) Does the result in (b) extend to ℓ^2 (assuming we reorder ℓ^2 to possess a two-dimensional indexing structure)?

22. Let $\Omega \subset \mathbb{R}^d$ be a domain with a Lipschitz boundary and let $w \in L^\infty(\Omega)$. Define

$$H_w(\Omega) = \{ f \in L^2(\Omega) : \nabla(wf) \in (L^2(\Omega))^d \}.$$

(a) Give reasonable conditions on w so that $H_w(\Omega) = H^1(\Omega)$.

(b) Prove that $H_w(\Omega)$ is a Hilbert space. What is the inner-product?

(c) Can you define the trace of f on $\partial\Omega$? [Hint: first consider the trace of wf.]

(d) Characterize $H_w(\mathbb{R})$ if $w(x)$ is the Heaviside function.

CHAPTER 8

Boundary Value Problems

The content we dwell on in this chapter is a first introduction to a vast area of the mathematical literature. It has roots in the studies of Leonhard Euler, Jean d'Alembert, Joseph-Louis Lagrange and Pierre-Simon Laplace, mostly in the 18th century. These folks derived and analysed what we would now designate the fundamental equations of mathematical physics. The study of these equations continued into the 19th century and involved some of the most gifted practitioners of the era. One of the burning questions in the second half of the 19th century was whether or not the German mathematician Bernard Riemann's solution of Laplace's equation complete with boundary conditions by way of Dirichlet's principle could be rigorously realized. This brought the question of boundary-value problems to the forefront. (The particular issue was resolved in the affirmative in the early 20th century by Hilbert.) Toward the end of the 19th century, it was realized, and enunciated explicitly for example by the French mathematician Henri Poincaré, that while these model equations originated in many different disciplines, the means of understanding them was universal. The study of boundary-value problems had found a permanent place in the mathematical firmament. As the 20th century progressed, such problems have arisen in other areas of mathematics, not arising from applications.

Considered in this chapter are boundary-value problems for some of these very simple partial differential equations (PDEs) important in science and engineering. The equations will be posed on a bounded Lipschitz domain $\Omega \subset \mathbb{R}^d$, where typically d is 1, 2, or 3. We also impose auxiliary conditions on the boundary $\partial\Omega$ of the domain, called *boundary conditions* (BCs). A PDE together with its BCs constitute a *boundary value problem* (BVP). We tacitly assume throughout most of this chapter that the underlying field $\mathbb{F} = \mathbb{R}$.

Here, we only touch the surface of this subject. Our object is to introduce students to the larger ideas in the subject and to show off the use of much of what has been developed heretofore. Indeed, historically, much of what has come before in this text was developed with exactly these sort of issues in mind.

DOI: 10.1201/9781003492139-8

It will be helpful to make the following remark before we begin. The divergence theorem, also called the Gauss theorem, implies that for vector $\psi \in (C^1(\bar{\Omega}))^d$ and scalar $\phi \in C^1(\bar{\Omega})$,

$$\int_\Omega \nabla \cdot (\phi\psi) \, dx = \int_{\partial\Omega} \phi\psi \cdot \nu \, d\sigma(x),$$

where ν is the unit outward normal vector (which is defined almost everywhere on the boundary of a Lipschitz domain) and $d\sigma$ is the $(d-1)$-dimensional measure on $\partial\Omega$. Since

$$\nabla \cdot (\phi\psi) = \nabla\phi \cdot \psi + \phi \, \nabla \cdot \psi,$$

the integration by parts formula in $\mathbb{R}^d$ says

$$\int_\Omega \phi \, \nabla \cdot \psi \, dx = -\int_\Omega \nabla\phi \cdot \psi \, dx + \int_{\partial\Omega} \phi\psi \cdot \nu \, d\sigma(x). \qquad (8.1)$$

By density, this formula extends immediately to the case where merely $\phi \in H^1(\Omega)$ and $\psi \in (H^1(\Omega))^d$. Note that the trace theorem (Theorem 7.36) gives meaning to the boundary integral.

8.1   SECOND ORDER LINEAR ELLIPTIC PDES

Let $\Omega \subset \mathbb{R}^d$ be some bounded Lipschitz domain. The general second order linear elliptic PDE in divergence form for the unknown function u is

$$-\nabla \cdot (a\nabla u + bu) + cu = f \quad \text{in } \Omega, \qquad (8.2)$$

where a is a $d \times d$ matrix, b is a d-vector, and c and f are functions. To be physically relevant and mathematically well-posed, it is often the case that $c \geq 0$, $|b|$ is not too large (in a sense to be made clear later), and the matrix a is uniformly positive definite, as defined next.

Definition. If $\Omega \subset \mathbb{R}^d$ is a domain and $a : \bar{\Omega} \to \mathbb{R}^{d \times d}$ is a matrix, then a is *positive definite* if for a.e. $x \in \bar{\Omega}$,

$$\xi^T a(x)\xi > 0 \quad \forall \xi \in \mathbb{R}^d, \; \xi \neq 0,$$

and a is merely *positive semidefinite* if only $\xi^T a(x)\xi \geq 0$. Moreover, a is *uniformly positive definite* if there is some constant $a_* > 0$ such that for a.e. $x \in \bar{\Omega}$,

$$\xi^T a(x)\xi \geq a_* |\xi|^2 \quad \forall \xi \in \mathbb{R}^d.$$

We remark that positive definiteness of a insures that

$$a\nabla u \cdot \nabla u \geq 0.$$

The positivity of this term can be exploited mathematically. It is also related to physical principles. In many applications, ∇u is the direction of a force and $a\nabla u$ is the direction of a response. Positive definiteness says that the response is generally in the direction of the force, possibly deflected a bit, but never more than $90°$.

8.1.1 Practical examples

Here are some examples of systems governed by (8.2).

Example (Steady-state conduction of heat). Let $\Omega \subset \mathbb{R}^3$ be a solid body, $u(x)$ the temperature of the body at $x \in \Omega$, and $f(x)$ an external source or sink of heat energy. The heat flux is a vector in the direction of the flow of heat, with magnitude given as the amount of heat energy that passes through an infinitesimal planar region orthogonal to the direction of flow divided by the area of the infinitesimal region, per unit time. Fourier's law of heat conduction says that the heat flux is $-a\nabla u$, where $a(x)$, the thermal conductivity of the body, is positive definite. Thus, heat flows generally from hot to cold. Finally, $s(x)$ is the specific heat of the body; it measures the amount of heat energy that can be stored per unit volume of the body per degree of temperature. The physical principle governing the system is energy conservation. If $V \subset \Omega$, then the total heat inside V is $\int_V su\,dx$. This total changes in time with the external heat added due to f minus the heat lost due to movement through ∂V; thus,

$$\frac{d}{dt}\int_V su\,dx = \int_V f\,dx - \int_{\partial V}(-a\nabla u)\cdot\nu\,d\sigma(x),$$

where, as always, ν is the outer unit normal vector. Applying the divergence theorem, the last term is

$$\int_{\partial V} a\nabla u\cdot\nu\,d\sigma(x) = \int_V \nabla\cdot(a\nabla u)\,dx,$$

and so, assuming the derivative may be moved inside the integral,

$$\int_V \left(\frac{\partial su}{\partial t} - \nabla\cdot(a\nabla u)\right)dx = \int_V f\,dx.$$

This holds for every $V \subset \Omega$ with a reasonable boundary. By a modification of Lebesgue's lemma, we conclude that, except on a set of measure zero,

$$\frac{\partial(su)}{\partial t} - \nabla\cdot(a\nabla u) = f.$$

In steady-state, the time derivative vanishes, and we have (8.2) with $b = 0$ and $c = 0$. But suppose that $f(x) = f(u(x), x)$ depends on the temperature itself; that is, the external world will add or subtract heat at x depending on the temperature found there. For example, a room Ω may have a thermostatically controlled heater/air conditioner $f = F(u, x)$. Suppose further that $F(u, x) = c(x)(u_{\text{ref}}(x) - u)$ for some $c \geq 0$ and reference temperature $u_{\text{ref}}(x)$. Then

$$\frac{\partial(su)}{\partial t} - \nabla\cdot(a\nabla u) = c(u_{\text{ref}} - u),$$

and, in steady-state, we have (8.2) with $b = 0$ and $f = cu_{\text{ref}}$. Note that if $c \geq 0$ and $u \leq u_{\text{ref}}$, then $F \geq 0$ and heat energy is added, tending to increase u. Conversely, if $u \geq u_{\text{ref}}$, u tends to decrease. In fact, in time, $u \to u_{\text{ref}}$. However, if $c < 0$, we have a potentially unphysical situation, in which hot areas (i.e., $u > u_{\text{ref}}$) tend to get even hotter and cold areas even colder. The steady-state configuration would be to have $u = +\infty$ in the hot regions and $u = -\infty$ in the cold regions! Thus $c \geq 0$ should be demanded on physical grounds (later it will be required on mathematical grounds as well).

Example (The electrostatic potential). Let u be the electrostatic potential, for which the electric flux is $-a\nabla u$ for some a that describes the electrostatic permitivity of the medium Ω. Conservation of charge over an arbitrary volume in Ω, the divergence theorem, and the Lebesgue lemma give (8.2) with $c = 0$ and $b = 0$, where f represents the electrostatic charges.

Example (Steady-state fluid flow in a porous medium). The equations of steady-state flow of a nearly incompressible, single phase fluid in a porous medium are similar to those for the flow of heat. In this case, u is the fluid pressure. Darcy's law gives the volumetric fluid flux (also called the *Darcy velocity*) as $-a(\nabla u - g\rho)$, where a is the permeability of the medium Ω divided by the fluid viscosity, g is the gravitational vector, and ρ is the fluid density. The total mass in volume $V \subset \Omega$ is $\int_V \rho\, dx$, and this quantity changes in time due to external sources (or sinks, if negative, such as wells) represented by f and *mass* flow through ∂V. The mass flux is given by multiplying the volumetric flux by ρ. That is, with t being time,

$$\frac{d}{dt}\int_V \rho\, dx = \int_V f\, dx - \int_{\partial V} -\rho a(\nabla u - g\rho) \cdot \nu\, d\sigma(x)$$
$$= \int_V f\, dx + \int_V \nabla \cdot [\rho a(\nabla u - g\rho)]\, dx,$$

and we conclude that, provided we can take the time derivative inside the integral,

$$\frac{\partial \rho}{\partial t} - \nabla \cdot [\rho a(\nabla u - g\rho)] = f.$$

Generally speaking, $\rho = \rho(u)$ depends on the pressure u through an equation-of-state, so this is a time dependent, nonlinear equation. If we assume steady-state flow, we can drop the first term. We might also simplify the equation-of-state if $\rho(u) \approx \rho_0$ is nearly constant (at least over the pressures being encountered). One choice uses

$$\rho(u) \approx \rho_0 + \gamma(u - u_0),$$

where γ and u_0 are fixed (note that these are the first two terms in a Taylor approximation of ρ about u_0). Substituting this in the equation above results in

$$-\nabla \cdot \{a[(\rho_0 + \gamma(u - u_0))\nabla u - g(\rho_0 + \gamma(u - u_0))^2]\} = f.$$

This is still nonlinear, so a further simplification would be to linearize the equation (i.e., assume $u \approx u_0$ and drop all higher order terms involving $u - u_0$). Since $\nabla u = \nabla(u - u_0)$, we obtain finally

$$-\nabla \cdot \{\rho_0 a[\nabla u - g(\rho_0 + 2\gamma(u - u_0))]\} = f,$$

which is (8.2) with a replaced by $\rho_0 a$, $c = 0$, $b = -2\rho_0 a g \gamma$, and f replaced by $f - \nabla \cdot [\rho_0 a g(\rho_0 - 2\gamma u_0)]$.

8.1.2 Boundary conditions (BCs)

In each of the previous examples, we determined the equation governing the behavior of the system, given the external forcing term f distributed over the domain Ω. However, the description of each system is incomplete, since we must also describe the external interaction with the world through its boundary $\partial\Omega$.

Boundary conditions generally take one of three forms, though others are possible depending on the system being modeled. Let Γ_D, Γ_N, and Γ_R be open subsets of $\partial\Omega$ that are are mutually disjoint (so $\Gamma_D \cap \Gamma_N = \Gamma_D \cap \Gamma_R = \Gamma_N \cap \Gamma_R = \emptyset$) and satisfy $\partial\Omega = \bar{\Gamma}_D \cup \bar{\Gamma}_N \cup \bar{\Gamma}_R$. The boundary conditions delineated here are

$$u = u_D \qquad \text{on } \Gamma_D, \tag{8.3}$$
$$-(a\nabla u + bu) \cdot \nu = g_N \qquad \text{on } \Gamma_N, \tag{8.4}$$
$$-(a\nabla u + bu) \cdot \nu = g_R(u - u_R) \quad \text{on } \Gamma_R, \tag{8.5}$$

where u_D, u_R, g_N, and g_R are functions with $g_R > 0$. One refers to (8.3) as a *Dirichlet* BC, (8.4) as a *Neumann* BC, and (8.5) as a *Robin* BC.

The Dirichlet BC fixes the value of the (trace of) the unknown function. In the heat conduction example, this would correspond to specifying the temperature on Γ_D.

The Neumann BC fixes the normal component of the flux $-(a\nabla u + bu) \cdot \nu$. The PDE controls the tangential component, as this component of the flux does not leave the domain in an infinitesimal sense. However, the normal component is the flux into or out of the domain, and so it may be fixed in certain cases. In the heat conduction example, $g_N = 0$ would represent a perfectly insulated boundary, as no heat flux may cross the boundary. If instead heat is added to (or taken away from) the domain through some external heater (or refrigerator), we would specify this through nonzero values of g_N.

The Robin BC is a combination of the first two types. It specifies that the flux is proportional to the deviation of u from u_R. If $u = u_R$, there is no flux; otherwise, the flux tends to drive u to u_R, since $g_R > 0$ and a is positive definite. This is a natural boundary condition for the heat conduction problem when the external world is held at a fixed temperature u_R and the body adjusts to it.

The PDE (8.2) and the BCs (8.3)–(8.5) constitute our boundary value problem (BVP). As we will see, this problem is *well-posed*, which means that there exists a unique solution to the system, and that it varies continuously in some norm with respect to changes in the data f, u_D, g_N, g_R and u_R. In the present development of the theory, we leave aside discussion of the Robin boundary condition and confine our attention to pure Dirichlet and pure Neumann BCs.

8.2 VARIATIONAL PROBLEMS AND MINIMIZATION OF ENERGY

For ease of exposition, let us consider the pure Dirichlet BVP

$$\begin{cases} -\nabla \cdot (a\nabla u) + cu = f & \text{in } \Omega, \\ \qquad\qquad\qquad u = u_D & \text{on } \partial\Omega, \end{cases} \tag{8.6}$$

where $b = 0$ and $\Gamma_D = \partial\Omega$. To make classical sense of this problem, we would expect $u \in C^2(\Omega) \cap C^0(\bar{\Omega})$, so we would need to require that $f \in C^0(\Omega)$, $a \in (C^1(\Omega))^{d \times d}$, $c \in C^0(\Omega)$, and $u_D \in C^0(\partial\Omega)$. Often in practice these functions are not quite that well behaved, so we attempt to interpret the problem in a weak or distributional sense.

If merely $f \in L^2(\Omega)$, $a \in (W^{1,\infty}(\Omega))^{d \times d}$, and $c \in L^\infty(\Omega)$, then we should expect $u \in H^2(\Omega)$. Moreover, then $u|_{\partial\Omega} \in H^{3/2}(\partial\Omega)$ is well-defined by the trace theorem. Thus the BVP has a mathematically precise and consistent meaning formulated as: if f, a, and c are as stated and $u_D \in H^{3/2}(\partial\Omega)$, then find $u \in H^2(\Omega)$ such that (8.6) holds. This is *not* an easy problem; fortunately, there is a better formulation using ideas of duality from distribution theory.

We first proceed formally: the calculations will be justified a bit later. First, multiply the PDE by a test function $v \in \mathcal{D}(\Omega)$, integrate in x, and integrate by parts. This yields the formula

$$\int_\Omega (-\nabla \cdot (a\nabla u) + cu)\, v \, dx = \int_\Omega (a\nabla u \cdot \nabla v + cuv)\, dx = \int_\Omega fv \, dx.$$

Notice that the required smoothness of u and v has been evened out. Both are now required to have only first-order derivatives. Now if we only ask that $f \in H^{-1}(\Omega)$, $a \in (L^\infty(\Omega))^{d \times d}$, and $c \in L^\infty(\Omega)$, then we would expect that $u \in H^1(\Omega)$; moreover, we merely need $v \in H_0^1(\Omega)$. This is much less restrictive than asking for $u \in H^2(\Omega)$, so it should be easier to find such a solution satisfying the PDE. Moreover, $u|_{\partial\Omega} \in H^{1/2}(\partial\Omega)$ is still a nice function, and only requires $u_D \in H^{1/2}(\partial\Omega)$.

Remark. Above we wanted to take cu in the same space as f, which was trivially achieved for $c \in L^\infty(\Omega)$. The Sobolev embedding theorem allows us to do better. For example, suppose indeed that $u \in H^1(\Omega)$ and that we want $cu \in L^2(\Omega)$ (to avoid negative index spaces). Then in fact $u \in L^q(\Omega)$ for any

$1 \leq q \leq 2d/(d-2)$ if $d \geq 2$ (but q finite) and $u \in C_B(\Omega) \subset L^\infty(\Omega)$ if $d = 1$. Thus we can take

$$c \in \begin{cases} L^2(\Omega) & \text{if } d = 1, \\ L^{2+\epsilon}(\Omega) & \text{if } d = 2 \text{ for any } \epsilon > 0, \\ L^d & \text{if } d \geq 3, \end{cases}$$

and obtain $cu \in L^2(\Omega)$ as desired.

With this reduced regularity requirement on u ($u \in H^1(\Omega)$, not $H^2(\Omega)$), the problem can be formulated rigorously as what is called a *variational problem* (VP) as outlined now.

The PDE (8.6) involves the *linear* operator

$$A \equiv -\nabla \cdot a\nabla + c : H^1(\Omega) \to H^{-1}(\Omega),$$

with range as specified by Corollary 7.11. We will transform A into a *bilinear* operator

$$B : H^1(\Omega) \times H^1_0(\Omega) \to \mathbb{R}.$$

Assume that $u \in H^1(\Omega)$ solves the PDE (we will show existence of a solution later), and take a test function $v \in H^1_0(\Omega)$. Then

$$\langle -\nabla \cdot (a\nabla u) + cu, v \rangle_{H^{-1}, H^1_0} = \langle f, v \rangle_{H^{-1}, H^1_0}.$$

Let $\{v_j\}_{j=1}^\infty \subset \mathcal{D}(\Omega)$ be a sequence converging to v in $H^1_0(\Omega)$. Then

$$\begin{aligned} \langle -\nabla \cdot (a\nabla u), v \rangle_{H^{-1}, H^1_0} &= \lim_{j \to \infty} \langle -\nabla \cdot a\nabla u, v_j \rangle_{H^{-1}, H^1_0} \\ &= \lim_{j \to \infty} \langle -\nabla \cdot a\nabla u, v_j \rangle_{\mathcal{D}', \mathcal{D}} \\ &= \lim_{j \to \infty} \langle a\nabla u, \nabla v_j \rangle_{\mathcal{D}', \mathcal{D}} \\ &= \lim_{j \to \infty} (a\nabla u, \nabla v_j)_{L^2(\Omega)} \\ &= (a\nabla u, \nabla v)_{L^2(\Omega)}, \end{aligned}$$

where the "$L^2(\Omega)$" inner-products are actually the one for $(L^2(\Omega))^d$. Thus

$$(a\nabla u, \nabla v)_{L^2(\Omega)} + (cu, v)_{L^2(\Omega)} = \langle f, v \rangle_{H^{-1}, H^1_0}.$$

If $B : H^1(\Omega) \times H^1(\Omega) \to \mathbb{R}$ is defined by

$$B(u, v) = (a\nabla u, \nabla v)_{L^2(\Omega)} + (cu, v)_{L^2(\Omega)} \quad \forall u, v \in H^1(\Omega),$$

and $F : H^1_0(\Omega) \to \mathbb{R}$ by

$$F(v) = \langle f, v \rangle_{H^{-1}, H^1_0},$$

then the PDE has been reduced to the variational problem:

Find $u \in H^1(\Omega)$ such that

$$B(u, v) = F(v) \quad \forall v \in H_0^1(\Omega).$$

What about the boundary condition? Recall that the trace operator γ_0 is continuous and linear with

$$\gamma_0 : H^1(\Omega) \xrightarrow{\text{onto}} H^{1/2}(\partial\Omega).$$

Thus there is some $\tilde{u}_D \in H^1(\Omega)$ such that $\gamma_0(\tilde{u}_D) = u_D \in H^{1/2}(\partial\Omega)$. It is therefore required that

$$u \in H_0^1(\Omega) + \tilde{u}_D,$$

so that $\gamma_0(u) = \gamma_0(\tilde{u}_D) = u_D$. For convenience, we no longer distinguish between u_D and its extension $\tilde{u}_D$. This construction is summarized below.

Theorem 8.1. *If $\Omega \subset \mathbb{R}^d$ is a domain with a Lipschitz boundary, and $a \in (L^\infty(\Omega))^{d \times d}$, $c \in L^\infty(\Omega)$, $f \in H^{-1}(\Omega)$, and $u_D \in H^1(\Omega)$, then the BVP for $u \in H^1(\Omega)$,*

$$\begin{cases} -\nabla \cdot (a\nabla u) + cu = f & \text{in } \Omega, \\ u = u_D & \text{on } \partial\Omega, \end{cases} \tag{8.7}$$

is equivalent to the variational problem:

Find $u \in H_0^1(\Omega) + u_D$ such that

$$B(u, v) = F(v) \quad \forall v \in H_0^1(\Omega), \tag{8.8}$$

where $B : H^1(\Omega) \times H^1(\Omega) \to \mathbb{R}$ is

$$B(u, v) = (a\nabla u, \nabla v)_{L^2(\Omega)} + (cu, v)_{L^2(\Omega)}$$

and $F : H_0^1(\Omega) \to \mathbb{R}$ is

$$F(v) = \langle f, v \rangle_{H^{-1}(\Omega), H_0^1(\Omega)}.$$

Actually, we showed that a solution to the BVP (8.7) gives a solution to the VP (8.8). By reversing the steps leading to this conclusion, the converse implication is validated. Note that along the way, the integration by parts formula (8.1) has been extended to the case where $\phi = v \in H_0^1(\Omega)$ and merely $\psi = -a\nabla u \in (L^2(\Omega))^d$.

The connection between the BVP (8.7) and the variational problem (8.8) is further illuminated by considering the following *energy functional*.

Definition. If a is symmetric (i.e., $a = a^T$), then the *energy functional* $J : H_0^1(\Omega) \to \mathbb{R}$ for (8.7) is

$$J(v) = \frac{1}{2}\left[(a\nabla v, \nabla v)_{L^2(\Omega)} + (cv, v)_{L^2(\Omega)}\right]$$
$$- \langle f, v \rangle_{H^{-1}(\Omega), H_0^1(\Omega)} + (a\nabla u_D, \nabla v)_{L^2(\Omega)} + (cu_D, v)_{L^2(\Omega)}. \tag{8.9}$$

The differential calculus in infinite dimensions and the calculus of variations will be studied in Chapters 9 and 10. This is presaged by the following simple computation here.

The present claim is that any solution of (8.7), minus u_D, minimizes the "energy" $J(v)$ just introduced. To see this might be a valid assertion, let $v \in H_0^1(\Omega)$ and compute

$$
\begin{aligned}
J(u - u_D + v) &- J(u - u_D) \\
&= (a\nabla u, \nabla v)_{L^2(\Omega)} + (cu, v)_{L^2(\Omega)} - \langle f, v \rangle_{H^{-1}(\Omega), H_0^1(\Omega)} \\
&\quad + \frac{1}{2}\left[(a\nabla v, \nabla v)_{L^2(\Omega)} + (cv, v)_{L^2(\Omega)}\right],
\end{aligned}
\tag{8.10}
$$

a calculation using the symmetry of a. If u satisfies (8.8), then

$$
J(u - u_D + v) - J(u - u_D) = \frac{1}{2}\left[(a\nabla v, \nabla v)_{L^2(\Omega)} + (cv, v)_{L^2(\Omega)}\right] \geq 0,
$$

provided that a is positive definite and $c \geq 0$. Thus every function in $H_0^1(\Omega)$ has "energy" at least as great as $u - u_D$.

Conversely, if $u - u_D \in H_0^1(\Omega)$ is to minimize the energy $J(v)$, then replacing in (8.10) v by ϵv for $\epsilon \in \mathbb{R}$, $\epsilon \neq 0$, we see that the difference quotient

$$
\begin{aligned}
\frac{1}{\epsilon}&\left[J(u - u_D + \epsilon v) - J(u - u_D)\right] \\
&= (a\nabla u, \nabla v)_{L^2(\Omega)} + (cu, v)_{L^2(\Omega)} - \langle f, v \rangle_{H^{-1}(\Omega), H_0^1(\Omega)} \\
&\quad + \frac{\epsilon}{2}\left[(a\nabla v, \nabla v)_{L^2(\Omega)} + (cv, v)_{L^2(\Omega)}\right],
\end{aligned}
$$

must be nonnegative if $\epsilon > 0$ and nonpositive if $\epsilon < 0$. Taking $\epsilon \to 0$ on the right-hand side shows that the first three terms must be both nonnegative and nonpositive, i.e., zero; thus, u must satisfy (8.8). Note that as $\epsilon \to 0$, the left-hand side is a kind of derivative of J at $u - u_D$. At the minimum, we have a critical point where the derivative vanishes.

Theorem 8.2. *If the hypotheses of Theorem 8.1 hold, and if $c \geq 0$ and a is symmetric and positive definite, then (8.7) and (8.8) are also equivalent to the minimization problem:*

Find $u \in H_0^1(\Omega) + u_D$ such that

$$
J(u - u_D) \leq J(v) \quad \forall v \in H_0^1(\Omega),
$$

where J is given above in (8.9).

Remark. As a consequence, the solution u to Laplace's equation $-\Delta u = 0$ with a Dirichlet BC can be obtained as the minimizer of an energy functional. This fact is called *Dirichlet's principle.*

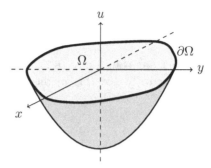

Figure 8.1 A thin membrane attached to a fixed frame represented by $\partial\Omega$ that sags under the influence of gravity.

The physical principles of conservation and energy minimization are equivalent in this context, and they are connected by the variational problem. First, the VP is the *weak form* of the BVP, given by multiplying by a test function, integrating, and integrating by parts to even out the number of derivatives on the solution and the test function. Second, the VP also defines a critical point of the energy functional where it is minimized.

Example. As another example of the use of energy functionals, consider a thin membrane stretched over a rigid frame in $\mathbb{R}^3$ of height zero, as depicted in Figure 8.1. Let $\Omega \subset \mathbb{R}^2$ be open in the xy-plane and let $u : \Omega \to \mathbb{R}$ be the height of the membrane, so it can be found at $\{(x, y, u(x, y)) \in \mathbb{R}^3 : (x, y) \in \Omega\}$. The membrane will assume the shape that minimizes the energy, subject to the constraint that it attaches to the rigid frame. The energy functional $E : H^1(\Omega) \to \mathbb{R}$ is a sum of the elastic energy and the gravitational potential energy, *viz.*

$$E(u) = \int_\Omega \left(\frac{1}{2} a |\nabla u|^2 + gu \right) dx,$$

where a is a constant related to the elasticity of the membrane and g is the gravitational constant. This energy E is minimized subject to the constraint that the trace of u, $\gamma_0(u)$, attaches to the frame, i.e., vanishes on the boundary of Ω. This minimization problem gives rise to the boundary value problem

$$\begin{cases} -\nabla \cdot a\nabla u = -g & \text{in } \Omega, \\ u = 0 & \text{on } \partial\Omega, \end{cases}$$

or, equivalently, its variational form.

8.3 THE CLOSED RANGE THEOREM AND LINEAR OPERATORS BOUNDED BELOW

The development continues with an abstract study of equation solvability. In this section, the ground field is not required to be real.

Definition. Let X be an NLS and $Z \subset X$. Then the *orthogonal complement* of Z is
$$Z^\perp = \{x^* \in X^* : \langle x^*, z \rangle_{X^*, X} = 0 \ \forall z \in Z\}.$$

Proposition 8.3. *Let X be an NLS and $Z \subset X$. Then*

(a) $Z^\perp$ *is closed in X^*, and*

(b) $Z \subset (Z^\perp)^\perp$.

Moreover, if $Z \subset X$ is a linear subspace and X is reflexive, then

(c) Z *is closed in X if and only if $Z = (Z^\perp)^\perp$.*

Of course, $(Z^\perp)^\perp \subset X^{**}$, but the natural inclusion $X \subset X^{**}$ has been used implicitly in the above statement.

Proof. For (a), suppose that we have a sequence $\{y_j\}_{j=1}^\infty \subset Z^\perp$ that converges in X^* to y. Then for any $z \in Z$,
$$0 = \langle y_j, z \rangle_{X^*, X} \to \langle y, z \rangle_{X^*, X},$$

so $y \in Z^\perp$ and $Z^\perp$ is closed. Result (b) is a direct consequence of the definitions: for $z \in Z \subset X \subset X^{**}$ it is simply asserted that $z \in (Z^\perp)^\perp$, i.e., that $\langle z, y \rangle_{X^{**}, X^*} = 0$ for all $y \in Z^\perp$. But $\langle z, y \rangle_{X^{**}, X^*} = \langle z, y \rangle_{X, X^*}$, which is 0 by definition.

Finally, for (c), that Z is closed follows from (a). For the other implication, suppose Z is closed. Because (b) is valid, it is only required to show that $(Z^\perp)^\perp \subset Z$. Suppose that there is some nonzero $x \in (Z^\perp)^\perp \subset X^{**} = X$ such that $x \notin Z$. The Hahn-Banach theorem, specifically Lemma 2.35, provides the existence of $f \in ((Z^\perp)^\perp)^*$ such that $f(x) \neq 0$ but $f(z) = 0$ for all $z \in Z$, since Z is linear. That is, $f \in Z^\perp$, so x cannot be in $(Z^\perp)^\perp$, a contradiction. □

Proposition 8.4. *Let X and Y be NLSs and $A : X \to Y$ a bounded linear operator. Then*
$$R(A)^\perp = N(A^*),$$

where $R(A)$ is the range of A and $N(A^*)$ is the null space of A^*.

Proof. Remark that $y \in R(A)^\perp$ if and only if for every $x \in X$,
$$0 = \langle y, Ax \rangle_{Y^*, Y} = \langle A^* y, x \rangle_{X^*, X},$$

which is true if and only if $A^* y = 0$. □

From these simple machinations, the following important theorem emerges.

Theorem 8.5 (closed range theorem). *Let X and Y be NLSs, $Y = Y^{**}$, and $A : X \to Y$ a bounded linear operator. Then $R(A)$ is closed in Y if and only if $R(A) = N(A^*)^{\perp}$.*

This theorem has implications for a class of operators that often arise. We saw them already in Lemma 4.17.

Definition. Let X and Y be NLSs and $A : X \to Y$ linear and bounded. We say that A is *bounded below* if there is some constant $\gamma > 0$ such that

$$\|Ax\|_Y \geq \gamma \|x\|_X \quad \forall x \in X.$$

A linear operator that is bounded below is one-to-one. If it also mapped onto Y, it would have a continuous inverse. We can determine whether $R(A) = Y$ by use of the closed range theorem.

Theorem 8.6. *Let X and $Y = Y^{**}$ be Banach spaces and $A : X \to Y$ a continuous linear operator. Then the following are equivalent:*

(i) *A is bounded below;*

(ii) *A is injective and $R(A)$ is closed;*

(iii) *A is injective and $R(A) = N(A^*)^{\perp}$.*

Proof. The closed range theorem gives the equivalence of (ii) and (iii), and we saw alredy that (i) implies (ii) in Lemma 4.17. So suppose (ii) holds. Then $R(A)$, being closed, is a Banach space itself. Thus $A : X \to R(A)$ is invertible, with continuous inverse by the open mapping theorem (Theorem 2.42). For $x \in X$, compute

$$\|x\|_X = \|A^{-1}Ax\|_X \leq \|A^{-1}\|\|Ax\|_Y,$$

which gives (i) with the constant $\gamma = 1/\|A^{-1}\|$. $\qquad \square$

Corollary 8.7. *Let X and $Y = Y^{**}$ be Banach spaces and $A : X \to Y$ a continuous linear operator. Then A is continuously invertible if and only if A is bounded below and $N(A^*) = \{0\}$ (i.e., A^* is injective).*

8.4   THE BABUŠKA-LAX-MILGRAM THEOREM

It is not difficult at this stage to prove existence of a unique solution to (8.7), or equivalently, (8.8), provided that a is symmetric and uniformly positive definite, $c \geq 0$, and both these functions are bounded. This is because $B(\cdot, \cdot)$ is then an inner-product on $H_0^1(\Omega)$, and this inner-product is equivalent to the usual one.

To see these facts, note that B is bilinear and symmetric (since a is symmetric), and $B(v,v) \geq 0$. For an upper bound, argue as follows:

$$B(v,v) = (a\nabla v, \nabla v) + (cv, v)$$
$$\leq \|a\|_{(L^\infty(\Omega))^{d\times d}} \|\nabla v\|^2_{L^2(\Omega)} + \|c\|_{L^\infty(\Omega)} \|v\|^2_{L^2(\Omega)} \leq C_1 \|v\|^2_{H^1_0(\Omega)},$$

for some constant C_1. A lower bound follows directly if c is *strictly* positive, i.e., bounded below by a positive constant. But one may allow merely $c \geq 0$ by using the following Poincaré-type inequality, which is a direct consequence of Corollary 7.21.

Theorem 8.8 (Poincaré inequality). *If $\Omega \subset \mathbb{R}^d$ is bounded, then there is some constant C_P such that*

$$\|v\|_{H^1_0(\Omega)} \leq C_P \|\nabla v\|_{L^2(\Omega)} \quad \forall v \in H^1_0(\Omega).$$

It follows that

$$B(v,v) = (a\nabla v, \nabla v) + (cv, v) \geq a_* \|\nabla v\|^2_{L^2(\Omega)} \geq (a_*/C_P^2) \|v\|^2_{H^1_0(\Omega)}.$$

Consequently, both $B(v,v) = 0$ implies $v = 0$ and the equivalence of norms is established.

Problem (8.8) becomes:

Find $w = u - u_D \in H^1_0(\Omega)$ such that

$$B(w,v) = F(v) - B(u_D, v) \equiv \tilde{F}(v) \quad \forall v \in H^1_0(\Omega).$$

The functonal $\tilde{F} : H^1_0(\Omega) \to \mathbb{R}$ is clearly linear and it is bounded since

$$|\tilde{F}(v)| \leq |F(v)| + |B(u_D, v)| \leq \left(\|F\|_{H^{-1}(\Omega)} + C\|u_D\|_{H^1(\Omega)}\right)\|v\|_{H^1_0(\Omega)},$$

where, again, C depends on the $L^\infty(\Omega)$-norms of a and c. Thus $\tilde{F} \in (H^1_0(\Omega))^* = H^{-1}(\Omega)$, and the problem devolves to seeking to represent $\tilde{F}$ as the inner product in B with some $w \in H^1_0(\Omega)$. The Riesz representation theorem (Theorem 3.15) gives us a unique such w. We have proved the following theorem.

Theorem 8.9. *If $\Omega \subset \mathbb{R}^d$ is a Lipschitz domain, $f \in H^{-1}(\Omega)$, $u_D \in H^1(\Omega)$, $a \in (L^\infty(\Omega))^{d\times d}$ is uniformly positive definite and symmetric on Ω, and $c \geq 0$ is in $L^\infty(\Omega)$, then there is a unique solution $u \in H^1(\Omega)$ to the BVP (8.7) and, equivalently, the variational problem (8.8). Moreover, there is a constant $C > 0$ such that*

$$\|u\|_{H^1(\Omega)} \leq C\left(\|F\|_{H^{-1}(\Omega)} + \|u_D\|_{H^1(\Omega)}\right). \tag{8.11}$$

This last inequality is a consequence of the facts that

$$\|u\|_{H^1(\Omega)} \leq \|u - u_D\|_{H^1_0(\Omega)} + \|u_D\|_{H^1(\Omega)}$$

and, for some $C > 0$ different from that in (8.11),

$$\|w\|^2_{H^1_0(\Omega)} \leq CB(w,w) = C\tilde{F}(w) \leq C\left(\|F\|_{H^{-1}(\Omega)} + \|u_D\|_{H^1(\Omega)}\right)\|w\|_{H^1_0(\Omega)}.$$

Remark. We leave it as an exercise to show that u is independent of the extension of u_D from $\partial\Omega$ to all of Ω. This extension is not unique. We have merely that once the extension for u_D is fixed, then w is unique. That is, w depends on the extension. The reader is asked to demonstrate that the sum $u = w + u_D$ does *not* depend on the extension chosen. In addition, since the extension operator is bounded,

$$\|u_D\|_{H^1(\Omega)} \le C\|u_D\|_{H^{1/2}(\partial\Omega)},$$

(8.11) can be modified, *viz.*

$$\|u\|_{H^1(\Omega)} \le C\big(\|F\|_{H^{-1}(\Omega)} + \|u_D\|_{H^{1/2}(\partial\Omega)}\big),$$

and thereby refer only to the raw data itself and not the extension.

For more general problems, where either a is not symmetric, or $b \ne 0$ in the original Dirichlet problem (8.6), B is no longer symmetric, so it won't be an inner-product. A generalization of the Riesz Representation theorem is needed to handle this case. In fact, we present this generalization for Banach spaces rather than restricting to Hilbert spaces.

Theorem 8.10 (Babuška-Lax-Milgram theorem). *Let $\mathcal{X}$ and Y be real Banach spaces, and suppose that Y is reflexive, $B : \mathcal{X} \times Y \to \mathbb{R}$ is bilinear, and $X \subset \mathcal{X}$ be a closed subspace. Assume also the following three conditions:*

(a) *B is continuous on $\mathcal{X} \times Y$, i.e., there is some $M > 0$ such that*

$$|B(x,y)| \le M\|x\|_{\mathcal{X}}\|y\|_Y \quad \forall x \in \mathcal{X}, \ y \in Y;$$

(b) *B satisfies the* inf-sup *condition on $X \times Y$, i.e., there is some $\gamma > 0$ such that*

$$\inf_{\substack{x \in X \\ \|x\|_{\mathcal{X}}=1}} \ \sup_{\substack{y \in Y \\ \|y\|_Y=1}} \ B(x,y) \ge \gamma > 0;$$

(c) *and B satisfies the nondegeneracy condition on X that*

$$\sup_{x \in X} B(x,y) > 0 \quad \forall y \in Y, \ y \ne 0.$$

If $x_0 \in \mathcal{X}$ and $F \in Y^*$, then there is a unique u solving the abstract variational problem:

Find $u \in X + x_0 \subset \mathcal{X}$ such that

$$B(u,v) = F(v) \quad \forall v \in Y. \tag{8.12}$$

Moreover, u admits the bound

$$\|u\|_{\mathcal{X}} \le \frac{1}{\gamma}\|F\|_{Y^*} + \left(\frac{M}{\gamma}+1\right)\|x_0\|_{\mathcal{X}}. \tag{8.13}$$

Remark. The inf-sup condition in (b) is often written equivalently as

$$\sup_{\substack{y \in Y \\ y \neq 0}} \frac{B(x,y)}{\|y\|_Y} \geq \gamma \|x\|_X \quad \forall x \in X,$$

and it is also called the *Ladyzhenskaya-Babuška-Brezzi* or *LBB* condition.

In our context, $\mathcal{X} = H^1(\Omega)$, $X = Y = H_0^1(\Omega)$, and $x_0 = u_D$. We call $\mathcal{X}$ the *trial space* (and a member a *trial function*), and Y is called the *test space* (of *test functions*).

Proof. Assume first that $x_0 = 0$. For each fixed $x \in X$, $B(x, \cdot)$ defines a linear functional on Y, since B is linear in each variable separately, so certainly the second. Let A represent the operator that takes x to $B(x, \cdot)$, i.e.,

$$\langle Ax, y \rangle = Ax(y) \equiv B(x, y) \quad \forall x \in X, \ y \in Y.$$

Since (a) implies that

$$|\langle Ax, y \rangle| = |B(x, y)| \leq (M\|x\|_X)\|y\|_Y,$$

Ax is a continuous linear functional and $A : X \to Y^*$. Moreover, A itself is linear, since B is linear in its first variable, and therefore A is a continuous linear operator with the bound

$$\|Ax\|_{Y^*} = \sup_{\|y\|_Y = 1} |\langle Ax, y \rangle| \leq M\|x\|_X.$$

Reformulate (8.12) in terms of A as the problem of finding $u \in X$ such that

$$Au = F.$$

The hypothesis (b) implies that

$$\|Ax\|_{Y^*} \geq \gamma \|x\|_X \quad \forall x \in X, \tag{8.14}$$

so A is bounded below and u, if it exists, must be unique (i.e., A is one-to-one). Since X is closed, it is a Banach space and we conclude that the range of A, $R(A)$, is closed in Y^* (Theorem 8.6). The closed range theorem (Theorem 8.5) now implies that $R(A) = N(A^*)^\perp$. We wish to show that $N(A^*) = \{0\}$, so that A maps onto Y^* (see Corollary 8.7). Suppose that for some $y \in Y = Y^{**}$, $y \in N(A^*)$; that is,

$$B(x, y) = \langle Ax, y \rangle = \langle x, A^*y \rangle = 0 \quad \forall x \in X.$$

Then (c) implies that $y = 0$, so A has a bounded inverse, with $\|A^{-1}\| \leq 1/\gamma$ by (8.14), and $u = A^{-1}F$ solves the problem.

Finally, compute as follows:

$$\|u\|_X = \|A^{-1}F\|_X \leq \|A^{-1}\|\|F\|_{Y^*} \leq \frac{1}{\gamma}\|F\|_{Y^*}.$$

The theorem is established when $x_0 = 0$.

When $x_0 \neq 0$, we can reduce to the previous case, since (8.12) is equivalent to:

Find $w \in X$ such that

$$B(w, v) = \tilde{F}(v) \quad \forall v \in Y,$$

where $u = w + x_0 \in X + x_0 \subset \mathcal{X}$ and

$$\tilde{F}(v) = F(v) - B(x_0, v).$$

As $\tilde{F} \in Y^*$ and

$$|\tilde{F}(v)| \leq |F(v)| + |B(x_0, v)| \leq \left(\|F\|_{Y^*} + M\|x_0\|_{\mathcal{X}} \right)\|v\|_Y.$$

the previous result gives

$$\|w\|_{\mathcal{X}} \leq \frac{1}{\gamma} \left(\|F\|_{Y^*} + M\|x_0\|_{\mathcal{X}} \right),$$

and so

$$\|u\|_{\mathcal{X}} \leq \|w + x_0\|_{\mathcal{X}} \leq \|w\|_{\mathcal{X}} + \|x_0\|_{\mathcal{X}}$$

provides the advertised bound. □

When $X = Y$ is a Hilbert space, things are a bit simpler.

Corollary 8.11 (Lax-Milgram theorem). *Let $\mathcal{H}$ be a real Hilbert space with closed subspace H. Let $B : \mathcal{H} \times \mathcal{H} \to \mathbb{R}$ be a bilinear functional satisfying the following two conditions:*

(i) *B is continuous on $\mathcal{H}$, i.e., there is some $M > 0$ such that*

$$|B(x, y)| \leq M\|x\|_{\mathcal{H}}\|y\|_{\mathcal{H}} \quad \forall x, y \in \mathcal{H};$$

(ii) *B is coercive (or elliptic) on H, which is to say, there is some $\gamma > 0$ such that*
$$B(x, x) \geq \gamma\|x\|_{\mathcal{H}}^2 \quad \forall x \in H.$$

If $x_0 \in \mathcal{H}$ and $F \in H^*$, then there is a unique u solving the abstract variational problem:

Find $u \in H + x_0 \subset \mathcal{H}$ such that

$$B(u, v) = F(v) \quad \forall v \in H.$$

Moreover,

$$\|u\|_{\mathcal{H}} \leq \frac{1}{\gamma}\|F\|_{H^*} + \left(\frac{M}{\gamma} + 1 \right)\|x_0\|_{\mathcal{H}}.$$

Proof. The corollary is just a special case of the theorem except that (ii) has replaced (b) and (c). We claim that (ii) implies both (b) and (c), so the corollary follows.

Easily, we have (c), since for any $y \in H$,

$$\sup_{x \in H} B(x, y) \geq B(y, y) \geq \gamma \|y\|_{\mathcal{H}}^2 > 0$$

whenever $y \neq 0$. Similarly, for any $x \in H$ with norm one,

$$\sup_{\substack{y \in H \\ \|y\|_{\mathcal{H}} = 1}} B(x, y) \geq B(x, x) \geq \gamma > 0,$$

so the infimum over all such x is bounded below by γ, which is (b). □

The Babuška-Lax-Milgram theorem gives the existence of a bounded linear solution operator $S : Y^* \times \mathcal{X} \to \mathcal{X}$ such that $S(F, x_0) = u \in X + x_0 \subset \mathcal{X}$ satisfies

$$B(S(F, x_0), v) = F(v) \quad \forall v \in Y.$$

The bound on S is given by (8.13). This bound shows that the solution varies continuously with the data. That is, by linearity,

$$\|S(F, x_0) - S(G, y_0)\|_{\mathcal{X}} \leq \frac{1}{\gamma} \|F - G\|_{X^*} + \left(\frac{M}{\gamma} + 1\right) \|x_0 - y_0\|_{\mathcal{X}}.$$

So if the data (F, x_0) is perturbed a bit to (G, y_0), then the solution $S(F, x_0)$ changes by a small amount to $S(G, y_0)$, where the magnitudes of the changes are measured in the norms as above.

8.5 APPLICATION TO LINEAR ELLIPTIC PDES

Consider again the BVP (8.7), in the form of the variational problem (8.8). To apply the Lax-Milgram theorem, set $\mathcal{H} = H^1(\Omega)$, $H = H_0^1(\Omega)$, and $x_0 = u_D$. Then $B : H^1(\Omega) \times H^1(\Omega) \to \mathbb{R}$ is continuous, since a and c are bounded, *viz.*

$$
\begin{aligned}
|B(u, v)| &= |(a\nabla u, \nabla v)_{L^2(\Omega)} + (cu, v)_{L^2(\Omega)}| \\
&\leq \|a\|_{(L^\infty(\Omega))^{d \times d}} \|\nabla u\|_{L^2(\Omega)} \|\nabla v\|_{L^2(\Omega)} + \|c\|_{L^\infty(\Omega)} \|u\|_{L^2(\Omega)} \|v\|_{L^2(\Omega)} \\
&\leq M \|u\|_{H^1(\Omega)} \|v\|_{H^1(\Omega)},
\end{aligned}
$$

by Hölder's inequality for some $M > 0$ depending on the bounds for a and c. Coercivity is more interesting. We will only assume that $c \geq 0$, since in practice often $c = 0$. Using that a is uniformly positive definite and Ω is bounded, compute

$$
\begin{aligned}
B(u, u) &= (a\nabla u, \nabla u)_{L^2(\Omega)} + (cu, u)_{L^2(\Omega)} \\
&\geq a_* (\nabla u, \nabla u)_{L^2(\Omega)} = a_* \|\nabla u\|_{L^2(\Omega)}^2 \geq (a_*/C_P^2) \|u\|_{H^1(\Omega)}^2,
\end{aligned}
$$

for some $C_P > 0$, by Poincaré's inequality (Theorem 8.8). Thus there exists a unique solution $u \in H_0^1(\Omega) + u_D$, and

$$\|u\|_{H^1(\Omega)} \leq \frac{C_P^2}{a_*}\|f\|_{H^{-1}(\Omega)} + \left(\frac{C_P^2 M}{a_*} + 1\right)\|u_D\|_{H^1(\Omega)}.$$

Note that the boundary condition $u = u_D$ on $\partial\Omega$ is enforced by our selection of the trial space $H_0^1(\Omega) + u_D$, i.e., the space within which we seek a solution has every member satisfying the boundary condition. Because of this, the Dirichlet BC is called an *essential* BC for this problem.

8.5.1 The general Dirichlet problem

Consider more generally the full elliptic equation (8.2) with a Dirichlet BC:

$$\begin{cases} -\nabla \cdot (a\nabla u + bu) + cu = f & \text{in } \Omega, \\ u = u_D & \text{on } \partial\Omega. \end{cases}$$

We leave it to the reader to show that an equivalent variational problem is:

Find $u \in H_0^1(\Omega) + u_D$ such that

$$B(u, v) = F(v) \quad \forall v \in H_0^1(\Omega),$$

where

$$B(u, v) = (a\nabla u, \nabla v)_{L^2(\Omega)} + (bu, \nabla v)_{L^2(\Omega)} + (cu, v)_{L^2(\Omega)},$$
$$F(v) = \langle f, v \rangle_{H^{-1}(\Omega), H_0^1(\Omega)}.$$

If $b \in (L^\infty(\Omega))^d$ (and a and c are bounded as before), then the bilinear form is bounded. For coercivity, assume again that $c \geq 0$ and a is uniformly positive definite. Then for $v \in H_0^1(\Omega)$,

$$\begin{aligned} B(v, v) &= (a\nabla v, \nabla v)_{L^2(\Omega)} + (bv, \nabla v)_{L^2(\Omega)} + (cv, v)_{L^2(\Omega)} \\ &\geq a_*\|\nabla v\|_{L^2(\Omega)}^2 - |(bv, \nabla v)_{L^2(\Omega)}| \\ &\geq \left(a_*\|\nabla v\|_{L^2(\Omega)} - \|b\|_{(L^\infty(\Omega))^d}\|v\|_{L^2(\Omega)}\right)\|\nabla v\|_{L^2(\Omega)}. \end{aligned}$$

Poincaré's inequality tells us that for some $C_P > 0$,

$$a_*\|\nabla v\|_{L^2(\Omega)} - \|b\|_{(L^\infty(\Omega))^d}\|v\|_{L^2(\Omega)} \geq \left(a_* - C_P\|b\|_{(L^\infty(\Omega))^d}\right)\|\nabla v\|_{L^2(\Omega)}.$$

To continue in the present context, we must assume that for some $\alpha > 0$,

$$a_* - C_P\|b\|_{(L^\infty(\Omega))^d} \geq \alpha > 0;$$

this restricts the size of b relative to a. In this case, it is confirmed that

$$B(v, v) \geq \alpha\|\nabla v\|_{L^2(\Omega)}^2 \geq \frac{\alpha}{C_P^2 + 1}\|v\|_{H^1(\Omega)}^2,$$

and the Lax-Milgram theorem provides a unique solution to the problem as well as a continuous dependence result. Note that in this more general case, if a is not symmetric or $b \neq 0$, then B is not symmetric, so B is not an inner-product. However, continuity and coercivity show that the diagonal of B (i.e., $u = v$) is equivalent to the square of the $H_0^1(\Omega)$-norm.

8.5.2 The Neumann problem with lowest order term

We turn now to the Neumann BVP

$$\begin{cases} -\nabla \cdot (a\nabla u) + cu = f & \text{in } \Omega, \\ -a\nabla u \cdot \nu = g & \text{on } \partial\Omega, \end{cases} \tag{8.15}$$

wherein we have set $b = 0$ for simplicity. This problem is more delicate than the Dirichlet problem, since for $u \in H^1(\Omega)$, we have no meaning in general for $a\nabla u \cdot \nu$. As before with the Dirichlet problem, proceed formally to derive a variational problem by assuming that the trial function u and the test function v are in, say $C^\infty(\bar{\Omega})$. Then the divergence theorem can be applied to obtain

$$-\int_\Omega \nabla \cdot (a\nabla u)\, v\, dx = \int_\Omega a\nabla u \cdot \nabla v\, dx - \int_{\partial\Omega} a\nabla u \cdot \nu v\, dx,$$

or, using the boundary condition and assuming that f and g are nice functions,

$$(a\nabla u, \nabla v)_{L^2(\Omega)} + (cu, v)_{L^2(\Omega)} = (f, v)_{L^2(\Omega)} - (g, v)_{L^2(\partial\Omega)}.$$

These integrals are well-defined on $H^1(\Omega)$, so a well defined variational problem emerges:

Find $u \in H^1(\Omega)$ such that

$$B(u, v) = F(v) \quad \forall v \in H^1(\Omega), \tag{8.16}$$

where $B : H^1(\Omega) \times H^1(\Omega) \to \mathbb{R}$ is

$$B(u, v) = (a\nabla u, \nabla v)_{L^2(\Omega)} + (cu, v)_{L^2(\Omega)}$$

and $F : H^1(\Omega) \to \mathbb{R}$ is

$$F(v) = \langle f, v \rangle_{(H^1(\Omega))^*, H^1(\Omega)} - \langle g, v \rangle_{H^{-1/2}(\partial\Omega), H^{1/2}(\partial\Omega)}. \tag{8.17}$$

It is clear we will require that $f \in (H^1(\Omega))^*$. In addition, for $v \in H^1(\Omega)$, its trace on the boundry lies in $H^{1/2}(\partial\Omega)$, so we merely require $g \in H^{-1/2}(\partial\Omega)$, the dual of $H^{1/2}(\partial\Omega)$. Note that the trace theorem (Theorem 7.37) implies that $F \in (H^1(\Omega))^*$, since $\|v\|_{H^{1/2}(\partial\Omega)} \leq C\|v\|_{H^1(\Omega)}$.

A solution of (8.16) will be called a *weak solution* of (8.15). These problems are not strictly equivalent, because of the boundary condition. For the PDE, consider u satisfying the variational problem. Restrict to test functions $v \in$

$\mathcal{D}(\Omega)$ to avoid $\partial\Omega$ and use the divergence theorem, as in the case of the Dirichlet boundary condition, to see that the differential equation in (8.15) is satisfied in the sense of distributions. This argument can be reversed to see that a solution in $H^1(\Omega)$ to the PDE gives a solution to the variational problem for $v \in \mathcal{D}(\Omega)$. By density, this extned to $v \in H_0^1(\Omega)$. The boundary condition will be satisfied only in some weak sense, i.e., only in the sense of the variational form.

If in fact the solution happens to be in, say, $H^2(\Omega)$, then $a\nabla u \cdot \nu \in H^{1/2}(\partial\Omega)$ and the argument above can be modified to show that indeed $-a\nabla u \cdot \nu = g$. Of course in this case, we must then have that $g \in H^{1/2}(\partial\Omega)$, and of course that $f \in L^2(\Omega)$. So suppose that $u \in H^2(\Omega)$ solves the variational problem (and f and g are as stated). Restrict now to test functions $v \in H^1(\Omega) \cap C^\infty(\bar\Omega)$ to show that

$$\begin{aligned} B(u,v) &= (a\nabla u, \nabla v)_{L^2(\Omega)} + (cu,v)_{L^2(\Omega)} \\ &= -(\nabla \cdot (a\nabla u), v)_{L^2(\Omega)} + (a\nabla u \cdot \nu, v)_{L^2(\partial\Omega)} + (cu,v)_{L^2(\Omega)} \\ &= F(v) = (f,v)_{L^2(\Omega)} - (g,v)_{L^2(\partial\Omega)}. \end{aligned}$$

Using test functions $v \in C_0^\infty$ shows again by the Lebesgue lemma that the PDE is satisfied. Thus, it transpires that

$$(a\nabla u \cdot \nu, v)_{L^2(\partial\Omega)} = -(g,v)_{L^2(\partial\Omega)},$$

and another application of the Lebesgue lemma (this time on $\partial\Omega$) shows that indeed $-a\nabla u \cdot \nu = g$ in $L^2(\partial\Omega)$, and therefore also in $H^{1/2}(\partial\Omega)$. That is, a smoother solution of (8.16) also solves (8.15). The converse can be shown to hold as well by reversing the steps above, up to the statement that indeed $u \in H^2(\Omega)$. But this latter fact follows from the elliptic regularity theorem (Theorem 8.13), which is presented at the end of this section to avoid interrupting the current development.

The next step is to apply the Lax-Milgram theorem to the variational problem (8.16) to obtain the existence and uniqueness of a solution. To this end, first of all remark that the bilinear form B is continuous if a and c are bounded functions. For coercivity, we require that a be uniformly positive definite and that c is uniformly positive, i.e., there exists $c_* > 0$ such that

$$c(x) \geq c_* > 0 \quad \text{for a.e. } x \in \Omega.$$

This assumption is needed rather than the weaker $c \geq 0$ since $H^1(\Omega)$ does *not* satisfy a Poincaré inequality. With these presumptions, a direct comptation reveals that

$$\begin{aligned} (a\nabla u, \nabla u)_{L^2(\Omega)} + (cu,u)_{L^2(\Omega)} &\geq a_* \|\nabla u\|_{L^2(\Omega)}^2 + c_* \|u\|_{L^2(\Omega)}^2 \\ &\geq \min(a_*, c_*) \|u\|_{H^1(\Omega)}^2, \end{aligned}$$

which is the desired coercivity of the form B. The conclusion is that there is a unique solution of the variational problem (8.16) which varies continuously with the data. Moreover, if the solution is more regular (i.e., $u \in H^2(\Omega)$), then (8.15) has a solution as well. (But is it unique?)

Note that the boundary condition $-a\nabla u \cdot \nu = g$ on $\partial\Omega$ is *not* enforced by the trial space $H^1(\Omega)$, since most elements of this space do *not* satisfy the boundary condition. Rather, the BC is imposed in a weak sense as noted above. In this case, the Neumann BC is said to be a *natural* BC. We obtain the bound

$$\|u\|_{H^1(\Omega)} \leq C\{\|f\|_{(H^1(\Omega))^*} + \|g\|_{H^{-1/2}(\partial\Omega)}\}.$$

8.5.3 The Neumann problem with no zeroth order term

In this subsection, it is presumed that Ω is connected. If it is not, consider each connected piece separately.

Often the Neumann problem (8.15) is posed with $c \equiv 0$, in which case the problem is *degenerate* in the sense that coercivity of B is lost. In that case, the solution cannot be unique. In fact any constant function solves the homogeneous problem (i.e., the problem for data $f = g = 0$).

The problem is that the kernel of the operator $\nabla \cdot a\nabla$ is larger than $\{0\}$, and this kernel intersects the kernel of the boundary operator $-a\partial/\partial\nu$. In fact, this intersection is

$$Z = \{v \in H^1(\Omega) : v \text{ is constant a.e. in } \Omega\},$$

which is a closed subspace isomorphic to $\mathbb{R}$. If we "mod out" by $\mathbb{R}$ in some way, we might be able to recover uniqueness. One way to do this is to insist that the solution has average zero. Let

$$\tilde{H}^1(\Omega) = \left\{u \in H^1(\Omega) : \int_\Omega u(x)\,dx = 0\right\},$$

which is isomorphic to the quotient space $H^1(\Omega)/Z$ or $H^1(\Omega)/\mathbb{R}$, i.e., $H^1(\Omega)$ modulo constant functions, and so is a Hilbert space. To establish coercivity of B on $\tilde{H}^1(\Omega)$, we need a variant of the Poincaré inequality, which is now addressed.

Theorem 8.12. *If $\Omega \subset \mathbb{R}^d$ is a bounded and connected domain, then there is some constant $C > 0$ such that*

$$\|v\|_{L^2(\Omega)} \leq C\|\nabla v\|_{(L^2(\Omega))^d} \quad \forall v \in \tilde{H}^1(\Omega).$$

Proof. Suppose not. Then there is a sequence $\{u_n\}_{n=1}^\infty \subset \tilde{H}^1(\Omega)$ such that

$$\|u_n\|_{L^2(\Omega)} = 1 \quad \text{and} \quad \|\nabla u_n\|_{L^2(\Omega)} < 1/n,$$

hence

$$\nabla u_n \to 0 \quad \text{strongly in } L^2(\Omega).$$

Furthermore, $\|u_n\|_{H^1(\Omega)} \leq \sqrt{2}$, so there is a subsequence (still denoted by u_n for convenience) that, by the Banach-Alaoglu theorem (Lemma 3.29), has

$$u_n \rightharpoonup u \quad \text{weakly in } H^1(\Omega)$$

and, by the Rellich-Kondrachov theorem (Theorem 7.23),

$$u_n \to u \quad \text{strongly in } L^2(\Omega).$$

Since $\nabla u_n \to 0$ and $\nabla u_n \to \nabla u$ as distributions, it must be that $\nabla u = 0$. Thus u is a constant (since Ω is connected) and has average zero, so $u = 0$. But this contradicts the fact that

$$1 = \|u_n\|_{L^2(\Omega)} \longrightarrow \|u\|_{L^2(\Omega)} = 0,$$

and the inequality claimed in the theorem must hold. □

On a connected domain, then, if $u \in \tilde{H}^1(\Omega)$

$$B(u,u) = (a\nabla u, \nabla u)_{L^2(\Omega)} \geq a_*\|\nabla u\|^2_{L^2(\Omega)} \geq C\|u\|^2_{H^1(\Omega)}$$

for some constant $C > 0$, which is the coercivity of B. The Lax-Milgram theorem again comes to the rescue and it is concluded that a solution exists and is unique for the variational problem:

Find $u \in \tilde{H}^1(\Omega)$ such that

$$B(u,v) = F(v) \quad \forall v \in \tilde{H}^1(\Omega),$$

where $B(u,v) = (a\nabla u, \nabla v)_{L^2(\Omega)}$ and F is defined in (8.17).
Note that $F \in (\tilde{H}^1(\Omega))^*$.

Often it is preferable to formulate this Neumann problem for test functions in $H^1(\Omega)$ rather than in $\tilde{H}^1(\Omega)$ (when designing a numerical scheme for an approximate solution, for example). In such a case, the variational problem becomes

Find $u \in \tilde{H}^1(\Omega)$ such that

$$B(u,v) = F(v) \quad \forall v \in H^1(\Omega).$$

In this situation, for any $\alpha \in \mathbb{R}$,

$$B(u, v+\alpha) = B(u,v),$$

so if we have a solution $u \in \tilde{H}^1(\Omega)$, then also

$$F(v) = B(u,v) = B(u, v+\alpha) = F(v+\alpha) = F(v) + F(\alpha)$$

implies that $F(\alpha) = 0$ is required. That is, $\mathbb{R} \subset \ker(F)$. Such a condition is called a *compatibility condition,* and it says that the kernel of $B(u, \cdot)$ is contained in the kernel of F; that is, f and g must satisfy

$$\langle f, 1 \rangle_{(H^1(\Omega))^*, H^1(\Omega)} - \langle g, 1 \rangle_{H^{-1/2}(\Omega), H^{1/2}(\Omega)} = 0,$$

which is to say

$$\int_\Omega f(x)\, dx = \int_{\partial\Omega} g(x)\, d\sigma(x),$$

provided that f and g are integrable.

The compatibility condition arises from enlarging the space of test functions from $\tilde{H}^1(\Omega)$ to $H^1(\Omega)$, which is needed to prove that the solution to the variational problem also solves the BVP. In abstract terms, we have the following situation. The problem is naturally posed for u and v in a Hilbert space X. However, there is nonuniqueness because the set $Z = \{u \in X : B(u, v) = 0 \ \forall v \in X\} = \{v \in X : B(u, v) = 0 \ \forall u \in X\}$ is contained in the kernel of the natural BC. But the problem is well behaved when posed over X/Z, which thus requires $F|_Z = 0$, i.e., the compatibility condition.

8.5.4 Elliptic regularity

We close this section with an important result from the theory of elliptic PDEs. See, e.g., [8] or [6] for a proof. This result can be used to prove the equivalence of the BVP and the variational problem in the case of Neumann BCs.

Theorem 8.13 (elliptic regularity theorem). *For the problem (8.2), suppose that $k \geq 0$ is an integer, $\Omega \subset \mathbb{R}^d$ is a bounded domain with a $C^{k+1,1}(\Omega)$-boundary, $a \in (W^{k+1,\infty}(\Omega))^{d \times d}$ is a uniformly positive definite matrix, $b \in (W^{k+1,\infty}(\Omega))^d$, and $c \in W^{k,\infty}(\Omega)$ is nonnegative. Let the bilinear form $B : H^1(\Omega) \times H^1(\Omega) \to \mathbb{R}$, be*

$$B(u, v) = (a\nabla u, \nabla v)_{L^2(\Omega)} + (bu, \nabla v)_{L^2(\Omega)} + (cu, v)_{L^2(\Omega)}.$$

(a) *If $f \in H^k(\Omega)$, $u_D \in H^{k+2}(\Omega)$, and B is continuous and coercive on $H_0^1(\Omega)$, then the Dirichlet problem:*

Find $u \in H_0^1(\Omega) + u_D$ such that

$$B(u, v) = (f, v)_{L^2(\Omega)} \quad \forall v \in H_0^1(\Omega),$$

has a unique solution $u \in H^{k+2}(\Omega)$ satisfying

$$\|u\|_{H^{k+2}(\Omega)} \leq C \big(\|f\|_{H^k(\Omega)} + \|u_D\|_{H^{k+3/2}(\partial\Omega)} \big)$$

for some $C > 0$ independent of f, u, and u_D. The specification $k = -1$ is allowed provided $c \in L^\infty(\Omega)$.

(b) If $f \in H^k(\Omega)$, $g \in H^{k+1/2}(\partial\Omega)$, and B is continuous and coercive on $H^1(\Omega)$, then the Neumann problem:

Find $u \in H^1(\Omega)$ such that

$$B(u,v) = (f,v)_{L^2(\Omega)} - (g,v)_{L^2(\partial\Omega)} \quad \forall v \in H^1(\Omega),$$

has a unique solution $u \in H^{k+2}(\Omega)$ satisfying

$$\|u\|_{H^{k+2}(\Omega)} \le C\big(\|f\|_{H^k(\Omega)} + \|g\|_{H^{k+1/2}(\partial\Omega)}\big).$$

for $C > 0$ independent of f, u, and g. The designation $k = -1$ is allowed provided $c \in L^\infty(\Omega)$ and we interpret $\|f\|_{H^k(\Omega)}$ as $\|f\|_{(H^1(\Omega))^*}$.

8.6   GALERKIN METHODS

In practice, simple approximations to solutions of these BVPs are needed, as closed form solutions are rare. This could be for computational purposes, to obtain an explicit approximation of the solution, perhaps with error estimates, or for theoretical purposes to indicate some general properties of solutions. We present here *Galerkin* methods, sometimes called *Bubnov-Galerkin* methods, which provide a framework for such approximations.

Theorem 8.14. *Suppose that H is a Hilbert space with closed subspaces $H_n \subset H$ such that for any $\phi \in H$, there are $\phi_n \in H_n$ such that $\phi_n \to \phi$ in H as $n \to \infty$. Suppose also that $B : H \times H \to \mathbb{R}$ is a continuous, coercive bilinear form on H and that $F \in H^*$. Then the variational problems, one for each n,*

Find $u_n \in H_n$ such that

$$B(u_n, v_n) = F(v_n) \quad \forall v_n \in H_n, \tag{8.18}$$

have unique solutions. The same problem posed on H also has a unique solution $u \in H$, and

$$u_n \to u \quad in \ H.$$

Moreover, if M and γ are respectively the continuity and coercivity constants for B, then for any n,

$$\|u - u_n\|_H \le \frac{M}{\gamma} \inf_{v_n \in H_n} \|u - v_n\|_H. \tag{8.19}$$

Furthermore, if B is symmetric, then for any n,

$$\|u - u_n\|_B = \inf_{v_n \in H_n} \|u - v_n\|_B, \tag{8.20}$$

where $\| \cdot \|_B = B(\cdot, \cdot)^{1/2}$ is the energy norm.

The hypothesis on H and its subspaces H_n is satisfied if the subspaces are *nested* (*viz.* $H_1 \subset H_2 \subset H_3 \subset \cdots \subset H$), and $\bigcup_{n=1}^{\infty} H_n$ is dense in H.

Remark. Estimate (8.19) says that the approximation of u by u_n in H_n is *quasi-optimal* in the H-norm; that is, up to the constant factor M/γ, u_n is the best approximation to u in H_n. When B is symmetric, $\|\cdot\|_B$ is indeed a norm equivalent to the H-norm by continuity and coercivity, as we have seen before. Estimate (8.20) says that the Galerkin approximation $u_n \in H_n$ is optimal in the energy norm.

Proof. Existence of unique solutions is given by the Lax-Milgram theorem. Thus there is in hand both

$$B(u_n, v_n) = F(v_n) \quad \forall v_n \in H_n,$$

for $n = 1, 2, \ldots$ and

$$B(u, v) = F(v) \quad \forall v \in H.$$

Since $H_n \subset H$, evaluate v at $v_n \in H_n$ in the relation defining u and subtract to obtain that

$$B(u - u_n, v_n) = 0 \quad \forall v_n \in H_n.$$

(Notice that when B provides an inner-product, this relation says that the error $u - u_n$ is B-orthogonal to H_n. Even when B does not present an inner-product, this relation is often referred to as *Galerkin orthogonality*.) Replace v_n by $(u - u_n) - (u - v_n) \in H_n$ for any $v_n \in H_n$ to obtain

$$B(u - u_n, u - u_n) = B(u - u_n, u - v_n) \quad \forall v_n \in H_n. \tag{8.21}$$

This leads to

$$\begin{aligned}
\gamma \|u - u_n\|_H^2 &\leq B(u - u_n, u - u_n) \\
&= B(u - u_n, u - v_n) \leq M \|u - u_n\|_H \|u - v_n\|_H,
\end{aligned}$$

and (8.19) follows. If B is symmetric, then B is an inner-product, and the Cauchy-Bunyakovsky-Schwarz inequality applied to (8.21) gives

$$\begin{aligned}
\|u - u_n\|_B^2 &= B(u - u_n, u - u_n) \\
&= B(u - u_n, u - v_n) \leq \|u - u_n\|_B \|u - v_n\|_B,
\end{aligned}$$

from which (8.20) follows.

Finally, there are $\phi_n \in H_n$ such that $\phi_n \to u$ in H as $n \to \infty$. Thus

$$\|u - u_n\|_H \leq \frac{M}{\gamma} \inf_{v_n \in H_n} \|u - v_n\|_H \leq \frac{M}{\gamma} \|u - \phi_n\|_H,$$

whence $u_n \to u$ in H as $n \to \infty$. $\qquad \square$

If (8.18) represents the equation for the critical point of an energy functional $J : H \to \mathbb{R}$, then for any n,

$$\inf_{v_n \in H_n} J(v_n) = J(u_n) \geq J(u) = \inf_{v \in H} J(v).$$

That is, we find the function with minimal energy in the space H_n to approximate u. In this minimization form, the method is called a *Ritz* method.

In the theory of *finite element methods*, one attempts to define explicitly the spaces $H_n \subset H$ in such a way that the equations (8.18) can be solved easily and so that the optimal error

$$\inf_{v_n \in H_n} \|u - v_n\|_H$$

is quantifiably small. Such Galerkin finite element methods are extremely effective for computing approximate solutions to elliptic BVPs, and for many other types of equations as well. Here is a very simple, but telling example.

Suppose that $\Omega = (0,1) \subset \mathbb{R}$ and $f \in L^2(0,1)$. Consider the BVP

$$\begin{cases} -u'' = f & \text{on } (0,1), \\ u(0) = u(1) = 0. \end{cases} \tag{8.22}$$

The equivalent variational problem is:

Find $u \in H_0^1(0,1)$ such that

$$(u', v')_{L^2} = (f, v)_{L^2} \quad \forall v \in H_0^1(0,1). \tag{8.23}$$

To implement a Galerkin approximation, one needs to construct a suitable finite element decomposition of $H_0^1(0,1)$. Let $n \geq 1$ be an integer, and define $h = h_n = 1/n$ and a grid $x_i = ih$ for $i = 0, 1, \ldots, n$ of uniform spacing $h > 0$. Let

$$H_n = V_h = \{v \in C^0(0,1) : v(0) = v(1) = 0 \text{ and } v(x) \text{ is a first degree}$$
$$\text{polynomial on } [x_{i-1}, x_i] \text{ for } i = 1, 2, \ldots, n\}.$$

Thus, V_h consists of continuous, piecewise linear functions. Note that $V_h \subset H_0^1(0,1)$, and V_h is a finite-dimensional vector space. We leave it to the reader to show that when $n = 2^N$ for some $N \in \mathbb{N}$, $V_1 \subset V_{1/2} \subset V_{1/4} \subset V_{1/8} \subset \cdots$ are nested and that $\bigcup_{m=0}^{\infty} V_{2^{-m}}$ is dense in $H_0^1(0,1)$. In fact, one can show (see [4] and an exercise at the end of this chapter) that there is a constant $C > 0$, independent of $h > 0$, such that for any $v \in H_0^1(0,1) \cap H^2(0,1)$,

$$\min_{v_h \in V_h} \|v - v_h\|_{H^1} \leq Ch \|v\|_{H^2}. \tag{8.24}$$

The Galerkin finite element approximation is:

Find $u_h \in V_h$ such that

$$(u_h', v_h')_{L^2} = (f, v_h)_{L^2} \quad \forall v_h \in V_h. \tag{8.25}$$

If u solves (8.23), then Theorem 8.14 implies that

$$\|u - u_h\|_{H^1} \le C \min_{v_h \in V_h} \|u - v_h\|_{H^1} \le Ch\|u\|_{H^2} \le Ch\|f\|_{L^2},$$

using elliptic regularity, at least when $n = 2^N$ (although in fact this restriction is not necessary). That is, the finite element approximations converge to the true solution linearly in the grid spacing h.

The problem (8.25) is easily solved by computer, since it reduces to an issue in linear algebra. For each $i = 1, 2, \ldots, n - 1$, let $\phi_{h,i} \in V_h$ be such that

$$\phi_{h,i}(x_j) = \begin{cases} 0 & \text{if } i \ne j, \\ 1 & \text{if } i = j. \end{cases}$$

The functions $\{\phi_{h,i}\}_{i=1}^{2^N - 1}$, which are uniquely detrmined by their values at the grid points, form a vector space basis for V_h, and so there are unique coefficients $\alpha_i \in \mathbb{R}$ such that

$$u_h(x) = \sum_{j=1}^{n-1} \alpha_j \phi_{h,j}(x).$$

Then, (8.25) reduces to

$$\sum_{j=1}^{n-1} \alpha_j (\phi'_{h,j}, \phi'_{h,i})_{L^2} = (f, \phi_{h,i})_{L^2}, \quad i = 1, 2, \ldots, n - 1,$$

since it is sufficient to test against the basis functions $\phi_{h,i}$. Let the $(n - 1) \times (n - 1)$ matrix M be defined by

$$M_{i,j} = (\phi'_{h,j}, \phi'_{h,i})_{L^2}$$

and the $(n - 1)$-vectors a and b by

$$a_j = \alpha_j \quad \text{and} \quad b_i = (f, \phi_{h,i})_{L^2}.$$

Then our problem is simply $Ma = b$, and the coefficients of u_h are given from the solution $a = M^{-1}b$ (why is this matrix invertible?). In fact M is tridiagonal (i.e., all the nonzero entries lie on the diagonal, subdiagonal, and superdiagonal), so the solution is easily and very efficiently computed.

8.7 GREEN'S FUNCTIONS

Let $\mathcal{L}$ be a linear partial differential operator, such as is given in (8.2). Often we can find a fundamental solution $E \in \mathcal{D}'$ satisfying

$$\mathcal{L}E = \delta_0,$$

wherein δ_0 is the Dirac delta function or point mass at the origin. If for the moment, $\mathcal{L}$ is taken to have constant coefficients, then the Malgrange-Ehrenpreis theorem (Theorem 5.29) asserts that such a fundamental solution exists. It is not unique, but for $f \in \mathcal{D}$, say, the equation $\mathcal{L}u = f$ has a solution $u = E * f$. However, u, defined this way, will generally fail to satisfy any imposed boundary condition. To resolve this difficulty, a special fundamental solution is developed in this section. For maximum generality, we will often proceed formally, assuming sufficient smoothness of all quantities involved to justify the calculations.

Let $\mathcal{B}$ denote a linear boundary condition operator (which generally involves the traces γ_0 and/or γ_1, and represents a Dirichlet, Neumann, or Robin boundary condition). For reasonable f and g, consider the BVP

$$\begin{cases} \mathcal{L}u = f & \text{in } \Omega, \\ \mathcal{B}u = g & \text{on } \partial\Omega. \end{cases} \tag{8.26}$$

Begin with the homogeneous case where $g = 0$.

Definition. Suppose $\Omega \subset \mathbb{R}^d$, $\mathcal{L}$ is a linear partial differential operator, and $\mathcal{B}$ is a homogeneous linear boundary condition. We call $G : \Omega \times \Omega \to \mathbb{R}$ a *Green's function* for $\mathcal{L}$ and $\mathcal{B}$ if, for any $f \in \mathcal{D}$, a weak solution u of (8.26) with $g = 0$ is given by

$$u(x) = \int_\Omega G(x, y) \, f(y) \, dy. \tag{8.27}$$

Here, it is assumed that $\partial\Omega$ is smooth enough to support the definition of the boundary condition.

Proposition 8.15. *The Green's function $G(\cdot, y) : \Omega \to \mathbb{R}$ is a fundamental solution for $\mathcal{L}$ with the point mass $\delta_y(\cdot) = \delta_0(\cdot - y)$, i.e., for a.e. $y \in \Omega$,*

$$\mathcal{L}_x G(x, y) = \delta_0(x - y) \quad \text{for } x \in \Omega$$

(wherein the notation $\mathcal{L}_x$ indicates that $\mathcal{L}$ acts on the variable x). What's more, $G(x, y)$ satisfies the homogeneous boundary condition

$$\mathcal{B}_x G(x, y) = 0 \quad \text{for } x \in \partial\Omega.$$

Proof. For any $f \in \mathcal{D}$, u defined by (8.27) solves $\mathcal{L}u = f$. We would like to proceed with the calculation

$$f(x) = \mathcal{L}u(x) = \mathcal{L} \int_\Omega G(x, y) \, f(y) \, dy = \int_\Omega \mathcal{L}_x G(x, y) \, f(y) \, dy,$$

which would indicate part of what we are after, but we need to justify moving $\mathcal{L}$ inside the integral. For $\phi \in \mathcal{D}(\Omega)$,

$$
\begin{aligned}
\int_\Omega f(x)\,\phi(x)\,dx &= \int_\Omega \mathcal{L}u(x)\,\phi(x)\,dx \\
&= \int_\Omega u(x)\,\mathcal{L}^*\phi(x)\,dx \\
&= \int_\Omega \int_\Omega G(x,y)\,f(y)\,\mathcal{L}^*\phi(x)\,dy\,dx \\
&= \int_\Omega \int_\Omega G(x,y)\,\mathcal{L}^*\phi(x)\,f(y)\,dx\,dy \\
&= \int_\Omega \langle \mathcal{L}_x G(\cdot,y), \phi \rangle\,f(y)\,dy,
\end{aligned}
$$

showing that

$$
\langle \mathcal{L}_x G(\cdot,y), \phi \rangle = \phi(y),
$$

that is, $\mathcal{L}_x G(x,y) = \delta_y(x)$.

That $G(x,y)$ satisfies a homogeneous Dirichlet condition in x is clear. Other boundary conditions involve normal derivatives, and it can be shown as above that G must satisfy them. □

Remark. For a fundamental solution E of a constant coefficient operator $\mathcal{L}$, $\mathcal{L}E = \delta_0$ and since us$\mathcal{L}$ commutes with translation, it follows that

$$
\mathcal{L}_x E(x-y) = \delta_y(x),
$$

which, as mentioned earlier, can be understood as giving the response $E(x-y)$ of the operator at $x \in \mathbb{R}^d$ to a point disturbance δ_y at $y \in \mathbb{R}^d$. Multiplying by the weight $f(y)$ and integrating (i.e., adding the responses) gives the solution $u = E * f$. When boundary conditions are imposed, a point disturbance at y is not necessarily translation equivalent to a disturbance at $\tilde{y} \neq y$. This is also true of nonconstant coefficient operators. Thus the more general form of the Green's function being a function of two variables is required; $G(x,y)$ is the response of the operator at $x \in \Omega$ to a point disturbance at $y \in \Omega$, subject also to the (homogeneous) boundary conditions.

Given a fundamental solution E that is sufficiently smooth outside the origin, we can construct the Green's function by solving a related BVP. For almost every $y \in \Omega$, solve

$$
\begin{cases}
\mathcal{L}_x w_y(x) = 0 & \text{for } x \in \Omega, \\
\mathcal{B}_x w_y(x) = \mathcal{B}_x E(x-y) & \text{for } x \in \partial\Omega,
\end{cases}
$$

and then

$$
G(x,y) = E(x-y) - w_y(x)
$$

is the Green's function. Note that indeed $\mathcal{L}_x G(x, y) = \delta_0(x - y)$ is a fundamental solution, and that this one is special in that $\mathcal{B}_x G(x, y) = 0$ on $\partial\Omega$.

It is generally difficult to find an explicit expression for the Green's function, except in special cases. However, its existence implies that the inverse operator of $(\mathcal{L}, \mathcal{B})$ is an integral operator, and thus has many important properties, such as compactness. When G can be found explicitly, it can be a powerful tool both theoretically and computationally.

Attention is now given to the nonhomogeneous BVP (8.26). Suppose that there is u_0 defined in Ω such that $\mathcal{B}u_0 = g$ on $\partial\Omega$. If $w = u - u_0$, then

$$\begin{cases} \mathcal{L}w = f - \mathcal{L}u_0 & \text{in } \Omega, \\ \mathcal{B}w = 0 & \text{on } \partial\Omega, \end{cases}$$

and this problem has a Green's function $G(x, y)$. Thus our solution is

$$u(x) = w(x) + u_0(x) = \int_\Omega G(x, y)\big(f(y) - \mathcal{L}u_0(y)\big)\, dy + u_0(x).$$

This formula has limited utility, since it is usually difficult to find such a u_0.

In some cases, the Green's function can be used to define a different integral operator involving an integral on $\partial\Omega$, thereby incorporating g directly. To illustrate, consider (8.26) with $\mathcal{L} = -\Delta + I$, where I is the identity operator. Now $\mathcal{L}_x G(x, y) = \delta_y(x)$, so this fact and integration by parts implies that

$$u(y) = \int_\Omega \mathcal{L}_x G(x, y)\, u(x)\, dx$$

$$= \int_\Omega [G(x, y)\, u(x) + \nabla_x G(x, y) \cdot \nabla u(x)]\, dx - \int_{\partial\Omega} \nabla_x G(x, y) \cdot \nu\, u(x)\, d\sigma(x)$$

$$= \int_\Omega G(x, y)\, \mathcal{L}u(x)\, dx + \int_{\partial\Omega} [G(x, y)\nabla u(x) \cdot \nu - \nabla_x G(x, y) \cdot \nu\, u(x)]\, d\sigma(x).$$

If $\mathcal{B}$ imposes the Dirichlet BC, so $u = u_D$, then since $\mathcal{L}u = f$ and $G(x, y)$ itself satisfies the homogeneous boundary conditions in x, there appears

$$u(y) = \int_\Omega G(x, y)\, f(x)\, dx - \int_{\partial\Omega} \nabla_x G(x, y) \cdot \nu\, u_D(x)\, d\sigma(x).$$

This is called the *Poisson integral formula*. If instead $\mathcal{B}$ imposes the Neumann BC, so $-\nabla u \cdot \nu = g$, then

$$u(y) = \int_\Omega G(x, y)\, f(x)\, dx - \int_{\partial\Omega} G(x, y)\, g(x)\, d\sigma(x).$$

Note that when $g = 0$, we have that

$$u(y) = \int_\Omega G(x, y)\, f(x)\, dx = \int_\Omega G(y, x)\, f(x)\, dx.$$

Since this formula holds for all $f \in \mathcal{D}$, it is concluded that $G(x, y) = G(y, x)$, i.e., G is symmetric. This is due to the fact that $\mathcal{L} = -\Delta + I$ is self-adjoint in Ω. When $\mathcal{L}$ is not self-adjoint, we would need to consider the Green's function for the operator $\mathcal{L}^*$, so that

$$u(y) = \int_\Omega \mathcal{L}_x^* G(x, y) \, u(x) \, dx = \int_\Omega G(x, y) \, \mathcal{L}u(x) \, dx + \text{boundary terms}$$

$$= \int_\Omega G(x, y) \, f(x) \, dx + \text{boundary terms}.$$

We remark that when a compatibility condition is required, it is not always possible to obtain the Green's function directly. For example, if $\mathcal{L} = -\Delta$ and one is posing a nonhomogeneous Neumann problem, then $\int_\Omega \delta_y(x) \, dx = 1 \neq 0$ as is required. So, instead, one solves the related problem

$$\begin{cases} -\Delta_x G(x, y) = \delta_y(x) - 1/|\Omega| & \text{in } \Omega, \\ -\nabla_x G(x, y) \cdot \nu = 0 & \text{on } \partial\Omega, \end{cases}$$

where $|\Omega|$ is the measure of Ω. Then our BVP (8.26) has the extra condition that the average of u vanishes. Thus, as above,

$$u(y) = -\int_\Omega \Delta_x G(x, y) \, u(x) \, dx$$

$$= \int_\Omega \nabla_x G(x, y) \cdot \nabla u(x) \, dx - \int_{\partial\Omega} \nabla_x G(x, y) \cdot \nu \, u(x) \, d\sigma(x)$$

$$= -\int_\Omega G(x, y) \, \Delta u(x) \, dx + \int_{\partial\Omega} G(x, y) \nabla u(x) \cdot \nu \, d\sigma(x)$$

$$= \int_\Omega G(x, y) \, f(x) \, dx - \int_{\partial\Omega} G(x, y) \, g(x) \, d\sigma(x).$$

8.8 EXERCISES

1. If A is a positive definite matrix, show that its eigenvalues are positive. Conversely, prove that if A is symmetric and has positive eigenvalues, then A is positive definite.

2. Suppose that the hypotheses of the Babuška-Lax-Milgram theorem (Theorem 8.10) are satisfied. Suppose also that $x_{0,1}$ and $x_{0,2}$ in $\mathcal{X}$ are such that the sets $X + x_{0,1} = X + x_{0,2}$. Prove that the solutions $u_1 \in X + x_{0,1}$ and $u_2 \in X + x_{0,2}$ of the abstract variational problem (8.12) agree (i.e., $u_1 = u_2$). What does this result say about Dirichlet boundary value problems?

3. Suppose that we wish to find $u \in H^2(\Omega)$ solving the nonlinear problem $-\Delta u + cu^2 = f \in L^2(\Omega)$, where $\Omega \subset \mathbb{R}^d$ is a bounded Lipschitz domain. For consistency, we would require that $cu^2 \in L^2(\Omega)$. Determine the smallest p

such that when $c \in L^p(\Omega)$, you can be certain that $cu^2 \in L^2(\Omega)$, if indeed it is possible. The answer depends on d.

4. Consider the boundary value problem for $u(x, y) : \mathbb{R}^2 \to \mathbb{R}$ such that

$$\begin{cases} - u_{xx} + e^y u = f & \text{for } (x, y) \in (0, 1)^2, \\ u(0, y) = 0, \ u(1, y) = \cos(y) & \text{for } y \in (0, 1). \end{cases}$$

Rewrite this as a variational problem and show that there exists a unique solution. Be sure to define your function spaces carefully and identify in which space f must lie.

5. Suppose that $\Omega \subset \mathbb{R}^d$ is a smooth, bounded, connected domain. Let

$$H = \left\{ u \in H^2(\Omega) : \int_\Omega u(x)\, dx = 0 \text{ and } \nabla u \cdot \nu = 0 \text{ on } \partial\Omega \right\}.$$

Show that H is a Hilbert space, and prove the Poincaré-type inequality that there exists $C > 0$ such that for any $u \in H$,

$$\|u\|_{H^1(\Omega)} \le C \sum_{|\alpha|=2} \|D^\alpha u\|_{L^2(\Omega)}.$$

6. Use the Lax-Milgram theorem to show that, for $f \in L^2(\mathbb{R}^d)$, there exists a unique solution $u \in H^1(\mathbb{R}^d)$ to the problem

$$-\Delta u + u = f \quad \text{in } \mathbb{R}^d.$$

Be careful to justify integration by parts. [Hint: $\mathcal{D}$ is dense in $H^1(\mathbb{R}^d)$.]

7. The *symmetric gradient tensor* $\epsilon(v)$ of a vector function $v \in (H^1(\Omega))^d$ is a $d \times d$ matrix of the form

$$\epsilon_{ij}(v) = \frac{1}{2}\left(\frac{\partial v_i}{\partial x_j} + \frac{\partial v_j}{\partial x_i} \right)$$

It arises often in structural mechanics problems. Assuming that $\Omega \subset \mathbb{R}^2$ is smooth and bounded, prove *Korn's inequality*, which says that there is some $\gamma > 0$ such that

$$\|\epsilon(v)\|_{(L^2(\Omega))^{2\times 2}} \ge \gamma \|v\|_{(H^1(\Omega))^2} \quad \forall v \in (H_0^1(\Omega))^2.$$

[Hint: you will need to show that all *rigid motions*, i.e., vector fields of the form $\alpha(x_2, -x_1) + (\beta, \gamma)$, vanish in $(H_0^1(\Omega))^2$.]

8. Suppose $\Omega \subset \mathbb{R}^d$ is a bounded, connected Lipschitz domain and $V \subset \Omega$ has positive measure. Let $H = \{u \in H^1(\Omega) : u|_V = 0\}$.

(a) Why is H a Hilbert space?

(b) Prove the following Poincaré inequality: there is some $C > 0$ such that

$$\|u\|_{L^2(\Omega)} \leq C\|\nabla u\|_{L^2(\Omega)} \quad \forall u \in H.$$

9. Let $\Omega \subset \mathbb{R}^d$ be a bounded domain with a Lipschitz boundary, $f \in L^2(\Omega)$, and $\alpha > 0$. Consider the Robin boundary value problem

$$\begin{cases} -\Delta u + u = f & \text{in } \Omega, \\ \dfrac{\partial u}{\partial \nu} + \alpha u = 0 & \text{on } \partial\Omega. \end{cases}$$

(a) For this problem, formulate a variational principle

$$B(u, v) = (f, v) \quad \forall v \in H^1(\Omega).$$

(b) Show that this problem has a unique weak solution.

10. For $\Omega = [0, 1]^d$, define

$$H_{\#}^1(\Omega) = \left\{ v \in H_{\text{loc}}^1(\mathbb{R}^d) : v \text{ is periodic of period } 1 \right.$$

$$\left. \text{in each direction and } \int_\Omega v \, dx = 0 \right\},$$

and consider the problem of finding a periodic solution $u \in H_{\#}^1(\Omega)$ of

$$-\Delta u = f \quad \text{on } \Omega,$$

where $f \in L^2(\Omega)$.

(a) Define precisely what it means for $v \in H_{\text{loc}}^1(\mathbb{R}^d)$ to be periodic of period 1 in each direction.

(b) Show that $H_{\#}^1(\Omega)$ is a Hilbert space.

(c) Show that there is a unique periodic solution to the partial differential equation.

11. Consider

$$B(u, v) = (a\nabla u, \nabla v)_{L^2(\Omega)} + (bu, \nabla v)_{L^2(\Omega)} + (cu, v)_{L^2(\Omega)}.$$

(a) Derive a condition on b to insure that B is coercive on $H^1(\Omega)$ when a is uniformly positive definite and c is uniformly positive.

(b) Suppose $b = 0$. If $c < 0$, is B not coercive? Show that this is correct on $H^1(\Omega)$, but that by restricting how negative c may be, B could still be coercive on $H_0^1(\Omega)$.

12. Suppose $\Omega \subset \mathbb{R}^d$ is a $C^{1,1}$ domain. Consider the biharmonic BVP

$$\begin{cases} \Delta^2 u = f & \text{in } \Omega, \\ \nabla u \cdot \nu = g & \text{on } \partial\Omega, \\ u = u_D & \text{on } \partial\Omega, \end{cases}$$

wherein $\Delta^2 u = \Delta\Delta u$ is the application of the Laplace operator twice.

(a) Determine appropriate Sobolev spaces within which the functions u, f, g, and u_D should lie, and formulate an appropriate variational problem for the BVP. Show that the two problems are equivalent.

(b) Show that there is a unique solution to the variational problem. [Hint: use the elliptic regularity theorem to prove coercivity of the bilinear form.]

(c) What would be the natural BCs for this partial differential equation?

(d) For simplicity, let u_D and g vanish and define the energy functional

$$J(v) = \int_\Omega \left(|\Delta v(x)|^2 - 2f(x)\,v(x) \right) dx.$$

Prove that minimization of J is equivalent to the variational problem.

13. Suppose $\Omega \subset \mathbb{R}^d$ is a bounded Lipschitz domain. Consider the Stokes problem for vector u and scalar p given by

$$\begin{cases} -\Delta u + \nabla p = f & \text{in } \Omega, \\ \nabla \cdot u = 0 & \text{in } \Omega, \\ u = 0 & \text{on } \partial\Omega, \end{cases}$$

where the first equation holds for each coordinate (i.e., $-\Delta u_j + \partial p/\partial x_j = f_j$ for each $j = 1, \ldots, d$). This problem is not a minimization problem; rather, it is a *saddle-point problem*, in that we minimize some energy subject to the *constraint* $\nabla \cdot u = 0$. However, if we work over the constrained space, we can handle this problem by the ideas of this chapter. Let

$$H = \{v \in (H_0^1(\Omega))^d : \nabla \cdot u = 0\}.$$

(a) Verify that H is a Hilbert space.

(b) Determine an appropriate Sobolev space for f, and formulate an appropriate variational problem for the constrained Stokes problem.

(c) Show that there is a unique solution u to the variational problem.

14. Let $\mathcal{H} = H_0^1(\Omega) \times H^1(\Omega)$ and consider the solution $(u, v) \in \mathcal{H}$ to the differential problem

$$\begin{cases} -\Delta u = f + a(v - u) & \text{in } \Omega, \\ v - \Delta v = g + a(u - v) & \text{in } \Omega, \\ u = 0 \quad \text{and} \quad \nabla v \cdot \nu = \gamma & \text{on } \partial\Omega, \end{cases}$$

where $a \in L^\infty(\Omega)$.

(a) Develop an appropriate weak or variational form for the problem. In what Sobolev spaces should f, g, and γ lie?

(b) Prove that there exists a unique solution to the problem, provided that $a \geq 0$.

15. Let $\Omega \subset \mathbb{R}^2$ be a bounded domain with a smooth boundary. The equations of linear elasticity can be formulated for the displacement vector $u = (u_1, u_2)^T$ using the symmetric gradient tensor $\epsilon(u)$ defined in Exercise 7, as

$$\sum_i \frac{\partial}{\partial x_i} \left[2\mu\epsilon_{ij}(u) + \lambda \nabla \cdot u\, \delta_{ij} \right] = f_j \quad \text{in } \Omega,$$

where μ and λ are positive constants. Impose the homgeneous Dirichlet boundary condition $u = 0$ on $\partial\Omega$.

(a) Develop an appropriate weak form for the equations that involve the integrals $(\nabla \cdot u, \nabla \cdot v)$ and $\sum_{i,j} \left(\epsilon_{ij}(u), \epsilon_{ij}(v) \right)$. For the latter, you will need to use the fact that $\epsilon(u)$ is indeed symmetric.

(b) In which spaces should u and $f = (f_1, f_2)^T$ lie?

(c) Show that there is a unique solution to the problem using the Lax-Milgram theorem. You will need to use Korn's inequality.

16. Let $\Omega \subset \mathbb{R}^2$ be a domain with a smooth boundary and consider the variational problem: find $u \in V$ such that

$$(au, v) + (\nabla \cdot u, \nabla \cdot v) = (f, \nabla \cdot v) \quad \text{for all } v \in V,$$

where u and v are vectors in $\mathbb{R}^2$, $a \in L^\infty(\Omega)$, $a(x) \geq a_* > 0$ for some constant a_*, $(au, v) = \int_\Omega a(x) \left(u_1(x) v_1(x) + u_2(x) v_2(x) \right) dx$, and $\nabla \cdot u = \text{div}\, u = \dfrac{\partial u_1}{\partial x_1} + \dfrac{\partial u_2}{\partial x_2}$.

(a) For the problem to make sense, define V and a space for f.

(b) Show that the hypotheses of the Lax-Milgram theorem hold for this problem. What norm do we use for V?

(c) Define $p = f - \nabla \cdot u$ and determine the strong form of the equation represented by the variational problem. What boundary condition should you impose on p?

17. Let H and W be real Hilbert spaces and let $V \subset H$ be a linear subspace. Let $A : H \to H$ and $B : V \to W$ be bounded linear operators, where we give V the graph norm $\|v\|_V = \|v\|_H + \|Bv\|_W$. For any $f \in V$ and $0 \leq \delta < 1$, consider the problem: find $(u, p) \in V \times W$ such that

$$\langle Au, v \rangle_H - \langle B^*p, v \rangle_H + \langle Bu, w \rangle_W + \langle p, w \rangle_W + \delta\langle Bu, Bv \rangle_W + \delta\langle p, Bv \rangle_W$$
$$= \langle f, v \rangle_H \quad \forall (v, w) \in V \times W.$$

Assume that A is coercive on V (i.e., $\gamma\|v\|_V^2 \le \langle Av, v\rangle_V$ for some $\gamma > 0$).

(a) Assuming there is a solution, find a bound on the norm of (u, p).

(b) Show that there is a unique solution for any $\delta \in (0, 1)$.

(c) Show that there is a unique solution for $\delta = 0$. [Hint: replace w by $w - \delta Bv$.]

18. Let $\Omega \subset \mathbb{R}^d$ be a domain and define the *potential* and *solenoidal* vector fields

$$H_{\mathrm{pot}} = \{\nabla\phi : \phi \in H_0^1(\Omega)\} \quad \text{and} \quad H_{\mathrm{sol}} = \{\psi \in (L^2(\Omega))^d : \nabla \cdot \psi = 0\}.$$

These are clearly linear subspaces of $(L^2(\Omega))^d$.

(a) Show that H_{pot} and H_{sol} are $(L^2(\Omega))^d$-orthogonal to each other.

(b) Show that both H_{pot} and H_{sol} are closed in $(L^2(\Omega))^d$.

(c) Prove that $(L^2(\Omega))^d = H_{\mathrm{pot}} \oplus H_{\mathrm{sol}}$, i.e., an L^2 vector can be written uniquely as the sum of a potential and solenoidal vector. This fact is called the *Helmholtz decomposition*.

(d) Show that for $u \in (L^2(\Omega))^d = \nabla\phi + \psi$,

$$\|u\|_{(L^2(\Omega))^d}^2 = \|\nabla\phi\|_{(L^2(\Omega))^d}^2 + \|\psi\|_{(L^2(\Omega))^d}^2.$$

(e) Suppose $u_j \in H_{\mathrm{pot}}$ and $\nabla \cdot u_j = f_j \in L^2(\Omega)$ for all j. If

$$\|u_j\|_{(L^2(\Omega))^d} + \|\nabla \cdot u_j\|_{(L^2(\Omega))^d} \le M < \infty \quad \forall j,$$

show that there is some $u \in H_{\mathrm{pot}}$ such that a subsequence, $\{u_{j_k}\}_{k=1}^{\infty}$, converges strongly to u in $(L^2(\Omega))^d$.

19. Let $\Omega \subset \mathbb{R}^d$ be a bounded Lipschitz domain. Using the Lax-Milgram theorem and the definition of the $H^{1/2}(\partial\Omega)$-norm (but *not* the elliptic regularity theorem), show that there is some constant $C > 0$ such that

$$\|u\|_{H^1(\Omega)} \le C\{\|\Delta u\|_{H^{-1}(\Omega)} + \|u\|_{H^{1/2}(\partial\Omega)}\} \quad \text{for all } u \in H^1(\Omega).$$

20. Let $\Omega \subset \mathbb{R}^d$ be a bounded domain with a Lipschitz boundary, and define

$$H(\mathrm{div}; \Omega) = \{v \in (L^2(\Omega))^d : \nabla \cdot v \in L^2(\Omega)\}.$$

(a) Show that $H(\mathrm{div}; \Omega)$ is a Hilbert space with the inner-product

$$(u, v)_{H(\mathrm{div};\Omega)} = (u, v)_{(L^2(\Omega))^d} + (\nabla \cdot u, \nabla \cdot v)_{L^2(\Omega)}.$$

(b) The trace theorem does not imply that $\partial_\nu v = v \cdot \nu$ exists on $\partial\Omega$. Nevertheless, show that

$$\partial_\nu : H(\mathrm{div};\Omega) \to H^{-1/2}(\partial\Omega) = (H^{1/2}(\partial\Omega))^*$$

is a well-defined bounded linear operator in the sense of integration by parts, *viz.*

$$\int_{\partial\Omega} v \cdot \nu\, \phi\, d\sigma(x) = \int_\Omega \nabla \cdot v\, \phi\, dx + \int_\Omega v \cdot \nabla\phi\, dx.$$

(c) Prove the following inf-sup condition: there exists $\gamma > 0$ such that

$$\inf_{w \in L^2(\Omega)} \sup_{v \in H(\mathrm{div};\Omega)} \frac{(w, \nabla \cdot v)_{L^2(\Omega)}}{\|w\|_{L^2(\Omega)}\|v\|_{H(\mathrm{div};\Omega)}} \geq \gamma > 0.$$

[Hint: solve $\Delta\varphi = w$ in $H_0^1(\Omega)$ and consider $v = \nabla\varphi$.]

21. Let $s, t \geq 0$ and define

$$\langle f, g \rangle = \int_{\mathbb{R}^2} \hat{f}(\eta, \zeta)\, \overline{\hat{g}(\eta, \zeta)}(1 + \eta^2)^s(1 + \zeta^2)^t\, d\eta\, d\zeta,$$
$$H^{s,t} = \{f \in L^2(\mathbb{R}^2) \,:\, \|f\|^2 = \langle f, f \rangle < \infty\}.$$

(a) Show that $\langle \cdot, \cdot \rangle$ is an inner-product on $H^{s,t}$.

(b) Show that $H^{s,t}$ is continuously imbedded in $H^{\min(s,t)}(\mathbb{R}^2)$ and that $H^{s+t}(\mathbb{R}^2)$ is continuously imbedded in $H^{s,t}(\mathbb{R}^2)$.

(c) Show that $H^{s,t}$ is complete and $\mathcal{S}(\mathbb{R}^2) \subset H^{s,t}$ is dense.

(d) Define the trace operator $\gamma_0 : C^2(\mathbb{R}^2) \to C^0(\mathbb{R})$ by $\gamma_0 f(x) = f(x, 0)$. Find reasonable hypotheses on s and t as well as an r so that γ_0 extends to a bounded linear operator $\gamma_0 : H^{s,t} \to H^r(\mathbb{R})$.

22. Modify the statement of Theorem 8.14 to allow for nonhomogeneous essential boundary conditions, and prove the result.

23. Let $\Omega \subset \mathbb{R}^d$ have a smooth boundary, V_n be the set of polynomials of degree up to n, for $n = 1, 2, \ldots$, and $f \in L^2(\Omega)$. Consider the problem: find $u_n \in V_n$ such that

$$(\nabla u_n, \nabla v_n)_{L^2(\Omega)} + (u_n, v_n)_{L^2(\Omega)} = (f, v_n)_{L^2(\Omega)} \quad \text{for all } v_n \in V_n.$$

(a) Show that there exists a unique solution for any n, and that

$$\|u_n\|_{H^1(\Omega)} \leq \|f\|_{L^2(\Omega)}.$$

(b) Show that there is $u \in H^1(\Omega)$ such that, for a subsequence, $u_n \rightharpoonup u$ weakly in $H^1(\Omega)$. Find a variational problem satisfied by u. Justify your answer.

(c) Show that $\|u - u_n\|_{H^1(\Omega)}$ decreases monotonically to 0 as $n \to \infty$.

(d) What can you say about u and $\nabla u \cdot \nu$ on $\partial\Omega$?

24. Consider the finite element method introduced in Section 8.6. Modify the method to account for:

(a) nonhomogeneous Neumann conditions;

(b) nonhomogeneous Dirichlet conditions.

25. Compute explicitly the finite element solution to (8.22) using $f(x) = x^2(1 - x)$ and $n = 4$. How does this approximation compare to the true solution?

26. Suppose that $u \in H^1(\Omega)$, where $\Omega \subset \mathbb{R}^d$ is a bounded, connected domain. Recall that the $H^1(\Omega)$-seminorm is $|u|_{H^1(\Omega)} = \left\{ \sum_{|\alpha|=1} \|D^\alpha u\|^2_{L^2(\Omega)} \right\}^{1/2}$.

(a) Show that there is some constant C_Ω, depending on Ω but not on u, such that
$$\inf_{c \in \mathbb{R}} \|u - c\|_{L^2(\Omega)} \le C_\Omega \, |u|_{H^1(\Omega)}.$$

(b) Let $\Omega = (0, h)^d$ for some $h > 0$. Show that there is a constant C, independent of h and u, such that
$$\inf_{c \in \mathbb{R}} \|u - c\|_{L^2(\Omega)} \le Ch \, |u|_{H^1(\Omega)}.$$

[Hint: change variables to integrate over $(0, 1)^d$, and use (a).]

(c) Let $\Omega = (0, 1)^d$ and let P be the set of piecewise discontinuous constants over the grid of spacing $h = 1/N$ for some positive integer N. Show that there is a constant C, independent of h and u, such that
$$\inf_{p \in P} \|u - p\|_{L^2(\Omega)} \le Ch \, |u|_{H^1(\Omega)}.$$

27. Let H_h be the set of continuous piecewise linear functions defined on the grid $x_j = jh$, where $h = 1/n$ for some integer $n > 0$. Let the interpolation operator $\mathcal{I}_h : H_0^1(0, 1) \to H_h$ be defined by
$$\mathcal{I}_h v(x_j) = v(x_j) \quad \forall j = 1, 2, \ldots, n - 1.$$

(a) Show that $\mathcal{I}_h$ is well-defined, and that it is continuous. [Hint: use the Sobolev embedding theorem.]

(b) Show that there is a constant $C > 0$ independent of h such that

$$\|v - \mathcal{I}_h v\|_{H^1(x_{j-1}, x_j)} \leq Ch \|v\|_{H^2(x_{j-1}, x_j)}.$$

[Hint: change variables so that the domain becomes $(0, 1)$, where the result is trivial by the Poincaré-Friedrichs inequality (Corollary 7.21) and Theorem 8.12.]

(c) Show that (8.24) holds.

28. Consider the problem (8.22).

(a) Find the Green's function.

(b) Instead impose Neumann BCs, and find the Green's function. [Hint: recall that now we require $-(\partial^2/\partial x^2)G(x, y) = \delta_y(x) - 1.$]

29. If $\mathcal{L}$ is self-adjoint on $L^2(\Omega)$ and has a Green's function G, show that $G(x, y) = G(y, x)$ for a.e. $x, y \in \Omega$.

CHAPTER 9

Differential Calculus in Banach Spaces

In this chapter, we move away from the rigid, albeit very useful confines of linear maps and contemplate maps $f : U \to Y$ which are not necessarily linear, where U is an open set in a Banach space X and Y is also a Banach space. Consideration of finding and classifying solutions to an equations like $f(u) = v$ in such a setting is a very natural generalization of questions arising in multivariable calculus in d-dimensional Euclidean space. Such infinite-dimensional equations arise in many applications where nonlinear effects cannot be ignored in the modeling.

The subject has precursors going back into the 19th century, but it is mostly a 20th century topic. One of the areas where it finds use is the calculus of variations, which is the subject of the next chapter. By the middle of the 20th century, most of the basic and even some of the more subtle aspects of multi-dimensional calculus had been sucessfully generalized to infinite dimensions. In the second half of the century, there arose a pure strain of nonlinear functional analysis aimed at more and more general settings, but often divorced from any real application. At the same time, the tools presented by the theory had worked their way into the everyday activities of many applied analysts.

As in finite-dimensional calculus, we begin the analysis of such functions by effecting a local approximation. In one-variable calculus, we are used to writing

$$f(x) \cong f(x_0) + f'(x_0)(x - x_0) \tag{9.1}$$

when $f : \mathbb{R} \to \mathbb{R}$ is continuously differentiable, say. This amounts to approximating f by an *affine* function, a translation of a linear mapping. This procedure allows the method of linear functional analysis to be brought to bear upon understanding a nonlinear function f.

DOI: 10.1201/9781003492139-9

9.1 DIFFERENTIATION

In attempting to generalize the notion of a derivative to more than one dimension, one realizes immediately that the one-variable calculus formula

$$f'(x) = \lim_{h \to 0} \frac{f(x+h) - f(x)}{h} \tag{9.2}$$

cannot be taken over intact. First, the quantity $1/h$ has no meaning in higher dimensions. Second, whatever $f'(x)$ might be, it is plainly not going to be a number. Instead, just as in multivariable calculus, it is a precise version of (9.1) that readily generalizes, and not (9.2). We start with some basic notions.

Definition. Suppose X, Y are NLSs, $U \subset X$ is open, $0 \in U$, and $f : U \to Y$. If

$$\frac{\|f(h)\|_Y}{\|h\|_X} \longrightarrow 0 \quad \text{as } h \to 0 \ (h \neq 0),$$

we say that as h tends to 0, f is "little oh" of h, and we denote this as

$$\|f(h)\|_Y = o(\|h\|_X),$$

or even $f(h) = o(h)$ when the norms are understood from context.

Definition. Let $f : U \to Y$ where $U \subset X$ is open and X and Y are normed linear spaces. Let $x \in U$. We say that f is *Fréchet differentiable* (or *strongly differentiable*) at x if there is an element $A \in B(X, Y)$ such that when

$$R(x, h) = f(x+h) - f(x) - Ah,$$

then

$$\frac{\|R(x, h)\|_Y}{\|h\|_X} \longrightarrow 0$$

as $h \to 0$ $(h \neq 0)$ in X, i.e.,

$$\|R(x, h)\|_Y = o(\|h\|_X).$$

When it exists, we call A the *Fréchet derivative* of f at x; it is denoted variously by

$$A = A_x = f'(x) = Df(x).$$

Notice that this definition generalizes the one-dimensional idea of being differentiable. Indeed, if $f \in C^1(\mathbb{R})$, then

$$R(x, h) = f(x+h) - f(x) - f'(x)h = \left[\frac{f(x+h) - f(x)}{h} - f'(x) \right] h,$$

and so

$$\frac{|R(x, h)|}{|h|} = \left| \frac{f(x+h) - f(x)}{h} - f'(x) \right| \longrightarrow 0$$

as $h \to 0$ ($h \neq 0$) in $\mathbb{R}$. Note that $B(\mathbb{R}, \mathbb{R}) \cong \mathbb{R}$, and thus that the product $f'(x)h$ may be viewed as the linear mapping that sends h to $f'(x)h$.

If f is Fréchet differentiable throughout U, then the map $Df(x)$ is a linear map from $U \subset X$ to Y. It is useful to think of Df as a mapping of $U \times X$ into Y via the correspondence

$$(x, h) \longmapsto Df(x)h.$$

In this viewpoint, the mapping D itself takes functions from U to Y and maps them to functions from $U \times X$ to Y. This latter map is linear.

Proposition 9.1. *If X, Y are NLSs, $f : U \to Y$ where $U \subset X$ is open, and f is Fréchet differentiable throughout U, then $Df(x)$ is unique and f is continuous at any $x \in U$. Moreover, if g is also Fréchet differentiable in U and $\alpha, \beta \in \mathbb{F}$, then*

$$D(\alpha f + \beta g)(\cdot) = \alpha Df(\cdot) + \beta Dg(\cdot).$$

Proof. Suppose $A, B \in B(X, Y)$ are such that

$$f(x + h) - f(x) - Ah = R_A(x, h)$$

and

$$f(x + h) - f(x) - Bh = R_B(x, h),$$

where

$$\frac{\|R_A(x, h)\|_Y}{\|h\|_X} \longrightarrow 0 \quad \text{and} \quad \frac{\|R_B(x, h)\|_Y}{\|h\|_X} \longrightarrow 0$$

as $h \to 0$ ($h \neq 0$) in X. It follows that

$$\begin{aligned}
\|A - B\|_{B(X,Y)} &= \frac{1}{\epsilon} \sup_{\|h\|_X = \epsilon} \|Ah - Bh\|_Y \\
&= \sup_{\|h\|_X = \epsilon} \frac{\|R_B(x, h) - R_A(x, h)\|_Y}{\|h\|_X} \\
&\leq \sup_{\|h\|_X = \epsilon} \frac{\|R_B(x, h)\|_Y}{\|h\|_X} + \sup_{\|h\|_X = \epsilon} \frac{\|R_A(x, h)\|_Y}{\|h\|_X},
\end{aligned}$$

and the right-hand side may be made as small as we like by taking ϵ small enough. Thus $A = B$.

Continuity of f at x is straightforward since

$$\begin{aligned}
\|f(x + h) - f(x)\|_Y &= \|Df(x)h + R(x, h)\|_Y \\
&\leq \|Df(x)\|_{B(X,Y)} \|h\|_Y + \|R(x, h)\|_Y,
\end{aligned}$$

and the right-hand side tends to 0 as $h \to 0$ in X.

The final result is left as an exercise. $\qquad\square$

In fact, we have much more than mere continuity. The following result is often useful. It says that when f is differentiable, it is locally Lipschitz.

Lemma 9.2 (local-Lipschitz lemma). *If $f : U \to Y$ is differentiable at $x \in U$, then given $\epsilon > 0$, there is a $\delta = \delta(x, \epsilon) > 0$ such that for all h with $\|h\|_X \le \delta$,*

$$\|f(x + h) - f(x)\|_Y \le \left(\|Df(x)\|_{B(X,Y)} + \epsilon \right) \|h\|_X. \qquad (9.3)$$

Proof. Simply write

$$f(x + h) - f(x) = R(x, h) + Df(x)h. \qquad (9.4)$$

Since f is differentiable at x, given $\epsilon > 0$, there is a $\delta > 0$ such that $\|h\|_X \le \delta$ and $h \ne 0$ implies

$$\frac{\|R(x, h)\|_Y}{\|h\|_X} \le \epsilon.$$

Then (9.4) implies the advertised result. $\qquad \Box$

Examples.

(1) If $f(x) = Ax$, where $A \in B(X, Y)$, then $f(x + h) - f(x) = Ah$, so f is Fréchet differentiable everywhere and

$$Df(x) = A$$

for all $x \in X$.

(2) Let $X = H$ be a Hilbert-space over $\mathbb{R}$. Let $f(x) = (x, Ax)_H$ where $A \in B(H, H)$. Then, $f : H \to \mathbb{R}$ and

$$f(x + h) - f(x) = (x, Ah)_H + (h, Ax)_H + (h, Ah)_H$$
$$= \left((A^* + A)x, h \right)_H + (h, Ah)_H.$$

Hence if we define, for $x, h \in X$,

$$Df(x)h = \left((A^* + A)x, h \right)_H,$$

then

$$\|f(x + h) - f(x) - Df(x)h\|_Y \le \|h\|_X^2 \|A\|_{B(X,Y)}.$$

Thus $Df(x) \in H^* = B(H, \mathbb{R})$ is the Riesz map associated with the element $(A^* + A)x$.

(3) Let $f : \mathbb{R}^n \to \mathbb{R}$ and suppose $f \in C^1(\mathbb{R}^n)$, to wit $\partial_i f = \partial f / \partial x_i$ exists and is continuous on $\mathbb{R}^n$, $1 \le i \le n$. Then $Df(x) \in B(\mathbb{R}^n, \mathbb{R})$ is defined by

$$Df(x)h = \nabla f(x) \cdot h.$$

(4) Let $f : \mathbb{R}^n \to \mathbb{R}^m$ and suppose $f \in C^1(\mathbb{R}^n, \mathbb{R}^m)$, which is to say each of the component functions of $f = (f_1, \ldots, f_m)$, as $\mathbb{R}$-valued functions, have all their first partial derivatives, and each of these is continuous. Then f is Fréchet differentiable and

$$Df(x)h = [\partial_j f_i(x)]h,$$

where the latter is matrix multiplication and the matrix itself is the usual Jacobian matrix. That is, $Df(x) \in B(\mathbb{R}^n, \mathbb{R}^m)$ is an $m \times n$ matrix, and the ith component of $Df(x)h$ is

$$\big(Df(x)h\big)_i = \sum_{j=1}^n \partial_j f_i(x) h_j.$$

(5) Let $\varphi \in L^p(\mathbb{R}^d)$, where $p \geq 1$, p an integer, and define

$$f(\varphi) = \int_{\mathbb{R}^d} \varphi^p(x)\, dx.$$

Then $f : L^p(\mathbb{R}^d) \to \mathbb{R}$, f is Fréchet differentiable and

$$Df(\varphi)h = p \int_{\mathbb{R}^d} \varphi^{p-1}(x)h(x)\, dx,$$

which is a bounded linear operator, since

$$\left| \int_{\mathbb{R}^d} \varphi^{p-1}(x)h(x)\, dx \right| \leq \|\varphi^{p-1}\|_{L^{p/(p-1)}} \|h\|_{L^p} = \|\varphi\|_{L^p}^{p-1} \|h\|_{L^p}.$$

To verify this formula, simply use the binomial theorem to deduce that

$$f(\varphi + h) - f(\varphi) = \int_{\mathbb{R}^d} [(\varphi + h)^p(x) - \varphi^p(x)]\, dx$$

$$= \int_{\mathbb{R}^d} \left[\varphi^p(x) + p\varphi^{p-1}(x)h(x) + \binom{p}{2}\varphi^{p-2}(x)h^2(x) + \cdots + h^p(x) - \varphi^p(x) \right] dx$$

$$= p \int_{\mathbb{R}^d} \varphi^{p-1}(x)\, h(x)\, dx + \int_{\mathbb{R}^d} \left[\binom{p}{2}\varphi^{p-2}(x)\, h^2(x) + \cdots + h^p(x) \right] dx.$$

There is a notion of differentiability weaker than Fréchet differentiability that is occasionally useful. In this conception, we only ask the function f to be differentiable in each specified direction $h \in X$, $h \neq 0$. That is, we consider the differentiability of $g(t) = f(x + th)$, a Y-valued function of the real variable t.

Definition. Suppose $f : U \to Y$, where X and Y are NLSs and $U \subset X$ is open. Then f is *Gâteaux differentiable* (or *weakly differentiable*) at $x \in U$ if there is an $A \in B(X, Y)$ such that for all $h \in X$, $h \neq 0$,

$$\frac{1}{t}\|f(x + th) - f(x) - tAh\| = \|h\| \left\| \frac{f(x + th) - f(x)}{t\|h\|} - A\frac{h}{\|h\|} \right\| \longrightarrow 0$$

as $t \to 0$, t real and nonzero. We call A the *Gâteaux derivative* of f.

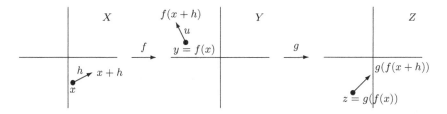

Figure 9.1   Notation used in the proof of the chain rule.

The reader should note that if the Gâteaux derivative exists, it is unique, and that the following result holds.

Proposition 9.3. *If f is Fréchet differentiable, then it is Gâteaux differentiable.*

Remark. The converse is not valid. The function $f : \mathbb{R}^2 \to \mathbb{R}$ given by

$$f(x) = \begin{cases} 0 & \text{if } x_2 = 0, \\ x_1^3/x_2 & \text{if } x_2 \neq 0, \end{cases}$$

is not continuous at the origin. For instance $f((t, t^3)) \to 1$ as $t \to 0$, but $f(0) = 0$. However, f is Gâteaux differentiable at $(0, 0)$ in every direction h since

$$\frac{f(th) - f(0)}{t} = \frac{f(th)}{t} = \begin{cases} 0 & \text{if } h_2 = 0, \\ t(h_1^3/h_2) & \text{if } h_2 \neq 0. \end{cases}$$

The limit as $t \to 0$ exists and is zero, whatever the value of h.

9.1.1   The chain rule

Just as in multi-variable calculus, the derivative of the composition $g \circ f = g(f)$ of two differentiable functions g and f can be expressed in terms of their individual derivatives. The outer function g in the composition must be restricted to being Fréchet differentiable, since the specified directions required in Gâteaux differentiability may be distorted by the inner function f.

Theorem 9.4 (chain rule). *Let X, Y, Z be NLSs, $U \subset X$ open, $V \subset Y$ open, $f : U \to Y$, and $g : V \to Z$. Let $x \in U$ and $y = f(x) \in V$. Suppose g is Fréchet differentiable at y and f is Gâteaux (respectively, Fréchet) differentiable at x. Then $g \circ f$ is Gâteaux (respectively, Fréchet) differentiable at x and*

$$D(g \circ f)(x) = Dg\big(f(x)\big) \, Df(x).$$

Proof. The proof is given for the case where both maps are Fréchet differentiable. The proof of the Gâteaux case is similar. Notation used in the proof are illustrated in Figure 9.1. Write

$$R_f(x, h) = f(x + h) - f(x) - Df(x)h$$

and

$$R_g(y, k) = g(y + k) - g(y) - Dg(y)k.$$

By assumption,

$$\frac{R_f(x, h)}{\|h\|} \xrightarrow{Y} 0 \quad \text{as } h \xrightarrow{X} 0 \tag{9.5}$$

and

$$\frac{R_g(y, k)}{\|k\|} \xrightarrow{Z} 0 \quad \text{as } k \xrightarrow{Y} 0. \tag{9.6}$$

Define

$$u = u(h) = f(x + h) - f(x) = f(x + h) - y$$

and note by continuity that $u(h) \to 0$ as $h \to 0$. Now consider the difference

$$
\begin{aligned}
(g \circ f)(x + h) - (g \circ f)(x) &= g(f(x + h)) - g(f(x)) \\
&= g(y + u) - g(y) \\
&= Dg(y)u + R_g(y, u) \\
&= Dg(y)[f(x + h) - f(x)] + R_g(y, u) \\
&= Dg(y)[Df(x)h + R_f(x, h)] + R_g(y, u) \\
&= Dg(f(x))Df(x)h + R(x, h),
\end{aligned}
$$

where

$$R(x, h) = Dg(y)R_f(x, h) + R_g(y, u).$$

To finish, it must be shown that $R(x, h) = o(\|h\|_X)$ as $h \to 0$. To this end, notice that

$$\frac{\|Dg(y)R_f(x, h)\|_Z}{\|h\|_X} \leq \|Dg(y)\|_{B(Y,Z)} \frac{\|R_f(x, h)\|_Y}{\|h\|_X} \longrightarrow 0 \quad \text{as } h \to 0$$

because of (9.5). The second term is slightly more interesting. We are trying to show

$$\frac{\|R_g(y, u)\|_Z}{\|h\|_X} \longrightarrow 0$$

as $h \to 0$ ($h \neq 0$). This does not follow immediately from (9.6). However, the local-Lipschitz property (Lemma 9.2) comes to our rescue.

If $u = 0$, then $R_g(y, u) = 0$. If not, then multiply and divide by $\|u\|_Y$ to reach

$$\frac{\|R_g(y, u)\|_Z}{\|h\|_X} = \frac{\|R_g(y, u)\|}{\|u\|_Y} \frac{\|u\|_Y}{\|h\|_X}. \tag{9.7}$$

Let $\epsilon > 0$ be given and suppose without loss of generality that $\epsilon \leq 1$. There is a $\eta > 0$ such that if $\|k\|_Y \leq \eta$, then

$$\frac{\|R_g(y, k)\|_Z}{\|k\|_Y} \leq \epsilon.$$

On the other hand, because of (9.3), there is a $\delta > 0$ such that $\|h\|_X \leq \delta$ implies

$$\|u(h)\|_Y = \|f(x+h) - f(x)\|_Y \leq (\|Df(x)\|_{B(X,Y)} + 1)\|h\|_X \leq \eta$$

(simply choose δ so that $\delta(\|Df(x)\|_{B(X,Y)} + 1) \leq \eta$ in addition to it satisfying the smallness requirement in Lemma 9.2). With this choice of δ, if $\|h\|_X \leq \delta$, then (9.7) implies

$$\frac{\|R_g(y,u)\|_Z}{\|h\|_X} \leq \epsilon(\|Df(x)\|_{B(X,Y)} + 1).$$

The result follows. □

Example. Let $A : L^2 \to L^2$ be a linear operator and define $f : L^2 \to \mathbb{R}$ by $f(\varphi) = \int \varphi(x)^2 \, dx$. Let $F : L^2 \to \mathbb{R}$ be the composition $F = f \circ A$. Then the reader can readily verify that $DF(\varphi)(h) = 2 \int Af(x) \, Ah(x) \, dx$.

9.1.2 The Mean-Value Theorem

The mean-value theorem relates the value of a function at two points to the way it changes between them, i.e., to the derivative of the function on a path joining the two points.

Proposition 9.5 (mean-value theorem for curves). *Let Y be an NLS and $\varphi : [a,b] \to Y$ be continuous, where $a < b$ are real numbers. Suppose $\varphi'(t)$ exists on (a,b) and that $\|\varphi'(t)\|_{B(\mathbb{R},Y)} \leq M$. Then*

$$\|\varphi(b) - \varphi(a)\|_Y \leq M(b-a). \tag{9.8}$$

Remark. Every bounded linear operator from $\mathbb{R}$ to Y is given by $t \mapsto ty$ for some fixed $y \in Y$, and we may identify $\varphi'(t)$ with this element y (i.e., $B(\mathbb{R}, Y) \cong Y$). In this case y can be obtained by the elementary limit

$$y = \varphi'(t) = \lim_{s \to 0} \frac{\varphi(t+s) - \varphi(t)}{s}.$$

Proof. Fix an $\epsilon > 0$ and suppose $\epsilon \leq 1$. For any $t \in (a,b)$, there is a $\delta_t = \delta(t, \epsilon)$ such that if $|s - t| < \delta_t$ and $s \in (a,b)$, then

$$\|\varphi(s) - \varphi(t)\|_Y \leq (M + \epsilon)|s - t|$$

by the local-Lipschitz lemma (Lemma 9.2). Let

$$S(t) = B_{\delta_t}(t) \cap (a,b),$$

which is open. Then if $a < \tilde{a} < \tilde{b} < b$,

$$[\tilde{a}, \tilde{b}] \subset \bigcup_{t \in [\tilde{a}, \tilde{b}]} S(t).$$

Hence by compactness, there is a finite sub-cover, of, say, N intervals, $S(\tilde{a})$, $S(t_2), S(t_4), \ldots, S(\tilde{b})$, where

$$\tilde{a} = t_0 < t_2 < \cdots < t_{2N} = \tilde{b},$$

such that also $S(t_{2k+2}) \cap S(t_{2k}) \neq \emptyset$ for all k. Choose points $t_{2k+1} \in S(t_{2k+2}) \cap S(t_{2k})$, enrich the partition to

$$\tilde{a} = t_0 < t_1 < t_2 < \cdots < t_{2N} = \tilde{b},$$

and note that

$$\|\varphi(t_{k+1}) - \varphi(t_k)\|_Y \leq (M + \epsilon)|t_{k+1} - t_k|$$

for all k. Hence

$$\|\varphi(\tilde{b}) - \varphi(\tilde{a})\|_Y \leq \sum_{k=1}^{2N} \|\varphi(t_k) - \varphi(t_{k-1})\|_Y$$

$$\leq (M + \epsilon) \sum_{k=1}^{2N} (t_k - t_{k-1}) = (M + \epsilon)(\tilde{b} - \tilde{a}).$$

By continuity, we may take the limit as $\tilde{b} \to b$ and $\tilde{a} \to a$, and the same inequality holds. Since $\epsilon > 0$ was arbitrary, (9.8) follows. □

Remark. The mean-value theorem for curves can be used to give reasonable conditions under which Gâteaux differentiability implies Fréchet differentiability.

Theorem 9.6 (mean-value theorem). *Let X, Y be NLSs and $U \subset X$ open. Let $f : U \to Y$ be Fréchet differentiable everywhere in U and suppose the line segment*

$$\ell = \{(1 - t)x_0 + tx_1 : 0 \leq t \leq 1\}$$

is contained in U. Then

$$\|f(x_1) - f(x_0)\|_Y \leq \sup_{x \in \ell} \|Df(x)\|_{B(X,Y)} \|x_1 - x_0\|_X.$$

Proof. Define $\varphi : [0, 1] \to Y$ by

$$\varphi(t) = f((1 - t)x_0 + tx_1) = f(x_0 + t(x_1 - x_0)) = f(\gamma(t)),$$

where $\gamma : [0, 1] \to X$. Certainly φ is differentiable on $[0, 1]$ by the chain rule. By the mean-value theorem for curves (Proposition 9.5),

$$\|f(x_1) - f(x_0)\|_Y = \|\varphi(1) - \varphi(0)\|_Y \leq \sup_{0 \leq t \leq 1} \|\varphi'(t)\|_Y.$$

But, the chain rule insures that

$$\varphi'(t) = Df(\gamma(t)) \circ \gamma'(t) = Df(\gamma(t))(x_1 - x_0),$$

so

$$\|\varphi'(t)\|_Y \leq \|Df(\gamma(t))\|_{B(X,Y)} \|x_1 - x_0\|_X$$

$$\leq \sup_{x \in \ell} \|Df(x)\|_{B(X,Y)} \|x_1 - x_0\|_X. \quad \square$$

9.1.3   Partial differentiation

One can immediately generalize the above discussion to partial Fréchet differentiability of $f : U \to Y$, as long as the NLS $X \supset U$ can be decomposed appropriately, which is to say that X can be written as the direct sum of the NLSs $X_1, \ldots, X_m$. Recall from Chapter 2 that we can view this direct sum as either external or internal. In the former case,

$$X = \{(x_1, \ldots, x_m) : x_i \in X_i \ \forall i\} = X_1 \times \cdots \times X_m,$$

and in the latter case, each X_i is a linear subspace of the larger NLS X and $X = X_1 + \cdots + X_m$ but written $X = X_1 \oplus \cdots \oplus X_m$. In either case, the requirement of a direct sum is that each $x \in X$ is *uniquely* decomposable into either $(x_1, \ldots, x_m)$ or $x_1 + \cdots + x_m$ for $x_i \in X_i$. Recall further that the norm on X is given for $p \in [1, \infty]$ by any of the equivalent functions

$$\|x\|_X = \left\|(\|x_1\|_{X_1}, \ldots, \|x_m\|_{X_m})\right\|_{\ell^p}, \tag{9.9}$$

where $\|\cdot\|_{X_i} = \|\cdot\|_X$ in the case of an internal direct sum.

To define a partial Fréchet derivative, it is more convenient to view the NLS X as an external direct sum, $X = X_1 \times \cdots \times X_m$ with the norm defined by (9.9). Let $U \subset X$ be open and $F : U \to Y$, Y an NLS. Let $x = (x_1, \ldots, x_m) \in U$ and fix an integer $k \in [1, m]$. For z near enough to x_k in X_k, the point $(x_1, \ldots, x_{k-1}, z, x_{k+1}, \ldots, x_m)$ lies in U, since U is open. Define

$$f_k(z) = F(x_1, \ldots, x_{k-1}, z, x_{k+1}, \ldots, x_m).$$

Then f_k maps an open subset of X_k into Y. If f_k has a Fréchet derivative at $z = x_k$, then we say F has a *kth partial derivative* at x and define

$$D_k F(x) = D f_k(x_k).$$

Notice that $D_k F(x) \in B(X_k, Y)$. In place of $D_k F$, one often writes $D_{X_k} F$, or even $D_{x_k} F$ when the context makes clear that $x_k \in X_k$.

Proposition 9.7. *Let $X = X_1 \times \cdots \times X_m$ be the (external) direct sum of NLSs, $U \subset X$ open, and $F : U \to Y$, another NLS. Suppose $D_j F(x)$ exists for $x \in U$ and $1 \le j \le m$, and that these linear maps are continuous as a function of x at $x_0 \in U$. Then F is Fréchet differentiable at x_0 and for $h = (h_1, \ldots, h_m) \in X$,*

$$DF(x_0)h = \sum_{j=1}^{m} D_j F(x_0) h_j. \tag{9.10}$$

Proof. The right-hand side of (9.10) defines a bounded linear map on X. Indeed, it may be written as

$$Ah = \sum_{j=1}^{m} D_j F(x_0) \circ \Pi_j h,$$

where $\Pi_j : X \to X_j$ is projection onto the jth component (which is easily seen to be a bounded linear map). So A is a sum of compositions of bounded linear operators and so is itself a bounded linear operator. Define

$$R(h) = F(x_0 + h) - F(x_0) - Ah.$$

It suffices to show that $R : X \to Y$ is such that

$$\frac{\|R(h)\|_Y}{\|h\|_X} \longrightarrow 0$$

as $h \to 0$ ($h \neq 0$).

Let $\epsilon > 0$ be given. Because F is partially Fréchet differentiable and A is linear, it follows immediately from the chain rule that R is partially Fréchet differentiable in h and

$$D_j R(h) = D_j F(x_0 + h) - D_j F(x_0).$$

Since the partial Fréchet derivatives are continuous as a function of x at x_0, it follows there is a $\delta > 0$ such that if $\|h\|_X \leq \delta$, then

$$\|D_j R(h)\|_{B(X_j, Y)} \leq \epsilon \quad \text{for } 1 \leq j \leq m. \tag{9.11}$$

On the other hand,

$$
\begin{aligned}
\|R(h)\|_Y \leq{} & \|R(h) - R(0, h_2, \dots, h_m)\|_Y \\
& + \|R(0, h_2, \dots, h_m) - R(0, 0, h_3, \dots, h_m)\|_Y \\
& + \cdots + \|R(0, \dots, 0, h_m) - R(0, \dots, 0)\|_Y.
\end{aligned}
\tag{9.12}
$$

If $\|h\|_X \leq \delta$, then by the mean-value theorem (Theorem 9.6) applied to the mappings

$$R_j(k) = R(0, \dots, 0, k, h_{j+1}, \dots, h_m),$$

it is determined on the basis of (9.11) that, for $1 \leq j \leq m$,

$$
\begin{aligned}
\|R_j(h_j) - R_j(0)\|_Y &\leq \sup_{t \in [0,1]} \|DR_j(th_j)\|_{B(X_j, Y)} \|h_j\|_{X_j} \\
&= \sup_{t \in [0,1]} \|D_j R(0, \dots, 0, th_j, h_{j+1}, \dots, h_m)\|_{B(X_j, Y)} \|h_j\|_{X_j} \\
&\leq \epsilon \|h_j\|_{X_j}.
\end{aligned}
$$

Choosing in (9.9) the ℓ^1-norm on X, it follows from (9.12) and the last inequalities that

$$\|R(h)\|_Y \leq \epsilon \sum_{j=1}^{m} \|h_j\|_{X_j} = \epsilon \|h\|_X,$$

provided that $\|h\|_X \leq \delta$. (If another ℓ^p-norm is used in (9.9), we merely get a fixed constant multiple of the right-hand side above.) The result follows. $\quad\square$

9.2 FIXED POINTS AND CONTRACTIVE MAPS

The discussion turns now to the problem of finding solutions to nonlinear equations. We begin by solving an equation of the form $G(x) = x$. It will prove to be surprisingly worthwhile to solve such an equation.

Definition. Let (X, d) be a metric space and $G : X \to X$. The mapping G is a *contraction* if there is a θ with $0 \le \theta < 1$ such that

$$d(G(x), G(y)) \le \theta d(x, y) \quad \text{for all } x, y \in X.$$

A *fixed point* of the mapping G is an $x \in X$ such that $x = G(x)$.

A contraction map is a Lipschitz map with Lipschitz constant less than 1. Of course, such maps are also continuous.

Theorem 9.8 (contraction mapping theorem of Banach). *If (X, d) is a complete metric space and G a contraction mapping of X, then there is a unique fixed point of G in X.*

Proof. If there were two fixed points x and y, then

$$d(x, y) = d(G(x), G(y)) \le \theta d(x, y),$$

and since $d(x, y) \ge 0$ and $0 \le \theta < 1$, it follows that $d(x, y) = 0$, whence $x = y$.

For existence of a fixed point, argue as follows. Fix an $x_0 \in X$ and let

$$x_1 = G(x_0), \quad x_2 = G(x_1), \quad x_3 = G(x_2), \quad \text{etc.}$$

We claim the sequence $\{x_n\}_{n=0}^{\infty}$ of iterates is a Cauchy sequence. If so, then since (X, d) is complete, there is an $x \in X$ such that $x_n \to x$. Furthermore, $G(x_n) \to G(x)$ by continuity. Since $G(x_n) = x_{n+1}$, it follows that $G(x) = x$.

To see $\{x_n\}_{n=0}^{\infty}$ is a Cauchy sequence, observe that

$$d(x_n, x_{n+1}) = d(G(x_{n-1}), G(x_n)) \le \theta d(x_{n-1}, x_n)$$

for $n = 1, 2, 3, \ldots$. In consequence, it follows by induction that

$$d(x_n, x_{n+1}) \le \theta^n d(x_0, x_1), \quad \text{for } n = 0, 1, 2, \ldots.$$

If $n \ge 0$ is fixed and $m > n$, then

$$
\begin{aligned}
d(x_n, x_m) &\le d(x_n, x_{n+1}) + d(x_{n+1}, x_{n+2}) + \cdots + d(x_{m-1}, x_m) \\
&\le \left(\theta^n + \cdots + \theta^{m-1} \right) d(x_0, x_1) \\
&= \theta^n \left(1 + \cdots + \theta^{m-n-1} \right) d(x_0, x_1) \\
&= \theta^n \frac{1 - \theta^{m-n}}{1 - \theta} d(x_0, x_1) \\
&\le \frac{\theta^n}{1 - \theta} d(x_0, x_1).
\end{aligned}
\tag{9.13}
$$

As $\theta < 1$, the right-hand side of the last inequality can be made as small as desired, independently of $m > n$, by taking n large enough. $\qquad\square$

Not only does this theorem provide existence and uniqueness, the proof is constructive. Indeed, the proof consists of generating a sequence of approximations to $x = G(x)$ that converge to the fixed point and satisfy the error bound (9.13). Taking a limit in (9.13) leads to the following result.

Corollary 9.9 (fixed point theorem). *Suppose that (X, d) is a complete metric space, G a contraction mapping of X with contraction constant θ, and $x_0 \in X$. If the sequence $\{x_n\}_{n=0}^{\infty}$ is defined successively by*

$$x_{n+1} = G(x_n) \quad for \ n = 0, 1, 2, \ldots,$$

then $x_n \to x$, where x is the unique fixed point of G in X. Moreover,

$$d(x_n, x) \leq \frac{\theta^n}{1 - \theta} d(x_0, x_1).$$

Example. Consider the initial value problem (IVP)

$$\begin{cases} u_t = \cos(u(t)), & t > 0, \\ u(0) = u_0. \end{cases}$$

We would like to obtain a solution to the problem, at least up to some final time $T > 0$, using the fixed point theorem. At the outset we require two things: a complete metric space within which to seek a solution, and a map on that space for which a fixed point is the solution to our problem. It is not easy to handle the differential operator directly in this context, so we remove it through integration, *viz.*

$$u(t) = u_0 + \int_0^t \cos(u(s)) \, ds.$$

Now it is natural to seek a continuous function as a solution, say in $X = C^0([0, T])$ for some as yet unknown $T > 0$. It is also natural to consider the function

$$G(u) = u_0 + \int_0^t \cos(u(s)) \, ds,$$

which clearly takes X to X and has a fixed point at the solution to our IVP. To see if G is contractive, consider two functions u and v in X and compute

$$\begin{aligned} \|G(u) - G(v)\|_{L^\infty} &= \sup_{0 \leq t \leq T} \left| \int_0^t \big(\cos(u(s)) - \cos(v(s)) \big) \, ds \right| \\ &= \sup_{0 \leq t \leq T} \left| \int_0^t (-\sin(w(s)))(u(s) - v(s)) \, ds \right| \\ &\leq T \|u - v\|_{L^\infty}, \end{aligned}$$

wherein we have used the ordinary mean-value theorem for functions of a real variable. Taking $T = 1/2$, say, a unique solution on the interval $[0, 1/2]$

obtains by the Banach contraction mapping theorem. Since T is a fixed number independent of the solution u, this process can be iterated, starting at $t = 1/2$ (with "initial condition" $u(1/2)$) to extend the solution uniquely to $[0, 1]$, and so on, to obtain a solution for all time; usually referred to as a *global* solution. Notice that this simple argument works because the contraction constant does not depend on the solution.

We next provide an extended example of the concepts. Let $\kappa \in L^1(\mathbb{R})$, $\varphi \in C_B^0(\mathbb{R})$ and consider the nonlinear operator

$$\Phi u(x, t) = \varphi(x) + \int_0^t \int_{-\infty}^\infty \kappa(x - y)\big(u(y, s) + u^2(y, s)\big)\, dy\, ds.$$

We claim that there exists $T = T(\|\varphi\|_{L^\infty}) > 0$ such that Φ has a fixed point in the space $X = C_B^0(\mathbb{R} \times [0, T])$.

Since κ is in $L^1(\mathbb{R})$, Φu makes sense. If $u \in X = C_B^0(\mathbb{R} \times [0, T])$, then it is an easy exercise to see $\Phi u \in X$. Indeed, Φu is C^1 in the temporal variable and continuous in x by the dominated convergence theorem. That is, $\Phi : X \to X$; however, Φ is *not* contractive on all of X.

Let $R > 0$ and B_R the closed ball of radius R about 0 in X. We want to show that if R and T are chosen well, $\Phi : B_R \to B_R$ is a contraction. Let $u, v \in B_R$ and consider

$$\|\Phi u - \Phi v\|_X = \sup_{(x,t)\in\mathbb{R}\times[0,T]} \left| \int_0^t \int_{-\infty}^\infty \kappa(x - y)(u - v + u^2 - v^2)\, dy\, ds \right|$$

$$\leq T \sup_{(x,t)\in\mathbb{R}\times[0,T]} \int_{-\infty}^\infty |\kappa(x - y)(u - v + u^2 - v^2)|\, dy$$

$$\leq T\|\kappa\|_{L^1}\big(\|u - v\|_X + \|u^2 - v^2\|_X\big)$$

$$\leq T\|\kappa\|_{L^1}\big(1 + \|u\|_X + \|v\|_X\big)\|u - v\|_X$$

$$\leq T\|\kappa\|_{L^1}(1 + 2R)\|u - v\|_X.$$

Let

$$\theta = T(1 + 2R)\|\kappa\|_{L^1},$$

choose $R = 2\|\varphi\|_{L^\infty}$ and then choose T so that $\theta = 1/2$. With these choices, Φ is contractive on B_R and if $u \in B_R$, then indeed

$$\|\Phi u\|_X \leq \|\Phi u - \Phi 0\|_X + \|\Phi 0\|_X$$

$$\leq \theta\|u - 0\|_X + \|\varphi\|_{L^\infty}$$

$$\leq \frac{1}{2}R + \frac{1}{2}R = R.$$

That is, $\Phi : B_R \to B_R$ and Φ is contractive. The set B_R, being closed, is a complete metric space, hence it is concluded that there exists a unique $u \in B_R$ such that

$$u = \Phi u.$$

Why do we care? Consider the partial differential equation

$$\frac{\partial u}{\partial t} + \frac{\partial u}{\partial x} + 2u\frac{\partial u}{\partial x} - \frac{\partial^3 u}{\partial x^2 \partial t} = 0, \tag{9.14}$$

that arises as a model in various wave propagation contexts. It is not elliptic, parabolic, or hyperbolic. It fits into the class of nonlinear, dispersive wave equations. Write it as

$$(1 - \partial_x^2)u_t = -u_x - 2uu_x \equiv f.$$

The left-hand side is a nice operator, at least from the point of view of the Fourier transform, as will be apparent in a moment, while the terms defining f are more troublesome. Take the Fourier transform on x to reach

$$(1 + \xi^2)\hat{u}_t = \hat{f}, \quad \text{i.e.,} \quad \hat{u}_t = \frac{1}{1 + \xi^2}\hat{f},$$

whence, by taking the inverse Fourier transform, it is formally deduced that

$$u_t = \tilde{\kappa} * f = -\tilde{\kappa} * (u_x + 2uu_x) = -\tilde{\kappa} * (u + u^2)_x,$$

where

$$\tilde{\kappa}(x) = \frac{1}{\sqrt{2\pi}}\mathcal{F}^{-1}\left(\frac{1}{1 + \xi^2}\right) = \frac{1}{2}e^{-|x|}.$$

Let $\kappa = -\tilde{\kappa}_x = \frac{1}{2}e^{-|x|}\,\text{signum}(x) \in L^1(\mathbb{R})$ to conclude

$$u_t(x, t) = \kappa * (u + u^2).$$

Now integrate over $[0, t]$ and use the fundamental theorem of calculus with constant of integration $u(x, 0) = \varphi(x) \in C_B^0(\mathbb{R})$ to reach

$$u(x, t) = \varphi(x) + \int_0^t \kappa * (u + u^2)\,ds,$$

which has the form with which we started the example. Thus our fixed point $\Phi u = u$ is formally a solution to (9.14), at least up to the time T, with the initial condition $u(x, 0) = \varphi(x)$.

In the extended example above, we used a corollary of the Banach contraction mapping theorem given as follows.

Corollary 9.10. *Let X be a Banach space and $G : X \to X$. Suppose there is $x_0 \in X$ and $r > 0$ such that both $G : \overline{B_r(x_0)} \to \overline{B_r(x_0)}$ and G is a contraction on $\overline{B_r(x_0)}$. Then G has a unique fixed point in $\overline{B_r(x_0)}$.*

9.3 NONLINEAR EQUATIONS

Developed next are some helpful techniques for understanding when a nonlinear equation has a solution and its basic properties. These include fixed point methods of Newton-type and the inverse and implicit function theorems.

9.3.1 Newton methods

The basic idea, already illustrated above, is to convert the equation into a fixed point problem. For example, when X, Y are Banach spaces and $f : X \to Y$, the problem of finding $x \in X$ for fixed $y \in Y$ in the equation

$$f(x) = y$$

is equivalent to finding a fixed point of

$$G(x) = x - T_x(f(x) - y),$$

where $T_x : Y \to X$ vanishes only at 0 for any $x \in X$. Often $T_x = T \in B(Y, X)$ is independent of x and injective. In that case, we have that

$$DG(x) = I - T \, Df(x),$$

so G is a contraction provided T can be chosen to make $\|I - T \, Df(x)\| < 1$, e.g. by the mean-value theorem (Theorem 9.6) or some other method.

Theorem 9.11 (simplified Newton method). *Let X, Y be Banach spaces, $f : X \to Y$ a differentiable mapping, and $x_0 \in X$. Suppose $A = Df(x_0)$ has a bounded inverse and*

$$\|I - A^{-1} Df(x)\| \leq \theta < 1 \qquad (9.15)$$

for all $x \in \overline{B_r(x_0)}$, for some $r > 0$. Let

$$\delta = \frac{(1 - \theta)r}{\|A^{-1}\|_{B(Y,X)}}.$$

Then the equation

$$f(x) = y$$

has a unique solution $x \in \overline{B_r(x_0)}$ whenever $y \in \overline{B_\delta(f(x_0))}$.

In more colloquial terms, the theorem says that given $x_0 \in X$ and $f(x_0) = y_0 \in Y$, if (9.15) holds in a neighborhood of x_0, then the problem $f(x) = y$ has a (unique) solution $x \in X$ near x_0 for all y near enough to y_0. Of course, as is usual in this chapter, this is a local result and f need not be defined on all of X.

Proof. Let $y \in \overline{B_\delta(f(x_0))}$ be given and define a mapping $G_y : X \to X$ by

$$G_y(x) = x - A^{-1}(f(x) - y). \qquad (9.16)$$

Notice that $G_y(x) = x$ if and only if $f(x) = y$. Note also that the chain rule implies

$$DG_y(x) = I - A^{-1} Df(x).$$

By assumption, $\|DG_y(x)\|_{B(X,X)} \leq \theta < 1$ for $x \in \overline{B_r(x_0)}$, so G_y is a contraction on $\overline{B_r(x_0)}$. Moreover, the mean-value theorem and the choice of y and δ implies that, for $x \in \overline{B_r(x_0)}$,

$$
\begin{aligned}
\|G_y(x) - x_0\|_X &\leq \|G_y(x) - G_y(x_0)\|_X + \|G_y(x_0) - x_0\|_X \\
&\leq \sup_{\xi \in \overline{B_r(x_0)}} \|DG_y(\xi)\|_{B(X,X)} \|x - x_0\|_X + \|A^{-1}(f(x_0) - y)\|_X \\
&\leq \theta r + \|A^{-1}\|_{B(Y,X)} \|f(x_0) - y\|_Y \\
&\leq \theta r + (1 - \theta)r = r.
\end{aligned}
$$

The hypotheses of Corollary 9.10 are verified, so G_y is a contractive map from $\overline{B_r(x_0)}$ to itself, and the conclusion follows. □

Remark. If f is continuously differentiable, i.e., $Df(x)$ is continuous as a function of x, then hypothesis (9.15) is true for r small enough. Thus another conclusion is that at any point x_0 where $Df(x_0)$ is boundedly invertible, there is an $r > 0$ and a $\delta > 0$ such that $f(\overline{B_r(x_0)}) \supset \overline{B_\delta(f(x_0))}$, and so f is a one-to-one map of $\overline{B_r(x_0)} \cap f^{-1}(\overline{B_\delta(f(x_0))})$ onto $\overline{B_\delta(f(x_0))}$.

Take note of the algorithm that is implied by the proof. Given y, start with a guess x_0 and form the sequence

$$
x_{n+1} = G_y(x_n) = x_n - A^{-1}(f(x_n) - y).
$$

If things are as in the theorem, the sequence converges to the solution of $f(x) = y$ in $\overline{B_r(x_0)}$. If x is the solution, then

$$
\begin{aligned}
\|x_n - x\|_X &= \|G_y(x_{n-1}) - G_y(x)\|_X \\
&\leq \theta \|x_{n-1} - x\|_X \\
&\leq \cdots \\
&\leq \theta^n \|x_0 - x\|_X,
\end{aligned}
$$

or, if one prefers the result from Corollary 9.9,

$$
\|x_n - x\|_X \leq \frac{\theta^n}{1 - \theta} \|x_0 - x_1\|_X.
$$

More can be shown. We leave the rather lengthy proof of the following result to the exercises at the end of the chapter.

Theorem 9.12 (Newton-Kantorovich method). *Let X, Y be Banach spaces and $f : X \to Y$ a differentiable mapping. Assume that there is $x_0 \in X$, $r > 0$, and $\kappa \geq 0$ such that*

(a) $A = Df(x_0)$ *has a bounded inverse, and*

(b) $\|Df(x_1) - Df(x_2)\|_{B(X,Y)} \leq \kappa \|x_1 - x_2\|$ *for all $x_1, x_2 \in B_r(x_0)$.*

Let $y \in Y$ and set

$$\epsilon = \|A^{-1}(f(x_0) - y)\|_X.$$

If ϵ is small enough that both

$$\epsilon \leq \frac{r}{2} \quad and \quad 4\epsilon\kappa\|A^{-1}\|_{B(Y,X)} \leq 1,$$

then the equation

$$y = f(x)$$

has a unique solution in $B_r(x_0)$. Moreover, the solution is obtained as the limit of the Newton iterates

$$x_{n+1} = x_n - Df(x_n)^{-1}(f(x_n) - y),$$

starting at x_0. The convergence is asymptotically quadratic; that is,

$$\|x_{n+1} - x_n\|_X \leq C\|x_n - x_{n-1}\|_X^2,$$

for n large, where $C \geq 0$ does not depend on n.

Theorem 9.13 (precursor to the inverse function theorem). *Suppose the hypotheses of the simplified Newton method (Theorem 9.11) hold. Then the inverse mapping $f^{-1} : \overline{B_\delta(f(x_0))} \rightarrow \overline{B_r(x_0)}$ is Lipschitz.*

Proof. Theorem 9.11 tells us that the map $f^{-1} : \overline{B_\delta(f(x_0))} \rightarrow \overline{B_r(x_0)}$ is well-defined. Let $y_1, y_2 \in \overline{B_\delta(f(x_0))}$ and let x_1, x_2 be the unique points in $\overline{B_r(x_0)}$ such that $f(x_i) = y_i$, for $i = 1, 2$. Fix a $y \in B_\delta(f(x_0))$, $y = y_0 = f(x_0)$ for example, and reconsider the mapping G_y defined in (9.16). As shown previously, G_y is a contraction mapping of $\overline{B_r(x_0)}$ into itself with Lipschitz constant $\theta < 1$. Then

$$\|f^{-1}(y_1) - f^{-1}(y_2)\|_X = \|x_1 - x_2\|_X$$
$$= \|G_y(x_1) - G_y(x_2) + A^{-1}(f(x_1) - f(x_2))\|_X$$
$$\leq \theta\|x_1 - x_2\|_X + \|A^{-1}\|_{B(Y,X)}\|y_1 - y_2\|_Y.$$

It follows that

$$\|x_1 - x_2\|_X \leq \frac{\|A^{-1}\|_{B(Y,X)}}{1 - \theta}\|y_1 - y_2\|_Y;$$

whence f^{-1} is Lipschitz with constant at most $\|A^{-1}\|_{B(Y,X)}/(1 - \theta)$. □

9.3.2 The Inverse Function Theorem

Earlier, we agreed that two Banach spaces X and Y are isomorphic if there is a map $T \in B(X, Y)$ which is one-to-one and onto (and hence with bounded inverse by the open mapping theorem). Isomorphic Banach spaces are indistinguishable as Banach spaces, up to the value of the norm. A local version of this idea is now introduced.

Definition. Let X, Y be Banach spaces and $U \subset X$, $V \subset Y$ open sets. Let $f : U \to V$ be one-to-one and onto. Then f is called a *diffeomorphism* on U and U is *diffeomorphic* to V if both f and f^{-1} are C^1, which is to say f and f^{-1} are Fréchet differentiable throughout U and V, respectively, and their derivatives are continuous on U and V, respectively. The latter condition means that the maps

$$x \longmapsto Df(x) \quad \text{and} \quad y \longmapsto Df^{-1}(y)$$

are continuous from U to $B(X, Y)$ and V to $B(Y, X)$, respectively.

The notion of diffeomorphism is stronger than that of homeomorphism.

Theorem 9.14 (inverse function theorem). *Let X, Y be Banach spaces. Let $x_0 \in X$ be such that f is C^1 in a neighborhood of x_0 and $Df(x_0)$ is an isomorphism. Then there is an open set $U \subset X$ with $x_0 \in U$ and an open set $V \subset Y$ with $f(x_0) \in V$ such that $f : U \to V$ is a diffeomorphism. Moreover, for $y \in V$, $x \in U$, and $y = f(x)$,*

$$D(f^{-1})(y) = (Df(x))^{-1}.$$

A little preparation is needed to establish this result. Let the general linear group $GL(X, Y)$ denote the set of all isomorphisms of X onto Y. Of course, $GL(X, Y) \subset B(X, Y)$. The theorem says that if we have the solution $f(x_0) = y_0 \in V$ when f is C^1 and $Df(x_0) \in GL(X, Y)$, then we can solve $f(x) = y$ for all $y \in V$, and uniquely for $x \in U$.

Lemma 9.15. *If X and Y are Banach spaces, then $GL(X, Y)$ is an open subset of $B(X, Y)$. If $GL(X, Y) \neq \emptyset$, then the mapping $J_{X,Y} : GL(X, Y) \to GL(Y, X)$ given by $J_{X,Y}(A) = A^{-1}$ is one-to-one, onto, and continuous.*

Proof. If $GL(X, Y) = \emptyset$, there is nothing to prove. Clearly $J_{Y,X} J_{X,Y} = I$ and $J_{X,Y} J_{Y,X} = I$, so $J_{X,Y}$ is both one-to-one and onto (but certainly *not* linear!). Let $A \in GL(X, Y)$ and $H \in B(X, Y)$. We claim that if $\|H\|_{B(X,Y)} < \theta / \|A^{-1}\|_{B(Y,X)}$ where $\theta < 1$, then $A + H \in GL(X, Y)$ also. To prove this, one need only show $A + H$ is one-to-one and onto.

Now, were it to exist,

$$(A + H)^{-1} = \left((I + HA^{-1})A \right)^{-1} = A^{-1}(I + HA^{-1})^{-1}.$$

We know that for any $|x| < 1$,

$$(1 + x)^{-1} = \sum_{n=0}^{\infty} (-x)^n,$$

so, as in the proof of Neumann series (Lemma 4.4), consider the operators

$$S_N = A^{-1} \sum_{n=0}^{N} (-HA^{-1})^n \in B(Y, X), \quad N = 1, 2, \ldots.$$

The sequence $\{S_N\}_{N=1}^{\infty}$ is Cauchy in $B(Y,X)$ since, for $M > N$,

$$\|S_M - S_N\|_{B(Y,X)} \leq \|A^{-1}\|_{B(Y,X)} \sum_{n=N+1}^{M} \|(HA^{-1})^n\|$$

$$\leq \|A^{-1}\|_{B(Y,X)} \sum_{n=N+1}^{M} (\|H\|_{B(X,Y)}\|A^{-1}\|_{B(Y,X)})^n \quad (9.17)$$

$$\leq \|A^{-1}\|_{B(Y,X)} \sum_{n=N+1}^{M} \theta^n \longrightarrow 0$$

as $N \to \infty$. Hence $S_N \to S$ in $B(Y,X)$ for some S. Observe that

$$(A + H)S = \lim_{N\to\infty} (A + H)S_N$$

$$= \lim_{N\to\infty} \sum_{n=0}^{N} \left[(-HA^{-1})^n - (-HA^{-1})^{n+1} \right]$$

$$= \lim_{N\to\infty} \left[I - (-HA^{-1})^{N+1} \right].$$

But as $\|HA^{-1}\| \leq \theta < 1$, $(HA^{-1})^N \to 0$ in $B(Y,Y)$. It is concluded that $(A + H)S = I$, and a similar calculation shows $S(A + H) = I$. Thus $A + H$ is one-to-one and onto, hence in $GL(X,Y)$. For use in a moment, notice that

$$\|S\|_{B(Y,X)} \leq \frac{\|A^{-1}\|_{B(Y,X)}}{1 - \theta},$$

by an argument similar to that given in (9.17).

For continuity, it suffices to take arbitrary $A \in GL(X,Y)$ and show that $(A+H)^{-1} \to A^{-1}$ in $B(Y,X)$ as $H \to 0$ in $B(X,Y)$. But, as $S = (A+H)^{-1}$, this amounts to showing $S - A^{-1} \to 0$ as $H \to 0$.

The calculation

$$S - A^{-1} = (SA - I)A^{-1} = (S(A + H) - SH - I)A^{-1} = -SHA^{-1}$$

gives the estimate

$$\|S - A^{-1}\|_{B(Y,X)} \leq \|S\|_{B(Y,X)}\|H\|_{B(X,Y)}\|A^{-1}\|_{B(Y,X)} \longrightarrow 0$$

as $H \to 0$ since $\|A^{-1}\|_{B(Y,X)}$ is fixed and $\|S\|_{B(Y,X)} \leq \|A^{-1}\|_{B(Y,X)}/(1 - \theta)$ is bounded independently of H. □

Proof of the inverse function theorem. Let $A = Df(x_0)$. Since f is a C^1-mapping, $Df(x) \to A$ in $B(X,Y)$ as $x \to x_0$ in X, so there is an $r' > 0$ such that

$$\left\|A^{-1}\big(Df(x_0) - Df(x)\big)\right\|_{B(X,X)} = \|I - A^{-1}Df(x)\|_{B(X,X)} \leq \frac{1}{2}$$

for all $x \in B_{r'}(x_0)$.

Because of Lemma 9.15, there is an r'' with $0 < r'' \le r'$ such that $Df(x)$ has a bounded inverse for all $x \in B_{r''}(x_0)$. It is further adduced that $Df(x)^{-1} \to A^{-1}$ as $x \to x_0$. In consequence, for some fixed r with $0 < r \le r''$ and for $x \in B_r(x_0)$,

$$\|Df(x)^{-1}\|_{B(Y,X)} \le 2\|A^{-1}\|_{B(Y,X)}.$$

Appealing now to the simplified Newton method, it is concluded that there is a $\delta > 0$ such that $f : U_* \to V_*$ is one-to-one, and onto, where

$$V_* = \overline{B_\delta(f(x_0))} \quad \text{with} \quad \delta = \frac{r}{2\|A^{-1}\|_{B(Y,X)}}$$

and

$$U_* = \overline{B_r(x_0)} \cap f^{-1}(V_*).$$

By the precursor to the inverse function theorem (Theorem 9.13), we know that f^{-1} is Lipschitz continuous. Take any open $U \subset U_*$ and let $V = f(U)$, which is also open.

It remains to establish that f^{-1} is a C^1 mapping with the indicated derivative. Suppose it is known that

$$Df^{-1}(y) = Df(x)^{-1} \quad \text{when } y = f(x), \tag{9.18}$$

where $x \in U$ and $y \in V$. In this case, the mapping from y to $Df^{-1}(y)$ is obtained in three steps, namely

$$Y \xrightarrow{f^{-1}} X \xrightarrow{Df} B(X,Y) \xrightarrow{J_{X,Y}} B(Y,X),$$

$$y \longmapsto f^{-1}(y) \longmapsto Df(f^{-1}(y)) \longmapsto \left(Df(f^{-1}(y))\right)^{-1} = Df^{-1}(y).$$

As all three of these components are continuous, so is the composite.

Thus it is only necessary to establish (9.18). To this end, fix $y \in V$ and let k be small enough that $y + k$ also lies in V. If $x = f^{-1}(y)$ and $h = f^{-1}(y+k) - x$, then

$$
\begin{aligned}
&\|f^{-1}(y+k) - f^{-1}(y) - Df(x)^{-1}k\|_X \\
&= \|h - Df(x)^{-1}[f(x+h) - f(x)]\|_X \\
&= \|Df(x)^{-1}[f(x+h) - f(x) - Df(x)h]\|_X \\
&\le 2\|A^{-1}\|_{B(Y,X)}\|f(x+h) - f(x) - Df(x)h\|_Y.
\end{aligned}
\tag{9.19}
$$

The right-hand side is $o(\|h\|)$ as $h \to 0$ in X since f is differentiable at x. But,

$$\|h\|_X = \|f^{-1}(y+k) - f^{-1}(y)\|_X \le M\|k\|_Y,$$

for some constant M since f^{-1} is Lipschitz. It follows that the expressions in (9.19) are $o(\|k\|)$, so f^{-1} is differentiable at $y = f(x)$ and

$$Df^{-1}(y) = Df(x)^{-1}.$$

The theorem is thereby established. $\qquad \square$

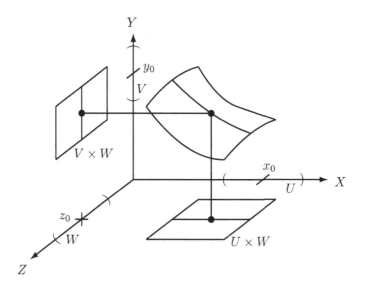

Figure 9.2 An illustration of the implicit function theorem. The cross section $\{z_0\} \times X \times Y$ is an illustration of the inverse function theorem.

9.3.3 The Implicit Function Theorem

In many applications, one has a map from X to Y that is parametrized by some parameter z taken from a Banach space Z. In that case, there is a generalization of the inverse function theorem, the implicit function theorem. It is illustrated in Figure 9.2.

Theorem 9.16 (implicit function theorem). *Let X, Y, Z be Banach spaces and suppose*

$$f : Z \times X \to Y$$

to be a C^1-mapping defined at least in a neighborhood of a point (z_0, x_0). If $D_x f(z_0, x_0) \in GL(X, Y)$, then there are open sets W, U, and V satisfying

$$z_0 \in W \subset Z, \quad x_0 \in U \subset X, \quad and \quad y_0 = f(z_0, x_0) \in V \subset Y,$$

and a unique mapping

$$g : W \times V \to U$$

such that

$$f(z, g(z, y)) = y \tag{9.20}$$

for all $(z, y) \in W \times V$. The function g is C^1 and, with $x = g(z, y)$,

$$Dg(z, y)(\eta, \zeta) = D_x f(z, x)^{-1}\big(\zeta - D_z f(z, x)\eta\big)$$

for all $(z, y) \in W \times V$ and $(\eta, \zeta) \in Z \times Y$.

Remark. If $Z = \{0\}$ is the trivial Banach space, this result recovers the inverse function theorem. Note that Proposition 9.7 on partial differentiation implies that the hypotheses are satisfied if $D_x f(z, x)$ and $D_z f(z, x)$ are continuous in (z, x) within a neighborhood of (z_0, x_0) and $D_x f(z_0, x_0)$ has a bounded inverse.

Proof. Define an auxiliary mapping $\hat{f}$ by

$$\hat{f}(z, x) = (z, f(z, x)).$$

Then $\hat{f} : Z \times X \to Z \times Y$ and $\hat{f}$ is C^1 since both its components are. Moreover, from Proposition 9.7 it is adduced that

$$D\hat{f}(z, x)(\eta, \varphi) = (\eta, D_z f(z, x)\eta + D_x f(z, x)\varphi)$$

for (z, x) in the domain of f and $(\eta, \varphi) \in Z \times X$. If $D_x f(z, x)$ is an invertible element of $B(X, Y)$, then $D\hat{f}$ is an invertible element of $B(Z \times X, Z \times Y)$ and its inverse is given by

$$D\hat{f}(z, x)^{-1}(\eta, \zeta) = (\eta, \varphi) = (\eta, D_x f(z, x)^{-1}(\zeta - D_z f(z, x)\eta)), \qquad (9.21)$$

as one checks immediately. For the point (z_0, x_0), the inverse function theorem (Theorem 9.14) implies $\hat{f}$ is a diffeomorphism from some open set $\hat{U}$ about (z_0, x_0) to an open set $\hat{V}$ containing (z_0, y_0). By continuity of the projections onto components in $Z \times Y$, there are open sets W and V in Z and Y, respectively, such that $W \times V \subset \hat{V}$. By construction,

$$\hat{f}^{-1}(z, y) = (z, g(z, y))$$

where g is a C^1-mapping. And, since

$$(z, y) = \hat{f}(\hat{f}^{-1}(z, y)) = \hat{f}(z, g(z, y)) = (z, f(z, g(z, y))),$$

g solves the equation (9.20). The derivative of g is the the second component of (9.21). □

Corollary 9.17. *Let f be as in Theorem 9.16. Then there is a unique C^1-branch of solutions of the equation*

$$f(z, x) = y_0$$

defined in a neighborhood of (z_0, x_0).

Proof. Let $h(z) = g(z, y_0)$ in the implicit function theorem. Then h is C^1, $h(z_0) = x_0$, and

$$f(z, h(z)) = y_0$$

for z near z_0. □

Example. The eigenvalues of an $n \times n$ matrix are given as the roots of the characteristic polynomial

$$p(A, \lambda) = \det(A - \lambda I).$$

In fact, p is a polynomial in λ and all the entries of A, so $p : \mathbb{C}^{n \times n} \times \mathbb{C} \to \mathbb{C}$ is a C^1 function. Fix A_0 and λ_0 such that λ_0 is a simple root of $p(A_0, \lambda)$. Then $D_2 p(A_0, \lambda_0) \neq 0$ (i.e., $D_2 p(A_0, \lambda_0) \in GL(\mathbb{C}, \mathbb{C})$), so every matrix A near A_0 has a unique eigenvalue

$$\lambda = g(A, 0) = \hat{g}(A),$$

where $\hat{g}$ is C^1. As we change A continuously from A_0, the eigenvalue λ_0 changes continuously until possibly it becomes a repeated eigenvalue, at which point a bifurcation may occur. A bifurcation cannot occur otherwise.

Example. Consider the ordinary differential initial value problem

$$\begin{cases} u' = 1 - u + \epsilon e^u, & 0 < t, \\ u(0) = 0. \end{cases}$$

If $\epsilon = 0$, this is a well-posed linear problem with solution

$$u_0(t) = 1 - e^{-t},$$

which exists for all time t. It is natural to ask if there is a solution for $\epsilon > 0$. Note that if ϵ is very large, then we have essentially the equation

$$w' = \epsilon e^w,$$

which has solution

$$w(t) = -\log(1 - \epsilon t) \longrightarrow \infty \quad \text{as } t \to 1/\epsilon.$$

Thus we do not have a solution w for all time. The implicit function theorem clarifies the situation. Our parameter space is $Z = \mathbb{R}$, and our function space is $X = \{f \in C_B^1(0, \infty) : f(0) = 0\}$. The mapping $T : Z \times X \to Y = C_B^0(0, \infty)$ defined by

$$T(\epsilon, u) = u' - 1 + u - \epsilon e^u$$

is C^1 with partial derivatives

$$D_Z T(\epsilon, u)(z) = -z e^u \quad \text{and} \quad D_X T(\epsilon, u)(v) = v' + v - \epsilon v e^u.$$

Now $D_X T(0, u)(v) = v' + v$ maps one-to-one and onto, since we can uniquely solve $v' + v = f$ by using an integrating factor. Thus the implicit function theorem gives us an $\epsilon_0 > 0$ such that for $|\epsilon| < \epsilon_0$, there exists a solution defined for all time. The solution is unique in a neighborhood of u_0 in X.

9.4   HIGHER DERIVATIVES

The discussion is now turned to higher order Fréchet derivatives. The development starts with some helpful preliminaries and ends with the Taylor series for nonlinear operators.

Definition. Let X, Y be vector spaces over $\mathbb{F}$ and n a positive integer. A multi-linear or n-linear operator is a function

$$f : \underbrace{X \times \cdots \times X}_{n \text{ components}} \longrightarrow Y$$

for which f is linear in each argument separately. The set of all n-linear maps from X to Y is denoted $\mathcal{B}^n(X, Y)$. By convention, we set $\mathcal{B}^0(X, Y) = Y$. If X and Y are NLSs, we say that $f \in \mathcal{B}^n(X, Y)$ is *bounded* if there is a constant M such that

$$\|f(x_1, \ldots, x_n)\|_Y \le M \|x_1\|_X \cdots \|x_n\|_X \quad \text{for all } x_1, \ldots, x_n \in X.$$

We denote by $B^n(X, Y)$ the subspace of $\mathcal{B}^n(X, Y)$ consisting of all bounded n-linear maps, and we set $B^0(X, Y) = Y$. Note that $B^1(X, Y)$ is simply $B(X, Y)$.

The following two results are left to the reader.

Proposition 9.18. *Let X, Y be NLSs and let $n \in \mathbb{N}$. The following are equivalent for $f \in \mathcal{B}^n(X, Y)$:*

(i) *f is continuous;*

(ii) *f is continuous at 0;*

(iii) *f is bounded.*

Proposition 9.19. *Let X, Y be NLSs and $n \in \mathbb{N}$. For $f \in B^n(X, Y)$, define*

$$\|f\|_{B^n(X,Y)} = \sup_{\substack{x_i \in X,\, \|x_i\| \le 1 \\ 1 \le i \le n}} \|f(x_1, \ldots, x_n)\|_Y.$$

Then $\|\cdot\|_{B^n(X,Y)}$ is a norm on $B^n(X, Y)$ and if Y is complete, so is $B^n(X, Y)$.

Proposition 9.20. *Let $k, \ell \in \mathbb{N}$ and X, Y be NLSs. Then $B^k(X, B^\ell(X, Y))$ is isomorphic to $B^{k+\ell}(X, Y)$ and the norms are the same.*

Proof. Let $n = k + \ell$ and define $J : B^k(X, B^\ell(X, Y)) \to B^n(X, Y)$ by

$$(Jf)(x_1, \ldots, x_n) = f(x_1, \ldots, x_k)(x_{k+1}, \ldots, x_n).$$

This makes sense because $f(x_1, \ldots, x_k) \in B^\ell(X, Y)$. Clearly $Jf \in \mathcal{B}^n(X, Y)$ and

$$\begin{aligned}
\|Jf\|_{B^n(X,Y)} &= \sup_{\|x_i\| \le 1, 1 \le i \le n} \|Jf(x_1, \ldots, x_n)\|_Y \\
&= \sup_{\|x_i\| \le 1, 1 \le i \le k} \|f(x_1, \ldots, x_k)\|_{B^\ell(X,Y)} \\
&= \|f\|_{B^k(X, B^\ell(X,Y))},
\end{aligned}$$

so $Jf \in B^n(X,Y)$ is a norm preserving and one-to-one map. For $g \in B^n(X,Y)$, define $\hat{g} \in \mathcal{B}^k(X, \mathcal{B}^\ell(X,Y))$ by

$$\hat{g}(x_1,\dots,x_k)(x_{k+1},\dots,x_n) = g(x_1,\dots,x_n).$$

A straightforward calculation shows that

$$\|\hat{g}\|_{\mathcal{B}^k(X,\mathcal{B}^\ell(X,Y))} \leq \|g\|_{B^n(X,Y)},$$

so $\hat{g} \in \mathcal{B}^k(X, \mathcal{B}^\ell(X,Y))$, and we conclude that $J\hat{g} = g$. Thus J is a one-to-one, onto, bounded linear map, so also has a bounded inverse and therefore is a Banach-space isomorphism. ∎

Definition. Let X, Y be Banach spaces and $f : X \to Y$. For $n = 2, 3, \dots$, define f to be n-times Fréchet differentiable in a neighborhood of a point x if f is $(n-1)$-times differentiable in a neighborhood of x and the mapping $x \mapsto D^{n-1}f(x)$ is Fréchet differentiable near x. Define

$$D^n f(x) = DD^{n-1}f(x), \quad n = 2, 3, \dots.$$

It is instructive to observe that

$$f : X \to Y,$$
$$Df : X \to B(X,Y),$$
$$D^2 f = D(Df) : X \to B(X, B(X,Y)) = B^2(X,Y),$$
$$\vdots$$
$$D^n f = D(D^{n-1}f) : X \to B(X, B^{n-1}(X,Y)) = B^n(X,Y).$$

When $n = 2$ and $f : X \to Y$, $D^2 f(x) \in B^2(X,Y)$ means

$$\|Df(x+h) - Df(x) - D^2 f(x)(h, \cdot)\|_{B(X,Y)} = o(\|h\|_X).$$

This is equivalent to

$$\|Df(x+h)k - Df(x)k - D^2 f(x)(h,k)\|_Y = o(\|h\|_X)\|k\|_X.$$

A similar remark holds for higher derivatives.

Proposition 9.21. *If X and Y are Banach spaces and $f : X \to Y$ has $n \geq 0$ derivatives, then $D^n f(x) = D^k D^\ell f(x)$ whenever $k + \ell = n$, $k \geq 0$ and $\ell \geq 0$.*

Proof. Use induction on n and Proposition 9.20. ∎

Examples.

(1) If $A \in B(X,Y)$, then $DA(x) = A$ for all x. Hence

$$D^2 A(x) = 0 \quad \text{for all } x.$$

This is because

$$DA(x+h) - DA(x) = 0 \quad \text{for all } x.$$

(2) Let $X = H$ be a real Hilbert space and $A \in B(H, H)$. Define $f : H \to \mathbb{R}$ by

$$f(x) = (x, Ax)_H.$$

We saw earlier that $Df(x) = \mathcal{R}((A + A^*)x)$, where $\mathcal{R}$ denotes the Riesz map. That is, $Df(x) \in B(H, \mathbb{R}) = H^*$, and for $y \in H$,

$$Df(x)(y) = (y, A^*x + Ax)_H.$$

To compute the second derivative, form the difference

$$[Df(x + h) - Df(x)]y = (y, (A + A^*)(x + h) - (A + A^*)x)$$
$$= (y, (A + A^*)h),$$

for $y \in H$. Thus it is determined that

$$D^2 f(x)(h, y) = (y, (A + A^*)h).$$

Note that $D^2 f(x)$ does not depend on x, so $D^3 f(x) = 0$.

(3) Let $K \in L^\infty(I \times I)$ where $I = [a, b] \subset \mathbb{R}$ is a bounded interval. Define $F : L^p(I) \to L^p(I)$ by

$$F(g)(x) = \int_I K(x, y) g^p(y) \, dy$$

for $p \in \mathbb{N}$ and $x \in I$. Then $DF(g) \in B(L^p(I), L^p(I))$ and

$$DF(g)h = p \int_I K(x, y) g^{p-1}(y) h(y) \, dy,$$

since the binomial theorem gives the expansion

$$F(g + h) - F(g) = \int_I K(x, y)[(g + h)^p - g^p] \, dy$$
$$= \int_I K(x, y)\left[pg^{p-1}(y)h(y) + \binom{p}{2} g^{p-2}(y)h^2(y) + \cdots\right] dy,$$

wherein all but the first term is higher order in h. Thus it follows readily that

$$DF(g + h)u - DF(g)u = p \int_I K(x, y) \left[(g + h)^{p-1}u - g^{p-1}u\right] dy$$
$$= p(p - 1) \int_I K(x, y)[g^{p-2}hu] \, dy + \text{ terms cubic in } h, u.$$

It follows formally, and can be verified under strict hypotheses, that

$$D^2 F(g)(h, k) = p(p - 1) \int_I K(x, y) g^{p-2}(y) h(y) k(y) \, dy.$$

Lemma 9.22 (Schwarz lemma). *Let X, Y be Banach spaces, U an open subset of X, and $f : U \to Y$ have two derivatives throughout U. Then $D^2 f(x)$ is a symmetric bilinear mapping.*

Proof. Consider the difference

$$g(h, k) = f(x + h + k) - f(x + h) - f(x + k) + f(x) - D^2 f(x)(k, h).$$

The first four terms on the right-hand side are invariant under $(h, k) \mapsto (k, h)$, so

$$
\begin{aligned}
\|D^2 f(x)(h, k) - D^2 f(x)(k, h)\|_Y &= \|g(h, k) - g(k, h)\|_Y \\
&\leq \|g(h, k) - g(0, k)\|_Y + \|g(0, k) - g(k, h)\|_Y \\
&= \|g(h, k) - g(0, k)\|_Y + \|g(0, h) - g(k, h)\|_Y
\end{aligned}
$$

since $g(0, k) = g(0, h) = 0$. But each term on the right-hand side of the last equality is bounded above by the mean-value theorem (Theorem 9.6) as

$$
\begin{aligned}
\|g(h, k) - g(0, k)\|_Y &\leq \sup \|D_1 g\|_{B(X,Y)} \|h\|_X, \\
\|g(k, h) - g(0, h)\|_Y &\leq \sup \|D_1 g\|_{B(X,Y)} \|k\|_X.
\end{aligned}
$$

Differentiate g partially with respect to the first variable h to obtain

$$
\begin{aligned}
D_1 g(h, k)\tilde{h} &= Df(x + h + k)\tilde{h} - Df(x + h)\tilde{h} - D^2 f(x)(k, \tilde{h}) \\
&= \big[Df(x + h + k)\tilde{h} - Df(x)\tilde{h} - D^2 f(x)(h + k, \tilde{h}) \big] \\
&\quad - \big[Df(x + h)\tilde{h} - Df(x)\tilde{h} - D^2 f(x)(h, \tilde{h}) \big].
\end{aligned}
$$

For $\|h\|_X$, $\|k\|_X$ small, it follows from the definition of the Fréchet derivative of Df that

$$\|D_1 g(h, k)\|_{B(X,Y)} = o(\|h\|_X + \|k\|_X).$$

In consequence, we have

$$\|D^2 f(x)(h, k) - D^2 f(x)(k, h)\|_Y = o(\|k\|_X + \|h\|_X) \, (\|h\|_X + \|k\|_X).$$

It follows that the norm of the difference of the two bilinear forms is zero, whence

$$D^2 f(x)(h, k) = D^2 f(x)(k, h)$$

for all $h, k \in X$ (simply replace (h, k) by $(\epsilon h, \epsilon k)$ and take $\epsilon \to 0$). $\quad\square$

Corollary 9.23. *Let f, X, Y and U be as in Lemma 9.22, but suppose f has $n \geq 2$ derivatives in U. Then $D^n f(x)$ is symmetric under permutation of its arguments. That is, if π is an $n \times n$ symmetric permutation matrix, then*

$$D^n f(x)(h_1, \ldots, h_n) = D^n f(x)(\pi(h_1, \ldots, h_n)).$$

Proof. The result follows by induction using the fact that $D^n f(x) = D^2(D^{n-2}f)(x)$ and Proposition 9.21. $\quad\square$

Theorem 9.24 (Taylor's theorem). *Let X, Y be Banach spaces, $U \subset X$ open and suppose $f : U \to Y$ has $n \geq 1$ derivatives throughout U. Then for $x \in U$ and h small enough that $x + h \in U$,*

$$f(x + h) = f(x) + Df(x)h + \frac{1}{2!}D^2 f(x)(h, h)$$
$$+ \cdots + \frac{1}{n!}D^n f(x)(h, \ldots, h) + R_n(x, h) \tag{9.22}$$

and

$$\frac{\|R_n(x, h)\|_Y}{\|h\|_X^n} \longrightarrow 0$$

as $h \to 0$, $h \neq 0$, in X. Another way of saying this is, $\|R_n(x, h)\|_Y = o(\|h\|_X^n)$.

Proof. First note in general that if $F \in B^m(X, Y)$ is symmetric and g is defined by

$$g(h) = F(h, \ldots, h),$$

then

$$Dg(h)k = mF(h, \ldots, h, k).$$

This follows by straightforward calculation as follows. For $m = 1$, F is just a linear map and the result is already known. For $m = 2$, for example, just compute

$$g(h + k) - g(h) - 2F(h, k)$$
$$= F(h + k, h + k) - F(h, h) - 2F(h, k) = F(k, k),$$

and

$$\|F(k, k)\|_Y \leq C\|k\|_X^2,$$

showing g is differentiable and that $Dg(h) = 2F(h, \cdot)$. The results for $m > 2$ follow similarly.

For the theorem, the case $n = 1$ just reproduces the definition of f being differentiable at x. We initiate an induction on n, supposing the result valid for all functions f satisfying the hypotheses for $k < n$, where $n \geq 2$. Let f satisfy the hypotheses for $k = n$. Define R_n as in (9.22) and notice that differentiation with respect to h yields

$$Df(x + h) = Df(x) + D^2 f(x)(h, \cdot)$$
$$+ \cdots + \frac{1}{(n - 1)!}D^n f(x)(h, \ldots, h, \cdot) + D_2 R_n(x, h),$$

which is the $(n-1)$st Taylor expansion of Df. By induction we conclude that

$$\frac{\|D_2 R_n(x, h)\|_{B(X,Y)}}{\|h\|_X^{n-1}} \longrightarrow 0$$

as $h \to 0$ ($h \neq 0$). On the other hand, by the mean-value theorem (Theorem 9.6), if $\|h\|_X$ is sufficiently small, then

$$\frac{\|R_n(x,h)\|_Y}{\|h\|_X^n} = \frac{\|R_n(x,h) - R_n(x,0)\|_Y}{\|h\|_X^n}$$

$$\leq \sup_{0 \leq \alpha \leq 1} \frac{\|D_2 R_n(x,\alpha h)\|_{B(X,Y)}}{\|h\|_X^{n-1}} \longrightarrow 0$$

as $h \to 0$. ◻

9.5   EXTREMA

Attention is now turned from solving operator equations to finding points that maximize or minimize operators mapping into the real line.

Definition. Let X be a set and $f : X \to \mathbb{R}$. A point $x_0 \in X$ is a *minimum* if $f(x_0) \leq f(x)$ for all $x \in X$; it is a *maximum* if $f(x_0) \geq f(x)$ for all $x \in X$. An *extremum* is a point which is a maximum or a minimum. If X has a topology, we say x_0 is a *relative* (or *local*) minimum if there is an open set $U \subset X$ with $x_0 \in U$ such that

$$f(x_0) \leq f(x) \quad \text{for all } x \in U.$$

Similarly, if

$$f(x_0) \geq f(x) \quad \text{for all } x \in U,$$

then x_0 is a *relative maximum*. The point x_0 is called a relative extremum when it is either a relative minimum or maximum. If equality is disallowed above when $x \neq x_0$, the (relative) minimum or maximum is said to be *strict*.

Theorem 9.25. *Let X be an NLS, U be an open set in X, and $f : U \to \mathbb{R}$ be differentiable. If $x_0 \in U$ is a relative extremum, then $Df(x_0) = 0$.*

Proof. We argue by contradiction, so suppose that $Df(x_0)$ is not the zero map. Then there is some $h \neq 0$ such that $Df(x_0)h \neq 0$. By possibly reversing the sign of h, we may assume that $Df(x_0)h > 0$. Let $t_0 > 0$ be small enough that $x_0 + th \in U$ for $|t| \leq t_0$ and consider for such t

$$\frac{1}{t}[f(x_0 + th) - f(x_0)] = \frac{1}{t}[Df(x_0)(th) + R_1(x_0,th)]$$

$$= Df(x_0)h + \frac{1}{t}R_1(x_0,th).$$

The quantity $R_1(x_0,th)/t \to 0$ as $t \to 0$. Hence for $t_1 \leq t_0$ small enough and $|t| \leq t_1$,

$$Df(x_0)h + \frac{1}{t}R_1(x_0,th) > 0.$$

It follows that

$$f(x_0 + th) = f(x_0) + t\left[Df(x_0)h + \frac{1}{t}R_1(x_0,th)\right]$$

can be made smaller or larger than $f(x_0)$, depending on the sign of t (for $0 < |t| < t_1$). This contradicts that x_0 is a relative extremum. □

Definition. A *critical point* of a mapping $f : U \to Y$, where U is open in X, is a point x_0 where $Df(x_0) = 0$. This is also referred to as a *stationary point* by some authors.

Corollary 9.26. *If $f : U \to \mathbb{R}$ is differentiable, then the relative extrema of f in U are critical points of f.*

Definition. Let X be a vector space over $\mathbb{R}$, $U \subset X$ a convex subset, and $f : U \to \mathbb{R}$. We say that f is *convex* if whenever $x_0, x_1 \in U$ and $\lambda \in (0, 1)$, then

$$f(\lambda x_1 + (1 - \lambda)x_0) \leq \lambda f(x_1) + (1 - \lambda)f(x_0).$$

We say that f is *concave* if the opposite inequality holds. We say that f is *strictly* convex or concave if equality is not allowed above when $\lambda \in (0, 1)$.

Proposition 9.27. *Linear functionals on X are both convex and concave (but not strictly so). If $a, b > 0$ and f, g are convex, then $af + bg$ is convex, and if at least one of f or g is strictly convex, then so is $af + bg$. Furthermore, f is (strictly) convex if and only if $-f$ is (strictly) concave.*

The proof is left as an easy exercise using only the definitions.

Proposition 9.28. *Let X be an NLS, U an open convex subset of X, and $f : U \to \mathbb{R}$ convex and differentiable. Then, for $x, y \in U$,*

$$f(y) \geq f(x) + Df(x)(y - x),$$

and, if $Df(x) = 0$, then x is a minimum of f in U. Moreover, if f is strictly convex, then for $x \neq y$,

$$f(y) > f(x) + Df(x)(y - x),$$

and $Df(x) = 0$ implies that f has a strict and therefore unique minimum in U.

Proof. By convexity, for $\lambda \in (0, 1]$,

$$\lambda f(y) + (1 - \lambda)f(x) \geq f(x + \lambda(y - x)),$$

whence

$$f(y) - f(x) \geq \frac{f(x + \lambda(y - x)) - f(x)}{\lambda}.$$

Take the limit as $\lambda \to 0$ on the right-hand side to obtain the desired result.

We leave the more delicate proof of the strictly convex case to the reader as an exercise at the end of the chapter. □

Examples.

(1) Quadratic functions are strictly convex. More precisely, if $a, b \in \mathbb{R}$, $a \neq b$, then for any $\lambda \in (0, 1)$,

$$
\begin{aligned}
\left(\lambda a + (1-\lambda)b\right)^2 &= \lambda^2 a^2 + (1-\lambda)^2 b^2 + 2\lambda(1-\lambda)\, ab \\
&= \lambda a^2 + (1-\lambda)b^2 - \lambda(1-\lambda)\,(a^2 + b^2 - 2ab) \\
&= \lambda a^2 + (1-\lambda)b^2 - \lambda(1-\lambda)\,(a-b)^2 \\
&< \lambda a^2 + (1-\lambda)b^2.
\end{aligned}
$$

(2) Let $\Omega \subset \mathbb{R}^d$, $f \in L^2(\Omega)$, and assume that the underlying field is real. Define $J : H_0^1(\Omega) \to \mathbb{R}$ by

$$
J(v) = \tfrac{1}{2}\|\nabla v\|_{L^2(\Omega)}^2 - (f, v)_{L^2(\Omega)}.
$$

We claim that $\|\nabla v\|_{L^2(\Omega)}^2$ is strictly convex. To verify this, fix $v, w \in H_0^1(\Omega)$ and $\lambda \in (0, 1)$, and consider $\left(\lambda \nabla v + (1-\lambda)\nabla w\right)^2$. An application of the result of the previous example and integration over Ω yields

$$
\|\nabla(\lambda v + (1-\lambda)w)\|_{L^2(\Omega)}^2 < \lambda\|\nabla v\|_{L^2(\Omega)}^2 + (1-\lambda)\|\nabla w\|_{L^2(\Omega)}^2,
$$

unless $v - w$ is identically constant on each connected component of Ω. In this case, since $v, w \in H_0^1(\Omega)$, $v = w = 0$ on $\partial\Omega$, and so $v = w = 0$ everywhere. That is, there is strict inequality whenever $v \neq w$, and so it is concluded that $\|\nabla v\|_{L^2(\Omega)}^2$ is strictly convex. Proposition 9.27 then implies that $J(v)$ is also strictly convex. A calculation shows that

$$
DJ(u, v) = (\nabla u, \nabla v)_{L^2(\Omega)} - (f, v)_{L^2(\Omega)}.
$$

It is concluded that $u \in H_0^1(\Omega)$ satisfies the boundary value problem

$$
(\nabla u, \nabla v)_{L^2(\Omega)} = (f, v)_{L^2(\Omega)}
$$

if and only if u minimizes the "energy functional" $J(v)$ over $H_0^1(\Omega)$:

$$
J(u) < J(v) \quad \text{for all } v \in H_0^1(\Omega), \ v \neq u.
$$

Strict convexity implies such a function u is unique.

Local convexity suffices to verify that a critical point is a relative extremum. More generally, we can examine the second derivative at the relevant point.

Theorem 9.29. *If X is an NLS and $f : X \to \mathbb{R}$ is twice differentiable at a relative minimum $x \in X$, then*

$$
D^2 f(x)(h, h) \geq 0 \quad \text{for all } h \in X.
$$

Proof. By Taylor's formula

$$f(x \pm \lambda h) = f(x) \pm Df(x)\lambda h + \tfrac{1}{2}\lambda^2 D^2 f(x)(h,h) + o(\lambda^2 \|h\|_X^2),$$

accordingly,

$$D^2 f(x)(h,h) = \lim_{\lambda \to 0} \left[\frac{f(x + \lambda h) + f(x - \lambda h) - 2f(x)}{\lambda^2} + \frac{o(\lambda^2 \|h\|_X^2)}{\lambda^2 \|h\|_X^2} \|h\|_X^2 \right]$$

$$= \lim_{\lambda \to 0} \frac{f(x + \lambda h) + f(x - \lambda h) - 2f(x)}{\lambda^2} \geq 0$$

when x is a local minimum. $\qquad\square$

In infinite dimensions, $Df(x) = 0$ and $D^2 f(x)(h,h) > 0$ for all $h \neq 0$ does *not* imply that x is a local minimum. For example consider the function $f : \ell^2 \to \mathbb{R}$ defined by

$$f(x) = \sum_{k=1}^{\infty} \left(\frac{1}{k} - x_k \right) x_k^2,$$

where $x = (x_k)_{k=1}^{\infty} \in \ell^2$. Note that f is well-defined on $\ell^2 \subset \ell^3$ (i.e., the sum converges). Direct calculation shows that

$$Df(x)(h) = \sum_{k=1}^{\infty} \left(\frac{2}{k} - 3x_k \right) x_k h_k,$$

$$D^2 f(x)(h,h) = \sum_{k=1}^{\infty} \left(\frac{2}{k} - 6x_k \right) h_k^2,$$

so $f(0) = 0$, $Df(0) = 0$, and $D^2 f(0)(h,h) > 0$ for all $h \neq 0$. However, let x^ℓ be the element of ℓ^2 such that x_k^ℓ is 0 if $k \neq \ell$ and $2/\ell$ if $k = \ell$. Compute that $f(x^\ell) < 0$, in spite of the fact that $x^\ell \to 0$ as $\ell \to \infty$. Thus 0 is not a local minimum of f.

Theorem 9.30 (second derivative test). *Let X be an NLS, and $f : X \to \mathbb{R}$ have two derivatives at a critical point $x \in X$. If there is some constant $c > 0$ such that*

$$D^2 f(x)(h,h) \geq c\|h\|_X^2 \quad \text{for all } h \in X,$$

then x is a strict local minimum point.

Proof. By Taylor's theorem, for any $\epsilon > 0$, there is $\delta > 0$ such that for $\|h\|_X \leq \delta$,

$$\left| f(x + h) - f(x) - \tfrac{1}{2}D^2 f(x)(h,h) \right| \leq \epsilon \|h\|_X^2,$$

since the Taylor remainder is $o(\|h\|_X^2)$. Thus,

$$f(x + h) - f(x) \geq \tfrac{1}{2}D^2 f(x)(h,h) - \epsilon \|h\|_X^2 \geq (\tfrac{1}{2}c - \epsilon)\|h\|_X^2,$$

and taking $\epsilon = c/4$, it follows that

$$f(x + h) \geq f(x) + \tfrac{1}{4} c \|h\|_X^2,$$

from which it is clear that f has a local minimum at x. □

This theorem is not as general as it appears. If we define the bilinear form

$$(h, k)_X = D^2 f(x)(h, k),$$

then with the assumption of the second derivative test, $(h, k)_X$ is an inner-product, which induces a norm equivalent to the original. Thus in fact X must be a pre-Hilbert space, and it makes no sense to attempt use of the theorem when X is known not to be pre-Hilbert.

9.6 EXERCISES

1. Let X, Y_1, Y_2, and Z be normed linear spaces and $P : Y_1 \times Y_2 \to Z$ be a continuous bilinear map (so P is a "product" between Y_1 and Y_2).

   (a) Show that for $y_i, \hat{y}_i \in Y_i$, $i = 1, 2$,

   $$DP(y_1, y_2)(\hat{y}_1, \hat{y}_2) = P(y_1, \hat{y}_2) + P(\hat{y}_1, y_2).$$

   (b) If $f : X \to Y_1 \times Y_2$ is differentiable, show that for $h \in X$,

   $$D(P \circ f)(x)\, h = P(Df_1(x)\, h, f_2(x)) + P(f_1(x), Df_2(x)\, h).$$

2. Let X be a real Hilbert space, $A_1, A_2 \in B(X, X)$, and define $f(x) = (x, A_1 x)_X A_2 x$. Show that $Df(x)$ exists for all $x \in X$ by finding an explicit expression for it.

3. Let $X = C([0, 1])$ be the space of bounded continuous functions on $[0, 1]$ and, for $u \in X$, define

   $$F(u)(x) = \int_0^1 K(x, y)\, f(u(y))\, dy,$$

   where $K : [0, 1] \times [0, 1] \to \mathbb{R}$ is continuous and f is a C^1-mapping of $\mathbb{R}$ into $\mathbb{R}$. Find the Fréchet derivative $DF(u)$ of F at $u \in X$. Is the map $u \mapsto DF(u)$ continuous?

4. Suppose X and Y are Banach spaces, and $f : X \to Y$ is differentiable with derivative $Df(x) \in B(X, Y)$ being a compact operator for any $x \in X$. Prove that f is also compact.

5. Set up and apply the contraction mapping theorem to show that the problem

   $$-u_{xx} + u - \epsilon u^2 = f(x), \quad x \in \mathbb{R},$$

   has a smooth bounded solution if $\epsilon > 0$ is small enough, where $f(x) \in \mathcal{S}(\mathbb{R})$.

6. Use the contraction mapping theorem to show that the Fredholm integral equation

$$f(x) = \varphi(x) + \lambda \int_a^b K(x, y) f(y) \, dy$$

has a unique solution $f \in C([a, b])$, provided that λ is sufficiently small, wherein $\varphi \in C([a, b])$ and $K \in C([a, b] \times [a, b])$.

7. Suppose that F is defined on a Banach space X, $x_0 = F(x_0)$ is a fixed point of F, $DF(x_0)$ exists, and 1 is *not* in the spectrum of $DF(x_0)$. Prove that x_0 is an isolated fixed point.

8. Consider the first-order differential equation

$$u'(t) + u(t) = \cos(u(t))$$

posed as an initial-value problem for $t > 0$ with initial condition $u(0) = u_0$.

(a) Use the contraction mapping theorem to show that there is exactly one solution u corresponding to any given $u_0 \in \mathbb{R}$.

(b) Prove that there is a number ξ such that $\lim_{t \to \infty} u(t) = \xi$ for any solution u, independent of the value of u_0.

9. Set up and apply the contraction mapping theorem to show that the boundary value problem

$$\begin{cases} -u_{xx} + u - \epsilon u^2 = f(x), & x \in (0, \infty), \\ u(0) = u(\infty) = 0, \end{cases}$$

has a smooth solution if $\epsilon > 0$ is small enough, where $f(x)$ is a smooth compactly supported function on $(0, \infty)$.

10. Consider the partial differential equation

$$\begin{cases} \dfrac{\partial u}{\partial t} - \dfrac{\partial^3 u}{\partial t \, \partial x^2} - \epsilon u^3 = f, & -\infty < x < \infty, \ t > 0, \\ u(x, 0) = g(x). \end{cases}$$

Use the Fourier transform and a contraction mapping argument to show that there exists a solution for small enough ϵ, at least up to some time $T < \infty$. In what spaces should f and g lie?

11. Let $\phi(x) \in C(\mathbb{R}) \cap L^\infty(\mathbb{R})$. Consider the nonlinear initial value problem

$$\begin{cases} \dfrac{\partial u}{\partial t} + \dfrac{\partial u}{\partial x} + 3u^2 \dfrac{\partial u}{\partial x} - \dfrac{\partial^3 u}{\partial x^2 \partial t} = 0, & x \in \mathbb{R}, \ t > 0, \\ u(x, 0) = \phi(x). \end{cases}$$

(a) Use the Fourier transform in x to show that the problem can be rewritten in the form

$$\begin{cases} \partial_t u = K * (u + u^3), & x \in \mathbb{R}, \ t > 0, \\ u(x,0) = \phi(x), \end{cases}$$

for some $K(x) \in L^1(\mathbb{R})$.

(b) Set up and apply the contraction mapping theorem to show that the initial value problem has a continuous and bounded solution $u = u(x,t)$, at least up to some time $T < \infty$.

12. Let X be a Banach space and $F : X \to X$ be a smooth map. Suppose that x_* is a simple root of F in the sense that $F(x_*) = 0$ and the derivative $DF(x_*)$ is invertible. Given any starting point x_0, for $k \geq 0$ consider the full Newton-Kantorovich method

$$x_{k+1} = G(x_k) \quad \text{where} \quad G(x) = x - DF(x)^{-1}F(x).$$

Complete the following outline to show that if x_0 is sufficiently close to x_*, then $x_k \to x_*$ as $k \to \infty$.

(a) Show that $G(x_*) = x_*$, $DG(x_*) = 0$, and that there is a closed ball B about x_* such that $\|DG(x)\| \leq \frac{1}{2}$ for all $x \in B$. [Hint: you do *not* need to compute $DG(x)$, only $DG(x_*)$.]

(b) Show that $G(x) \in B$ for all $x \in B$.

(c) Show that $G : B \to B$ is a contraction.

(d) Prove that $x_k \to x_*$ as $k \to \infty$ for any $x_0 \in B$.

(e) If $F \in C^3$, show that, for some $C < \infty$,

$$\|x_{k+1} - x_*\| \leq C\|x_k - x_*\|^2.$$

That is, if the Newton-Kantorovich iteration converges, it does so quadratically in the error. [Hint: note that $x_{k+1} - x_* = G(x_k) - G(x_*)$ and apply the mean value theorem twice.]

13. Surjective mapping theorem: let X and Y be Banach spaces, $U \subset X$ be open, $f : U \to Y$ be C^1, and $x_0 \in U$. If $Df(x_0)$ has a bounded right inverse, then $f(U)$ contains a neighborhood of $f(x_0)$.

(a) Prove this theorem from the inverse function theorem. [Hint: let R be the right inverse of $Df(x_0)$ and consider $g : V \to Y$ where $g(y) = f(x_0 + Ry)$ and $V = \{y \in Y : x_0 + Ry \in U\}$.]

(b) Prove that if $y \in Y$ is sufficiently close to $f(x_0)$, there is at least one solution to $f(x) = y$.

14. Suppose $f \in C^0([0,1])$ and that we want to solve

$$\frac{1}{1 + \epsilon u^2} u' = f(x), \quad x \in (0,1), \quad \text{and} \quad u(0) = 0.$$

We note that there is a unique solution $u_0(x)$ when $\epsilon = 0$.

(a) Use the implicit function theorem to show that there is a continuously differentiable solution for ϵ small enough.

(b) Use the Banach contraction mapping theorem to show that there is a unique continuous solution in a closed ball about u_0 in an appropriate Banach space for ϵ sufficiently small.

15. Let X and Y be Banach spaces.

(a) Let F and G mapping X to Y be C^1 on X, and let $H(x,\epsilon) = F(x) + \epsilon G(x)$ for $\epsilon \in \mathbb{R}$. If $H(x_0, 0) = 0$ and $DF(x_0)$ is invertible, show that there exists $x \in X$ such that $H(x,\epsilon) = 0$ for ϵ sufficiently close to 0.

(b) Let $f \in C^0([0,1])$ and $\epsilon \in \mathbb{R}$. Use the previous result to show that, for sufficiently small $|\epsilon|$, there is a solution $w \in C^2([0,1])$ to

$$w'' = f + w + \epsilon w^2, \quad w(0) = w(1) = 0.$$

16. Prove that for sufficiently small $\epsilon > 0$, there is at least one solution to the functional equation

$$f(x) + \sin x \int_{-\infty}^{\infty} f(x-y) f(y) \, dy = \epsilon e^{-|x|^2}, \quad x \in \mathbb{R},$$

such that $f \in L^1(\mathbb{R})$.

17. Let X and Y be Banach spaces, and let $U \subset X$ be open and convex. Let $F : U \to Y$ be an n-times Fréchet differentiable operator. Let $x \in U$ and $h \in X$. Prove that in Taylor's theorem, the remainder is actually bounded as

$$\|R_{n-1}(x,h)\|$$
$$= \left\| F(x+h) - F(x) - DF(x)h + \cdots + \frac{1}{(n-1)!} D^{n-1} F(x)(h, \ldots, h) \right\|$$
$$\leq \sup_{0 \leq \alpha \leq 1} \|D^n F(x + \alpha h)\| \, \|h\|^n.$$

18. Prove that if X is an NLS, U an open convex subset of X, and $f : U \to \mathbb{R}$ is strictly convex and differentiable, then, for $x, y \in U$, $x \neq y$,

$$f(y) > f(x) + Df(x)(y - x),$$

and $Df(x) = 0$ implies that f has a strict and therefore unique minimum.

19. Let $\Omega \subset \mathbb{R}^d$ have a smooth boundary, and let $g(x)$ be real with $g \in H^1(\Omega)$. Consider the BVP

$$\begin{cases} -\Delta u + u = 0 & \text{in } \Omega, \\ u = g & \text{on } \partial\Omega. \end{cases}$$

(a) Write this as a variational problem.

(b) Define an appropriate energy functional $J(v)$ and find $DJ(v)$.

(c) Relate the BVP to a constrained minimization of $J(v)$.

20. Let $\Omega \subset \mathbb{R}^n$ have a smooth boundary, $A(x)$ be an $n \times n$ real matrix with components in $L^\infty(\Omega)$, and let $c(x)$ and $f(x)$ be real with $c \in L^\infty(\Omega)$ and $f \in L^2(\Omega)$. Consider the BVP

$$\begin{cases} -\nabla \cdot A\nabla u + cu = f & \text{in } \Omega, \\ u = 0 & \text{on } \partial\Omega. \end{cases}$$

(a) Write this as a variational problem.

(b) Assume that A is symmetric and uniformly positive definite and c is uniformly positive. Define the energy functional $J : H_0^1 \to \mathbb{R}$ by

$$J(v) = \frac{1}{2} \int_\Omega \{|A^{1/2}\nabla v|^2 + c|v|^2 - 2fv\}dx.$$

Find $DJ(v)$.

(c) Prove that for $u \in H_0^1$, the following are equivalent: (i) u is the solution of the BVP; (ii) $DJ(u) = 0$; (iii) u minimizes $J(v)$.

21. Let X and Y be Banach spaces, $U \subset X$ an open set, and $f : U \to Y$ Fréchet differentiable. Suppose that f is *compact*, in the sense that for any $x \in U$, if $\overline{B_r(x)} \subset U$, then $f(B_r(x))$ is precompact in Y. If $x_0 \in U$, prove that $Df(x_0)$ is a compact linear operator.

22. Suppose we wish to find the surface $u(x, y)$ above the square $Q = [-1, 1]^2$, with $u = 0$ on ∂Q, that encloses the greatest volume, subject to the constraint that the surface area is fixed at $s > 4$.

(a) Formulate the problem, and reformulate it incorporating the constraint as a Lagrange multiplier (see Chapter 10). [Hint: the surface area is $\iint_Q \sqrt{1 + |\nabla u|^2}\, dx\, dy$.]

(b) Using the definition of the Fréchet derivative, find the conditions for a critical point.

(c) Find a partial differential equation that u must satisfy to be a critical point of this problem. [Remark: a solution of this differential equation, that also satisfies the area constraint, gives the solution to our problem.]

CHAPTER 10

The Calculus of Variations

A common problem in science and engineering applications is to find extrema of a real-valued functional that involves an integral of a function. Such problems lie in the realm of the calculus of variations, a term coined by Leonhard Euler in the mid-18th century. The subject got its start with a problem of least resistance introduced by Newton in the later 17th century followed by the famous brachistochrone problem raised by Johann Bernoulli at the end of the century (see later in this chapter). Joesph-Louis Lagrange was influenced by the work of Euler and at age 19, put forward an analytic approach to these sorts of problems. His ideas were immediately championed by Euler and the subject was off and running. Many of the well known 19th century mathematicians contributed to its development and this has carried over to the 20th and 21st century where the subject is still alive and well.

Our consideration of these types of issues will start more or less where the subject itself started, with the following problem. Let $a < b$,

$$f : [a, b] \times \mathbb{R}^d \times \mathbb{R}^d \to \mathbb{R},$$

and $F : C^1([a, b]; \mathbb{R}^d) \to \mathbb{R}$ be defined by

$$F(y) = \int_a^b f(x, y(x), y'(x)) \, dx.$$

With α and β given in $\mathbb{R}^d$, let

$$C^1_{\alpha,\beta}([a, b]; \mathbb{R}^d) = \{v : [a, b] \to \mathbb{R}^d \mid v \text{ has a continuous}$$

$$\text{first derivative and } v(a) = \alpha, \ v(b) = \beta\}.$$

Only the one-sided derivatives are considered at the endpoints of the interval $[a, b]$. The goal is to find $y \in C^1_{\alpha,\beta}([a, b]; \mathbb{R}^d)$ such that

$$F(y) = \min_{v \in C^1_{\alpha,\beta}([a,b];\mathbb{R}^d)} F(v).$$

DOI: 10.1201/9781003492139-10

This is also written for the set of all such points y as

$$\{y\} = \underset{v \in C^1_{\alpha,\beta}([a,b];\mathbb{R}^d)}{\arg\min} F(v).$$

Example. Find $y(x) \in C^1([a,b];\mathbb{R})$ such that $y(a) = \alpha > 0$ and $y(b) = \beta > 0$ and the surface of revolution of the graph of y about the x-axis has minimal area. Recall that a differential of arc length is given by

$$ds = \sqrt{1 + (y'(x))^2}\, dx,$$

so our area as a function of the curve y is

$$A(y) = \int_a^b 2\pi y(x)\sqrt{1 + (y'(x))^2}\, dx, \tag{10.1}$$

since clearly a solution $y(x)$ will be non-negative for all $x \in [a,b]$.

10.1 THE EULER-LAGRANGE EQUATIONS

If α and β are both zero, $C^1_{0,0}([a,b];\mathbb{R}^d) = C^1_0([a,b];\mathbb{R}^d)$ is a Banach space with its usual norm $\|y\| = \|y\|_\infty + \|y'\|_\infty$ and the minimum is found at a critical point. However, in general $C^1_{\alpha,\beta}([a,b];\mathbb{R}^d)$ is not a linear vector space. Rather it is an affine space, a translate of a vector space. More precisely, let

$$\ell(x) = \frac{1}{b-a}[\alpha(b-x) + \beta(x-a)]$$

be the linear function connecting (a,α) to (b,β). Then

$$C^1_{\alpha,\beta}([a,b];\mathbb{R}^d) = C^1_0([a,b];\mathbb{R}^d) + \ell.$$

To solve this problem, then, one needs to consider any fixed element of $C^1_{\alpha,\beta}$, such as $\ell(x)$, and all possible "admissible variations" of it that lie in C^1_0; that is, we minimize if possible $F(v)$ by searching among all possible "competing functions" $v = \ell + h \in C^1_{\alpha,\beta}$, where $h \in C^1_0$ (the *admissible variations*). On C^1_0, we can compute the derivative of $F(\ell+h)$ as a function of h, and thereby restrict our search to the critical points. Such a point $y = \ell + h$ is called a *critical point* for F defined on $C^1_{\alpha,\beta}$. Here is a general result about the derivative of F of the form considered in this section.

Theorem 10.1. *If $f \in C^1([a,b] \times \mathbb{R}^d \times \mathbb{R}^d)$ and*

$$F(y) = \int_a^b f(x, y(x), y'(x))\, dx,$$

then $F : C^1([a,b]) \to \mathbb{R}$ is continuously differentiable and

$$DF(y)(h) = \int_a^b [D_2 f(x, y(x), y'(x))\, h(x) + D_3 f(x, y(x), y'(x))\, h'(x)]\, dx$$

for all $h \in C^1([a,b])$.

Proof. Let A be defined by

$$Ah = \int_a^b [D_2f(x, y(x), y'(x))\, h(x) + D_3f(x, y(x), y'(x))\, h'(x)]\, dx,$$

which is clearly a bounded linear functional on C^1, since the norm of any $v \in C^1$ can be taken to be

$$\|v\| = \max(\|v\|_{L^\infty}, \|v'\|_{L^\infty}).$$

Calculate as follows:

$$F(y + h) - F(y) = \int_a^b \int_0^1 \frac{d}{dt} f(x, y + th, y' + th')\, dt\, dx$$

$$= \int_a^b \int_0^1 [D_2f(x, y + th, y' + th')\, h + D_3f(x, y + th, y' + th')\, h']\, dt\, dx,$$

so

$$|F(y + h) - F(y) - Ah|$$

$$\leq \int_a^b \int_0^1 \big|[D_2f(x, y + th, y' + th') - D_2f(x, y, y')]\, h\big|\, dt\, dx$$

$$+ \int_a^b \int_0^1 \big|[D_3f(x, y + th, y' + th') - D_3f(x, y, y')]\, h'\big|\, dt\, dx.$$

Since D_2f and D_3f are uniformly continuous on compact sets, the right-hand side is $o(\|h\|)$, and thus $DF(y) = A$.

It remains to show that $DF(y)$ is continuous. This follows from uniform continuity of D_2f and D_3f, and from the computation

$$|DF(y + h)k - DF(y)k|$$

$$\leq \int_a^b \big|[D_2f(x, y + h, y' + h') - D_2f(x, y, y')]\, k\big|\, dx$$

$$+ \int_a^b \big|[D_3f(x, y + h, y' + h') + D_3f(x, y, y')]\, k'\big|\, dx,$$

which tends to 0 as $\|h\| \to 0$ for any $k \in C^1([a, b])$ with $\|k\| \leq 1$. □

Theorem 10.2. *Suppose $f \in C^1([a, b] \times \mathbb{R}^d \times \mathbb{R}^d)$, $y \in C^1_{\alpha,\beta}([a, b])$, and*

$$F(y) = \int_a^b f(x, y(x), y'(x))\, dx.$$

Then y is a critical point for F if and only if the curve $x \mapsto D_3f(x, y(x), y'(x))$ is $C^1([a, b])$ and y satisfies the Euler-Lagrange equations

$$D_2f(x, y, y') - \frac{d}{dx} D_3f(x, y, y') = 0.$$

In component form, the Euler-Lagrange Equations are

$$\frac{\partial f}{\partial y_k} = \frac{d}{dx}\frac{\partial f}{\partial y'_k}, \quad k = 1, \ldots, d,$$

or

$$f_{y_k} = \frac{d}{dx} f_{y'_k}, \quad k = 1, \ldots, d.$$

The converse implication of the theorem is easily shown from the previous result after integrating by parts, since $h \in C_0^1$. The direct implication follows from the previous result and the following lemma, which can be proved by classical methods, but is also trivial to prove from the Lebesgue lemma (Lemma 5.7). The details are left to the reader as an exercise.

Lemma 10.3 (Dubois-Reymond lemma). *Let φ and ψ lie in $C^0([a,b]; \mathbb{R}^d)$. Then*

(a) $\displaystyle\int_a^b \varphi(x) \cdot h'(x)\, dx = 0$ *for all $h \in C_0^1([a,b]; \mathbb{R}^d)$ if and only if φ is*

identically constant;

(b) $\displaystyle\int_a^b [\varphi(x) \cdot h(x) + \psi(x) \cdot h'(x)]\, dx = 0$ *for all $h \in C_0^1([a,b]; \mathbb{R}^d)$ if and only*

if $\psi \in C^1$ and $\psi' = \varphi$.

Proof. Both converse implications are trivial after integrating by parts. For the direct implication of (a), let

$$\bar{\varphi} = \frac{1}{b-a} \int_a^b \varphi(x)\, dx,$$

and note that

$$0 = \int_a^b \varphi(x) \cdot h'(x)\, dx = \int_a^b (\varphi(x) - \bar{\varphi}) \cdot h'(x)\, dx.$$

Take

$$h(x) = \int_a^x (\varphi(s) - \bar{\varphi})\, ds \in C_0^1([a,b]; \mathbb{R}^d),$$

so that $h' = \varphi - \bar{\varphi}$. Combining these two observations shows that

$$\|\varphi - \bar{\varphi}\|_{L^2} = 0,$$

so that $\varphi = \bar{\varphi}$ (almost everywhere, but both functions are continuous, so everywhere).

For the direct implication of (b), let

$$\Phi = \int_a^x \varphi(s)\, ds,$$

so that $\Phi' = \varphi$. Then the hypothesis of (b) and integration by parts show that

$$0 = -\int_a^b (\varphi\, h + \psi\, h') = \int_a^b [\Phi - \psi] \cdot h' \, dx,$$

since h vanishes at a and b. Part (a) then implies that $\Phi - \psi$ is constant. Since Φ is C^1, so is ψ, and $\psi' = \Phi' = \varphi$. □

Definition. Solutions of the Euler-Lagrange equations are called *extremals*.

Example. We illustrate the theory by finding the shortest path between two points. Suppose $y(x)$ is a path in $C^1_{\alpha,\beta}([a,b])$, which connects (a,α) to (b,β). The goal is to minimize the length functional

$$L(y) = \int_a^b \sqrt{1 + (y'(x))^2} \, dx$$

over all such y. The integrand is

$$f(x, y, y') = \sqrt{1 + (y'(x))^2},$$

so the Euler-Lagrange equations become simply

$$(D_3 f)' = 0.$$

This yields that

$$\frac{y'(x)}{\sqrt{1 + (y'(x))^2}} = c.$$

for some constant c. Thus,

$$y'(x) = \pm\sqrt{\frac{c^2}{1 - c^2}},$$

if $c^2 \neq 1$, and there is no solution otherwise. In any case, $y'(x)$ is constant, so the only critical paths are lines, and there is a unique such line in $C^1_{\alpha,\beta}([a,b])$. Since $L(y)$ is convex, this path is necessarily a minimum, and the well-known maxim emerges: "the shortest distance between two points is a straight line" (to which the mathematician adds "in Euclidean space").

Example. Many problems have no solutions. For example, consider the problem of minimizing the length of the curve $y \in C^1([0,1])$ such that $y(0) = y(1) = 0$ and $y'(0) = 1$. The previous example shows that extremals would have to be lines. But there is no line satisfying the three boundary conditions, so there are no extremals. Clearly the minimum approaches 1, but is never attained by a C^1-function.

It is generally not easy to solve the Euler-Lagrange equations. They constitute a nonlinear second order ordinary differential equation for $y(x)$. In more detail, suppose that f and y have continuous second order derivatives and use the chain rule to compute

$$D_2 f = (D_3 f)' = D_1 D_3 f + D_2 D_3 f\, y' + D_3^2 f\, y''.$$

Observe that $D_3^2 f(x, y, y') \in B^2(\mathbb{R}^d, \mathbb{R})$, which is isomorphic to $B(\mathbb{R}^d, \mathbb{R}^d)$, so if $D_3^2 f(x, y, y')$ is invertible,

$$y'' = (D_3^2 f)^{-1}(D_2 f - D_1 D_3 f - D_2 D_3 f\, y').$$

Definition. If y is an extremal and $D_3^2 f(x, y, y')$ is invertible for all $x \in [a, b]$, then we call y a *regular* extremal.

Proposition 10.4. *If $f \in C^2([a, b] \times \mathbb{R}^d \times \mathbb{R}^d)$ and $y \in C^1([a, b])$ is a regular extremal, then $y \in C^2([a, b])$.*

The proof should be clear. In the case of a regular extremal, the problem can be reduced to first order when f does not depend explicitly on x.

Theorem 10.5. *If $f \in C^2([a, b] \times \mathbb{R}^d \times \mathbb{R}^d)$, $f(x, y, z) = f(y, z)$ only, and $y \in C^1([a, b])$ is a regular extremal, then $D_3 f\, y' - f$ is constant.*

Proof. Simply compute

$$\begin{aligned}(D_3 f\, y' - f)' &= D_3 f\, y'' + (D_3 f)' y' - f' \\ &= D_3 f\, y'' + D_2 f\, y' - (D_2 f\, y' + D_3 f\, y'') = 0,\end{aligned}$$

using the Euler-Lagrange equation for the extremal and partial differentiation of f. □

Example. Reconsider the problem of finding $y(x) \in C^1([a, b])$ such that $y(a) = \alpha > 0$ and $y(b) = \beta > 0$ and the surface of revolution of the graph of y about the x-axis has minimal area. The area as a function of the curve is given in (10.1), so

$$f(y, y') = 2\pi y(x)\sqrt{1 + (y'(x))^2}.$$

Remark that

$$D_3 f(y, y') = \frac{2\pi y(x)\, y'(x)}{\sqrt{1 + (y'(x))^2}} \quad \text{and} \quad D_3^2 f(y, y') = \frac{2\pi y(x)}{(1 + (y'(x))^2)^{3/2}} \neq 0,$$

unless $y(x) = 0$. Clearly $y(x) > 0$, so our extremals are regular, and we can use the theorem to find them. For some constant C,

$$\frac{2\pi y\, (y')^2}{(1 + (y')^2)^{1/2}} - 2\pi y(1 + (y')^2)^{1/2} = 2\pi C,$$

which implies that

$$y' = \frac{1}{C}\sqrt{y^2 - C^2}.$$

Applying separation of variables leads to the relation

$$\frac{dy}{\sqrt{y^2 - C^2}} = \frac{dx}{C},$$

which, for some constant λ, gives the solution

$$y(x) = C\cosh(x/C + \lambda),$$

which is called a *catenary*. Suppose that $a = 0$, so that $C = \alpha/\cosh\lambda$ and

$$y(b) = \beta = \frac{\alpha}{\cosh\lambda}\cosh\left(\frac{\cosh\lambda}{\alpha}b + \lambda\right).$$

That is, we determine C once we have λ, which must solve the above equation. There may or may not be solutions λ (i.e., there may not be regular extremals). It is a fact, which we will not prove (see [22, pp. 62ff.]), that the minimal area is given either by a regular extremal or the *Goldschmidt solution*, which is the piecewise graph that uses straight lines to connect the points $(0, \alpha)$ to $(0, 0)$, $(0, 0)$ to $(b, 0)$, and finally $(b, 0)$ to (b, β). This is not a C^1 curve, so it is technically inadmissible, but it has area $A_G = \pi(\alpha^2 + \beta^2)$. If there are no extremals, then, given $\epsilon > 0$, we have C^1 curves approximating the Goldschmidt solution such that the area is greater than but within ϵ of A_G.

Example (The brachistochrone problem with a free end). Sometimes one does not impose a condition at one end. An example is the brachistochrone problem. Consider a particle moving under the influence of gravity in the xy-plane, where y points upwards. We assume that the particle starts from rest at the position $(0, 0)$ and slides frictionlessly along a curve $y(x)$, moving in the x-direction a distance $b > 0$ and falling an unspecified distance (see Figure 10.1). We wish to minimize the total *travel time*. Let the final position be (b, β), where $\beta < 0$ is unspecified. We assume that the curve $y \in C^1_{0,*}([0, b])$, where

$$C^1_{0,*}([a, b]) = \{v \in C^1([a, b]) : v(a) = 0\}.$$

The steeper the curve, the faster it will move; however, it must convert some of this speed into motion in the x-direction to travel distance b. To derive the travel time functional $T(y)$, note first that Newton's law implies that for a mass m traveling on the arc s with angle θ from the downward direction,

$$m\frac{d^2s}{dt^2} = -mg\cos\theta = -mg\frac{dy}{ds},$$

where g is the gravitational constant. The mass cancels and

$$\frac{1}{2}\frac{d}{dt}\left(\frac{ds}{dt}\right)^2 = \frac{d^2s}{dt^2}\frac{ds}{dt} = -g\frac{dy}{dt},$$

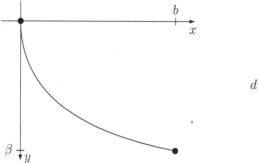 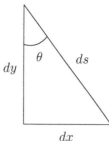

Figure 10.1 The brachistochrone problem (left) and the arc length differential ds for an arc at angle θ (right).

so there is some constant C for which

$$\left(\frac{ds}{dt}\right)^2 = -2gy + C.$$

But at $t = 0$, both the speed and $y(0)$ are zero, so $C = 0$, and

$$\frac{ds}{dt} = \sqrt{-2gy}.$$

Now, the travel time is given by

$$T(y) = \int dt = \int \frac{ds}{\sqrt{-2gy}} = \int_0^b \sqrt{\frac{1 + (y'(x))^2}{-2gy(x)}}\, dx.$$

We interrupt the discussion to establish a general result to deal with the free end.

Theorem 10.6. *If $y \in C^2([a, b])$ minimizes*

$$F(y) = \int_a^b f(x, y(x), y'(x))\, dx$$

subject only to the single constraint that $y(a) = \alpha \in \mathbb{R}$, then y must satisfy the Euler Lagrange equations and $D_3 f(b, y(b), y'(b)) = 0$.

Proof. For $y \in C^1_{0,*}([a, b]) + \alpha$ and $h \in C^1_{0,*}([a, b])$, just compute

$$DF(y)\,h = \int_a^b (D_2 f\, h + D_3 f\, h')\, dx$$

$$= \int_a^b (D_2 f\, h - (D_3 f)'\, h)\, dx + D_3 f(b, y(b), y'(b))\, h(b).$$

If $h \in C^1_{0,0}([a, b])$, we derive the Euler-Lagrange equations, and otherwise we obtain the second condition at $x = b$. $\qquad\square$

Example (The brachistochrone problem with a free end, continued). Since we are looking for a minimum, the factor $\sqrt{2g}$ can be dropped. So attention is given to

$$f(y, y') = \sqrt{\frac{1 + (y'(x))^2}{-y(x)}}.$$

This is independent of x, so we solve

$$y'D_3f - f = C_1 = \frac{1}{\sqrt{-y}} \left(\frac{(y')^2}{\sqrt{1 + (y')^2}} - \sqrt{1 + (y')^2} \right),$$

or

$$\int \sqrt{\frac{-y}{C_1^{-2} + y}} \, dy = x - C_2.$$

This in turn can be solved using a trigonometric substitution, *viz.*

$$y = -C_1^{-2} \sin^2(\phi/2) = -(1 - \cos\phi)/2C_1^2,$$

where $0 \le \phi \le \pi$, and then

$$x = -(\phi - \sin\phi)/2C_1^2 + C_2.$$

Applying the initial condition ($\phi = 0$), the curve is determined to be

$$(x, y) = C(\phi - \sin\phi, 1 - \cos\phi)$$

for some constant C. This is a *cycloid*. Now C is determined by the auxiliary condition

$$0 = D_3f(y(b), y'(b)) = \frac{1}{\sqrt{-y(b)}} \frac{y'(b)}{\sqrt{1 + (y'(b))^2}},$$

which requires

$$0 = y'(b) = \frac{dy}{d\phi} \left(\frac{dx}{d\phi} \right)^{-1} = \frac{\sin\phi(b)}{1 - \cos\phi(b)}.$$

Thus $\phi(b) = \pi$ (since $\phi \in [0, \pi]$), so $C = b/\pi$ and we have found the only extremal. Proving that this extremal is actually a minimum is a bit involved, and we omit the details.

10.2 CONSTRAINED EXTREMA AND LAGRANGE MULTIPLIERS

When discussing the Euler-Lagrange equations, we came upon the problem of finding relative extrema of a nonlinear functional in $C^1_{\alpha,\beta}$, which is an affine translate of a Banach space. This can be phrased differently: search for extrema in the Banach space C^1 subject to the linear constraint that the function agrees with α and β at its endpoints. This segues into the more general problem of finding relative extrema of a nonlinear functional subject to a possibly nonlinear constraint.

Let X be a Banach space, $U \subset X$ open, and $f : U \to \mathbb{R}$. To describe our constraint, we assume that there are functions $g_i : X \to \mathbb{R}$ for $i = 1, \ldots, m$ that define the set $M \subset U$ by

$$M = \{x \in U : g_i(x) = 0 \text{ for all } i\}.$$

Our problem is to find the relative extrema of f restricted to M. Note that M is not necessarily open, so we must discuss what happens on ∂M. To rephrase our problem: find the relative extrema of $f(x)$ on U subject to the constraints

$$g_1(x) = \cdots = g_m(x) = 0. \tag{10.2}$$

It can be difficult to solve a constrained problem. Rather than finding the relative extrema of $f(x)$ on U subject to the constraints (10.2), one can instead solve an unconstrained problem, albeit in more dimensions. Define $H : X \times \mathbb{R}^m \to \mathbb{R}$ by

$$H(x, \lambda) = f(x) + \lambda_1 g_1(x) + \cdots + \lambda_m g_m(x). \tag{10.3}$$

The critical points of H are given by solving for a root of the system of equations defined by the partial derivatives

$$D_1 H(x, \lambda) = Df(x) + \lambda_1 Dg_1(x) + \cdots + \lambda_m Dg_m(x),$$
$$D_2 H(x, \lambda) = g_1(x),$$
$$\vdots$$
$$D_{m+1} H(x, \lambda) = g_m(x).$$

Such a critical point satisfies the m constraints and an additional condition which is necessary for an extremum, as is shown next.

Theorem 10.7 (Lagrange multiplier theorem). *Let X be a Banach space, $U \subset X$ open, and $f, g_i : U \to \mathbb{R}$, $i = 1, \ldots, m$, be continuously differentiable. If $x \in M$ is a relative extremum for $f|_M$, where*

$$M = \{x \in U : g_i(x) = 0 \text{ for all } i\},$$

then there is a nonzero $\lambda = (\lambda_0, \ldots, \lambda_m) \in \mathbb{R}^{m+1}$ such that

$$\lambda_0 Df(x) + \lambda_1 Dg_1(x) + \cdots + \lambda_m Dg_m(x) = 0. \tag{10.4}$$

This says that, to find a local extremum in M, one need only consider points that satisfy (10.4). So, for example, one could search through the unconstrained space U for points x satisfying (10.4), and then check which ones of these lie in M. Two possibilities arise for $x \in U$. If $\{Dg_i(x)\}_{i=1}^m$ is linearly independent, the only nontrivial way to satisfy (10.4) is to take $\lambda_0 \neq 0$. Otherwise, $\{Dg_i(x)\}_{i=1}^m$ is linearly dependent, and (10.4) is satisfied for a nonzero λ with $\lambda_0 = 0$.

Our method of search is then clear.

(1) Find the critical points of H as defined above in (10.3). These points automatically satisfy both (10.4) (with $\lambda_0 = 1$) and $x \in M$, so they are potential relative extrema.

(2) Find points $x \in U$ where $\{Dg_i(x)\}_{i=1}^m$ is linearly dependent. Then (10.4) is satisfied (with $\lambda_0 = 0$), so we must further check to see if indeed $x \in M$, i.e., each $g_i(x) = 0$. If so, x is also a potential relative extremum.

(3) Determine if the potential relative extrema are indeed extrema or not.

(4) If extrema are desired in $\bar{M}$, then all points on ∂M must also be considered.

Often, the constraints are chosen so that $\{Dg_i(x)\}_{i=1}^m$ is always linearly independent, and the second step does not arise.

Proof of the Lagrange multiplier theorem. Suppose that x is a local minimum of $f|_M$; the case of a local maximum is similar. Then there is an open set $V \subset U$ such that $x \in V$ and

$$f(x) \leq f(y) \quad \text{for all } y \in M \cap V.$$

Define $F : V \to \mathbb{R}^{m+1}$ by

$$F(y) = (f(y), g_1(y), \ldots, g_m(y)).$$

Since x is a local minimum on M, for any $\epsilon > 0$,

$$(f(x) - \epsilon, 0, \ldots, 0) \neq F(y) \quad \text{for all } y \in V.$$

Thus, the function F does *not* map V onto an open neighborhood of $F(x) = (f(x), 0, \ldots, 0) \in \mathbb{R}^{m+1}$.

Suppose that $DF(x)$ maps X onto $\mathbb{R}^{m+1}$. Then construct a space $\tilde{X} = \text{span}\{v_1, \ldots, v_{m+1}\} \subset X$, where the v_i are chosen such that $DF(x)(v_i) = e_i$, the standard unit vector in the ith direction in $\mathbb{R}^{m+1}$. Let $\tilde{V} = \{v \in \tilde{X} : x + v \in V\}$, and define the function $h : \tilde{V} \to \mathbb{R}^{m+1}$ by $h(v) = F(x + v)$. Now $Dh(0) = DF(x)$ maps $\tilde{X}$ onto $\mathbb{R}^{m+1}$ and so is invertible. The inverse function theorem (Theorem 9.14) implies that h maps an open subset S of $\tilde{V}$ containing 0 onto an open subset of $\mathbb{R}^{m+1}$ containing $h(0) = F(x)$. But then $x + S \subset V$ is an open set that contradicts our previous conclusion regarding F.

Thus $DF(x)$ cannot map onto all of $\mathbb{R}^{m+1}$, and so it maps onto a proper subspace. There is then some nonzero vector $\lambda \in \mathbb{R}^{m+1}$ orthogonal to $DF(x)(X)$. Then, for any $y \in X$,

$$\lambda_0 Df(x)(y) + \lambda_1 Dg_1(x)(y) + \cdots + \lambda_m Dg_m(x)(y) = 0,$$

and one conclude that this linear combination of operators must vanish, i.e., (10.4) holds. ☐

Note that this theorem is especially useful when the function F and constraints G_i are given as integral operators, say of the form

$$F(y) = \int_a^b f(x, y, y') \, dx \quad \text{and} \quad G_i(y) = \int_a^b g_i(x, y, y') \, dx.$$

In that case,

$$H(y, \lambda) = \int_a^b h_\lambda(x, y, y') \, dx,$$

where

$$h_\lambda(x, y, y') = f(x, y, y') + \sum_{i=1}^m \lambda_i g_i(x, y, y'),$$

and the Euler-Lagrange equations can be used to find the extrema, *viz.*

$$D_y h_\lambda(x, y, y') = \frac{d}{dx} D_{y'} h_\lambda(x, y, y').$$

In the previous section we had boundary conditions. It is sometimes best to impose such point constraints directly, as in the following example.

Example (The isoperimetric problem). The isoperimetric problem can be stated as follows: among all *rectifiable* curves (i.e., curves with well-defined lengths) in $\mathbb{R}_+^2$ from $(-1, 0)$ to $(1, 0)$ with length ℓ, find the one enclosing the greatest area. We need to maximize the functional

$$A(u) = \int_{-1}^1 u(t) \, dt$$

subject to the constraint

$$L(u') = \int_{-1}^1 \sqrt{1 + (u'(t))^2} \, dt = \ell$$

over the set $u \in C_{0,0}^1([-1, 1])$ with $u \geq 0$. Let

$$H(u, \lambda) = A(u) + \lambda[L(u') - \ell] = \int_{-1}^1 h_\lambda(u, u') \, dt,$$

where

$$h_\lambda(u, u') = u + \lambda\left(\sqrt{1 + (u'(t))^2} - \ell/2\right).$$

To find a critical point of the system, we need to find both $D_u H$ and $D_\lambda H$. For the former, it is given by considering λ fixed and solving the Euler-Lagrange equations $D_2 h_\lambda = (D_3 h_\lambda)'$. That is,

$$1 = \lambda \frac{d}{dt}\left(\frac{u'}{\sqrt{1 + (u'(t))^2}}\right),$$

so for some constant C_1,

$$t = \lambda \frac{u'}{\sqrt{1 + (u'(t))^2}} + C_1.$$

Solving for u' yields

$$u'(t) = \frac{t - C_1}{\sqrt{\lambda^2 - (t - C_1)^2}}.$$

Another integration gives a constant C_2 and

$$u(t) = \sqrt{\lambda^2 - (t - C_1)^2} + C_2,$$

or, rearranging, there obtains

$$(u(t) - C_2)^2 + (t - C_1)^2 = \lambda^2,$$

which is the equation of a circular arc with center (C_1, C_2) and radius λ. The partial derivative $D_\lambda H$ simply recovers the constraint that the arc length is ℓ, and the requirement that $u \in C_{0,0}([-1, 1])$ says that it must go through the points $u(-1) = (-1, 0)$ and $u(1) = (1, 0)$. We leave it to the reader to complete the example by showing that these conditions uniquely determine $C_1 = 0$, C_2, and $\lambda = \sqrt{1 + C_2^2}$, where C_2 satisfies the transcendental equation

$$\sqrt{1 + C_2^2} = \frac{\ell}{2[\pi - \tan^{-1}(1/C_2)]}.$$

Additionally, the reader may justify that a maximum is obtained at this critical point. We also need to check the condition $DL(u') = 0$. Again the Euler-Lagrange equations allow us to find these points easily. The result, left to the reader, is that for some constant C of integration,

$$u' = \frac{C}{\sqrt{1 - C}},$$

which means that u is a straight line. The fixed ends imply that $u \equiv 0$, and so we do not satisfy the length constraint unless $\ell = 2$, a trivial case to analyze.

As a corollary, among curves of fixed lengths, the circle encloses the region of greatest area.

10.3   LOWER SEMICONTINUITY AND EXISTENCE OF MINIMA

Whether there exists a minimum of a functional is an important question. If a minimum exists, we can locate it by analyzing critical points. Perhaps the simplest criterion for the existence of a minimum is to consider convex functionals, as we have done previously. Next simplest is perhaps to note that a continuous function on a compact set attains its minimum.

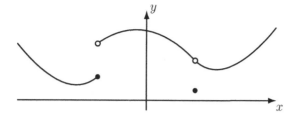

Figure 10.2   An example of a function $f : \mathbb{R} \to \mathbb{R}$ that is lower semicontinuous.

However, in an infinite-dimensional Banach space X, bounded sets are not compact; that is, compact sets are very small. This observation suggests that, at least when X is reflexive, we consider using the weak topology, since then the Banach-Alaoglu theorem (Theorem 2.55) implies that bounded sets are weakly compact. The problem now is that many interesting functionals are not weakly continuous, such as the norm itself. For the norm, it is easily seen that

$$\text{if } u_n \rightharpoonup u, \text{ then } \liminf_{n \to \infty} \|u_n\| \geq \|u\|,$$

with inequality possible. We are lead to consider a weaker notion of continuity.

Definition. Let X be a topological space. A function $f : X \to (-\infty, \infty]$ is said to be *lower semicontinuous* (l.s.c.) if whenever $\lim_{n \to \infty} x_n = x$, then

$$\liminf_{n \to \infty} f(x_n) \geq f(x).$$

Clearly a continuous function is lower semicontinuous. See Figure 10.2 for a more illuminating example.

Proposition 10.8. *Let X be a topological space and $f : X \to (-\infty, \infty]$. Then f is lower semicontinuous if and only if the sets*

$$A_\alpha = \{x \in X : f(x) \leq \alpha\} = f^{-1}((-\infty, \alpha])$$

are closed for all $\alpha \in \mathbb{R}$. Equivalently, $f^{-1}((\alpha, \infty])$ is open for all $\alpha \in \mathbb{R}$.

Proof. Suppose f is l.s.c. Let $x_n \in A_\alpha$ be such that $x_n \to x \in X$. Then

$$f(x) \leq \liminf_{n \to \infty} f(x_n) \leq \alpha,$$

so $x \in A_\alpha$ and A_α is closed.

Suppose now each A_α is closed. Then

$$A_\alpha^c = \{x \in X : f(x) > \alpha\}$$

is open. Let $x_n \to x \in X$, and suppose that $x \in A_\alpha^c$ for some α (i.e., $f(x) > \alpha$). Then there is some $N_\alpha > 0$ such that for all $n \geq N_\alpha$,

$x_n \in A_\alpha^c$, and so $\liminf_{n\to\infty} f(x_n) \geq \alpha$. In other words, whenever $f(x) > \alpha$, $\liminf_{n\to\infty} f(x_n) \geq \alpha$, so we conclude that

$$\liminf_{n\to\infty} f(x_n) \geq \sup\{\alpha : f(x) > \alpha\} = f(x). \qquad \square$$

Theorem 10.9. *If M is compact and $f : M \to (-\infty, \infty]$ is lower semicontinuous, then f is bounded below and takes on its minumum value.*

Proof. Let

$$\beta = \inf_{x\in M} f(x) \in [-\infty, \infty].$$

If $\beta = -\infty$, choose a sequence $x_n \in M$ such that $f(x_n) \leq -n$ for all $n \geq 1$. Since M is compact, there is $x \in M$ such that, for some subsequence, $x_{n_i} \to x$ as $i \to \infty$. But

$$f(x) \leq \liminf_{i\to\infty} f(x_{n_i}) = -\infty,$$

contradicting that f maps into $(-\infty, \infty]$. Thus $\beta > -\infty$, and f is bounded below.

Now choose a sequence $x_n \in M$ such that $f(x_n) \leq \beta + 1/n$, and again extract a convergent subsequence $x_{n_i} \to x \in M$ as $i \to \infty$. Then, compute

$$\beta \leq f(x) \leq \liminf_{i\to\infty} f(x_{n_i}) \leq \liminf_{i\to\infty} \left(\beta + \frac{1}{n_i}\right) = \beta,$$

and conclude that $f(x) = \beta$ attains its minimum at x. $\qquad \square$

The previous results apply to general topological spaces. For reflexive Banach spaces, we have both the strong (or norm) and weak topologies.

Theorem 10.10. *Let M be a weakly closed subspace of a reflexive Banach space X. If $f : M \to (-\infty, \infty]$ is weakly lower semicontinuous and, for some α, $A_\alpha = \{x \in X : f(x) \leq \alpha\}$ is bounded and nonempty, then there is some $x_0 \in M$ such that*

$$f(x_0) = \min_{x\in M} f(x) > -\infty.$$

Proof. By the Banach-Alaoglu theorem, and specifically the generalized Heine-Borel theorem (Theorem 2.60, extended to nonseparable spaces if necessary), $\bar{A}_\alpha$ is weakly compact, so $f|_{\bar{A}_\alpha}$ attains its minimum at, say, $x_0 \in \bar{A}_\alpha$. But for $x \in M \setminus \bar{A}_\alpha$, $f(x) > \alpha \geq \min_{x\in\bar{A}_\alpha} f(x)$, and the theorem follows. $\qquad \square$

As is apparent from the ideas just introduced, it is important to determine when a function is weakly lower semicontinuous. The following requirement is left to the reader as an exercise at the end of the chapter. Its near converse follows.

Proposition 10.11. *If X is a Banach space and $f : X \to (-\infty, \infty]$ is weakly lower semicontinuous, then f is strongly lower semicontinuous.*

Theorem 10.12. *Suppose X is a Banach space and $f : X \to (-\infty, \infty]$. If $V = \{x \in X : f(x) < \infty\}$ is a subspace of X, and if f is both convex on V and strongly lower semicontinuous, then f is weakly lower semicontinuous.*

Proof. For $\alpha \in \mathbb{R}$, let $A_\alpha = \{x \in X : f(x) \le \alpha\}$ be as usual. Since f is strongly l.s.c., Proposition 10.8 implies that A_α is closed in the strong (i.e., norm) topology. But f being convex on V implies that A_α is also convex. A strongly closed convex set is weakly closed (see Corollary 2.63), so we conclude that f is weakly l.s.c. $\qquad\square$

Lemma 10.13. *Let $f : \mathbb{C} \to [0, \infty)$ be convex, Ω a domain in $\mathbb{R}^d$, and $1 \le p < \infty$. Then $F : L^p(\Omega) \to [0, \infty]$, defined by*

$$F(u) = \int_\Omega f(u(x))\, dx,$$

is strongly and weakly l.s.c.

Proof. Since F is convex, it is enough to prove the norm l.s.c. property. Let $u_n \to u$ in $L^p(\Omega)$ and choose a subsequence such that

$$\lim_{i \to \infty} F(u_{n_i}) = \liminf_{n \to \infty} F(u_n)$$

and $u_{n_i}(x) \to u(x)$ for almost every $x \in \Omega$. Then $f(u_{n_i}(x)) \to f(u(x))$ for a.e. x, since f being convex is also continuous. Fatou's lemma finally implies that

$$F(u) \le \liminf_{i \to \infty} F(u_{n_i}) = \liminf_{n \to \infty} F(u_n). \qquad\square$$

Corollary 10.14. *If Ω is a domain in $\mathbb{R}^d$ and $1 \le p, q < \infty$, then the $L^q(\Omega)$-norm is weakly l.s.c. on $L^p(\Omega)$.*

We close this section with two examples that illustrate the concepts. For the first example, let $f \in C_0^\infty(\mathbb{R}^d)$ and consider the nonlinear partial differential equation

$$-\Delta u + u|u| + u = f.$$

Let us show that there is a solution. Define the functional F as

$$F(u) = \int_{\mathbb{R}^d} \left(\tfrac{1}{2}|\nabla u|^2 + \tfrac{1}{3}|u|^3 + \tfrac{1}{2}|u|^2 - fu \right) dx,$$

which may be $+\infty$ for some u. Now if $v \in C_0^\infty(\mathbb{R}^d)$,

$$DF(u)(v) = \int_{\mathbb{R}^d} \left(\nabla u \cdot \nabla v + |u|uv + uv - fv \right) dx$$

$$= \int_{\mathbb{R}^d} \left(-\Delta u + u|u| + u - f \right) v\, dx$$

which vanishes for all v if and only if the differential equation is satisfied. Since F is clearly convex, there will be a solution to the differential equation if F takes on its minimum.

Calclate a lower bound for F, *viz.*

$$F(u) \geq \tfrac{1}{2}\|u\|_{H^1(\mathbb{R}^d)}^2 - \|f\|_{L^2(\mathbb{R}^d)}\|u\|_{L^2(\mathbb{R}^d)} \geq \tfrac{1}{4}\|u\|_{H^1(\mathbb{R}^d)}^2 - \|f\|_{L^2(\mathbb{R}^d)}^2,$$

so the set $\{u \in H^1(\mathbb{R}^d) : F(u) \leq 1\}$ is bounded by $4(1 + \|f\|_{L^2(\mathbb{R}^d)}^2)$, and nonempty (since it contains $u \equiv 0$). The proof will be complete if it can be shown that F is l.s.c.

The last term of F is weakly continuous, and the second and third terms are weakly l.s.c., since they are norm l.s.c. and the space is convex. For the first term, let $u_n \rightharpoonup u$ in L^2. Then

$$
\begin{aligned}
\|\nabla u\|_{L^2} &= \sup_{\psi \in (C_0^\infty)^d,\ \|\psi\|_{L^2}=1} |(\psi, \nabla u)_{L^2}| \\
&= \sup_{\psi \in (C_0^\infty)^d,\ \|\psi\|_{L^2}=1} |(\nabla \cdot \psi, u)_{L^2}| \\
&\leq \sup_{\psi \in (C_0^\infty)^d,\ \|\psi\|_{L^2}=1} \lim_{n\to\infty} |(\nabla \cdot \psi, u_n)_{L^2}| \\
&= \sup_{\psi \in (C_0^\infty)^d,\ \|\psi\|_{L^2}=1} \lim_{n\to\infty} |(\psi, \nabla u_n)_{L^2}| \\
&\leq \liminf_{n\to\infty} \|\nabla u_n\|_{L^2}
\end{aligned}
$$

by Cauchy-Bunyakovsky-Schwartz. Thus the first term is l.s.c. as well.

In the final example, we compute a geodesic. Let $M \subset \mathbb{R}^d$ be closed and let $\gamma : [0,1] \to M$ be a *rectifiable* curve (i.e., γ is continuous and γ', as a distribution, is in $L^1([0,1]; \mathbb{R}^d)$). The *length* of γ is

$$L(\gamma) = \int_0^1 |\gamma'(s)|\, ds.$$

Theorem 10.15. *Suppose $M \subset \mathbb{R}^d$ is closed. If $x, y \in M$ and there is at least one rectifiable curve $\gamma : [0,1] \to M$ with $\gamma(0) = x$ and $\gamma(1) = y$, then there exists a rectifiable curve $\tilde{\gamma} : [0,1] \to M$ such that $\tilde{\gamma}(0) = x$, $\tilde{\gamma}(1) = y$, and*

$$L(\tilde{\gamma}) = \inf \big\{ L(\gamma) \mid \gamma : [0,1] \to M \text{ is rectifiable and } \gamma(0) = x,\ \gamma(1) = y \big\}.$$

Such a minimizing curve is called a geodesic.

Note that a geodesic is the shortest path on some manifold M (i.e., surface in $\mathbb{R}^d$) between two points. One exists provided only that the two points can be joined within M. Note that a geodesic may not be unique (e.g., consider joining points $(-1,0)$ and $(1,0)$ within the unit circle).

Proof. It would be natural to use Theorem 10.10; however, L^1 is *not* reflexive. We need two key ideas to resolve this difficulty. We expect that γ' is constant along a geodesic, so define

$$E(\gamma) = \int_0^1 |\gamma'(s)|^2 \, ds$$

and let us try to minimize E in $L^2([0,1])$. This is the first key idea.

Define

$$Y = \left\{ f \in L^2([0,1]; \mathbb{R}^d) : \right.$$
$$\left. \gamma_f(s) \equiv x + \int_0^s f(t) \, dt \in M \text{ for all } s \in [0,1] \text{ and } \gamma_f(1) = y \right\}.$$

These are the derivatives of rectifiable curves going from x to y. The map $f \mapsto \int_0^s f(t) \, dt$ is a continuous linear functional, so Y is weakly closed in $L^2([0,1]; \mathbb{R}^d)$. Since $\gamma_f' = f$, define $\tilde{E} : Y \to [0, \infty)$ by

$$\tilde{E}(f) = E(\gamma_f) = \int_0^1 |f(s)|^2 \, ds.$$

Clearly $|\cdot|$ is convex, so $\tilde{E}$ is weakly l.s.c. by Lemma 10.13. Let

$$A_\alpha = \{ f \in Y : \tilde{E}(f) \le \alpha \},$$

so that by definition A_α is bounded for any α. If A_α is not empty for some α, then there is a minimizer f_0 of $\tilde{E}$, by Theorem 10.10.

Now we need the second key idea. Given any rectifiable γ, define its *geodesic reparametrization* γ^* by

$$T(s) = \frac{1}{L(\gamma)} \int_0^s |\gamma'(t)| \, dt \in [0,1] \quad \text{and} \quad \gamma^*(T(s)) = \gamma(s),$$

which is well-defined since T is nondecreasing and $T(s)$ is constant only where γ is also constant. But

$$\gamma'(s) = \left(\gamma^*(T(s)) \right)' = \gamma^{*'}(T(s)) \, T'(s) = \gamma^{*'}(T(s)) \frac{|\gamma'(s)|}{L(\gamma)},$$

so

$$|\gamma^{*'}(s)| = L(\gamma)$$

is constant. Moreover, $L(\gamma^*) = L(\gamma)$, and so

$$E(\gamma^*) = L(\gamma^*)^2.$$

Now at least one γ exists by hypothesis, so the reparametrized γ^* has $E(\gamma^*) < \infty$. Thus, for some α, A_α is nonempty, and we conclude that there is a minimizer f_0 of $\tilde{E}$.

Finally, for any rectifiable curve,

$$E(\gamma) \geq L(\gamma)^2 = L(\gamma^*)^2 = E(\gamma^*).$$

Thus a curve of minimal energy E must have $|\gamma'|$ constant. So, for any rectifiable $\gamma = \gamma_f$ (where $f = \gamma'$),

$$L(\gamma) = E(\gamma^*)^{1/2} = \tilde{E}(f)^{1/2} \geq \tilde{E}(f_0)^{1/2} = E(\gamma_{f_0})^{1/2} = L(\gamma_{f_0}),$$

and γ_{f_0} is our geodesic. $\qquad\qquad\qquad\qquad\qquad\qquad\qquad\qquad\Box$

10.4   EXERCISES

1. Prove Theroem 10.2.

2. Let $F(u) = \displaystyle\int_{-1}^{5} [(u'(x))^2 - 1]^2\, dx$.

   (a) Find all extremals in $C^1([-1,5])$ such that $u(-1) = 1$ and $u(5) = 5$.

   (b) Decide if any extremal from (a) is a minimum of F. Consider $u(x) = |x|$.

3. Consider the functional

$$F(y) = \int_0^1 [(y(x))^2 - y(x)\, y'(x)]\, dx,$$

   defined for $y \in C^1([0,1])$.

   (a) Find all extremals.

   (b) If we require $y(0) = 0$, show by example that there is no minimum.

   (c) If we require $y(0) = y(1) = 0$, show that the extremal is a minimum. [Hint: note that $y\,y' = (\frac{1}{2}y^2)'$.]

4. Find all extremals of

$$\int_0^{\pi/2} \left[(y'(t))^2 + (y(t))^2 + 2\,y(t) \right] dt$$

   under the condition $y(0) = y(\pi/2) = 0$.

5. Suppose that we wish to minimize

$$F(y) = \int_0^1 f(x, y(x), y'(x), y''(x))\, dx$$

over the set of $y(x) \in C^2([0,1])$ such that $y(0) = \alpha$, $y'(0) = \beta$, $y(1) = \gamma$, and $y'(1) = \delta$. That is, with $C_0^2([0,1]) = \{u \in C^2([0,1]) : u(0) = u'(0) = u(1) = u'(1) = 0\}$, $y \in C_0^2([0,1]) + p(x)$, where p is the cubic polynomial that matches the boundary conditions.

(a) Find a differential equation, similar to the Euler-Lagrange equation, that must be satisfied by the minimum (if it exists).

(b) Apply your equation to find the extremal(s) of

$$F(y) = \int_0^1 (y''(x))^2 \, dx,$$

where $y(0) = y'(0) = y'(1) = 0$ but $y(1) = 1$, and justify that each extremal is a (possibly nonstrict) minimum.

6. Prove the theorem: If f and g map $\mathbb{R}^3$ to $\mathbb{R}$ and have continuous partial derivatives up to second order, and if $u \in C^2([a,b])$, $u(a) = \alpha$ and $u(b) = \beta$, minimizes

$$\int_a^b f(x, u(x), u'(x)) \, dx,$$

subject to the constraint

$$\int_a^b g(x, u(x), u'(x)) \, dx = 0,$$

then there is a nontrivial linear combination $h = \mu f + \lambda g$ such that $u(x)$ satisfies the Euler-Lagrange equation for h.

7. Consider the functional

$$\Phi(x, y, y') = \int_a^b F(x, y(x), y'(x)) \, dx.$$

(a) If $F \in C^2$ and $F = F(y, y')$ only, and if we assume that $y \in C^2$, prove that in this case the Euler-Lagrange equations reduce to

$$\frac{d}{dx}(F - y'F_{y'}) = 0.$$

(b) Among all C^2 curves $y(x)$ joining the points $(0,1)$ and $(1, \cosh(1))$, find the one which generates the minimum area when rotated about the x-axis. Recall that this area is

$$A = 2\pi \int_0^1 y\sqrt{1 + (y')^2} \, dx.$$

$$\left[\text{Hint: } \int \frac{dt}{\sqrt{t^2 - C^2}} = \ln(t + \sqrt{t^2 - C^2}).\right]$$

8. Consider the functional

$$J[x, y] = \int_0^{\pi/2} [(x'(t))^2 + (y'(t))^2 + 2x(t)y(t)] \, dt$$

and the boundary conditions

$$x(0) = y(0) = 0 \quad \text{and} \quad x(\pi/2) = y(\pi/2) = 1.$$

(a) Find the Euler-Lagrange equations for the functional.

(b) Find all extremals.

(c) Find a global minimum, if it exists, or show it does not exist.

(d) Find a global maximum, if it exists, or show it does not exist.

9. Consider the problem of finding a C^1 curve that minimizes

$$\int_0^1 (y'(t))^2 \, dt$$

subject to the conditions that $y(0) = y(1) = 0$ and

$$\int_0^1 (y(t))^2 \, dt = 1.$$

(a) Remove the integral constraint by incorporating a Lagrange multiplier, and find the Euler-Lagrange equations.

(b) Find all extremals to this problem.

(c) Find the solution to the problem.

(d) Use your result to find the best constant C in the inequality

$$\|y\|_{L^2(0,1)} \le C \|y'\|_{L^2(0,1)}$$

for functions that satisfy $y(0) = y(1) = 0$.

10. Find the C^2 curve $y(t)$ that minimizes the functional

$$\int_0^1 [(y(t))^2 + (y'(t))^2] \, dt$$

subject to the endpoint constraints

$$y(0) = 0 \quad \text{and} \quad y(1) = 1$$

and the constraint

$$\int_0^1 y(t) \, dt = 0.$$

11. Find the form of the curve in the plane (*not* the curve itself), of minimal length, joining $(0,0)$ to $(1,0)$ such that the area bounded by the curve, the x and y axes, and the line $x = 1$ has area $\pi/8$.

12. Solve the constrained brachistochrone problem: in a vertical plane, find a C^1-curve joining $(0,0)$ to (b, β), b and β positive and given, such that if the curve represents a track along which a particle slides without friction under the influence of a constant gravitational force of magnitude g, the time of travel is minimal. Note that this travel time is given by the functional

$$T(y) = \int_0^b \sqrt{\frac{1 + (y'(x))^2}{2g(\beta - y(x))}}\, dx.$$

13. Consider a stream between the lines $x = 0$ and $x = 1$, with speed $v(x)$ in the y-direction. A boat leaves the shore at $(0,0)$ and travels with constant speed $c > v(x)$. The problem is to find the path $y(x)$ of minimal crossing time, where the terminal point $(1, \beta)$ is unspecified.

   (a) Find conditions on y so that it satisfies the Euler-Lagrange constraint. $\left[\text{Hint: the crossing time is } t = \int_0^1 \frac{\sqrt{c^2(1 + (y')^2) - v^2} - vy'}{c^2 - v^2}\, dx.\right]$

   (b) What free endpoint constraint is required?

   (c) If v is constant, find y.

14. For $y \in C^1([0,1])$ with $y(0) = 0$, consider minimizing

$$F(y) = \int_0^1 \left[xy(x) + (y'(x))^2\right] dx.$$

   (a) Show that $F(y)$ is strictly convex.

   (b) Find the minimum if $y(1) = 0$.

   (c) Find the minimum if $y(1) = 0$ and $\int_0^1 y(x)\, dx = 1$.

   (d) Suppose we apply the same constraint, $\int_0^1 y(x)\, dx = 1$, but we do *not* constrain y at $x = 1$. First derive conditions for the minimum, including a free end condition, and then find the minimum.

15. Given $I = [0, b]$, consider the problem of finding $u : I \to \mathbb{R}$ such that

$$\begin{cases} u'(t) = g(t)f(u(t)) & \text{for a.e. } t \in I, \\ u(0) = \alpha, \end{cases}$$

where $\alpha \in \mathbb{R}$ is a given constant, $g \in L^p(I)$, $p \geq 1$, and $f : \mathbb{R} \to \mathbb{R}$ are given functions. We suppose that f is Lipschitz continuous and satisfies $f(0) = 0$. Consider the functional

$$F(u) = \alpha + \int_0^t g(s)\, f(u(s))\, ds.$$

(a) Show that F maps $C^0(I)$ into $C^0(I) \cap W^{1,p}(I)$. Moreover, show that $u \in C^0(I) \cap W^{1,p}(I)$ solves the problem if and only if it is a fixed point of F.

(b) Show that there exists b small enough, not depending on α, such that F has a unique fixed point in $C^0(I)$.

(c) Show that the problem has a unique solution $u \in C^0(I) \cap W^{1,p}(I)$ for any $g \in L^p(I)$ and $b > 0$.

16. Let X be a topological space and let f and g map X to $(-\infty, \infty]$ be lower semicontinuous.

(a) Show that $f + g$ is l.s.c.

(b) Show that $\inf(f, g)$ and $\sup(f, g)$ are l.s.c.

(c) Show by example that fg need not be l.s.c. [Hint: the functions can have both positive and negative values.]

(d) Show that if f and g map X to $[0, \infty]$, then fg is l.s.c.

17. Prove Proposition 10.11.

Bibliography

[1] R. A. Adams. *Sobolev Spaces*. Academic Press, New York, 1975.

[2] A. Boggess and F. J. Narcowich. *A First Course in Wavelets with Fourier Analysis*. Prentice Hall, Englewood Cliffs, New Jersey, 2001.

[3] E. W. Cheney. *Analysis for Applied Mathematics*. Number 208 in Graduate Texts in Mathematics. Springer, New York, 2001.

[4] Ph. G. Ciarlet. *Linear and Nonlinear Functional Analysis with Applications*. Society for Industrial and Applied Mathematics, Philadelphia, 2013.

[5] L. Debnath and P. Mikusiński. *Introduction to Hilbert Spaces with Applications*. Academic Press, New York, 1990.

[6] G. B. Folland. *Introduction to Partial Differential Equations*. Princeton University Press, Princeton, New Jersey, 1976.

[7] I. M. Gelfand and S. V. Fomin. *Calculus of Variations*. Prentice Hall, Englewood Cliffs, New Jersey, 1963.

[8] D. Gilbarg and N. S. Trudinger. *Elliptic Partial Differential Equations of Second Order*. Spring-Verlag, 1983.

[9] M. W. Hirsch. *Differential Topology*. Springer-Verlag, New York, 1976.

[10] J. Jost and X. Li-Jost. *Calculus of Variations*. Number 64 in Cambridge Studies in Advanced Mathematics. Cambridge University Press, Cambridge, 1998.

[11] J. L. Kelley. *General Topology*. Springer-Verlag, New York, 1955.

[12] E. Kreyszig. *Introductory Functional Analysis with Applications*. Wiley Classics Library. Wiley, New York, 1989 edition, 1978.

[13] E. H. Lieb and M. Loss. *Analysis*, volume 14 of *Graduate Studies in Mathematics*. American Mathematical Society, Providence, Rhode Island, 1997.

[14] J. R. Munkres. *Topology: A First Course*. Prentice-Hall, Englewood Cliffs, New Jersey, 1975.

[15] J. Nečas. *Direct Methods in the Theory of Elliptic Equations*. Springer-Verlag, Berlin and Heidelberg, 2012.

[16] J. T. Oden and L. F. Demkowicz. *Applied Functional Analysis*. Textbooks in Mathematics. CRC Press, Boca Raton, Florida, third edition, 2018.

[17] M. Reed and B. Simon. *Methods of Modern Physics. I: Functional Analysis*. Academic Press, San Diego, California, revised and enlarged edition, 1980.

[18] H. L. Royden. *Real Analysis*. MacMillan Publishing Co, New York, third edition, 1988.

[19] W. Rudin. *Principles of Mathematical Analysis*. McGraw-Hill, New York, third edition, 1976.

[20] W. Rudin. *Real and Complex Analysis*. McGraw-Hill, New York, third edition, 1987.

[21] W. Rudin. *Functional Analysis*. McGraw-Hill, New York, second edition, 1991.

[22] H. Sagan. *Introduction to the Calculus of Variations*. Dover Publications, New York, 1969.

[23] R. E. Showalter. *Hilbert Space Methods in Partial Differential Equations*. Dover Publications, Mineola, New York, 2010.

[24] E. Stein. *Singular Integrals and Differentiability Properties of Functions*. Princeton University Press, Princeton, New Jersey, 1970.

[25] E. Stein and G. Weiss. *Introduction to Fourier Analysis on Euclidean Spaces*. Princeton University Press, Princeton, New Jersey, 1971.

[26] L. Tartar. *An Introduction to Sobolev Spaces and Interpolation Spaces*. Springer-Verlag, Berlin, 2007.

[27] H. F. Weinberger. *A First Course in Partial Differential Equations: with Complex Variables and Transform Methods*. Dover Publications, New York, 1965.

[28] K. Yosida. *Functional Analysis*. Classics in Mathematics. Springer-Verlag, Berlin, sixth edition, 1995.

Index

$B(X, Y)$, 43
$B(\mathbb{R}^{d_1}, \mathbb{R}^{d_2})$, 43
$B_r(x)$, 4, 38
$\mathcal{B}^n(X, Y)$, 358
$\mathbb{C}, \mathbb{C}^d$, 1, 2, 36
$C(X, Y)$, 136
$C([a, b])$, 37
$C^0(\Omega)$, 181
$C_0^1([a, b]; \mathbb{R}^d)$, 373
$C_{\alpha, \beta}^1([a, b]; \mathbb{R}^d)$, 372
$C^\infty(\Omega)$, 181
$C^n(\Omega)$, 181
$C^{m,1}$-boundary, 261
$C^{m,1}(\Omega)$, 261
c_0, 51
$C_0(\Omega)$, 22, 58
$C_0^\infty(\Omega)$, 179–181
$C_B^n(\Omega)$, 182, 270
$C_{\mathrm{v}}(\mathbb{R}^d)$, 218
$C_{\mathrm{loc}}^{0,1}(\mathbb{R}^d)$, 291
$\mathcal{D}'(\Omega)$, 184
$D(T)$, 130
$\mathcal{D}(\Omega)$, 180, 181, 183
$\mathcal{D}_K$, 181
diam, 93
dist, 25, 66, 68
ess sup, 54
$\mathcal{F}$, 216
$\mathbb{F}, \mathbb{F}^d$, 36, 37
f_0, 51, 100
$GL(X), GL(X, Y)$, 134, 352
$H^m(\Omega)$, 255
$H^s(\Omega)$, 279
$H^s(\mathbb{R}^d)$, 276
$\mathcal{I}$, 2
inf, 17
$\ker(T)$, 97, 112, 130
$L_{\mathrm{loc}}^1(\Omega)$, 185

$\ell^2(\mathcal{I})$, 115
ℓ^p, 50
$L^p(\Omega)$, 53
lim inf, 17
lim sup, 17
$N(T)$, 97, 112, 130
$o(\cdot)$, 335
Ω, 16, 53, 181
$PV\frac{1}{x}$, 187
$\mathbb{Q}$, 67
$\mathbb{R}, \mathbb{R}^d$, 1, 2, 36
$R(T)$, 130
$\mathbb{R}^+$, 36
Res, 30
$\mathcal{S}$, 221
$\mathcal{S}'$, 231
$S_r(x)$, 45
σ-algebra, 12
sup, 17
τ_y, 96, 192, 217
$\mathrm{tr}(T)$, 174
$W^{-m,p}(\Omega)$, 257
$W^{m,p}(\Omega)$, 253
$W_0^{m,p}(\Omega)$, 257
$W^{s,p}(\Omega)$, 289
X^*, 47
X^{**}, 79
$\mathbb{Z}$, 1
$*$, 193, 211, 219, 234
$\|\cdot\|$, 36
$\cong$, 43
$\hat{\ }$, 216
$\hookrightarrow$, 231
$\overset{c}{\hookrightarrow}$, 273
$\langle \cdot, \cdot \rangle_{X^*, X}$, 186
$(\cdot, \cdot)_H$, 104
$\langle \cdot, \cdot \rangle_H$, 104
$\leq$, 147

$\oplus$, 74, 108, 343
$\perp$, 106, 108
$\preceq$, 61
$\rightarrow$, 83
$\xrightarrow{w}$, 83
$\xrightarrow{w^*}$, 83
$\sqrt{T}$, 147
$\subset\subset$, 10, 181
$\times$, 74

a.e., *see* almost every/everywhere
Abel's theorem, 161
absolute value of an operator, 149
accumulation point, 5
adjoint
    Hilbert, *see* Hilbert adjoint
    operator, *see* dual, operator
admissible variation, 373
affine
    function, 334
    space, 373
almost every/everywhere, 21
analytic, *see* holomorphic
annihilator, 97
approximation to the identity, 196,
        197
Ascoli-Arzelà theorem, 153
axiom of choice, 62

Babuška-Lax-Milgram theorem, 306,
        308
Baire category theorem, 71
balanced set, 68
ball, 4, 38
Banach contraction mapping
        theorem, 345
Banach space, 35, 39
    isomorphic, 74
Banach-Alaoglu theorem, 85
    for separable Hilbert spaces, 123
Banach-Saks theorem, 89
Banach-Steinhaus theorem, 77, 196
base for a topology, 3
basis, 47
    dual, 91

Hamel, 51
    orthonormal, 113, 117, 119
    Schauder, 51, 95
BC, *see* boundary condition
Bessel potential, 276
Bessel's inequality, 115
best approximation, 107, 124, 319
best approximation theorem, 107
bifurcation, 357
biharmonic equation/operator, 212,
        250, 328
bilinear, 99, 104, 301
Borel set, 14
boundary, 5
    $C^{m,1}$, 261
    Lipschitz, 261
    locally $C^{m,1}$, 261
    locally Lipschitz, 261
boundary condition, 158, 214, 295,
        299
    Dirichlet, 299
    essential, 312
    natural, 315
    Neumann, 299
    Robin, 299
boundary value problem, 158, 295
bounded, 54
    below, 140, 305, 306
    bilinear operator, 99
    essentially, 54
    linear functional, 47
    linear operator, 43
    multi-linear operator, 358
    n-linear operator, 358
    operator, 42, 43
    set, 39, 93
brachistochrone problem, 378, 380,
        393
Bubnov-Galerkin method, *see*
        Galerkin method
BVP, *see* boundary value problem

calculus of variations, 372
Cantor diagonalization argument,
        136

catenary, 378
Cauchy sequence, 39
  in $\mathcal{D}(\Omega)$, 183
  weakly, 101
Cauchy's infinitely differentiable
    function, 181
Cauchy's theorem, 27
Cauchy-Bunyakovsky-Schwarz
    inequality, 48, 104
Cauchy-Schwarz inequality, *see*
    Cauchy-Bunyakovsky-
    Schwarz inequality
chain, *see* ordered set, totally
chain rule, 339
characteristic function, 17, 186
Chebyshev's inequality, 23
closed graph theorem, 75
closed operator, 75
closed range theorem, 305, 306
closed set, 2
closest point, 107
closure of a set, 4
coercive, 310
compact
  embedding of Sobolev spaces,
    273
  operator, 127, 129, 134, 136,
    149, 371
  sequentially, 11, 86
  set, 10
  weak-∗, 81
  weakly, 81
compact support, 22, 58
compactly contained in, 10, 181
comparable elements, 61
compatibility condition, 317
complete, 39
  of $\mathcal{D}'(\Omega)$, 196
  weakly, 101
completely continuous linear
    operator, 137
completion, 255
complex differentiable, 24
concave
  function, 364

strictly, 364
conjugate
  exponent, 47, 48, 54
  linear, 104
  operator, *see* dual, operator
  symmetric, 103
continuous
  embedding, 231
  function, 7, 43
  linear operator, 43
  Lipschitz, 261
  sequentially, 7
contour, 26
contour integration, 24
contraction, 345
contraction mapping theorem, 345
converge, 6
convergence
  in $\mathcal{D}'(\Omega)$, 197
  in $\mathcal{D}(\Omega)$, 183
  of distributions, 196
  weak, 81, 83, 122
  weak-∗, 83
convex
  function, 364
  set, 41
  strictly, 364
convex hull, 94
convolution
  of distributions, 211
  of functions, 193, 219
  with a distribution, 193, 234
coordinate mapping, 47
coset, 96, 279, 294
cosine series, 169
countably additive, 13
critical point, 364, 373
curve, 26
cycloid, 380

Darcy's law, 298
degenerate problem, 315
delta function, 179, 187
dense, 5
derivative, 24

diameter, 93
diffeomorphic, 352
diffeomorphism, 352
differentiation, 335
    complex, 24
    Fréchet, 335
    Gâteaux, 338
    higher order Fréchet, 358
    of distribution, 189
    partial Fréchet, 343
dilation operator, 192, 193
Dirac
    delta function, 179, 187
    distribution, 187
    measure/mass, 33, 187
direct sum, 74, 108, 343
    orthogonal, 110
Dirichlet boundary condition, 299
Dirichlet principle, 252, 303
Dirichlet problem, 312
discrete topology, 3, 32
dispersive wave equations, 348
distance function, *see* metric
distribution, 178, 179, 184, 186
    compact support, 211
    convergence of, 197
    derivative of, 189
    order of, 184
    regular, 186
    singular, 186
    tempered, 231
divergence theorem, 33, 296
domain, 181
domain of dependence, 206
domain of influence, 206
dominated convergence theorem, 23
double dual space, 79
dual
    algebraic, 99
    basis, *see* basis, dual
    double, 79
    Hilbert adjoint, 113
    operator, 89
    space, 47, 103, 111
duality pairing, 180, 186

Dubois-Reymond lemma, 375

Eberlein-Šmulian theorem, 88
eigenfunction, 131
eigenvalue, 131
eigenvector, 131
electrostatic potential, 298
elliptic, 310
elliptic equation, 296, 311
elliptic regularity theorem, 317
energy functional, 302
energy norm, 318
entire function, 26
equicontinuous, 153, 196
equivalence relation, 293
Euclidean norm, 47, 94
Euler operator, 176
Euler's formula, 26
Euler-Lagrange equations, 373, 374
evaluation map, 79
extension
    from dense subset, 228
    linear, 62, 63
    operator, 259, 263
extremal, 376
    regular, 377
extremum, 363
    constrained, 380

Fatou's lemma, 23
filter, 241
    causal, 245
    linear, 241
    translation invariant, 241
finite element method, 320
finite-dimensional space, 37, 47
first category, 72
fixed point, 345
fixed point theorem, 346
formal adjoint, 156
formally self-adjoint, 157
Fourier coefficients, 119
Fourier series, 122, 169, 213
Fourier transform, 213
    in $\mathcal{S}'(\mathbb{R}^d)$, 231, 234

in $\mathcal{S}(\mathbb{R}^d)$, 221
in $L^1(\mathbb{R}^d)$, 216
in $L^2(\mathbb{R}^d)$, 228–230
Fourier's law of heat conduction, 297
Fréchet derivative, 335
  higher order, 358
  partial, 343
Fréchet differentiable, 335
Fréchet space, 183
Fredholm alternative for compact
    operators, 142
Fredholm integral equation, 368
Fubini's theorem, 23
function
  affine, 334
  characteristic, 17
  entire, 26
  integrable, 19
  locally Lipschitz, 291, 337
  measurable, 16
  of rapid decrease, 221
  simple, 17
  slowly increasing, 233
  test, 179, 181, 309
  trial, 309
functional, 46
  linear, 47
fundamental solution, *see* solution,
    fundamental
fundamental theorem of calculus, 24

Galerkin method, 318
Galerkin orthogonality, 319
Gâteaux derivative, 338
Gâteaux differentiable, 338
Gauss theorem, 33, 296
Gaussian kernel, *see* heat kernel
general linear group, 134, 352
generalized function, 178, *see*
    distribution
geodesic, 388
  reparametrization, 389
Goldschmidt solution, 378
Gram-Schmidt orthogonalization,
    119

graph, 75
graph norm, 75
Green's function, 158, 159, 321, 322
ground field, 36

Hahn-Banach theorem, 59, 65
  for general vector spaces, 63
  for normed linear spaces, 64
  for real vector spaces, 62
Hamel basis, 51
Hausdorff, 6
Hausdorff-Young inequality, 246
heat conduction, 297
heat equation, 213, 238
heat kernel, 239
heat operator, 238
Heaviside function, 178
Heine-Borel theorem, 10, 49, 81
  generalized, 87, 88
Helmholtz decomposition, 330
Helmholtz operator, 211
Hermitian form, 104
Hermitian operator, 143
Hilbert adjoint, 113
Hilbert space, 103, 107
  orthonormal basis, 117, 119
  isomorphic, 119
Hilbert-Schmidt operator, 174
Hilbert-Schmidt theorem, 150
Hölder's inequality, 48
  generalized, 266
  in ℓ^p, 50
  in L^p, 54
holomorphic, 24
homeomorphic, 8
homeomorphism, 8

idempotent, 108
implicit function theorem, 355
incomparable elements, 61
induced
  metric, 38
  norm, 104
  topology, 8
inf-sup condition, 308

infimum, 17
infinite-dimensional space, 37
inherited topology, 8
initial value problem, 156
inner-product, 103
inner-product space, 104
integrable function, 19
integral operator, 153
integration by parts, 33, 179, 296, 302
interior of a set, 4
interior point, 5
interpolation inequalities, 292
inverse function theorem, 351, 352
    precursor, 351
IPS, *see* inner-product space
isolated point, 5
isometric, 80
isometry, 80
    partial, 173
isomorphic
    Banach spaces, 74
    Hilbert spaces, 119
    isometrically, 74, 119
    vector spaces, 43
isomorphism, 43
isoperimetric problem, 383
IVP, *see* initial value problem

kernel of an operator, 130
Klein-Gordon equation, 250
Korn's inequality, 326
Kuratowski-Zorn lemma, *see* Zorn's lemma

l.s.c, *see* lower semicontinuous
Ladyzhenskaya-Babuška-Brezzi condition, 309
Lagrange multiplier, 380
Lagrange multiplier theorem, 381
Laplace equation/operator, 206
Laplacian, 206
Laurent series, 29
Lax-Milgram theorem, 310
LBB condition, *see* Ladyzhenskaya-Babuška-Brezzi condition

Lebesgue integration, 12, 18
Lebesgue lemma, 185
Lebesgue measurable set, 15
Lebesgue measure, 12, 15
Lebesgue space, 53
Leibniz rule, 191
limit, 6
    lim inf, 17
    lim sup, 17
linear differential operator, 199
linear elasticity equations, 329
linear operator, 42
linear space, *see* vector, space
linearly independent, 113
Lipschitz boundary, *see* boundary, Lipschitz
Lipschitz constant, 261
Lipschitz continuous, 261
little-ℓ^p space, 50, 115
little-o notation, 335
local-Lipschitz lemma, 337
locally Lipschitz function, 291, 337
lower semicontinuous, 384, 385
Lusin's theorem, 22

Malgrange-Ehrenpreis theorem, 204
matrix, 43
maximal element, 62
maximal orthonormal set, 117
maximum, 363
    relative, 363
    strict, 363
Mazur separation lemma, 66, 68
mean-value theorem, 341, 342
    for curves, 341
measurable
    function, 16
    set, 15
measure, 13
    complex, 13
    Lebesgue, 12, 15
    of a set, 13
    positive, 13
measure space, 14
meromorphic, 30

metric, 2
metric space, 2, 38
metrizable, 32
minimization of energy, 300
minimum, 363
    local, 363
    relative, 363
    strict, 363
Minkowski functional, 68
Minkowski's inequality, 55
monotone convergence theorem, 23
multi-index, 181
multi-linear operator, 358
multiplier, 243
    operator, 243, 250, 275

n-linear operator, 358
neighborhood, 4
Neumann boundary condition, 299
Neumann problem, 313, 315
Neumann series, 133
Newton method, 349
    Newton-Kantorovich, 350
    simplified, 349
Newton-Kantorovich method, 350,
        369
NLS, see normed linear space
nonlinear equations, 348
nonmeasurable, 16
norm, 36
    equivalent, 40
    topology, 38, 81
normal derivative, 286
normal operator, 175
normed linear space, 35, 36
nowhere dense, 72
null space, 130

ON, see orthonormal
open cover, 10
open mapping theorem, 71, 72
open set, 2, 38
operator
    bilinear, 99, 104, 301
    bounded, 42, 43

bounded below, 140, 305, 306
bounded linear, 43
closed, 75
compact, 127, 129, 134, 136, 149
completely continuous linear,
        137
continuous
    in $\mathcal{D}(\Omega)$, 184
    linear, 43
dilation, 192, 193
dual, 89
extension, 259, 263
Hermitian, 104, 143
Hilbert-Schmidt, 174
idempotent, 108
integral, 153
linear, 42
multi-linear, 358
multiplier, 243, 250, 275
n-linear, 358
normal, 175
positive, 147
projection, 108
self-adjoint, 143, 149
translation, 96, 192, 193, 217
unitary, 125
order
    of differential operator, 199
    of distribution, 184
    of pole, 29
ordered set
    partially, 61
    totally, 61
ordering, 61
Orlicz space, 98
orthogonal, 106, 113
orthogonal complement, 305
orthogonal projection, 107, 108
orthonormal, 113
    basis, 113, 117, 119

Paley-Wiener theorem, 221
parallelogram law, 106
parametrized, 26
Parseval identity, 117, 228

partial differential equation/operator
  biharmonic, 212, 250, 328
  dispersive wave, 348
  elliptic, 296, 311
  Fredholm, 368
  heat, 213, 238
  Helmholtz, 211
  Klein-Gordon, 250
  Laplace, 206
  Laplacian, 206
  linear, 199
  linear elasticity, 329
  Poisson, 206
  Schrödinger, 239
  Stokes, 328
  telegrapher's, 249
  wave, 203
partial isometry, 173
partition of unity, 262
  locally finite, 256, 262
PDE, *see* partial differential equation
perp-space, 108
Picard's theorem, 156
Plancherel theorem
  in $\mathcal{S}'(\mathbb{R}^d)$, 235
  in $\mathcal{S}(\mathbb{R}^d)$, 226
  in $L^2(\mathbb{R}^d)$, 229
plane wave, 216
Plemelij-Sochozki formula, 210
Poincaré inequality, 270, 307, 315, 327
Poincaré-Friedrichs inequality, 270
point measure, 33
point-set topology, *see* topology
Poisson equation, 206
Poisson integral formula, 324
pole, 29
porous medium, 298
positive
  definite, 104, 296
  operator, 147
  semidefinite, 296
  square root operator, 147
  uniformly, 296
potential vector field, 330

power series, 25
  representable by, 25
pre-Hilbert space, 104
precompact set, 134
primitive, 199
principal value, 188
principle of superposition, 204
product space, 97
product topology, 9
projection, 108
  map, 9
  orthogonal, 107, 108
Pythagorean law, 106

quasi-optimal, 319
quotient space, 96, 279

radius of convergence, 25
Radon-Nikodym theorem, 58
rapidly decreasing function, 221
Rayleigh quotient, 145
rectifiable, 383, 388
reflexive space, 80
Rellich-Kondrachov theorem, 273
residue, 30
residue theorem, 30
resolvent, 130, 131
  operator, 131
reverse triangle inequality, 39
Riemann integral, 12
Riemann-Lebesgue lemma, 218
Riesz map, 112
Riesz representation theorem, 111
Riesz-Fischer map, 116
Riesz-Fischer theorem, 118
rigid motion, 326
Ritz method, 320
Robin boundary condition, 299

saddle-point problem, 328
Schauder basis, 51
Schrödinger equation, 239
Schrödinger kernel, 240
Schwartz space, 221
Schwarz lemma, 361
second category, 72

second derivative test, 366
self-adjoint operator, 143, 149
self-similar, 239
seminorm, 59
separable space, 67
separate points, 66, 102, 183
separating hyperplane theorem, 69
separation of variables, 169
sequentially
    closed, 38
    compact, 11, 86
    continuous, 7
sesquilinear form, 104
Shannon-Whittaker sampling
        theorem, 245
signal, 241
    band limited, 245
signal processing, 241
simple closed curve, 26
simple function, 17
simplified Newton method, 349
sine series, 168
singularity
    essential, 29
    isolated, 29
    removable, 29
SL, *see* Sturm-Liouville problem
slowly increasing function, 233
Sobolev embedding theorem, 265,
        271
Sobolev inequality, 265, 294
Sobolev space, 252, 253, 255, 257,
        275, 276, 279, 289
solenoidal vector field, 330
solution
    classical, 199
    distributional, 199
    fundamental, 203, 204
    weak, 199, 313
space, 1
    C^n, 181
    L^1_{loc}, 185
    $\mathcal{D}_K$, 181
    affine, 373
    Banach, 35, 39

distributions, $\mathcal{D}'$, 184
double dual, 79
dual, 47, 103, 111
Fréchet, 183
functions that vanish at infinity,
        C_v, 218
Hilbert, 103, 107
inner-product, 104
Lebesgue, 53
Lipschitz continuous, $C^{m,1}$, 261
little-ℓ^p, 50, 115
measure, 14
metric, 2, 38
normed linear, 35, 36
Orlicz, 98
pre-Hilbert, 104
product, 97
quotient, 96, 279
Schwartz, $\mathcal{S}$, 221
Sobolev, 252, 253, 255, 257, 275,
        276, 279, 289
tempered L^p, 233
tempered distributions, $\mathcal{S}'$, 231
test functions, $\mathcal{D}$, 180, 181, 183
topological, 2
vector, 36
spectral mapping theorem, 171
spectral theorem
    for compact operators, 142
    for compact self-adjoint
        operators, 151
    for self-adjoint operators, 144,
        146
spectral theory, 129, 132
spectrum, 130, 131
    continuous, 131
    point, 131
    residual, 131
square root operator, 147
stationary point, *see* critical point
Stokes equations, 328
strict limit point, 6
stronger topology, 8
strongly differentiable, *see* Fréchet
        differentiable

Sturm-Liouville problem, 155, 158
  applications, 168
  Bessel, 158
  Chebyshev, 158
  Hermite, 158
  Laguerre, 158
  Legendre, 158
  regular, 158
  singular, 158
  spectral properties, 164
sublinear, 59
subspace, 9, 38
support, 181
supremum, 17
  essential, 54
Surjective mapping theorem, 369
symbol of multiplier operator, 275
symmetric, 103
symmetric gradient, 326

Taylor series, 25, 362
Taylor's theorem, 362
telegrapher's equation, 249
tempered L^p, 233
tempered distribution, 231
test function, 179, 181, 309
test space, 180, 309
topological space, 2
topology, 1, 2
  base, 3
  discrete, 3, 32
  induced, 8
  inherited, 8
  norm, 38, 81
  product, 9
  strong, *see* topology, norm
  stronger, 8
  trivial, 3, 32
  weak, 82, 84, 102
  weak-∗, 82, 84
  weaker, 8
trace, 174, 252, 281, 284
trace theorem, 281, 285, 286
transfinite induction, 61
translation invariant, 15, 16, 241

translation operator, 96, 192, 193, 217
trial function, 309
trial space, 309
triangle inequality, 2, 36, 59
trivial topology, 3, 32
Tychonoff theorem, 11

uniform boundedness principle, 77
unitary
  matrix, 172
  operator, 125
unitary isomorphism, 229
unmeasurable, *see* nonmeasurable
upper bound, 62

vanish at infinity, 218
variational problem, 300, 301
vector, 36
  space, 36
VP, *see* variational problem

wave equation/operator, 203
weak convergence, 81, 122
weak form, 304
weak solution, 199, 313
weak topology, 82, 84, 102
weak-∗ compact, 81
weak-∗ topology, 82, 84
weaker topology, 8
weakly compact, 81
weakly differentiable, *see* Gâteaux
    differentiable
Weierstrass approximation theorem,
    127
well-defined, 16
well-posed, 300
Wronskian, 161

Young's inequality, 220
  for products, 47
  generalized, 220, 246

Zorn's lemma, 62

Printed in the United States
by Baker & Taylor Publisher Services